INTRODUCTORY TOPOLOGICAL ANALYSIS OF ELECTRICAL NETWORKS

Holt, Rinehart and Winston Series in Electrical Engineering, Electronics, and Systems

OTHER BOOKS IN THE SERIES:

George R. Cooper and Clare D. McGillem,
METHODS OF SIGNAL AND SYSTEM ANALYSIS

Samuel Seely, ELECTRONIC CIRCUITS

Mohammed S. Ghausi and John J. Kelly,
INTRODUCTION TO DISTRIBUTED-PARAMETER NETWORKS: WITH APPLICATION TO INTEGRATED CIRCUITS

Benjamin J. Leon, LUMPED SYSTEMS

INTRODUCTORY TOPOLOGICAL ANALYSIS OF ELECTRICAL NETWORKS

SHU-PARK CHAN

Professor of Electrical Engineering, University of Santa Clara

HOLT, RINEHART AND WINSTON, INC.

New York Chicago San Francisco Atlanta
Dallas Montreal Toronto London Sydney

Library of Congress Catalog Card Number: 69-11392
03-071415-X
Printed in the United States of America
1 2 3 4 5 6 7 8 9

To my late parents,
General and Mrs. Chi-Tong Chan;

To my wife,
Stella;

And to my children,
Charlene and Yau-Gene

PREFACE

This text is, to a great extent, based on the class notes prepared by the author for two courses, modern network analysis and network graph theory, offered at the University of Santa Clara. After teaching these two courses for the past several years, the author has collected the following noteworthy observations:

(a) Since the digital computer has recently become a very powerful tool for solving engineering problems, the application of matrix theory to the analysis of networks and systems constitutes an essential portion of any modern analysis course in networks and systems.

(b) The application of graph theory to the analysis of networks and systems has yielded many interesting and significant results in recent years. Examples are the application of network topology in the formulation of state equations and the use of the so-called "k-tree" topological techniques as part of the computer programs developed for the analysis and design of passive and active networks.

(c) There have been recent developments in the theory and the application of flow graphs in circuit theory and control systems, which are recognized as handy tools for analyzing linear networks and systems.

These observations prompted the author to combine his revised lecture notes for the two network courses to produce this text. It not only gives an introduction to the topological study of electrical networks, which, in the author's opinion, is the foundation for modern network theory, but also includes some of the significant results recently developed in topological analysis of active networks and flow-graph applications to linear systems.

In order to make the book self-contained, the discussions of determinants, matrices, and linear equations, as well as the methods of proofs are included in the Appendixes. The only prerequisite to this text is an introductory course in modern network analysis in which Laplace transform and the concept of the complex frequency are discussed. Thus, it can be used as a text for an advanced undergraduate or a first-year graduate course. If the student has already had a course in matrix algebra and linear transformations, he can easily begin with Chapter 1 on elementary network topology; otherwise, the coverage of matrices included in the Appendixes should provide the student with a good enough working knowledge about the subject to proceed with the main text.

The first three chapters constitute an introduction to the theory of linear graphs, and discuss basic definitions and theorems, network matrices, and duality in detail. Chapter 4 is devoted to the matrix formulation of network equations, namely the loop-basis, the node-basis, and the state equations.

Chapters 5 and 6 form the classical topological analysis of passive networks without mutual inductances. Chapter 7 discusses various types of flow graphs as applied to the analysis of linear systems and networks, including some of the latest developments. In Chapter 8, topological formulas for active networks with mutual magnetic coupling are developed and the relationships between the flow-graph formulas (discussed in Chapter 7) and the k-tree formulas are established. Thus, Chapters 5 through 8 constitute a study of the topological analysis (k-tree and flow-graph) techniques for linear systems. Some of the topological formulas that have been developed recently for the digital computer are also included in these four chapters.

Chapter 9 provides a discussion of the topological applications in switching theory with emphasis on the single-contact networks; and, finally, in Chapter 10, a brief introduction to topological synthesis and a survey of other topological applications are presented.

In order to make the discussion more understandable, and make the concepts simple enough for the student to grasp, adequate examples are cited throughout the text whenever applicable. Also, whenever a new concept is to be discussed, an example is used to introduce it in an informal manner so as to give the reader a feeling of the terms involved before they are rigorously defined. In addition, a selected set of problems is included at the end of each chapter to illustrate important ideas introduced throughout the discussions and, in many instances, to introduce new concepts or terms that are closely related to the topics discussed in that chapter. Hence, it is important that the reader work a sufficient number of problems in each chapter to secure an understanding of important concepts, and then correlate these concepts to see the whole picture of the subject.

During the writing of the book, the author deeply appreciated the encouragement and assistance given him by colleagues and students at the University of Santa Clara, especially Dr. A. Burzio, who read the entire manuscript critically and made many helpful suggestions; Mr. H. N. Mai, whose original contributions have been included as part of Chapter 7 and whose numerous problems are distributed throughout the first seven chapters; Mr. R. Woc, who did all the original sketches for the final manuscript with enthusiasm; and Messrs. F. F. Lin and C. C. Hsiech, who did an excellent job of proofreading all the galleys and page proof. The author is indebted to Professor R. W. Newcomb of Stanford University, M. E. Van Valkenburg of Princeton University, B. J. Leon of Purdue University, W. Mayeda of the University of Illinois, G. J. Thaler and S. G. Chan of the U.S. Naval Postgraduate School, W. K. Chen of Ohio University, and S. Y. King of the University of Hong Kong for their valuable reviews and many constructive comments.

The author is also most grateful to Professor S. Y. Chan of the the University of California at Berkeley and San Diego State College for his critical reading of the entire manuscript and numerous helpful discussions, and to Professor P. M. Lin of Purdue University for his generous contribution of over fifty excellent class-tested problems as well as for his constant encouragement and many critical comments during the writing of the book. He also wishes to thank Mrs. M. McKenna, Mrs. M. Mahaffey, and Mrs. K. Vandendool for their efficient and careful typing of the manuscript. To Dr. Robert J. Parden, Dean of the School of Engineering, and Dr. Richard C. Dorf, Chairman of the Electrical Engineering Department, the author wishes to express his appreciation for their encouragement of his efforts. The author would also like to acknowledge his appreciation for the assistance given him by Holt, Rinehart and Winston, Inc., and especially thank Miss Lois Wernick for her help in the production of this book.

Finally, the author's special thanks go to his wife, Stella, for her encouragement and cooperation in making the writing of this book possible.

Shu-Park Chan

Santa Clara, California

November 1968

CONTENTS

CHAPTER 1

ELEMENTARY NETWORK TOPOLOGY

1-1 INTRODUCTION

Topology may be described as a study of certain geometrical properties of figures or objects. Network topology is, therefore, a study of (electrical) networks in connection with their nonmetric geometrical or topological properties through the investigation of their graphs. Such a study of graphs is known as the theory of linear graphs, a name first suggested by D. König in his famous book, *Théorieder Graphen*, published in 1936.

Linear graph theory is merely a part of topology and the application of such theory to electrical networks can be dated back to the nineteenth century[1] when, in 1847, Kirchhoff applied the topological concepts to formulate the loop-basis and node-basis network equations. However, not much significant progress has been made with respect to the formulation of the so-called "modern graph theory" until 1911 when O. Veblen's book, *Analysis Situs*, was published.

During the years 1930 to 1936, Kuratowski, Foster, and Whitney made important contributions in the development of graph theory by introducing concepts such as planarity, nullity, 2-isomorphism, cutsets, and so forth. The application of topological methods to the analysis and synthesis of electrical networks had not made much significant advancement until the last

[1] Strictly speaking, the earliest known paper on graph theory had been written in 1736 by L. Euler [30] in connection with the solution of the celebrated Königsberger bridge problem. See also Seshu and Reed [76], p. 19, and Seshu [74] in the References.

decade. During that period, network topology was gradually made known and recognized as the foundation for the theory of electrical networks. In recent years, the availability of computers made the application of topological techniques a powerful tool in network analysis and design.

1-2 AN INFORMAL DISCUSSION

In this section, we shall introduce a number of terms that are considered basic definitions in graph theory in the sense that they are the building blocks to all the lemmas, theorems, and corollaries developed throughout the entire text. Some of these definitions, such as circuits, series and parallel elements, and so forth, are already familiar to the reader; but some of them, such as cutset and isomorphism, may be entirely new to him. We shall first discuss the basic terms, in an informal manner, by means of an example through which the reader may acquire an intuitive understanding of the terms introduced. After this informal introduction, we shall define the terms again in a rigorous manner and, at the same time, introduce new ones as we progress.

Consider the network shown in Figure 1-1(a).

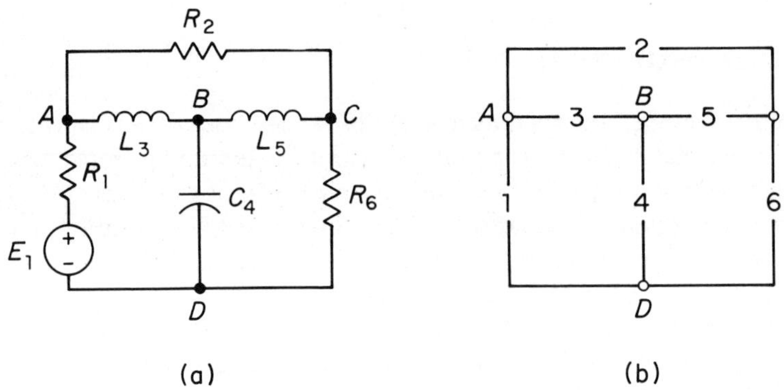

Figure 1-1 (a) An electrical network. (b) The graph of the network.

In this network, there are four nodes — A, B, C, and D — and six branches — E_1 and R_1, R_2, L_3, C_4, L_5, and R_6. If each of the six branches is replaced by an arc or a line segment with the four nodes remaining unchanged, we obtain a graphical representation of the network as shown in Figure 1-1(b). This graphical representation is known as the *linear graph* (or, simply, the graph) of the network, and each of the six line segments is called an *edge* of the graph, corresponding to a branch in the network. The four nodes in the graph are called the *vertices* of the graph.

Note that each branch of the network, regardless of what type of electrical element it is, becomes a line segment in the graph. Thus, it will be impossible to tell from the graph what types of electrical elements these edges represent in the original network. In other words, it is easy to construct the graph from a given network, but it is not possible to reconstruct the network from its graph. Therefore, the purpose of drawing the graph is merely to show the interconnections between the nodes and the branches of the network. The electrical properties of the network obviously will not be part of our discussions when we confine our study only to the theory of graphs.

We might point out that even though the graph in (b) and the network in (a) of Figure 1-2 are quite similar in shape, it will be incorrect to assume that

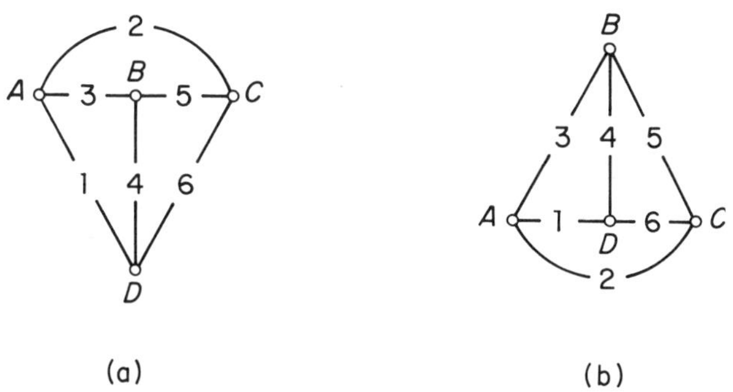

Figure 1-2 Graphs of the network of Figure 1-1(a).

the shape of the edges of a graph should affect the (topological) properties of the graph. In fact, both of the graphs of Figure 1-2 are graphs of the network of Figure 1-1(a).

This, then, brings up the question of uniqueness of the graphs of a given network. Also, since both graphs (a) and (b) of Figure 1-2, as well as graph (b) of Figure 1-1, are graphs of the same network, the reader may wonder whether there exist any relationships among these three graphs. Such questions and many others will be answered once we discuss isomorphism of graphs in a later section.

At this point, a few observations are appropriate. In Figure 1-2(a), we see that a number of edges are connected together at a vertex and each of the two endpoints of an edge is always terminated at a vertex. Since a graph is formed by connecting together in a prescribed manner the edges by joining their endpoints (namely, the vertices of the graph) and since network topology is a study of certain geometrical properties of graphs, it is easy to realize that the interconnections between the edges and the vertices play a very important role in the theory. We shall describe graph (a) of Figure 1-2 as follows:

Since three edges, 1, 2, and 3, are connected to vertex A, we say that vertex A is of *degree* 3. In fact, all the four vertices of the graph are of degree 3 since there are exactly three edges connected to or *incident with* each of the four vertices (this fact being just a coincidence in this example).

Now, if edge 2 is removed from graph (a), a graph as shown in Figure 1-3 results. Figure 1-3 is a portion of the whole graph of Figure 1-2(a) and is called a *subgraph* of the graph — analogous to the term subset of a set in set theory.

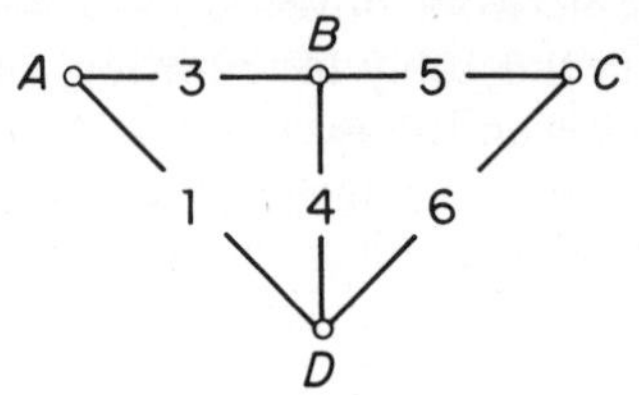

Figure 1-3 A subgraph of the graph of Figure 1-2(a).

Next, we observe that edges 1, 3, and 4 form a closed path or a loop in the graph of Figure 1-2(a), and the corresponding branches form a circuit in the original network as shown in Figure 1-1(a). Thus, we call the closed path formed by these three edges a *loop* or *circuit* in the graph.

Furthermore, if edges 3 and 5 are also removed from the graph, a subgraph is formed by the remaining three edges (1, 4, and 6) as shown in Figure 1-4(a).

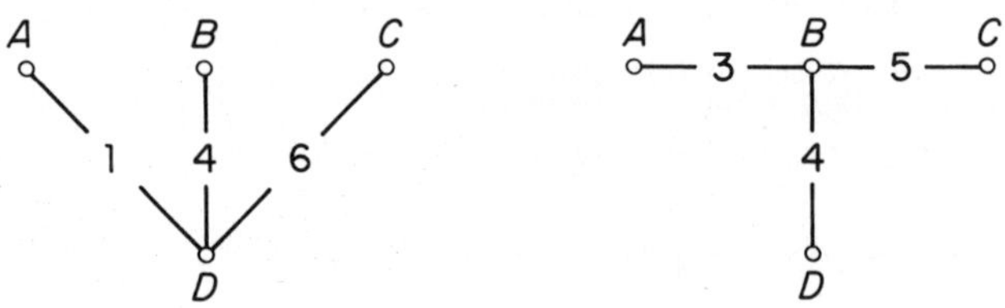

(a) (b)

Figure 1-4 Trees of graph of Figure 1-2(a).

This subgraph, containing only three edges, still connects all the four vertices in one piece and is called a *tree* of the graph of Figure 1-2(a), and the edges contained in the tree are referred to as *branches* of the tree. Another tree of the same graph is shown in Figure 1-4(b). Note that this subgraph also has the same properties as the one shown in Figure 1-4(b). We shall now point out some of the common properties of the two trees considered here.

(a) Both subgraphs contain the same number of edges (three in this example).
(b) They both contain all four vertices of the whole graph [Figure 1-2(a)].
(c) Each of the two subgraphs is in one piece by itself.
(d) None of the two subgraphs contains any circuits.

There are, of course, additional properties of a tree. It will be a good exercise for the reader to draw a few other trees of the same graph [Figure 1-2(a)] to see whether he can discover other properties that have not yet been mentioned.

It should be noted that the graph of Figure 1-2(a) is called a *connected graph* — namely, a graph that is in one piece by itself. Similarly, a tree of a connected graph is a connected subgraph of the graph [property (c) of a tree]. If the graph we obtain from a given network has more than one piece (part, or component), the graph is called an *unconnected graph.* In such a case, the graph will not have a tree but a *forest* — a collection of the trees of all the components of the graph. We shall define these basic terms formally in the next section.

1-3 BASIC DEFINITIONS

DEFINITION 1-1. Linear Graph and Its Edges and Vertices. A *linear graph* (or, simply, a *graph*) is a set of line segments called *edges* and points called *vertices*, which are the endpoints of the edges, interconnected in such a way that the edges are connected only to (or *incident* with) the vertices.

We should note that a linear graph may contain an infinite number of edges (and vertices). Such a graph is called an *infinite graph.* The term "linear" simply means that a linear graph is one consisting of line segments. It does not imply, however, that there also exist nonlinear graphs in the theory of graphs in opposition to the cases of linear systems versus nonlinear systems in systems theory, or in the case of linear networks versus nonlinear networks in network theory. In fact, we have only linear graphs. Also, we shall limit our discussions to finite graphs only.

The symbol G (sometimes with subscripts such as G_1 and G_2 if more than one graph is considered at the same time) will be used to denote a graph; symbols such as $e_1, e_2, \ldots,$ will be used to identify the edges; and $v_1, v_2, \ldots,$ (or $A, B, \ldots$) to represent vertices of a graph. Thus, the symbol $e_k(v_i,v_j)$ denotes the kth edge, whose two vertices (endpoints) are v_i and v_j in a given graph.

DEFINITION 1-2. Degree of a Vertex. The *degree* of a vertex of a graph is the number of edges incident with that vertex.

DEFINITION 1-3. Subgraph, Proper Subgraph, and Complement of a Subgraph. A portion G_i containing a subset of the edges and vertices of a given graph G is called a *subgraph* of G. If G_i does not contain all the edges of G, it is a *proper subgraph* of G. The complement G_i' of G_i containing all the edges of G that are not in G_i is called the *complement subgraph* of the subgraph G_i of G.

In Definition 1-3, the terms "subgraph" and "proper subgraph" are used in the same sense as "subset" and "proper subset" in set theory. We note in set theory that a set is also a subset of itself. This same idea applies in graph theory; that is, if a subgraph G_i of a graph G is not a proper subgraph of G, G_i must be the graph G itself. In such a case, the complement G_i' of G_i is a *null graph*, a subgraph of G containing no edges or vertices, which is analogous to an empty set in set theory.

DEFINITION 1-4. Path and Path Set. If, in any graph G, a set $P_{v_1,v_{i+1}}$ of edges $e_1, e_2, \ldots, e_i$ can be ordered in the form $e_1(v_1,v_2), e_2(v_2,v_3), \ldots, e_i(v_i,v_{i+1})$, where the vertices $v_1, v_2, \ldots, v_i, v_{i+1}$ are distinct, then the set $P_{v_1,v_{i+1}}$ is said to form a *path* with v_1 and v_{i+1} as its terminals, and $v_2, v_3, \ldots, v_i$ as its *interior vertices*. $P_{v_1,v_{i+1}}$ is said to be connected between its terminals through its interior vertices. A *path set* is the set of all the edges contained in a path.

DEFINITION 1-5. Circuit (Loop) and Circuit Set. If the two terminals of a path coincide and the interior vertices are distinct, the path is a closed path and is called a *circuit* (or *loop*). The set of all the edges contained in a circuit is called a *circuit set*.

We might point out that a circuit may also be defined as a path, every vertex of which is of degree two *with respect to the path itself*, in light of Definitions 1-2 and 1-5.

DEFINITION 1-6. Connected and Unconnected Graphs. A graph is said to be a *connected graph* if there is at least one path between *every* pair of its vertices; otherwise, the graph is an *unconnected graph*.

DEFINITION 1-7. Edges Connected in Series (or Series Edges). If all the interior vertices of a path P in a graph G are of degree two *with respect to G*, the edges in P are said to be connected in *series*.

DEFINITION 1-8. Edges Connected in Parallel (or Parallel Edges). If, in a set C of edges of a graph G, any two of them form a circuit, the edges in C are said to be connected in *parallel*.

We shall now illustrate, by the following example, the definitions introduced thus far.

EXAMPLE 1-1. The graph G shown in Figure 1-5(a) is a linear graph consisting of seven edges ($e_1, e_2, \ldots, e_7$) and five vertices ($v_1, v_2, \ldots, v_5$). Vertex v_1 is of degree 2, since two edges (e_1, e_2) are incident with it. Similarly, vertices v_2, v_3, v_4, and v_5 are all of degree 3. Edges e_2, e_1, and e_4 together form a path between vertices v_3 and v_4 with the two vertices as terminals. Vertices v_1 and v_2 are the interior vertices of that path. The set $P_1\{e_1, e_2, e_4\}$ is a path set. In Figure 1-5(b), subgraph G_1 is shown consisting of edges e_1, e_2, and e_4.

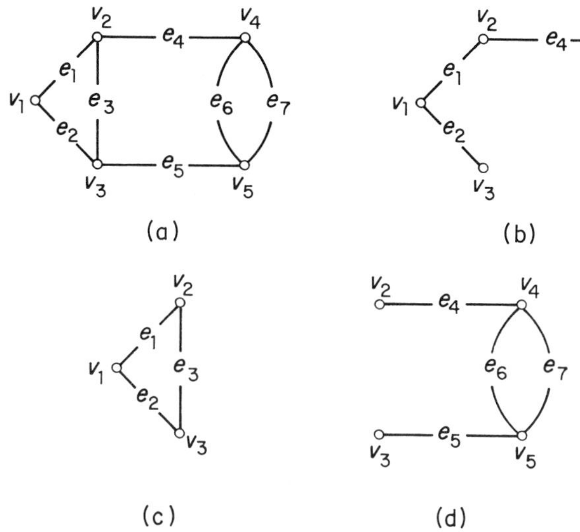

Figure 1-5 (a) Graph G. (b) Subgraph G_1. (c) Subgraph G_2. (d) Subgraph G_2'.

It is the path we just described. Note that, with respect to the path itself, the two interior vertices v_1 and v_2 are both of degree 2; but, with respect to G, v_1 is of degree 2 and v_2 is of degree 3. Graph G obviously is a connected, finite, linear graph — *connected* because there exists at least one path between any two vertices (Definition 1-6), and *finite* because the number of edges is finite. Subgraph G_2 in Figure 1-5(c), consisting of edges e_1, e_2, and e_3, is a circuit since the edges form a closed path. Again, with respect to G_2, all the vertices, v_1, v_2, and v_3, are of degree 2; but, with respect to G, v_1 is of degree 2, and v_2 and v_3 are both of degree 3. The set $C_1\{e_1, e_2, e_3\}$ is a circuit set, by Definition 1-5. Note that e_1 and e_2 in G are connected in series since, in the path formed by these two edges, the (only) interior vertex v_1 is of degree 2 with respect to G (Definition 1-7). Edges e_6 and e_7 in G are in parallel because both of them together form a circuit (Definition 1-8). Now, let us consider subgraph G_2 of G. The complement G_2' of G_2 in G is the subgraph containing edges e_4, e_5, e_6, and e_7, which form the set of all the edges that are not in G_2, as shown in Figure 1-5(d). All three subgraphs, G_1, G_2, and G_2', are proper subgraphs of G since none of the three subgraphs contains all the edges of G (Definition 1-3). But, G_2 and G_2' together form a subgraph (which, in fact, is the graph G itself). Hence, any subgraph formed by taking a subgraph and the complement of that subgraph is equal to the entire graph; and, by Definition 1-3, this resulting subgraph is not a proper subgraph.

DEFINITION 1-9. Separable and Nonseparable Graphs. A connected graph G is said to be *separable* if a subgraph G_i of G is connected with its complement G_i' through only one vertex v_k. Vertex v_k is called a *cut-vertex* (or separation vertex). Any connected graph that has no cut-vertices is *nonseparable*.

DEFINITION 1-10. Disjoint Graphs. Two graphs are *disjoint* if they have no vertices in common.

Since the graphs discussed in this text are those obtained from electrical networks, the direction of current flow and/or the voltage polarity marks are also of interest to us. In fact, we can assign arbitrarily an orientation to each branch of the network by either attaching an arrow to that branch to indicate the *assigned* positive direction of the current flow in that branch, or by putting a positive (+) polarity mark at one of the two terminals of that branch to denote the *arbitrarily assigned* positive terminal of the voltage across that branch. When such orientation is assigned to each of the branches, the graph of this *oriented* network will also carry the same orientations in the corresponding edges, and the graph becomes an oriented graph. Thus we state the following definitions.

DEFINITION 1-11. Oriented and Nonoriented Graphs. A graph is *oriented* if in each edge there is an orientation, either in the form of an arrow (indicating the assigned current flow in the corresponding branch of the network) or with a "+" sign at one of its two vertices (to denote the assigned positive terminal of the voltage across the corresponding branch of the network). If no orientations are assigned to its edges, the graph is a *nonoriented graph.*

In order to simplify our notation, we shall establish certain rules about the assignments of a positive direction of current flow and a positive voltage polarity mark in a branch. Consider a branch Z_i of a certain network as shown in Figure 1-6(a). We have assigned a positive current flow from terminal

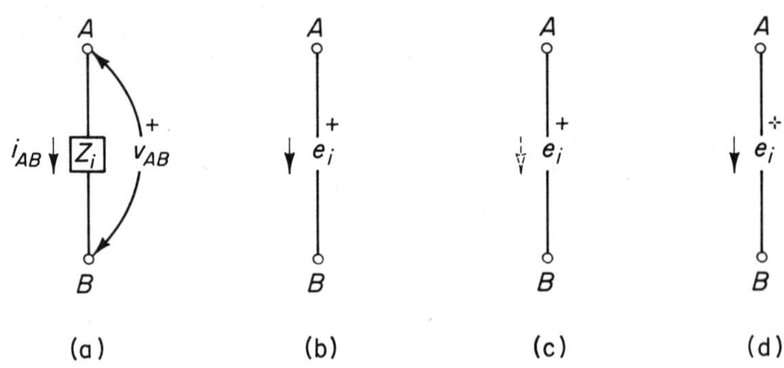

Figure 1-6 (a) An oriented branch. (b) The corresponding oriented edge. (c) The same edge with "+" mark only. (d) The same edge with arrow only.

A to terminal *B* through the branch denoted by Z_i. Also, the voltage V_{AB} across the two terminals is positive at *A* with respect to *B*. Accordingly, the edge is oriented in the same manner as shown in Figure 1-6(b). Since the assignments of positive current flow and positive voltage polarity are entirely arbitrary, these two assignments are independent from each other; that is, in Figure 1-6(a) and (b) we can just as well keep the "+" mark at A and reverse the sense of the arrow so that the tail of the arrow is at B instead of A. However, this will serve no purpose other than to confuse the issue. In fact, we shall establish a convention that only one set of signs — either arrows or "+" marks — is necessary in order to give the same information about the orientation of the edges, such as the case where both sets of signs are used. The convention is now stated as follows:

(a) If a "+" mark is attached to a terminal of an edge to indicate the positive voltage polarity, it also implies that the positive sense of the current flow is the one from the marked terminal to the unmarked terminal through the edge. That is, we shall visualize an arrow being assigned to the edge in such a way that the tail of the arrow is with the marked terminal.

(b) If an arrow is assigned to an edge to denote the positive sense of the current flow, the voltage polarity is also automatically assumed to be fixed in such a manner that the "+" mark is placed at the terminal identified by the tail of the arrow, even though it is not explicitly shown.

The idea stated in (a) is illustrated by Figure 1-6(c), and that in (b) is shown in Figure 1-6(d).

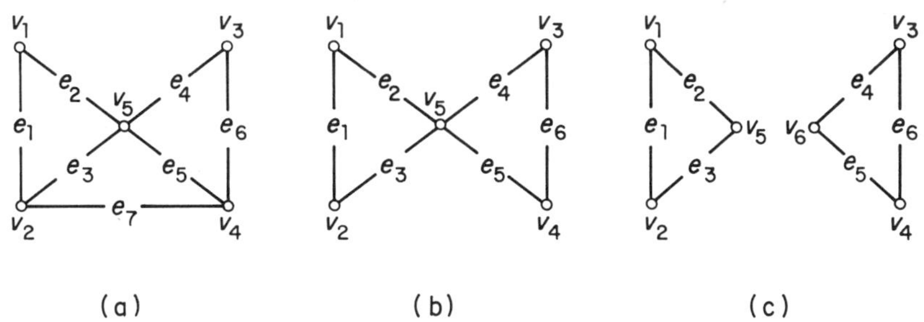

Figure 1-7 (a) Graph G. (b) Graph G_1. (c) Graph G_2.

EXAMPLE 1-2. Let us consider the graph G shown in Figure 1-7(a), which is a finite, connected, nonseparable, nonoriented graph. The fact that G is nonoriented is quite evident since no orientations are assigned to its edges. That it is nonseparable can be easily seen by observing that there exist no cut-vertices in G (Definition 1-9). In fact, we should note that each subgraph of G has at least two common vertices with its complement; this characteristic sometimes is used as a formal definition of a nonseparable graph. Now, suppose we remove edge e_7 from G. We then obtain a new graph G_1, which, of course, is a proper subgraph of G as shown in Figure 1-7(b). The new graph is a separable subgraph because, with e_7 removed, v_5 in G_1 now becomes a cut-vertex. This can be seen by realizing that the subgraph containing e_1,

e_2, and e_3 is connected at the cut-vertex v_5 with its complement, the subgraph of G_1, which does not contain any of the three edges, e_1, e_2, and e_3 (Definition 1-9).

Next, if vertex v_5 is "split," the resulting graph G_2 is a nonconnected graph with two disjoint subgraphs, each of which is a nonseparable, connected graph as shown in Figure 1-7(c). The two disjoint subgraphs are the one containing e_1, e_2, and e_3, and the one consisting of e_4, e_5, and e_6. Note that these two subgraphs are disjoint with respect to one another since they have no vertices in common, but each of the two subgraphs is a connected, nonseparable graph by itself. These two subgraphs are said to be *edge-disjoint* in G_1 of Figure 1-7(b) before the vertex was split. They are edge-disjoint simply because they have no edges in common.

Graph G of Figure 1-7(a) becomes an oriented graph when a set of arrows is assigned to all of its edges. The resulting graph is denoted by G_3 and is shown in Figure 1-8(a). If "+" marks are used instead of arrows, graph G_4 results, as shown in Figure 1-8(b). Finally, if both arrows and "+" marks are used, we get G_5, as illustrated in Figure 1-8(c).

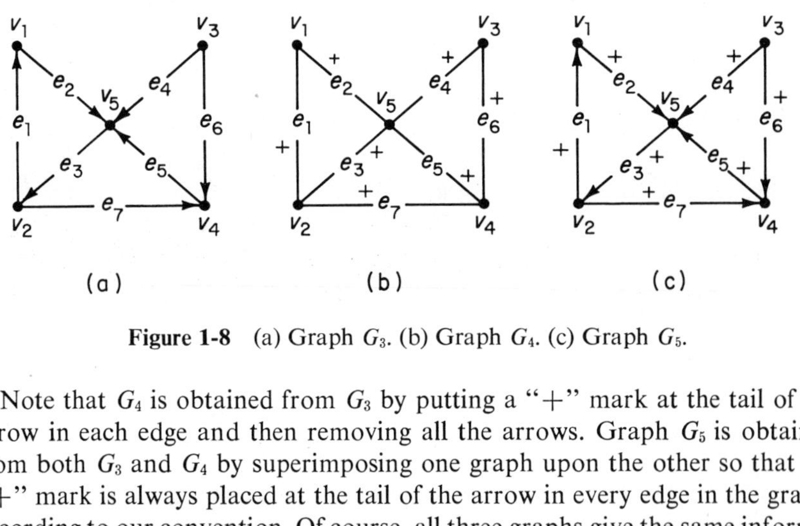

Figure 1-8 (a) Graph G_3. (b) Graph G_4. (c) Graph G_5.

Note that G_4 is obtained from G_3 by putting a "+" mark at the tail of the arrow in each edge and then removing all the arrows. Graph G_5 is obtained from both G_3 and G_4 by superimposing one graph upon the other so that the "+" mark is always placed at the tail of the arrow in every edge in the graph, according to our convention. Of course, all three graphs give the same information about edge orientations.

It is worth noting that when two graphs are vertex-disjoint, they are necessarily edge-disjoint since, if the two graphs have some common edges, the vertices that are the endpoints of these common edges are also the common vertices to both graphs. Hence, we shall use the term "disjoint" to mean *vertex-disjoint;* thus, if two graphs are *disjoint* they must have neither vertices nor edges in common, since vertex-disjoint also implies edge-disjoint.

In the same manner, the terms "disjoint" and "edge-disjoint" may be applied to circuits, and so forth. For example, if two circuits are disjoint, they are completely separated and have no common vertices or edges.

We shall next define one of the most important basic concepts in graph theory — a tree.

DEFINITION 1-12. Tree and Branch. A *tree* of a connected graph G is a connected subgraph that contains all the vertices of G but no circuits. The edges contained in a tree are called the *branches* of that tree.

DEFINITION 1-13. Chord Set and Chord. If T_i denotes a tree of a graph G, then the complement subgraph T_i' of T_i is called the *chord set* (or *cotree*) with respect to T_i and the edges contained in T_i' are called the *chords* (or *links*).

From Definitions 1-12 and 1-13, we see that for any connected G, if T_i is a tree of G and T_i' is the chord set with respect to T_i, then the two subgraphs T_i and T_i' are edge-disjoint but not (vertex) disjoint. This is because G is a connected graph and, consequently, T_i and T_i' must have some common vertices.

DEFINITION 1-14. Cutset and Vertex Cutset. A *cutset* of a connected graph is a minimal set of edges whose removal will separate the remainder of the graph into two disjoint subgraphs, one (or both) of which may be an *isolated vertex*. If one isolated vertex results, the cutset is called a *vertex cutset.*

DEFINITION 1-15. Incidence Set. An *incidence set* is a set of all edges incident with a common vertex.

In Definition 1-14, the term "a minimal set" simply means that no *proper* subset of it possesses the same property as stated in the definition.

We might point out also that an incidence set may or may not be a cutset whereas a vertex cutset is always an incidence set. This point, as well as others, will be illustrated by the next example.

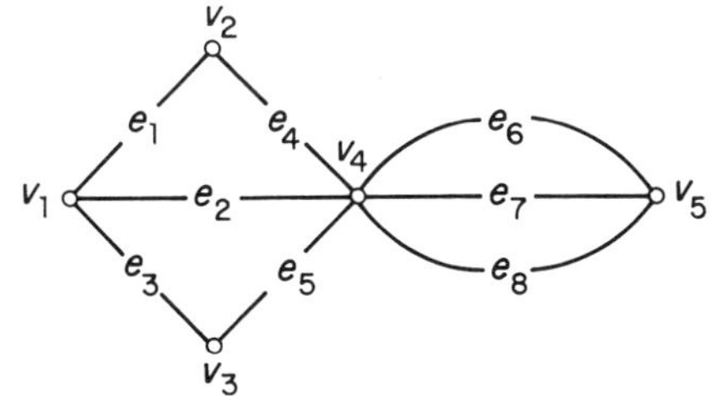

Figure 1-9 Graph G.

EXAMPLE 1-3. Consider the graph G of Figure 1-9. We see that G is a nonoriented, connected, separable graph containing eight edges and five vertices. Two trees of G are shown in Figure 1-10.

The chord sets T_a' and T_b' corresponding to T_a and T_b, respectively, are shown in Figure 1-11. Three cutsets of G are shown in Figure 1-12.

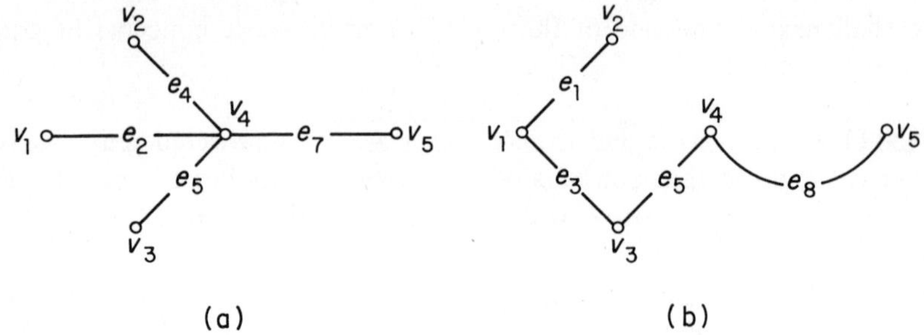

Figure 1-10 (a) Tree $T_a\{e_2,e_4,e_5,e_7\}$. (b) Tree $T_b\{e_1,e_3,e_5,e_8\}$.

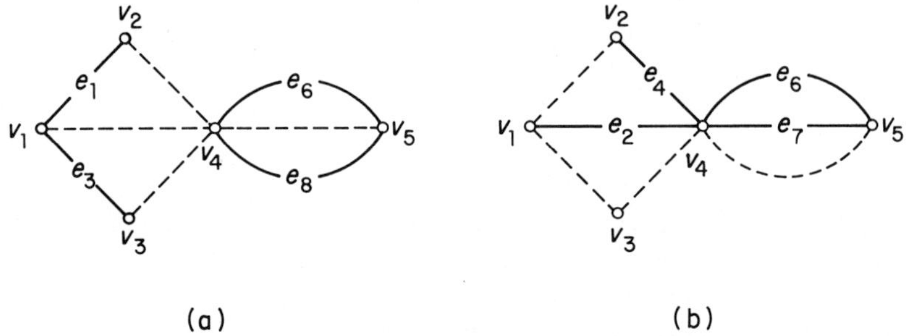

Figure 1-11 (a) Chord set $T'_a\{e_1,e_3,e_6,e_8\}$. (b) Chord set $T'_b\{e_2,e_4,e_6,e_7\}$.

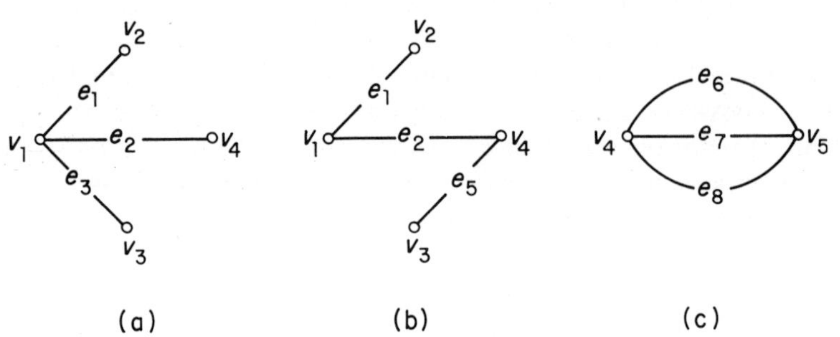

Figure 1-12 (a) Cutset $K_1\{e_1,e_2,e_3\}$. (b) Cutset $K_2\{e_1,e_2,e_5\}$. (c) Cutset $K_3\{e_6,e_7,e_8\}$.

In Figure 1-13, the remaining graph is shown after the cutset is removed from G in each case. We see that each of the three remaining subgraphs is separated into two parts — namely, two disjoint subgraphs. In Figure 1-13(a), one part is the *isolated vertex* v_1, and the other part is the subgraph containing edges e_4, e_5, e_6, e_7, and e_8; in Figure 1-13(b), one part is the edge e_3 and the other part is the subgraph containing edges e_4, e_6, e_7, and e_8; and, finally, in Figure 1-13(c), one part is the subgraph containing edges e_1, e_2, e_3, e_4, and e_5 and the other part is the isolated vertex v_5.

Evidently, cutsets K_1 and K_3 are vertex cutsets, whereas cutset K_2 is not. Also, cutsets K_1 and K_3 are incidence sets, since all the edges in each of the

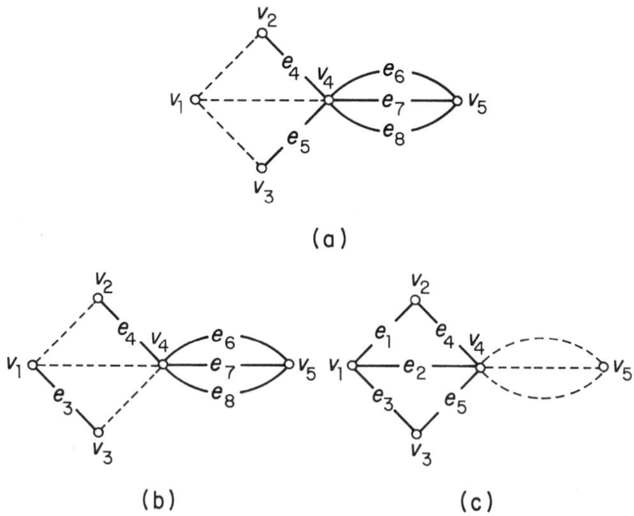

Figure 1-13 (a) Subgraph $K_1'\{e_4,e_5,e_6,e_7,e_8\}$. (b) Subgraph $K_2'\{e_3,e_4,e_6,e_7,e_8\}$. (c) Subgraph $K_3'\{e_1,e_2,e_3,e_4,e_5\}$.

two sets are incident with a common vertex — v_1 for K_1 and v_5 for K_3. The only incidence set *that is not a cutset* in G is the set of all edges incident with v_4 (a cut-vertex) — that is, $S_4\{e_2, e_4, e_5, e_6, e_7, e_8\}$. The fact that S_4 is not a cutset can be readily seen by observing that K_3 is a proper subset of S_4. Consequently, S_4 is not a minimal set as stated in Definition 1-14, since there exists a proper subset of S_4 (namely, K_3) which possesses the property "the removal of the set will separate the remainder of the graph into two disjoint subgraphs, one (or both) of which may be an isolated vertex" as stated in the definition.

In the next section, we shall define a set of operations among the subgraphs of a graph. Some of these operations are borrowed from set theory.

1-4 OPERATIONS AMONG SUBGRAPHS

There are four operations to discuss in this section: *union, intersection, ring sum*, and *difference*. We shall first define each of them and then illustrate these operations through examples.

DEFINITION 1-16. Union (∪). The *union* of two sets S_1 and S_2 of objects (which are called *elements* of the sets) is the set $S_1 \cup S_2$ of all objects that are in S_1, or in S_2, or in both.

If G_1 and G_2 are two subgraphs of a graph G, then the union $G_1 \cup G_2$ is the subgraph containing all the edges (and vertices) that are in G_1, or in G_2, or in both G_1 and G_2.

DEFINITION 1-17. Intersection ($\cap$). The *intersection* of two sets S_1 and S_2 is the set $S_1 \cap S_2$, consisting of all the elements that are common to both sets.

Hence, by the intersection of two circuits, for example, we mean the set of all edges common to both circuits.

DEFINITION 1-18. Ring Sum ($\oplus$). The *ring sum* of two sets S_1 and S_2 is the set, $S_1 \oplus S_2$, of all the edges that are either in S_1 or in S_2 but *not in both.*

DEFINITION 1-19. Difference ($-$). The *difference* between two sets, S_1 and S_2 (where S_2 is a proper subset of S_1) in the order S_1, S_2 is the set, $S_1 - S_2$, of all edges that are in S_1 but not in S_2.

We may now interrelate the four operations by means of the following equation:

$$S_1 \oplus S_2 = \{S_1 \cup S_2\} - \{S_1 \cap S_2\} \tag{1-1}$$

where S_1 and S_2 are two sets.

Equation (1-1) may be illustrated by the diagrams of Figure 1-14. The two sets, S_1 and S_2, are shown in Figure 1-14(a) with S_1 represented by the area

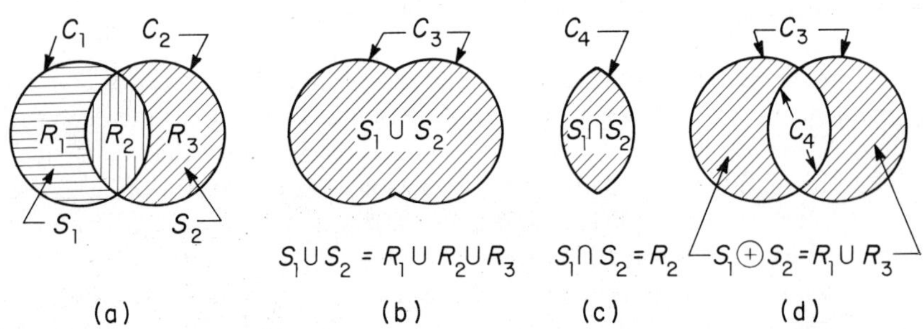

Figure 1-14 (a) Sets S_1 and S_2. (b) $S_1 \cup S_2$. (c) $S_1 \cap S_2$. (d) $S_1 \bigcirc S_2$.

bounded by the closed curve C_1, and S_2 identified by the area bounded by the closed curve C_2. The total area bounded by C_1 and C_2 together can be divided into three regions, R_1, R_2, and R_3, as shown in Figure 1-14(a). The union of the two sets, $S_1 \cup S_2$, may now be represented by the three regions joined together as depicted by Figure 1-14(b). Mathematically, we write

$$S_1 \cup S_2 = R_1 \cup R_2 \cup R_3. \tag{1-2}$$

The intersection of the two sets, $S_1 \cap S_2$, is the area bounded by C_4 and is equal to the region R_2 as indicated in Figure 1-14(c). Thus,

$$S_1 \cap S_2 = R_2 \tag{1-3}$$

The ring sum of the two sets, $S_1 \oplus S_2$, is the area bounded by the interior of C_3 and the exterior of C_4 as shown in Figure 1-14(d). Hence,

$$S_1 \oplus S_2 = R_1 \cup R_3$$

Now, if we take the difference, $\{S_1 \cup S_2\} - \{S_1 \cap S_2\}$ (where $S_1 \cap S_2$ is a proper subset of $S_1 \cup S_2$), by subtracting Equation (1-3) from Equation (1-2), or by inspection of Figure 1-14, we have

$$S_1 \oplus S_2 = \{S_1 \cup S_2\} - \{S_1 \cap S_2\} = R_1 \cup R_3$$

as stated in Equation (1-1).

The next example will illustrate how the foregoing operations are applied to graphs.

EXAMPLE 1-4. Consider the graph G of Figure 1-9. As given in Example 1-3, two of the trees are $T_a\{e_2, e_4, e_5, e_7\}$ and $T_b\{e_1, e_3, e_5, e_8\}$. The reader can easily verify the following results:

$$T_a \cup T_b = \{e_1, e_2, e_3, e_4, e_5, e_7, e_8\} \tag{1-4}$$

$$T_a \cap T_b = \{e_5\} \tag{1-5}$$

$$T_a \oplus T_b = \{e_1, e_2, e_3, e_4, e_7, e_8\} \tag{1-6}$$

From Equations (1-4), (1-5), and (1-6), we see that

$$T_a \oplus T_b = \{T_a \cup T_b\} - \{T_a \cap T_b\} = \{e_1, e_2, e_3, e_4, e_7, e_8\}$$

Next, performing the operations between tree T_a and the corresponding chord set T_a', we find

$$T_a \cup T_a' = \{e_1, e_2, e_3, e_4, e_5, e_6, e_7, e_8\} = G \tag{1-7}$$

and

$$T_a \cap T_a' = \varnothing \tag{1-8}$$

where $\varnothing$ is an empty set (or null set) — that is, a set containing no edges. Also,

$$T_a \oplus T_a' = \{e_1, e_2, e_3, e_4, e_5, e_6, e_7, e_8\} = G \tag{1-9}$$

Hence, in the case where $T_a \cap T_a' = \varnothing$,

$$T_a \oplus T_a' = T_a \cup T_a' \tag{1-10}$$

This should be quite obvious since T_a and T_a' are complements to each other by definition.

Now let us consider the two cutsets of G — K_1 and K_3 — both of which were shown in Figure 1-12, and $K_5\{e_2, e_4, e_5\}$. For convenience, graph G and cutsets K_1 and K_3 are repeated in Figure 1-15(a), (b), and (c), respectively. Cutset K_5 is shown in Figure 1-15(d). We see that

$$K_3 \cup K_5 = S_4 = \{e_2, e_4, e_5, e_6, e_7, e_8\}$$

which is the same incidence set S_4 discussed in Example 1-3.

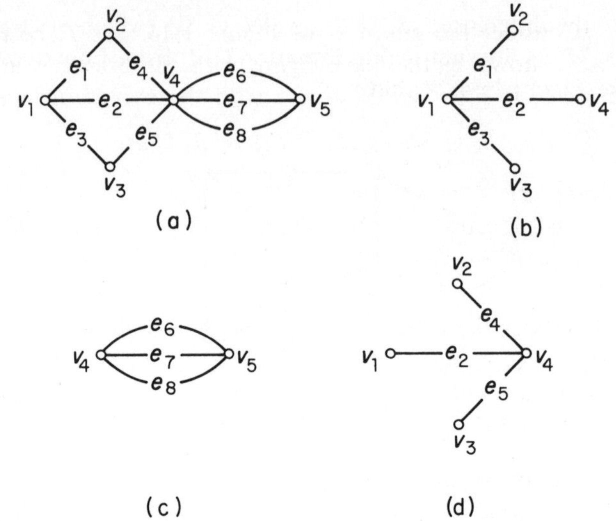

Figure 1-15 (a) Graph G. (b) Cutset K_1. (c) Cutset K_3. (d) Cutset K_5.

Next, taking the union of K_1 and K_3, we get

$$K_1 \cup K_3 = S_5 = \{e_1, e_2, e_3, e_6, e_7, e_8\}$$

which is a disjoint union of two cutsets.

In Example 1-4, two interesting results are suggested:

(a) If S_1 and S_2 are two sets, the equation

$$S_1 \oplus S_2 = S_1 \cup S_2 \tag{1-11}$$

holds if and only if $S_1 \cap S_2 = \varnothing$. Thus, if S_1 and S_2 are two subgraphs of a given graph, Equation 1-11 implies that S_1 and S_2 are at least edge-disjoint (they might even be vertex-disjoint).

(b) An incidence set is either a cutset or an edge-disjoint union of cutsets.

1-5 SOME BASIC THEOREMS

We shall define a few more terms before attempting to state and prove some of the basic theorems in graph theory.

DEFINITION 1-20. Fundamental Circuit (f-circuit). A *fundamental circuit* with respect to a given tree T of a connected graph is a unique circuit containing one chord and the unique tree path formed by some or all of the branches of T between the two vertices of the chord.

DEFINITION 1-21. Fundamental Cutset (f-cutset). A *fundamental cutset* with respect to a given tree T of a connected graph is a (unique) cutset consisting of one branch of T and possibly some (or all) chords.

EXAMPLE 1-5. Consider the graph G of Figure 1-16. Let T_1 be the tree $T_1\{e_1, e_4, e_5, e_8, e_{10}\}$, shown in the same figure with the branches in heavy lines. The chord set T_1' is $T_1'\{e_2, e_3, e_6, e_7, e_9\}$ with the chords in light lines.

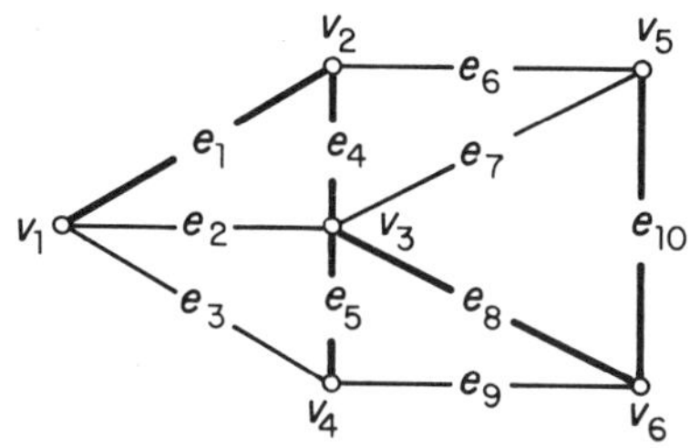

Figure 1-16 A connected graph G.

Since each fundamental circuit contains one distinct chord, and since there are five chords, there will be five fundamental circuits. They are:

$$C_{e_2}\{e_1, e_2, e_4\} \qquad C_{e_3}\{e_1, e_3, e_4, e_5\} \qquad C_{e_6}\{e_4, e_6, e_8, e_{10}\}$$

$$C_{e_7}\{e_7, e_8, e_{10}\} \qquad C_{e_9}\{e_5, e_8, e_9\}$$

Also, each fundamental cutset contains a unique branch (along with some chords, possibly) and there are five branches; we conclude, therefore, the number of fundamental cutsets is also five. They are:

$$K_{e_1}\{e_1, e_2, e_3\} \qquad K_{e_4}\{e_2, e_3, e_4, e_6\} \qquad K_{e_5}\{e_3, e_5, e_9\}$$

$$K_{e_8}\{e_6, e_7, e_8, e_9\} \qquad K_{e_{10}}\{e_6, e_7, e_{10}\}$$

From this example, we see that

(a) The number of f-circuits is equal to the number of elements contained in a chord set.

(b) The number of f-cutsets is equal to the number of elements contained in a tree.

For any tree T_i and the corresponding chord set T_i', we know that $T_i \cup T_i' = G$. In other words, the number of edges in T_i plus the number of edges in T_i' is equal to the total number of edges in G. Hence, if we know what the number of edges in a tree must be for a given graph, the number of f-circuits and f-cutsets can easily be determined.

We shall state and prove the following two theorems about the trees of a graph:

THEOREM 1-1. There exists at least one tree in every connected graph.

Proof: Let G be a connected graph. Then either (A) G contains some circuits or (B) none at all.

(A) If G does not contain any circuit, it is a tree by itself since it contains all its vertices (being a connected graph by assumption) and no circuit, thus satisfying the definition of a tree.

(B) If G contains some circuits, then consider one of the circuits of G. This circuit will be destroyed when one of its edges is removed from G; but the remaining subgraph will still be connected, since there exist exactly two paths between any two vertices in a circuit. Repeat this circuit-destroying process until there is no circuit contained in the remaining subgraph. Since the application of each of the circuit-destroying steps will result in a connected subgraph, the final one that contains no circuits is a connected subgraph of G, and, hence, is a tree by definition. The theorem is now proved.

It is worth noting that the method of proof used in proving Theorem 1-1 is a direct proof. We shall now state the next theorem and prove it by mathematical induction.

THEOREM 1-2. If a connected graph G contains v vertices, a tree of G contains $v - 1$ edges.

Proof:

(A) For the case when $v = 1$ — that is, when G is a connected graph consisting of only one vertex — there can be no edges contained in G since an edge must have two vertices.[2] This means that G is just a vertex. Hence a tree of G is the vertex also, containing $1 - 1 = 0$ or no edge. Therefore, the theorem is true for $v = 1$.

(B) Let the theorem be true for $v = k$, some positive integer. That is, we assume that the following statement is true: "If G_1 contains k vertices, then a tree of G_1 contains $k - 1$ edges."

Let us identify this statement by P and consider any graph G_2 containing $k + 1$ vertices. G_2 must contain a vertex v_i that is not a cut-vertex[3] (see Definition 1-9). Then the incidence set K_i of v_i is a cutset since the removal of K_i from G_2 separates v_i from the remaining subgraph, say G_3. But G_3 is a connected graph containing $(k + 1) - 1 = k$ vertices with vertex v_i subtracted from a set of $k + 1$ vertices. Hence, by statement P, a tree T_3 of G_3 must contain $k - 1$ edges. Now, adding any member of the cutset K_i to the tree T_3 of G_3 will yield a circuitless connected subgraph T_2 containing all the $k + 1$ vertices of G_2. Hence, by definition, T_2 is a tree of G_2. But, since T_3 contains $k - 1$ edges, and since T_2 is obtained by adding one edge (in the cutset K_i) to T_3, it follows that T_2 contains $(k - 1) + 1 = k$ edges. Thus, by virtue of mathematical induction, the theorem is true for any positive integer v.

As suggested in an earlier example, every tree of a connected graph G of v vertices and e edges has the following four properties:

[2]Clearly, we exclude the trivial case of a "self-loop" — a loop that contains only one edge.
[3]We shall leave the proof of this statement to the reader.

(a) It is a connected subgraph of G.
(b) It contains all the v vertices.
(c) It contains exactly $v - 1$ of the e edges.
(d) It contains no circuits.

The next theorem may be considered as another definition of a tree.

THEOREM 1-3. A graph G is a tree if and only if there exists exactly one path between any two vertices of G.

Proof: The proof of this theorem consists of two parts: The necessity and the sufficiency.

(A) For the necessity [the "only if" portion ($\Rightarrow$)], let G be a tree. Then, by Definition 1-12, it is a connected graph and hence there exists *at least one* path between any two of its vertices. Moreover, since G also contains no circuits, according to Definition 1-12 there can be *no more than one* path between any pair of vertices of G, otherwise it will contain some circuits. Thus the necessary condition is established.

(B) For the sufficiency [the "if" portion ($\Leftarrow$)], let there be exactly one path between any two vertices of G. Then, obviously, G is connected. Furthermore, since there is only one path between any two vertices of G, there can be no circuits in G, keeping in mind that for a circuit containing some two vertices to exist, there must be two paths between those two vertices. This completes the proof of the sufficiency part of the theorem.

Parts (A) and (B) together constitute a deductive proof of Theorem 1-3.

We might mention here that any three (or certain two) of the four properties (a), (b), (c), and (d) given earlier define a tree of a connected graph of v vertices. For example, we may state the following theorem:

THEOREM 1-4. Let G be a connected graph of v vertices, and let G_t be a subgraph of G containing $v - 1$ edges and no circuits [thus G_t has properties (c) and (d)]; then G_t is a tree of G.

The proof of Theorem 1-4 is left to the reader as an exercise (Problem 1-4). In fact, the reader certainly will gain a great deal if he states and proves, as additional theorems, statements with other combinations of the four properties of a tree, taken three at a time.

Since a tree of a connected graph G of v vertices contains exactly $v - 1$ edges, the next theorem is almost self-evident.

THEOREM 1-5. For any given tree T of a connected graph G of v vertices and e edges, there are exactly $v - 1$ branches and $e - v + 1$ chords with respect to T.

Proof: Since T contains $v - 1$ edges, these $v - 1$ edges of G are the branches of T. But G contains a total of e edges by assumption. The number of edges which are not contained in T is $e - (v - 1) = e - v + 1$. By Definition 1-13, these $e - v + 1$ edges of G form a chord set with respect to T and, therefore, there are $e - v + 1$ chords. Hence the theorem.

It is now obvious that, for a given tree T, the number of f-circuits is $e - v + 1$ and the number of f-cutsets is $v - 1$. This is due to the fact that each of the $e - v + 1$ chords determines a unique circuit (along with some of the $v - 1$ tree branches), and each of the $v - 1$ branches determines a unique cutset (with possibly some of the $e - v + 1$ chords).

We next state and prove a theorem which exhibits certain properties of a graph with regard to its cutsets and circuits.

THEOREM 1-6. The number of common edges between a cutset and a circuit in a connected graph is always even.

Proof: Let C be a circuit and K be a cutset of a connected graph G. Then either C and K have some common edges or none at all. These two cases will now be considered separately.

(A) Let C and K have some common edges and let l be the "cut line" which intersects all the edges of K as shown in Figure 1-17. Let $e_{k_1}, e_{k_2}, \ldots, e_{k_i}$ be

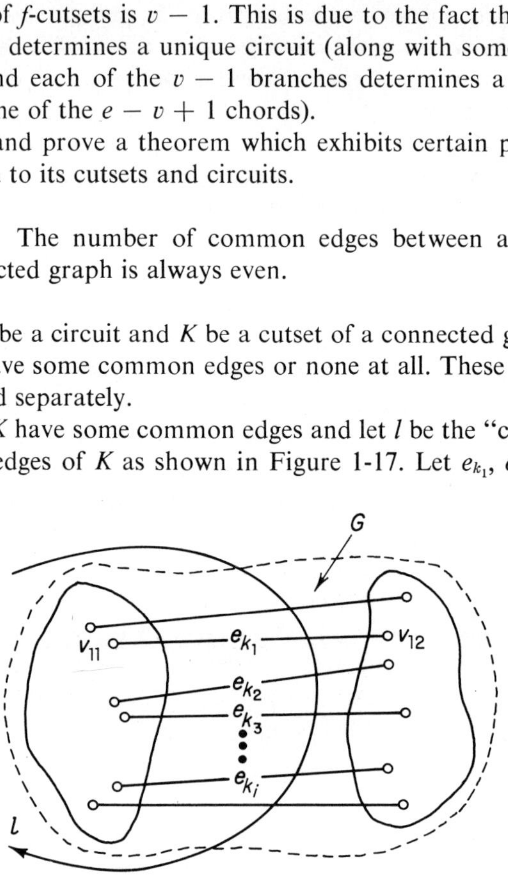

Figure 1-17 Common edges between a circuit C and a cutset K.

the common edges between C and K. Since cut line l intersects all the edges of K, it must also intersect edges e_{k_1} through e_{k_i}. Also, the endpoints of edges e_{k_1} through e_{k_i} are such that one endpoint of each edge is located to the left of l and the other to the right of l. Let us now trace circuit C from left to right starting at vertex v_{11} of e_{k_1}, through e_{k_1} to the other vertex v_{12} of e_{k_1} as shown in Figure 1-17. In order to complete the circuit, we must, at a certain stage of the tracing, cross l again from right to left (and thus intersect another common edge, say e_{k_2}) since our starting point v_{11} is assumed to be on the left-hand side of l. If this completes the circuit, then the number of common edges will be exactly two. But, if the tracing is not complete, we must then cross l once again from left to right through another common edge, say e_{k_3}, and back from right to left through still another common edge, say e_{k_4}, in order to come back to v_{11} to complete the circuit. This argument can be repeated, and for every cycle — from left to right and from right back to left — two edges will be intersected from the set of edges $e_{k_1}, e_{k_2}, \ldots, e_{k_i}$. Thus the number of common edges between C and K is the even number $2n$ where n is the number of such "round trips" or cycles crossing line l from left to right and back to left.

(B) If C and K have no common edges, then the number of common edges between C and K is zero, which is considered an even number in this trivial case. The proof of the theorem is now complete.

Finally, we shall state the following theorem about the ring-sum operation between circuits.

THEOREM 1-7. The ring sum of two circuits of a connected graph G is either a circuit or an edge-disjoint union of circuits of G.

The proof of Theorem 1-7 may be accomplished by considering the two cases separately: (A) The two circuits have some common edges. (B) They have no common edges. (This approach is similar to the one used in the proof of Theorem 1-6.) The execution of these two steps will be left as a problem (Problem 1-23).

1-6 CONGRUENCE AND EQUIVALENCE

Consider the three graphs, G_1, G_2, and G_3, of Figure 1-18. We note that all three graphs have six edges and four vertices. Furthermore, each vertex is of degree three. We also observe that graphs G_1 and G_3 are similar in appearance, whereas G_2 is quite different from the other two in shape.

Now let us consider graph G_2 in Figure 1-18(b). We note that if vertex v_4' is "pushed" toward the center of the triangle identified by edges e_1', e_3', and

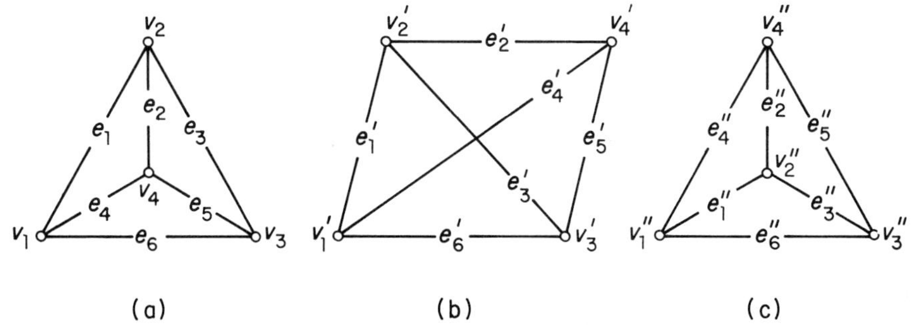

Figure 1-18 (a) Graph G_1. (b) Graph G_2. (c) Graph G_3.

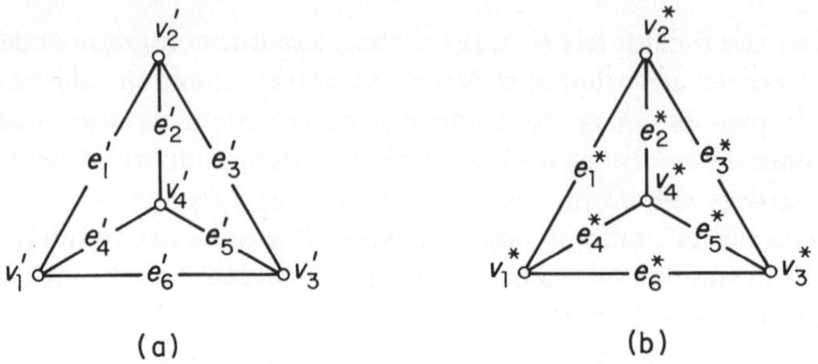

Figure 1-19 (a) Graph G_4. (b) Graph G_5.

e_6', a new graph, G_4, will result, as shown in Figure 1-19(a). Graph G_4 is exactly the same graph as G_1 if the primes in the vertices and edges of G_4 are omitted.

Next, consider G_3 of Figure 1-18(c). We see that this graph can also be made identical to G_1 provided the following changes of the labels are performed with respect to the vertices and edges.

For the vertices: $v_1'' \to v_1^*$, $v_2'' \to v_4^*$, $v_3'' \to v_3^*$, $v_4'' \to v_2^*$

For the edges: $e_1'' \to e_4^*$, $e_2'' \to e_2^*$, $e_3'' \to e_5^*$, $e_4'' \to e_1^*$, $e_5'' \to e_3^*$, $e_6'' \to e_6^*$

The new graph, G_5, is shown in Figure 1-19(b). Evidently, if all the asterisks are dropped, G_5 becomes exactly the same as G_1.

The above discussion has been intended to illustrate the idea of *congruent* (isomorphic or 1-isomorphic) graphs, namely, graphs that are topologically identical in that the interconnections between the edges and vertices are exactly the same — provided, of course, that the relabeling of the vertices and edges is allowed when necessary. All five graphs, G_1 through G_5, are examples of congruent graphs. We shall now formally define congruence.

DEFINITION 1-22. Congruence. Two graphs are said to be *congruent* (*1-isomorphic* or simply *isomorphic*) if the following conditions are satisfied:

(a) They have the same number of vertices.
(b) They have the same number of edges.
(c) A one-to-one correspondence between their vertices and a one-to-one correspondence between their edges can be established such that the incidence relationships between the vertices and edges are preserved.

Obviously, in Definition 1-22, condition (c) implies both conditions (a) and (b). But, if any of the first two conditions is not satisfied, there will be no need to go any further to examine the third condition in order to determine whether two graphs are congruent. Since conditions (a) and (b) can be checked very easily, it pays to count the number of vertices and edges in both graphs before considering condition (c); for if the two graphs do not have the same number of vertices and edges, they cannot be congruent.

On the other hand, even if both conditions, (a) and (b), are satisfied, we will still have to see whether condition (c) is also satisfied before we can determine whether the two graphs are congruent. In fact, the third condition must be satisfied by the two graphs as the necessary and sufficient condition for congruence, whereas the first two conditions are merely two of the necessary conditions for congruence.

EXAMPLE 1-6. Let us consider graphs G_1 and G_3 of Figure 1-18. We see that both conditions, (a) and (b), are satisfied since both graphs have four vertices and six edges. Condition (c) is also satisfied by the two graphs — as shown by Table 1-1, which reveals the incidence relationships between the vertices and edges of the two graphs.

The table is self-explanatory. For example, in the first row, it is shown that (a) vertex v_1 of G_1 is incident with edges e_1, e_4, and e_6 of G_1, and vertex

	G_1		G_3	
row 1	v_1	e_1 ↔	e''_4	v''_1
		e_4 ↔	e''_1	
		e_6 ↔	e''_6	
row 2	v_2	e_1 ↔	e''_4	v''_4
		e_2 ↔	e''_2	
		e_3 ↔	e''_5	
row 3	v_3	e_3 ↔	e''_5	v''_3
		e_5 ↔	e''_3	
		e_6 ↔	e''_6	
row 4	v_4	e_2 ↔	e''_2	v''_2
		e_4 ↔	e''_1	
		e_5 ↔	e''_3	

Table 1-1 The incidence relationships between the edges and vertices of Graph G_1 and G_3 of Figure 1-18.

v_1'' of G_3 is incident with edges e_4'', e_1'', and e_6'' of G_3; that (b) edge e_1 of G_1 corresponds to edge e_4'' of G_3, e_4 corresponds to e_1'', and e_6 corresponds to e_6''; and that (c) v_1 of G_1 corresponds to v_1'' of G_3. Similar remarks may be applied to the three other rows. Thus, by Definition 1-22, G_1 and G_3 are congruent.

From this example we see that the "incidence relationships" requirement implies that the one-to-one correspondence between the vertices of two congruent graphs should be set up in such a way that two corresponding vertices must have the same degree. This implication being a necessary condition, however, is not a sufficient one for condition (c) of Definition 1-22 to be satisfied, as the next example will show.

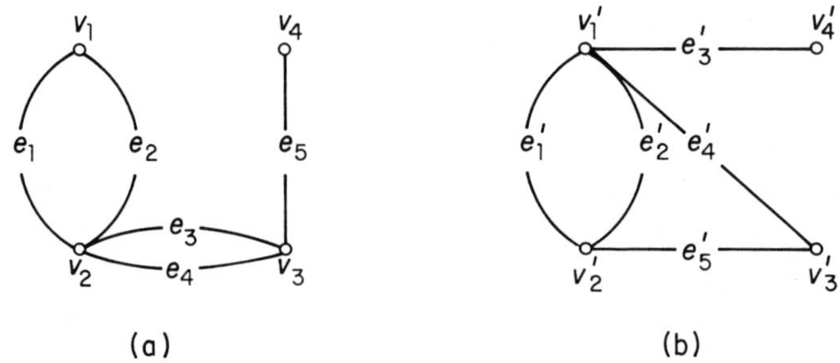

Figure 1-20 (a) Graph G_a. (b) Graph G_b.

EXAMPLE 1-7. Both graphs shown in Figure 1-20 have the same number of vertices (4) and the same number of edges (5). Let the vertices of the two graphs be paired off in the manner indicated in Figure 1-21.

In the first row, we see that v_1 of G_a is made correspondent to v_3' of G_b and both vertices are of degree two. Similarly, v_2 and v_1', v_3 and v_2', and v_4 and v_4' are paired off, each pair of vertices having the same degree as indicated in Figure 1-21. However, G_a and G_b are not congruent even though they satisfy the first two conditions of Definition 1-22 as well as the degree requirement

	G_a	G_b	Degree
row 1	v_1 ⟷	v_3'	(vertices of degree 2)
row 2	v_2 ⟷	v_1'	(vertices of degree 4)
row 3	v_3 ⟷	v_2'	(vertices of degree 3)
row 4	v_4 ⟷	v_4'	(vertices of degree 1)

Figure 1-21 One-to-one correspondence between the vertices of G_a and G_b.

in the vertex-matching operation because they do not satisfy condition (c) in the definition. This can be seen easily by noting that two of the edges, e_1 and e_2, incident with v_2 in G_a also are incident with vertex v_1, which is of degree two. This incidence relationship does not exist in G_b, since at v_1' of G_b (the vertex corresponding to v_2 of G_a) there is only one edge (e_4') incident with v_3' — the vertex of degree two, corresponding to v_1 of G_a. Hence the incidence relationship in G_a is not preserved in G_b; consequently, condition (c) of Definition 1-22 cannot be satisfied.

We shall next discuss a class of graphs called *equivalent graphs*, which are not congruent but can be made congruent by the (repeated) application of certain operations. We shall introduce the idea of equivalent graphs by means of an example and then define it formally.

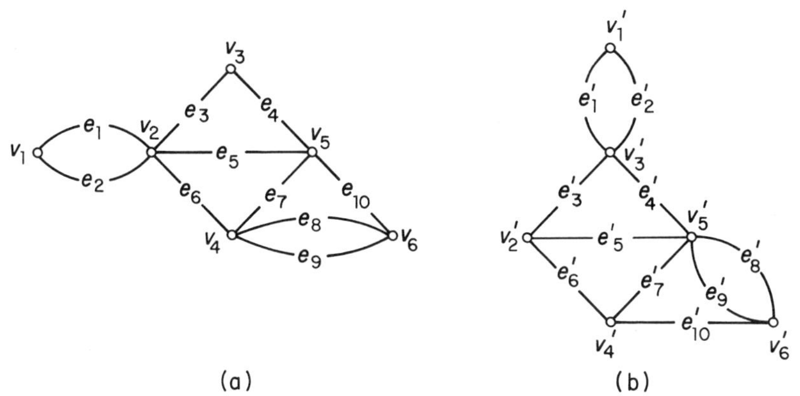

Figure 1-22 Two equivalent graphs: (a) Graph G_a. (b) Graph G_b.

EXAMPLE 1-8. Consider the two graphs depicted in Figure 1-22. The two graphs can be made congruent by transforming both graphs into two new ones through (a) the separation of a graph at a cut-vertex and (b) the turning

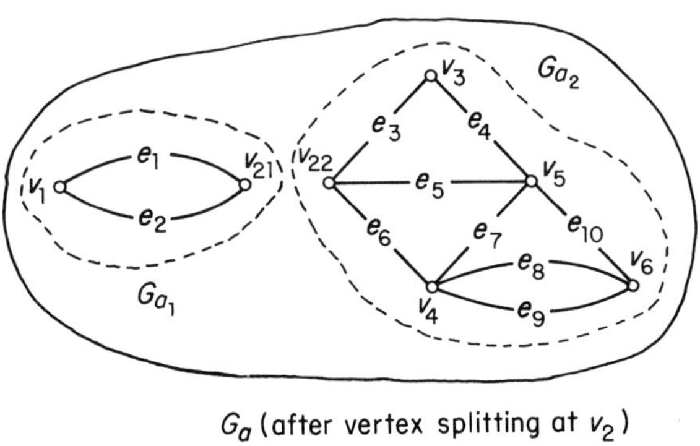

Figure 1-23 Separation of graph G_a at a cut-vertex v_2.

around of a subgraph about two vertices. We shall deal with the two graphs separately and then compare the two final transformed graphs.

(A) For graph G_a: Separate the graph into two connected pieces by splitting the cut-vertex v_2 and then identifying the two connected pieces by G_{a_1} and G_{a_2}, as indicated in Figure 1-23.

(B) For graph G_b:

Step 1: Separate the graph into two connected pieces at the cut-vertex v_3' and then identify the two subgraphs by G_{b_1} and G_{b_2} as shown in Figure 1-24(a).

Step 2: Turn around the subgraph consisting of edges e_8', e_9', e_{10}' in G_{b_2} about vertices v_4' and v_5', as shown in Figure 1-24, so that after the turning around, edges e_8' and e_9' are incident with vertex v_4' and edge e_{10}' is incident with vertex v_5'.

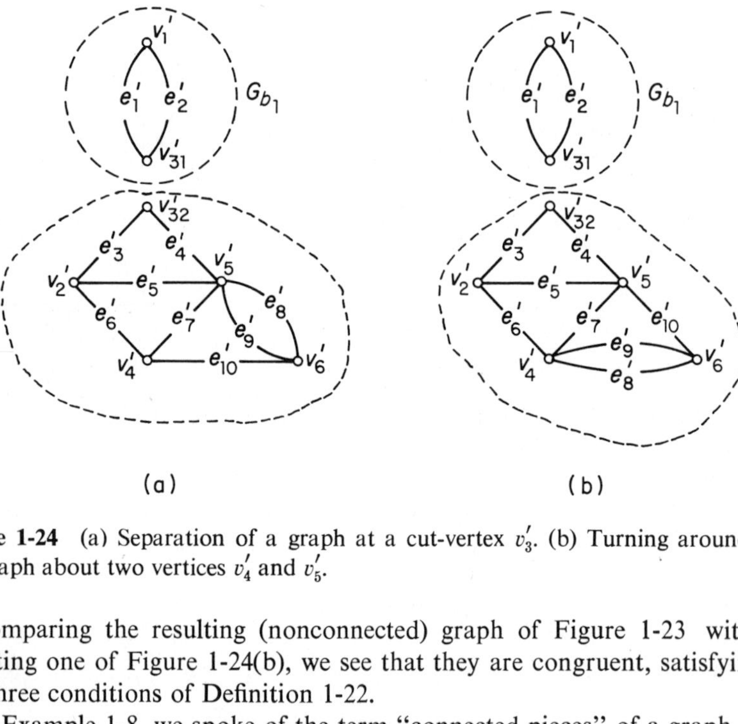

Figure 1-24 (a) Separation of a graph at a cut-vertex v_3'. (b) Turning around of a subgraph about two vertices v_4' and v_5'.

Comparing the resulting (nonconnected) graph of Figure 1-23 with the resulting one of Figure 1-24(b), we see that they are congruent, satisfying all the three conditions of Definition 1-22.

In Example 1-8, we spoke of the term "connected pieces" of a graph as an undefined term assuming that it would be clearly understood by the reader; but a connected piece can be defined as a maximally connected subgraph as follows.

DEFINITION 1-23. Maximally Connected Subgraph. A *maximally connected subgraph* of an unconnected graph G is a connected subgraph G_m of G such that the addition of any edge from its complement, G_m' to G_m, will result in an unconnected subgraph of G.

To illustrate, let us consider once again the (unconnected) graph G_a of Figure 1-23. We see that the two subgraphs, G_{a_1} and G_{a_2}, are complements. Also, both subgraphs are connected pieces, or maximally connected subgraphs of G_a as defined in Definition 1-23 because if we add any edge, say e_6, of G_{a_2} to G_{a_1}, a subgraph consisting of edges e_1, e_2, and e_6 results, which obviously is not a connected subgraph. Similarly, if either e_1 or e_2 (or both) of G_{a_1} is added to G_{a_2}, an unconnected subgraph will result.

Now let us formally define equivalence.

DEFINITION 1-24. Equivalence. Two graphs G_a and G_b are *equivalent*[4] if they can be made congruent by the (repeated) application of either or both of the following two operations:

(a) Separation of a graph into components by splitting the cut-vertices.

(b) Turning around a subgraph G_i of a graph G about two vertices, v_a and v_b, provided that G_i and its complement G_i' are connected *only* at vertices v_a and v_b and have no other common vertices.

Note that in Example 1-8, subgraph G_{a_1} of Figure 1-23 and subgraph G_{b_1} of Figure 1-24 are congruent after the application of operation (a) of Definition 1-24 at the cut-vertices; and the subgraph G_{a_2} of Figure 1-23 and subgraph G_{b_3} of Figure 1-24(b) are made congruent after the application of operation (b) of Definition 1-24 by turning around the subgraph consisting of edges e_8', e_9', and e_{10}' about vertices v_4' and v_5' in G_{b_2} of Figure 1-24(a).

We should also point out that sometimes it may be necessary to apply more than once either or both of the two operations described in Definition 1-24 before the two graphs can be made congruent.

1-7 REMARKS

In this chapter, we introduced a set of basic definitions that are essential for the discussions in the subsequent chapters. It is hoped that the reader will not try to memorize the definitions word by word. Instead, he should study these terms carefully and repeatedly, if necessary, in order to understand fully their meaning, since they are the basic vocabulary of the language in network topology.

Some basic theorems, which include the properties of trees, cutsets, and circuits, were also stated in this chapter. Some of the methods of proof (discussed in Appendix B2) were used in the proofs of the theorems in Section 1-5 as illustrations of these methods. The procedure of proving a theorem of the "if-and-only-if" type was demonstrated in the proof of Theorem 1-3. The meaning of a necessary and sufficient condition as discussed in detail in the

[4]Equivalent graphs are also called 2-isomorphic graphs in other texts. For example, see Seshu and Reed [76], Definition 3-4, p. 38.

Appendices should be fully understood by the reader at this point. If there is any doubt about the meaning of the phrase "if and only if," it is now time to review it again since many lemmas, theorems and corollaries of the "if-and-only-if" category will be discussed in the following chapters throughout the text.

We have defined both nonoriented and oriented graphs. The reader may wonder why we are studying both types. It seems quite logical to believe that the investigation of the oriented graphs alone will be sufficient, since an oriented graph contains all the information that is contained in the corresponding nonoriented graph. We must admit, also, that we can accomplish the same goal in the study of network graphs by omitting entirely the discussions on the nonoriented graphs. However, we do not choose such an alternative for two good reasons: First, we must realize that when we deal with oriented graphs, our discussions will be more complicated because we have to take edge orientations into consideration. Therefore, if we can solve our problems with a nonoriented graph, there is no reason to do it the hard way by taking the oriented graph, even though we finally will obtain the same

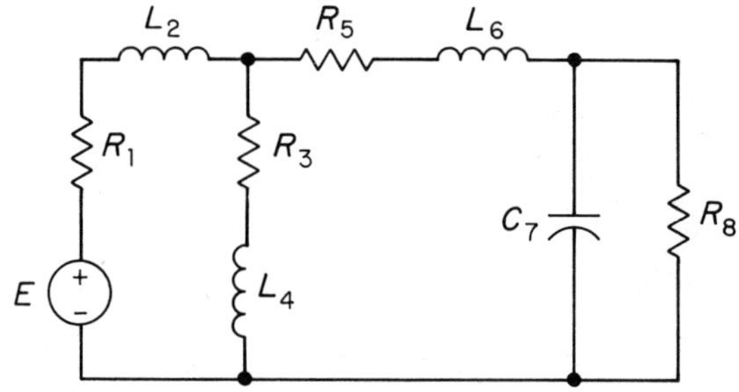

(a)

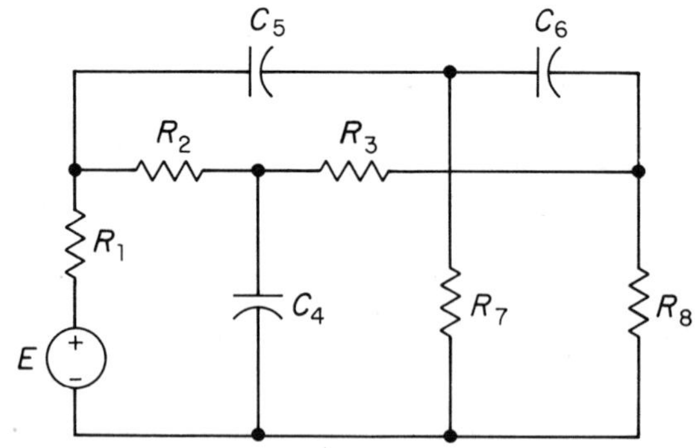

(b)

Figure P1-1

result. Second, most of the properties developed for the nonoriented graph can easily be extended to the oriented one. Hence, it will be a great saving, both in time and effort, to develop our theory with the nonoriented graphs and then make our extensions with just a little effort to include the oriented graphs.

PROBLEMS

1-1 Obtain a graph from each of the networks shown in Figure P1-1. Label the vertices and the edges in the graph.

1-2 List all the circuits of each of the graphs shown in Figure P1-2.

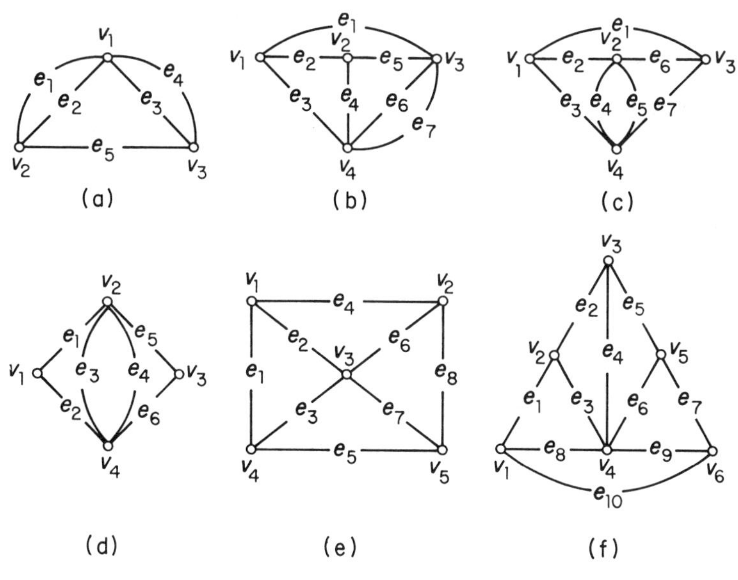

Figure P1-2

1-3 Refer to Figure P1-2. List all the trees for each of the graphs in the figure.

1-4 Some of the properties of a tree T of a connected graph G consisting of v vertices and e edges are as follows:
(a) T contains all the v vertices of G.
(b) T contains no circuits.
(c) T contains $v - 1$ of the e edges.
Show that any two of the three properties imply the third.

1-5 State two different systematic procedures for enumerating all the trees of a connected graph G. Make your methods as simple as you can.

1-6 Let G_s be a subgraph of a connected graph G.

(a) State the necessary and sufficient conditions for the existence of a tree of G that contains G_s.
(b) Propose an algorithm for enumerating all the trees that contain G_s.
(c) In Figure P1-6, it is desired to find all the trees that contain edges e_1 and e_2. Solve the problem using the method proposed in (b).

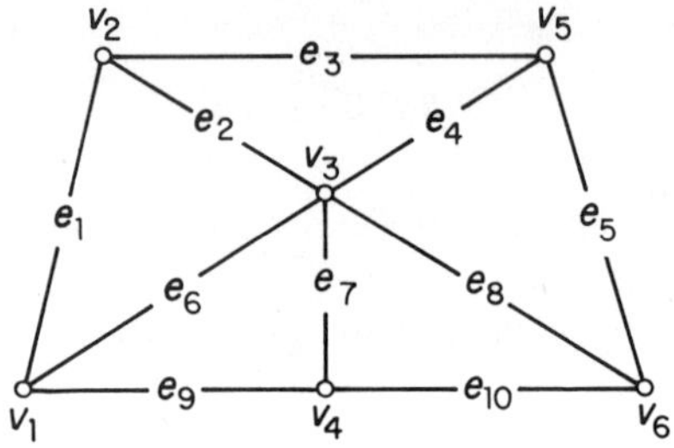

Figure P1-6

1-7 A graph G is called a *complete graph* if there exists an edge between every pair of vertices of G. Determine
(a) The total number of edges
(b) The total number of trees
(c) The total number of circuits
in a complete graph of v vertices.

1-8 A graph is called a *full graph* of order m if there exist m edges between every pair of vertices of G. A complete graph, defined in Problem 1-7, is a special case of a full graph with $m = 1$. Repeat Problem 1-7 for a full graph of order m with v vertices.

1-9 Show, as many as you can, graphs with the following three properties:
(1) The ring sum of *any* two (distinct) circuits is also a circuit.
(2) No edges are connected in series.
(3) Every edge is contained in some circuit.
(*Note:* If a vertex is of degree 2, then the two edges having this vertex in common are said to be connected in series. Condition 2 is the same as saying that no vertex of the graph is of degree 2.)

1-10 Prove that the number of circuits in a complete graph with n vertices is as follows:

(a) $n = 3,\ N_{\text{circuit}} = \dfrac{3\cdot 2\cdot 1}{2\cdot 3} = 1$

(b) $n = 4,\ N_{\text{circuit}} = \dfrac{4\cdot 3\cdot 2}{2\cdot 3} + \dfrac{4\cdot 3\cdot 2\cdot 1}{2\cdot 4} = 7$

(c) More generally, for $n = 3$,

$$N_{\text{circuit}} = \frac{n(n-1)(n-2)}{2\cdot 3} + \frac{n(n-1)(n-2)(n-3)}{2\cdot 4} + \cdots$$

$$= \sum_{r=3}^{n} \frac{n(n-1)\cdots(n-r+1)}{2r}$$

1-11 Let G_s be a subgraph of a connected graph G.

(a) State the necessary and sufficient conditions for the existence of a tree of G whose branches do not contain any of the edges of G_s.

(b) Propose a method for enumerating all the trees of G that do not contain any edges of G_s.

(c) For the graph of Problem 1-6, find all trees not containing edges e_3, e_4, and e_7. Use the method you proposed in part (b).

1-12 Find all the paths between vertices V_1 and V_2 of each of the graphs shown in Figure P1-12.

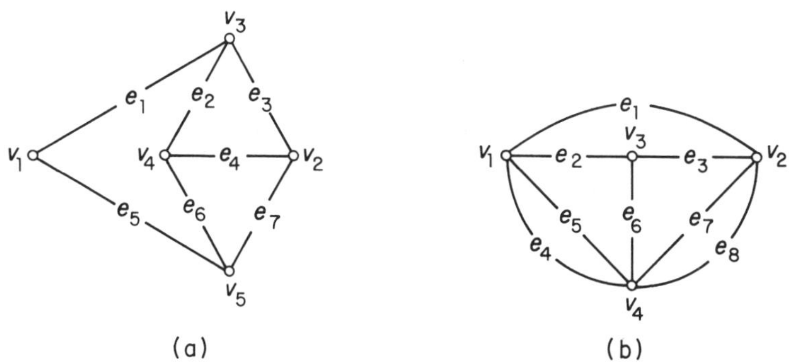

Figure P1-12

1-13 For each of the graphs in Problem 1-12, find all the cutsets.

1-14 Prove that for a complete graph of v vertices, the number of cutsets equals $2^{v-1} - 1$. (Refer to Problem 1-7 for the definition of a complete graph.)

1-15 For the graph shown in Figure P1-15,

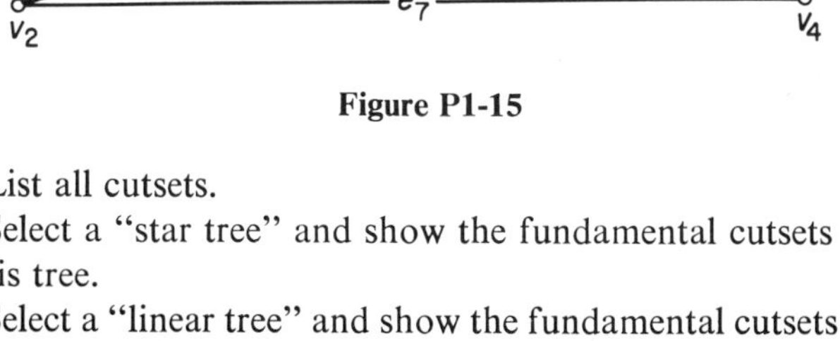

Figure P1-15

(a) List all cutsets.

(b) Select a "star tree" and show the fundamental cutsets with respect to this tree.

(c) Select a "linear tree" and show the fundamental cutsets with respect to this tree.

(d) Express all the cutsets in part (a) as the ring sum of some, or all, cutsets in part (b).

(e) Repeat part (d) using the cutsets in part (c).
(*Note:* In plain language, a *star tree* has one vertex common to all tree branches, and a *linear tree* is a tree that may be stretched out as a straight line.)

1-16 Construct a connected graph G with nine edges, $e_1, e_2, \ldots e_8, e_9$, such that each of the following sets of edges forms a cutset of G:

$$K_1 = \{e_2, e_3, e_4\} \qquad K_2 = \{e_3, e_5, e_6\}$$

$$K_3 = \{e_1, e_3, e_5, e_7\} \qquad K_4 = \{e_2, e_3, e_5, e_9\}$$

1-17 Construct a graph G with nine edges, $e_1, e_2, \ldots, e_8, e_9$, such that each of the following sets of edges forms a circuit of G:

$$C_1 = \{e_1, e_4, e_7, e_8\} \qquad C_2 = \{e_3, e_6, e_7, e_9\}$$

$$C_3 = \{e_2, e_3, e_4, e_8\} \qquad C_4 = \{e_4, e_5, e_6, e_7\}$$

1-18 For each of the following cases, determine whether a graph can be constructed to meet the requirements:
(a) Edges: e_1, e_2, e_3, e_4, e_5 Circuits: $\{e_1, e_2, e_5\}$, $\{e_1, e_3, e_5\}$, $\{e_1, e_2, e_4\}$.
(b) Edges: e_1, e_2, e_3, e_4, e_5 Circuits: $\{e_1, e_3, e_5\}$, $\{e_1, e_4, e_5\}$, $\{e_2, e_3, e_4, e_5\}$.

1-19 State a general systematic procedure for solving problems of the type similar to Problem 1-18.

1-20 (a) For the nonoriented graph shown in Figure P1-20, find the fundamental set of circuits with respect to the tree T: $\{e_1, e_2, e_3, e_4\}$ by inspection of the graph.

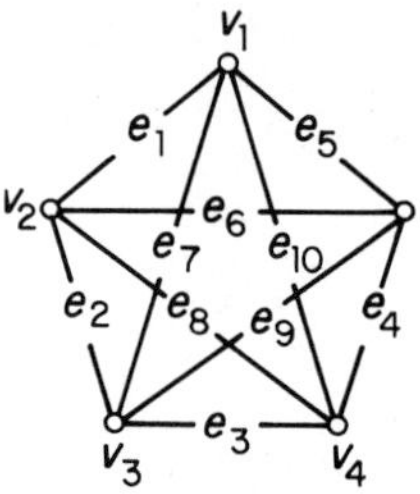

Figure P1-20

(b) The problem of part (a) is to be solved by the use of digital computer. The information about the connection of the edges is completely contained in the incidence matrix[5] of the graph. The computer cannot see the graph, but can perform matrix manipulations. How would you program the computer to obtain the fundamental set of circuits with respect to the tree T? In this relatively simple case, you are asked to carry out the details of the computation without the actual use of a computer. The use of modulo 2 algebra greatly simplifies the solution.

[5]Refer to Definition 2-1 for the definition of the incidence matrix of a graph.

1-21 For the graph of Figure P1-20, find the set of all fundamental cutsets with respect to the tree T: $\{e_1, e_2, e_3, e_4\}$.

1-22 The definition of a cutset as stated in Definition 1-14 is for a connected graph only. State a more general definition of a cutset so that it applies to an unconnected graph as well as a connected one.

1-23 Prove Theorem 1-7.

1-24 Are the two graphs, G_1 and G_2, of Figure P1-24

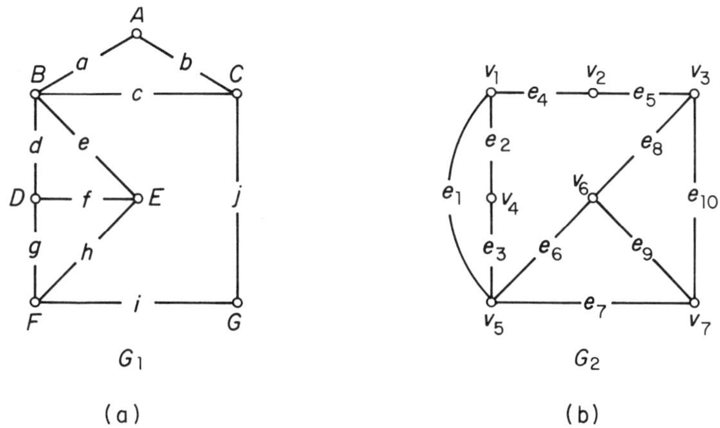

Figure P1-24

(a) Congruent?
(b) Equivalent (2-isomorphic)?
(c) Neither congruent nor equivalent?
Show your work.

1-25 Repeat Problem 1-24 for the two graphs, G_3 and G_4, of Figure P1-25.

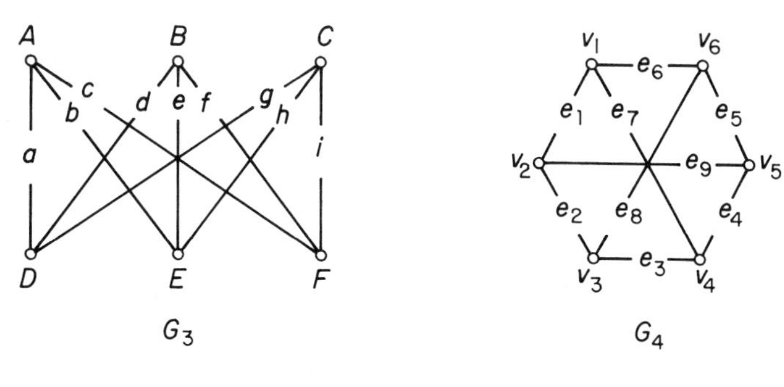

Figure P1-25

1-26 Suggest a systematic procedure for determining the total number of (a) trees, (b) circuits, and (c) cutsets of a given connected graph without actually listing them.

CHAPTER 2

MATRICES OF A GRAPH

2-1 INTRODUCTION

In this chapter we shall discuss some matrices of a graph. First we shall define each of these matrices for a nonoriented graph and point out the various properties of each matrix. Then we shall derive the interrelationships between these matrices. Other matrices, such as the path matrix of a graph, that are perhaps less basic will also be discussed, but only very briefly.

Next, the results obtained from the discussions on the matrices of a nonoriented graph will be extended for an oriented graph. This will be done by first pointing out the differences between the matrices of a nonoriented graph and those of the corresponding oriented graph. In the case of an oriented graph, modifications can be made to extend the theorems obtained in the nonoriented case. The proofs of the theorems for an oriented graph are, in fact, quite similar to those of the corresponding theorems for the corresponding nonoriented graph. These similarities, as well as the differences in the proofs, will be pointed out by comparing the two proofs of the same theorem in both cases. After such comparison has been made, the reader will then be expected to follow a similar pattern and complete the task of extending the remaining theorems and corollaries from the nonoriented to the oriented case.

Finally, we shall summarize the results and definitions discussed in the first two chapters in a tabulated or itemized form, giving a clear comparison between the properties of the matrices for the nonoriented case and those of the matrices for the oriented case.

2-2 THE INCIDENCE MATRIX

In this section, we shall introduce a matrix A_a, called the incidence matrix of a graph, which plays an important role in network graph theory.

DEFINITION 2-1. Incidence Matrix. The *incidence matrix* of a nonoriented graph G of v vertices and e edges is the matrix A_a of order $v \times e$ with each row identified by a vertex and each column by an edge; that is,

$$A_a = [a_{ij}]_{v\times e}$$

where

$$a_{ij} = 1 \text{ if edge } e_j \text{ is incident with vertex } v_i$$

$$a_{ij} = 0 \text{ if edge } e_j \text{ is not incident with vertex } v_i$$

EXAMPLE 2-1. Let G_1 be the nonoriented, connected graph shown in Figure 2-1.

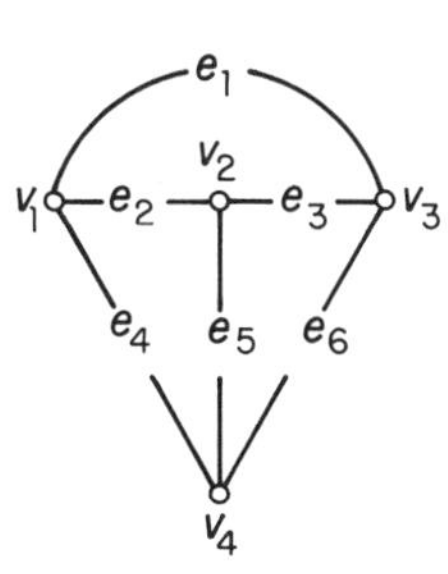

Figure 2-1 A nonoriented, connected graph G_1.

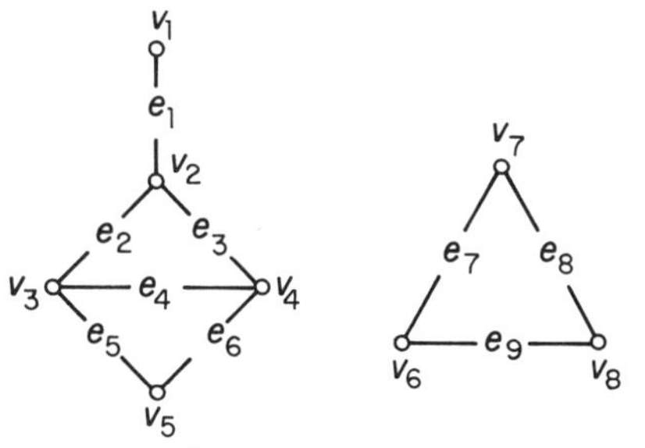

Figure 2-2 A nonoriented, nonconnected graph G_2.

The incidence matrix A_{a_1} of G_1 is given by

$$\mathrm{A}_{a_1} = [a_{ij}]_{4\times 6} = \begin{array}{c} \\ v_1 \\ v_2 \\ v_3 \\ v_4 \end{array} \begin{array}{c} \begin{array}{cccccc} e_1 & e_2 & e_3 & e_4 & e_5 & e_6 \end{array} \\ \begin{bmatrix} 1 & 1 & 0 & 1 & 0 & 0 \\ 0 & 1 & 1 & 0 & 1 & 0 \\ 1 & 0 & 1 & 0 & 0 & 1 \\ 0 & 0 & 0 & 1 & 1 & 1 \end{bmatrix} \end{array} \tag{2-1}$$

EXAMPLE 2-2. Consider the nonoriented, unconnected graph G_2 depicted by Figure 2-2. The incidence matrix A_{a_2} of G_2 is

$$A_{a_2} = [a_{ij}]_{8\times 9} = \begin{array}{c} \\ v_1 \\ v_2 \\ v_3 \\ v_4 \\ v_5 \\ v_6 \\ v_7 \\ v_8 \end{array} \begin{array}{c} \begin{array}{ccccccccc} e_1 & e_2 & e_3 & e_4 & e_5 & e_6 & e_7 & e_8 & e_9 \end{array} \\ \left[\begin{array}{cccccc|ccc} 1 & 0 & 0 & 0 & 0 & 0 & 0 & 0 & 0 \\ 1 & 1 & 1 & 0 & 0 & 0 & 0 & 0 & 0 \\ 0 & 1 & 0 & 1 & 1 & 0 & 0 & 0 & 0 \\ 0 & 0 & 1 & 1 & 0 & 1 & 0 & 0 & 0 \\ 0 & 0 & 0 & 0 & 1 & 1 & 0 & 0 & 0 \\ \hline 0 & 0 & 0 & 0 & 0 & 0 & 1 & 0 & 1 \\ 0 & 0 & 0 & 0 & 0 & 0 & 1 & 1 & 0 \\ 0 & 0 & 0 & 0 & 0 & 0 & 0 & 1 & 1 \end{array}\right] \end{array} \tag{2-2}$$

From these two examples, we observe the following properties of the incidence matrix A_a:

(a) There are two types of elements in the matrix: 0's and 1's.

(b) In each column there are exactly two 1's.[1]

(c) The number of 1's in each row indicates the degree of the vertex corresponding to that row.

(d) There can be no row in the matrix without a 1 in it for a connected graph. (For a graph that is not connected, we must then consider the possibility of *isolated vertices* — vertices that are not incident with any edges, in which case a row of zeros results.)

(e) If there is only one 1 in a row, the edge that identifies the column that the 1 is in is a *noncircuit edge* — an edge that is not in any of the circuits of the graph.

(f) If the matrix can be partitioned in the form (A_{a_2} in Example 2-2)

$$A_a = \begin{bmatrix} A_{11} & 0 \\ 0 & A_{22} \end{bmatrix}$$

such that minor-diagonal submatrices in the partitioned form are zero matrices, the graph is an unconnected graph with at least two maximally connected subgraphs [connected pieces (Definition 1-23)] upon possible interchange of rows and columns.

All of these properties are quite obvious. For instance, property (a) simply states that the possible element values in the matrix are 1 and 0 with 1 meaning "in" and 0 meaning "out". That is, if $a_{ij} = 1$, it implies that e_j is connected to vertex v_i; otherwise, $a_{ij} = 0$. Since each edge is connected to two vertices, each column of the matrix must contain exactly two 1's, as stated in property (b). The reader is urged to supply explanations for the other properties as exercises.

It is evident that the requirements for a matrix to be an incidence matrix of a nonoriented graph are:

(a) Each of the entries of the matrix is either 1 or 0.

[1]Here we have excluded self-loops. A self-loop may be defined as a circuit containing only one edge and one vertex.

(b) There are exactly two 1's in each column,[1] as suggested in the two foregoing examples.

We shall restate this in the following theorem.

THEOREM 2-1. A matrix is the incidence matrix of some nonoriented graph if and only if (a) its elements are either 1's or 0's; and (b) there are exactly two 1's in each column.[1]

Proof: Let A be the incidence matrix of some nonoriented graph. Then, by Definition 2-1, A must satisfy the two requirements, (a) and (b), stated in this theorem.

If A does possess these two properties, a nonoriented graph G can be constructed with each vertex identified by a distinct row of A and each edge by a distinct column of A. That is, the ith row of A corresponds to vertex v_i of G and the jth column of A to edge e_j of G. Furthermore, if the two 1's of the kth column of A are located in the mth and the nth row, then edge e_k is connected between vertices v_m and v_n. Hence A is the incidence matrix of G, and the proof of the theorem is complete.

It is interesting to note that at least one row of the incidence matrix A_a is dependent upon the others. This can be readily seen from the fact that there are always two 1's in each column of the matrix. Thus, if any row of A_a is omitted, this deleted row can be reconstructed by placing a 0 in a column if there are already two 1's in that column, and placing a 1 in the column if there is only one 1 in it.

EXAMPLE 2-3. Consider the incidence matrix A_{a_1} of graph G_1 in Example 2-1. The matrix is shown in Equation (2-1). If the third row of this matrix is deleted, a matrix A_1 results, as shown in Equation (2-3).

$$A_1 = \begin{array}{c} \\ v_1 \\ v_2 \\ v_4 \end{array}\!\!\begin{array}{c} \begin{array}{cccccc} e_1 & e_2 & e_3 & e_4 & e_5 & e_6 \end{array} \\ \begin{bmatrix} 1 & 1 & 0 & 1 & 0 & 0 \\ 0 & 1 & 1 & 0 & 1 & 0 \\ 0 & 0 & 0 & 1 & 1 & 1 \end{bmatrix} \end{array} \tag{2-3}$$

Now, suppose we were given A_1 as a submatrix of the incidence matrix A_{a_1} of some graph G, and if we were to obtain A_{a_1} from A_1, we would observe that each column of A_1 has at least one 1, but not more than two 1's. Consequently, if we placed a 1 in the fourth row of column 1, a 0 in the fourth row of column 2, and so on, the resulting matrix would be

$$A'_{a1} = \begin{array}{c} \\ v_1 \\ v_2 \\ v_4 \\ v_3 \end{array}\!\!\begin{array}{c} \begin{array}{cccccc} e_1 & e_2 & e_3 & e_4 & e_5 & e_6 \end{array} \\ \begin{bmatrix} 1 & 1 & 0 & 1 & 0 & 0 \\ 0 & 1 & 1 & 0 & 1 & 0 \\ 0 & 0 & 0 & 1 & 1 & 1 \\ 1 & 0 & 1 & 0 & 0 & 1 \end{bmatrix} \end{array} \tag{2-4}$$

which is the same matrix as A_{a_1} of Equation (2-1) with the third and fourth row interchanged.

Therefore, it is evident that both A_{a_1} of Equation (2-1) and A'_{a_1} of Equation (2-4) correspond to the same graph G_1 of Figure 2-1. In fact, any one of the three matrices, A_{a_1} of Equation (2-1), A'_{a_1} of (2-4), and A_1 of (2-3) fully represents graph G_1 of Figure 2-1 in that each of them gives the same (sufficient) information about the graph such that the graph can be uniquely determined from it.

We next define an incidence submatrix.

DEFINITION 2-2. Incidence Submatrix. An *incidence submatrix* A is a matrix obtained from the incidence matrix A_a by deleting (any) one of its rows. The following lemma is a result of the above discussion on the incidence matrix.

LEMMA 2-1. There are *at most* $v - 1$ independent rows in the incidence matrix A_a of a connected graph G.

The proof of Lemma 2-1 is left as an exercise (Problem 2-4).

The incidence matrix A_a and the submatrix A are special types of matrices, namely, matrices with 1's and 0's as elements. We see that if a row of such a matrix is added to another row, it is quite possible to have a 2 as one of the elements in the resulting row. Since the value 2 is excluded from being an element of a matrix of this kind, we must define our rules of operations so that the law of closure[2] will be valid. In other words, we would like to define a set of rules of operations such that the resulting matrix from such operations must also be a matrix of the same kind. (It must be a matrix consisting of only 1's and 0's as elements.)

First we define the "addition" table and the "multiplication" table for the set of two elements, 0 and 1, as shown in Figure 2-3.

$$\left\{\begin{array}{l} 0+0=0 \\ 0+1=1 \\ 1+0=1 \\ 1+1=0 \end{array}\right. \qquad \begin{array}{l} 0\cdot 0=0 \\ 0\cdot 1=0 \\ 1\cdot 0=0 \\ 1\cdot 1=1 \end{array}$$

(a) (b)

Figure 2-3 (a) Addition table. (b) Multiplication table for the set of two elements: 0 and 1.

[2]The law of closure states that if a and b are two elements of a set S and if $\circ$ stands for an operation, then $a \circ b$ is also an element of S.

We shall define the so-called "addition" of two rows as the addition of the corresponding elements in the two rows in the way we do for any matrix except that now the addition of elements must be governed by the addition table defined in Figure 2-3(a). That is, in adding a row of a matrix of the two-element kind to another row of the same matrix, if there is a 0 in some column of the first row and a 0 in the same column of the second row, then the element in that column of the resulting row is 0, since $0 + 0 = 0$. But, if the two 0's are replaced by two 1's, the result is still zero, since $1 + 1 = 0$. Finally, if one 0 is replaced by 1 and the other remains 0, the result is 1, because $0 + 1 = 1 + 0 = 1$.

In the case of an incidence matrix A_a (or a submatrix A) of a graph G, the addition of two rows defined above actually is the ring-sum operation defined in Definition 1-18! This can be seen by noting that a 1 in the (i, j) position of A_a (or A) means that edge e_j of G is incident with vertex v_i. Now, if there is again a 1 in the (k, j) position of the same matrix, it means that e_j is also incident with vertex v_k. Thus, adding the ith row to the kth row will result in a row whose element in the jth column is 0. This is precisely the ring-sum operation, since edge e_j is in both the ith and the kth row of the same column; e_j must then be removed from the same column in the resulting row if the operation is to be identified by the ring sum.

The multiplication table of Figure 2-3(b) shows that the so-called "multiplication" here is identical to the multiplication of the set of elements 1 and 0 in the ordinary algebra of scalars.

We now show how we can reduce an incidence matrix to normal form in order to determine its rank.

EXAMPLE 2-4. Let us consider the incidence matrix A'_{a1} of Equation (2-4). Upon performing elementary transformations on the rows, A'_{a1} may be reduced to normal form as shown in Equation (2-5).

$$A'_{a1} = \begin{array}{c} \begin{matrix} e_1 & e_2 & e_3 & e_4 & e_5 & e_6 \end{matrix} \\ \begin{bmatrix} 1 & 1 & 0 & 1 & 0 & 0 \\ 0 & 1 & 1 & 0 & 1 & 0 \\ 0 & 0 & 0 & 1 & 1 & 1 \\ 1 & 0 & 1 & 0 & 0 & 1 \end{bmatrix} \end{array} \sim \underset{\text{(first equivalent)}}{\begin{array}{c} \begin{matrix} e_1 & e_2 & e_3 & e_4 & e_5 & e_6 \end{matrix} \\ \begin{bmatrix} 1 & 1 & 0 & 1 & 0 & 0 \\ 0 & 1 & 1 & 0 & 1 & 0 \\ 0 & 0 & 0 & 1 & 1 & 1 \\ 0 & 1 & 1 & 1 & 0 & 1 \end{bmatrix} \end{array}}$$

$$\sim \underset{\text{(second equivalent)}}{\begin{array}{c} \begin{matrix} e_1 & e_2 & e_3 & e_4 & e_5 & e_6 \end{matrix} \\ \begin{bmatrix} 1 & 1 & 0 & 1 & 0 & 0 \\ 0 & 1 & 1 & 0 & 1 & 0 \\ 0 & 0 & 0 & 1 & 1 & 1 \\ 0 & 0 & 0 & 1 & 1 & 1 \end{bmatrix} \end{array}} \sim \underset{\text{(third equivalent)}}{\begin{array}{c} \begin{matrix} e_1 & e_2 & e_3 & e_4 & e_5 & e_6 \end{matrix} \\ \begin{bmatrix} 1 & 1 & 0 & 1 & 0 & 0 \\ 0 & 1 & 1 & 0 & 1 & 0 \\ 0 & 0 & 0 & 1 & 1 & 1 \\ 0 & 0 & 0 & 0 & 0 & 0 \end{bmatrix} \end{array}}$$

$$\sim \left[\begin{array}{ccc:ccc} e_1 & e_2 & e_4 & e_3 & e_5 & e_6 \\ 1 & 1 & 1 & 0 & 0 & 0 \\ 0 & 1 & 0 & 1 & 1 & 0 \\ 0 & 0 & 1 & 0 & 1 & 1 \\ \hdashline 0 & 0 & 0 & 0 & 0 & 0 \end{array}\right] \tag{2-5}$$

(fourth equivalent)

The first equivalent of A'_{a1} in Equation (2-5) is obtained by adding row 1 of A'_{a1} to row 4; the second equivalent is obtained from the first by adding the second row of the first equivalent to its fourth row; the third equivalent is obtained from the second by adding row 3 to row 4; and the fourth one is obtained from the third by interchanging columns 3 and 4. By inspection of this last matrix, it is clear that the rank of A'_{a1} is 3.

We next state and prove the following lemma.

LEMMA 2-2. The number of independent rows in the incidence matrix A_a of a connected graph G is *at least* $v - 1$.

Proof: We shall employ the method of indirect proof. Assume that the number of independent rows of an incidence matrix A_a is less than $v - 1$. Then a set of $v - 1$ rows of A_a constitutes a dependent set. If we represent these $v - 1$ rows by $R_1, R_2, \ldots, R_{v-1}$, then, by definition of linear dependence (see Appendixes), there exists a set of constants, $k_1, k_2, \ldots, k_{v-1}$, not all zero, such that

$$k_1R_1 + k_2R_2 + \cdots + k_{v-1}R_{v-1} \equiv 0 \tag{2-6}$$

Since the addition operation is interpreted as ring sum with all elements equal to either 0 or 1, Equation (2-6) implies that there exist at least two nonzero constants[3] (which must be 1 since we have only two values, 1 and 0, in the set) such that the equation holds identically. This, in turn, means that a nonempty subset (which may or may not be a proper subset) of the set of $v - 1$ rows will add to yield a row of zeros. But this is an impossible situation, and the reason is as follows.

Note that each row of A_a must contain at least one 1 for a connected graph. Also, each column must contain exactly two 1's. Let us consider a 1 in any row, say R_i, located in the jth column. Then there must be another 1 in that column but in some other row. Now, if row R_i is deleted and the other $v - 1$ rows are added together, the resulting row must have a 1 in the jth column, and, hence, it cannot be a row of zeros. Since R_i can be any of the v rows of A_a, we conclude that no proper subset of the v rows of A_a can be added to produce a row of zeros. Thus our assumption that "the number of independent rows of A_a is less than $v - 1$" leads to a contradiction. This implies that the number of independent rows of A_a must be at least $v - 1$. Hence the lemma.

[3]This is true since a row of zeros cannot be a row of A_a. (Why?)

We are now ready to prove the next theorem with the help of Lemmas 2-1 and 2-2.

THEOREM 2-2. The rank of the incidence matrix A_a of a connected graph G of v vertices is $v - 1$; that is,

$$R(A_a) = v - 1 \tag{2-7}$$

Proof: We observe that for both lemmas to hold at the same time, it is inevitable to conclude that the number of independent rows in A_a is exactly $v - 1$. Also, it is easy to show that a square matrix is nonsingular if and only if all its rows are linearly independent. This implies that there exists a nonsingular submatrix of A_a of order $v - 1$. Hence the theorem.

As we mentioned earlier, if the graph G is not a connected graph, we have a forest instead of a tree for G, and we shall define this as follows:

DEFINITION 2-3. Forest. Let G be a nonconnected graph with k maximally connected subgraphs $G_1, G_2, \ldots, G_k$. Then a *forest* of G is a set of k connected subgraphs $T_1, T_2, \ldots, T_k$, containing all the vertices of G and no circuits, such that T_i is a tree of G_i $(i = 1, 2, \ldots, k)$.

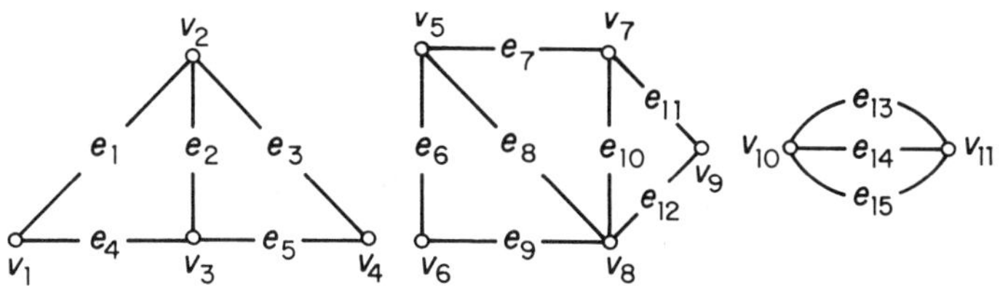

Figure 2-4 A nonconnected graph G.

EXAMPLE 2-5. In the nonconnected graph G of Figure 2-4, there are 11 vertices and 15 edges. It contains three maximally connected subgraphs G_1, G_2, G_3. A forest of G is formed by taking a set of three trees — one from each maximally connected subgraph. One of the forests of G is shown in Figure 2-5.

In the case of a nonconnected graph, Theorem 2-2 should be revised as stated in the next theorem.

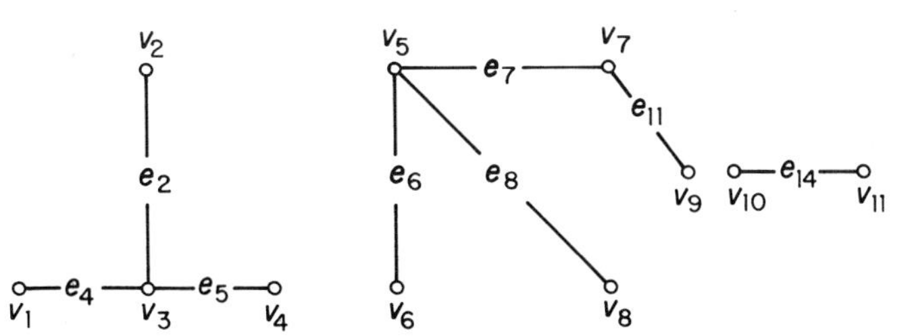

Figure 2-5 A forest of graph G of Figure 2-4.

THEOREM 2-3. For a nonconnected graph G of v vertices with k maximally connected subgraphs, the rank of the incidence matrix A_a is $v - k$; that is,

$$R(A_a) = v - k \tag{2-8}$$

Proof: The proof of this theorem can easily be established by following the same pattern as shown in the proof of Theorem 2-2, keeping in mind that graph G now is composed of k individually connected pieces (k maximally connected subgraphs), say $G_1, G_2, \ldots, G_k$. It is simple to show (try it!) that

$$R(A_a) = R(A_{a_1}) + R(A_{a_2}) + \cdots + R(A_{a_k}) \tag{2-9}$$

and

$$v = v_1 + v_2 + \cdots + v_k \tag{2-10}$$

where $A_{a_1}, A_{a_2}, \ldots, A_{a_k}$ are the incidence matrices and $v_1, v_2, \ldots v_k$ are the number of vertices of $G_1, G_2, \ldots, G_k$, respectively.

From Equation (2-9), and Theorem 2-2, it is evident that

$$\begin{aligned} R(A_a) &= (v_1 - 1) + (v_2 - 1) + \cdots + (v_k - 1) \\ &= (v_1 + v_2 + \cdots + v_k) - \underbrace{(1 + 1 + \cdots + 1)}_{\text{a sum of } k \text{ terms}} \end{aligned}$$

or

$$R(A_a) = (v_1 + v_2 + \cdots + v_k) - k \tag{2-11}$$

Combining Equations (2-10) and (2-11), we get

$$R(A_a) = v - k$$

This completes the proof.

Note that Equation (2-8) is true for a graph G which can be either connected or unconnected, because if G is a connected graph, then $k = 1$ and Equation (2-8) will be reduced to Equation (2-7). We might also point out that even though we have included the unconnected graph in our discussion here, we shall be concerned mainly with the connected graph in the remainder of this chapter as well as in the chapters that follow, since most of the networks with which we are concerned have connected graphs.

Finally, we should emphasize the following point. We have introduced the incidence matrix A_a and the incidence submatrix A of a connected graph G of v vertices and e edges in this section; and we have used subscript a in A_a to mean that *all* of the v vertices have been included in the incidence matrix A_a. On the other hand, the symbol A without a subscript denotes an incidence submatrix that includes some $v - 1$ of the v rows of A_a. Thus, one of the vertices of G has been left out. This omitted vertex is referred to as the *reference* vertex of G when A is being considered, and it may be any one of the v vertices of G. We should also note that for a matrix to be an incidence submatrix, it must be a submatrix of A_a of G containing (any) $v - 1$ of the v rows of A_a. Thus, for example, a submatrix of order $(v - 2) \times e$ of A_a *cannot* be called an incidence submatrix.

2-3 THE CIRCUIT MATRIX

Another matrix of great importance in the study of network topology is the so-called *circuit matrix*. We shall devote this entire section to the discussion of its properties.

DEFINITION 2-4. Circuit matrix. The *circuit matrix* of a nonoriented graph G of v vertices and e edges is the matrix B_a of order $c \times e$ with each row identified by a circuit and each column by an edge such that

$$B_a = [b_{ij}]_{c \times e}$$

where

$$b_{ij} = 1 \text{ if edge } e_j \text{ is in circuit } c_i$$

$$b_{ij} = 0 \text{ if edge } e_j \text{ is not in } c_i$$

c is the total number of circuits in G

EXAMPLE 2-6. Let us consider graph G_1 of Figure 2-1. The circuit matrix B_{a_1} of this graph is

$$B_{a_1} = \begin{array}{c} \\ c_1 \\ c_2 \\ c_3 \\ c_4 \\ c_5 \\ c_6 \\ c_7 \end{array} \begin{array}{c} \begin{array}{cccccc} e_1 & e_2 & e_3 & e_4 & e_5 & e_6 \end{array} \\ \begin{bmatrix} 0 & 1 & 0 & 1 & 1 & 0 \\ 0 & 0 & 1 & 0 & 1 & 1 \\ 1 & 1 & 1 & 0 & 0 & 0 \\ 1 & 0 & 0 & 1 & 0 & 1 \\ 1 & 0 & 1 & 1 & 1 & 0 \\ 1 & 1 & 0 & 0 & 1 & 1 \\ 0 & 1 & 1 & 1 & 0 & 1 \end{bmatrix} \end{array} \tag{2-12}$$

Graph G_1 is repeated in Figure 2-6 for convenience, and the set of all seven circuits is shown in Figure 2-7.

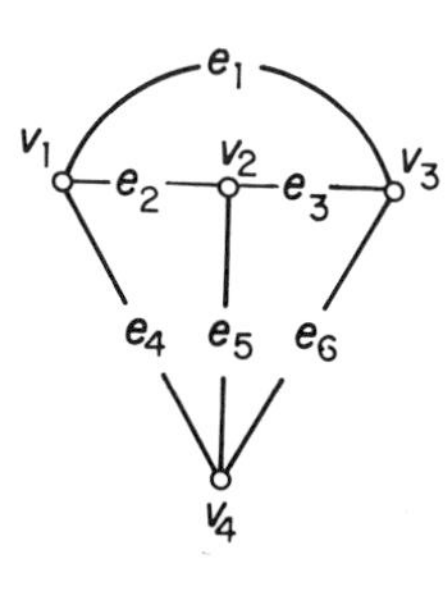

Figure 2-6 A nonoriented connected graph G_1.

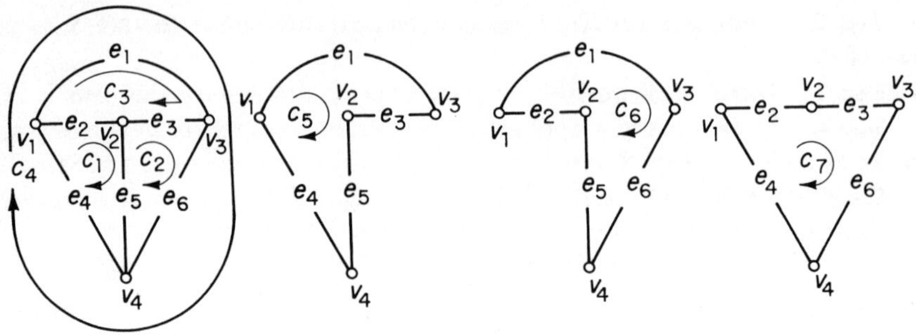

Figure 2-7 A set of all seven circuits of G_1 of Figure 2-6.

From Definition 2-4 and Example 2-6, we see that the number of rows in a circuit matrix cannot be readily determined, since it is quite apparent that if the graph is a complicated one, it will be difficult to determine all possible circuits by inspection. Therefore, the reader would wonder whether or not there is a systematic way of determining the total number c of circuits, and, consequently, finding out what these circuits are. One thing, however, is obvious: If such a method does exist, it must be in the form of an algorithm if it is to be of any value in determining, step by step, the complete set of all circuits with no possibility of missing any one of the circuits in the set. Needless to say, the method will be of greater value if it is a simple one.

Before attempting to derive such a method, let us recall Theorem 1-7, which states that the ring sum of two circuits of a graph is either a circuit or a edge-disjoint union of circuits. This suggests that if we can choose a subset S_i of (a sufficient number of) independent circuits from the set S of all circuits and take all possible ring sums among the circuits in S_i, then we obtain a set S^* that contains all of the circuits in S and some of the edge-disjoint unions of circuits as a result of such ring-sum operations. Thus, if we eliminate all of the edge-disjoint unions of circuits in S^*, the remaining set must be the set S.

Now, consider the circuit matrix B_a of a graph G. Each row of B_a represents a distinct circuit in G. If we can determine the rank r_B of B_a, a set of r_B independent rows of B_a can be chosen. These r_B rows of B_a will be the set S_i of independent circuits we want in order to generate the rest of the circuits, which are identified by the other rows of B_a after the edge-disjoint unions of circuits have been deleted.

Thus our algorithm is stated in the following theorem:

THEOREM 2-4. Let B_a be the circuit matrix of a given graph G. The set of all rows of B_a can be generated from a subset S_i of (r_B) independent rows of B_a by the following steps:

Step 1: Find the rank r_B of B_a.

Step 2: Choose a set S_i of r_B independent rows from the set S of all rows of B_a.

Step 3: Perform all possible ring-sum operations among the rows in S_i.

Step 4: Eliminate the *redundant* rows (that is, rows that are edge-disjoint unions of some rows in S_i resulting from the ring-sum operations in Step 3).

Step 5: The set that contains all of the rows in S_i and the new rows after the elimination of redundant ones is the set S of all rows in B_a.

EXAMPLE 2-7. Let us consider graph G_1 of Figure 2-6. The circuit matrix B_{a_1} of G_1 is given in Equation (2-12).

Step 1: To determine the rank of B_{a_1}, we shall reduce the matrix to triangular form as follows:

$$B_{a_1} \sim \begin{array}{c} \\ c_4 \\ c_1 \\ c_2 \\ c_3 \\ c_5 \\ c_6 \\ c_7 \end{array} \begin{array}{c} \begin{array}{cccccc} e_1 & e_2 & e_3 & e_4 & e_5 & e_6 \end{array} \\ \left[\begin{array}{cccccc} 1 & 0 & 0 & 1 & 0 & 1 \\ 0 & 1 & 0 & 1 & 1 & 0 \\ 0 & 0 & 1 & 0 & 1 & 1 \\ 1 & 1 & 1 & 0 & 0 & 0 \\ 1 & 0 & 1 & 1 & 1 & 0 \\ 1 & 1 & 0 & 0 & 1 & 1 \\ 0 & 1 & 1 & 1 & 0 & 1 \end{array}\right] \end{array} \sim \begin{array}{c} \\ c_1'(= c_4) \\ c_2'(= c_1) \\ c_3'(= c_2) \\ c_4'(= c_4 \oplus c_3) \\ c_5'(= c_4 \oplus c_5) \\ c_6'(= c_4 \oplus c_6) \\ c_7'(= c_7) \end{array} \begin{array}{c} \begin{array}{cccccc} e_1 & e_2 & e_3 & e_4 & e_5 & e_6 \end{array} \\ \left[\begin{array}{cccccc} 1 & 0 & 0 & 1 & 0 & 1 \\ 0 & 1 & 0 & 1 & 1 & 0 \\ 0 & 0 & 1 & 0 & 1 & 1 \\ 0 & 1 & 1 & 1 & 0 & 1 \\ 0 & 0 & 1 & 0 & 1 & 1 \\ 0 & 1 & 0 & 1 & 1 & 0 \\ 0 & 1 & 1 & 1 & 0 & 1 \end{array}\right] \end{array}$$

$$\sim \begin{array}{c} \\ c_1''(= c_1') \\ c_2''(= c_2') \\ c_3''(= c_3') \\ c_4''(= c_2' \oplus c_4') \\ c_5''(= c_5') \\ c_6''(= c_2' \oplus c_6') \\ c_7''(= c_2' \oplus c_7') \end{array} \begin{array}{c} \begin{array}{cccccc} e_1 & e_2 & e_3 & e_4 & e_5 & e_6 \end{array} \\ \left[\begin{array}{cccccc} 1 & 0 & 0 & 1 & 0 & 1 \\ 0 & 1 & 0 & 1 & 1 & 0 \\ 0 & 0 & 1 & 0 & 1 & 1 \\ 0 & 0 & 1 & 0 & 1 & 1 \\ 0 & 0 & 1 & 0 & 1 & 1 \\ 0 & 0 & 0 & 0 & 0 & 0 \\ 0 & 0 & 1 & 0 & 1 & 1 \end{array}\right] \end{array}$$

$$\sim \begin{array}{c} \\ c_1'' \\ c_2'' \\ c_3'' \\ c_3'' \oplus c_4'' \\ c_3'' \oplus c_5'' \\ c_6'' \\ c_3'' \oplus c_7'' \end{array} \begin{array}{c} \begin{array}{cccccc} e_1 & e_2 & e_3 & e_4 & e_5 & e_6 \end{array} \\ \left[\begin{array}{cccccc} 1 & 0 & 0 & 1 & 0 & 1 \\ 0 & 1 & 0 & 1 & 1 & 0 \\ 0 & 0 & 1 & 0 & 1 & 1 \\ 0 & 0 & 0 & 0 & 0 & 0 \\ 0 & 0 & 0 & 0 & 0 & 0 \\ 0 & 0 & 0 & 0 & 0 & 0 \\ 0 & 0 & 0 & 0 & 0 & 0 \end{array}\right] \end{array} \qquad (2\text{-}13)$$

In Equation (2-13), the row operations performed to transform one equivalent matrix of B_{a_1} into another have clearly been indicated by the expression in front of each row of the matrix. The rank of B_{a_1} is seen to be three from the last equivalent matrix of B_{a_1} in Equation (2-13).

Step 2: Choose the independent rows, c_1, c_2, and c_4, to form S_i. The fact that these three rows are linearly independent can be easily seen by noting the leading unit matrix of order three in the first equivalent matrix B_{a_1} in Equation (2-13).

Step 3: The execution of the ring-sum operations among the three rows is as follows:

$$\begin{array}{rcl} & & \begin{array}{cccccc} e_1 & e_2 & e_3 & e_4 & e_5 & e_6 \end{array} \\ c_1 \oplus c_2 &=& [\begin{array}{cccccc} 0 & 1 & 1 & 1 & 0 & 1 \end{array}] = c_7 \\ c_1 \oplus c_4 &=& [\begin{array}{cccccc} 1 & 1 & 0 & 0 & 1 & 1 \end{array}] = c_6 \\ c_2 \oplus c_4 &=& [\begin{array}{cccccc} 1 & 0 & 1 & 1 & 1 & 0 \end{array}] = c_5 \\ c_1 \oplus c_2 \oplus c_4 &=& [\begin{array}{cccccc} 1 & 1 & 1 & 0 & 0 & 0 \end{array}] = c_3 \end{array}$$

Step 4: Since none of the new rows generated in Step 3 is a redundant row, no elimination process is necessary.

Step 5: Collecting rows c_1, c_2, and c_4 of S_i and the new rows c_3, c_5, c_6, c_7 generated in Step 3, we get the set $S = \{c_1, c_2, c_3, c_4, c_5, c_6, c_7\}$, which includes all the rows of B_a.

From Example 2-7, we see that the algorithm works quite nicely and produces the desired results. Every step in the procedure is straightforward and easy to carry out. However, there is one little thing (perhaps not so little!) that might have bothered the reader. We note that in order to determine the number of independent circuits in S_i, we must first find the rank r_B of B_a. In the example, we found the rank of B_a by reducing it to triangular form. But, if the circuit matrix B_a were available to begin with, it would not have been necessary to go through all these steps just to find out something we already knew — the set of all circuits in G that corresponds to the set of all rows in B_a!

This difficulty will be overcome if we can somehow relate the rank of B_a to the number v of vertices and number e edges of G. If this can be done, the number r_B can be determined without knowing B_a, since the numbers v and e can be found simply by counting the vertices and edges of G, respectively.

Thus, our next concern is to determine the rank r_B of B_a in terms of v and e of G. We shall accomplish this in four steps. The first step is to establish the relationship between A_a and B_a of a given graph G, as stated in the following theorem.

THEOREM 2-5. Let A_a and B_a be the incidence matrix and the circuit matrix, respectively, of a nonoriented, connected graph G, and let the columns of the two matrices be in the same element order, so that column i, for example, of both matrices corresponds to the same edge e_i of G. Then,

$$A_a B_a^T = 0 \tag{2-14}$$

or

$$B_a A_a^T = 0 \tag{2-15}$$

since

$$(A_a B_a^T)^T = (B_a^T)^T A_a^T = B_a A_a^T = 0$$

Proof: Let the product $A_a B_a^T$ be represented by P, where $P = A_a B_a^T = [p_{ij}]$. Then the (i,j) element, p_{ij}, of P is obtained by multiplying the ith row of A_a and the jth column of B_a^T; or

$$p_{ij} = \sum_{k=1}^{e} a_{ik} b_{kj}^T = a_{i1} b_{1j}^T + a_{i2} b_{2j}^T + \cdots + a_{ie} b_{ej}^T \tag{2-16}$$

where b_{kj}^T is the (k,j) element of B_a^T and e is the number of edges in G.

We note that the ith row of A stands for vertex v_i, and the jth column of B_a^T represents the circuit c_j of G. We also note that the number of edges common to any vertex v_i and any circuit c_j is either 2 or 0 (that is, 2 if vertex v_i is in circuit c_j; and 0 otherwise). This means that in Equation (2-16) there are either exactly two nonzero terms in the form $1 \cdot 1$, corresponding to the two edges incident with vertex v_i in circuit c_j, or no nonzero terms, corresponding to the case where v_i is not in c_j. But, in either case, the sum in Equation (2-16) is equal to zero. Therefore, the product $P = A_a B_a^T$ is a zero matrix and the proof of the theorem is complete.

EXAMPLE 2-8. Again, consider graph G_1 of Figure 2-6. The incidence matrix of this graph is

$$A_{a_1} = \begin{array}{c} \\ v_1 \\ v_2 \\ v_3 \\ v_4 \end{array} \begin{array}{c} \begin{array}{cccccc} e_1 & e_2 & e_3 & e_4 & e_5 & e_6 \end{array} \\ \begin{bmatrix} 1 & 1 & 0 & 1 & 0 & 0 \\ 0 & 1 & 1 & 0 & 1 & 0 \\ 1 & 0 & 1 & 0 & 0 & 1 \\ 0 & 0 & 0 & 1 & 1 & 1 \end{bmatrix} \end{array}$$

The circuit matrix of G_1 is given by B_{a_1} of Equation (2-12). Thus,

$$A_{a_1} B_{a_1}^T = \begin{array}{c} \\ v_1 \\ v_2 \\ v_3 \\ v_4 \end{array} \begin{array}{c} \begin{array}{cccccc} e_1 & e_2 & e_3 & e_4 & e_5 & e_6 \end{array} \\ \begin{bmatrix} 1 & 1 & 0 & 1 & 0 & 0 \\ 0 & 1 & 1 & 0 & 1 & 0 \\ 1 & 0 & 1 & 0 & 0 & 1 \\ 0 & 0 & 0 & 1 & 1 & 1 \end{bmatrix} \end{array} \cdot \begin{array}{c} \\ e_1 \\ e_2 \\ e_3 \\ e_4 \\ e_5 \\ e_6 \end{array} \begin{array}{c} \begin{array}{ccccccc} c_1 & c_2 & c_3 & c_4 & c_5 & c_6 & c_7 \end{array} \\ \begin{bmatrix} 0 & 0 & 1 & 1 & 1 & 1 & 0 \\ 1 & 0 & 1 & 0 & 0 & 1 & 1 \\ 0 & 1 & 1 & 0 & 1 & 0 & 1 \\ 1 & 0 & 0 & 1 & 1 & 0 & 1 \\ 1 & 1 & 0 & 0 & 1 & 1 & 0 \\ 0 & 1 & 0 & 1 & 0 & 1 & 1 \end{bmatrix} \end{array}$$

$$= \begin{bmatrix} 0 & 0 & 0 & 0 & 0 & 0 & 0 \\ 0 & 0 & 0 & 0 & 0 & 0 & 0 \\ 0 & 0 & 0 & 0 & 0 & 0 & 0 \\ 0 & 0 & 0 & 0 & 0 & 0 & 0 \end{bmatrix} \tag{2-17}$$

or $B_{a1}A_{a1}^T = 0$:

$$B_{a1}A_{a1}^T = \begin{array}{c} \\ c_1 \\ c_2 \\ c_3 \\ c_4 \\ c_5 \\ c_6 \\ c_7 \end{array}\!\!\begin{array}{c} \begin{matrix} e_1 & e_2 & e_3 & e_4 & e_5 & e_6 \end{matrix} \\ \begin{bmatrix} 0 & 1 & 0 & 1 & 1 & 0 \\ 0 & 0 & 1 & 0 & 1 & 1 \\ 1 & 1 & 1 & 0 & 0 & 0 \\ 1 & 0 & 0 & 1 & 0 & 1 \\ 1 & 0 & 1 & 1 & 1 & 0 \\ 1 & 1 & 0 & 0 & 1 & 1 \\ 0 & 1 & 1 & 1 & 0 & 1 \end{bmatrix} \end{array} \cdot \begin{array}{c} \\ e_1 \\ e_2 \\ e_3 \\ e_4 \\ e_5 \\ e_6 \end{array}\!\!\begin{array}{c} \begin{matrix} v_1 & v_2 & v_3 & v_4 \end{matrix} \\ \begin{bmatrix} 1 & 0 & 1 & 0 \\ 1 & 1 & 0 & 0 \\ 0 & 1 & 1 & 0 \\ 1 & 0 & 0 & 1 \\ 0 & 1 & 0 & 1 \\ 0 & 0 & 1 & 1 \end{bmatrix} \end{array}$$

$$= \begin{bmatrix} 0 & 0 & 0 & 0 \\ 0 & 0 & 0 & 0 \\ 0 & 0 & 0 & 0 \\ 0 & 0 & 0 & 0 \\ 0 & 0 & 0 & 0 \\ 0 & 0 & 0 & 0 \\ 0 & 0 & 0 & 0 \end{bmatrix} \qquad (2\text{-}18)$$

Hence, $A_{a1}B_{a1}^T = 0$, or $B_{a1}A_{a1}^T = 0$ as stated in Theorem 2-5.

It should be emphasized that the assumption in Theorem 2-5 that the columns of A_a and B_a of G be of the same edge order is a necessary and sufficient condition for Equations (2-14) and (2-15) to hold true, and our proof is based on this assumption. To appreciate this point, the reader should try to form the product $A_{a1}B_{a1}^T$ (or $B_{a1}A_{a1}^T$) for the same graph G_1 in Example 2-8 by rearranging the columns of A_{a1}, but keeping the column order of B_{a1} unchanged, and see what effects this would cause in either of the two products, $A_{a1}B_{a1}^T$ and $B_{a1}A_{a1}^T$.

We should, perhaps, point out also that it would be incorrect to combine Equations (2-14) and (2-15) to conclude that $A_aB_a^T = B_aA_a^T$. The reason for this is quite obvious, for one is the transpose of the other and $A_{a1}B_{a1}^T = B_{a1}A_{a1}^T$ only if the product $A_{a1}B_{a1}^T$ happens to be square, as a special case.

Our second step toward the determination of the rank of B_a in terms of v and e of G is stated in the following lemma:

LEMMA 2-3. The rank of the circuit matrix B_a of a nonoriented, connected graph G of v vertices and e edges is not less than $e - v + 1$; that is,

$$R(B_a) \geqq e - v + 1 \qquad (2\text{-}19)$$

Proof: The proof of this lemma is very simple. Recall that for a given tree T_i of G there exists a set of $e - v + 1$ (independent) fundamental circuits, each of which contains a unique chord from the chord set T_i' with respect to T_i (Definition 1-20). Since the rows of B_a correspond to the set of all circuits in G, these $e - v + 1$ fundamental circuits may then be identified by some $e - v + 1$ rows of B_a. Furthermore, since these $e - v + 1$ rows of B_a correspond to $e - v + 1$ independent fundamental circuits, they constitute a set of $e - v + 1$ independent rows of B_a; and, consequently, the rank of B_a can be no less than $e - v + 1$. Hence the lemma.

As our third step in expressing the rank of B_a in terms of v and e of G, we shall show that the rank of B_a is no greater than $e - v + 1$ for any nonoriented, connected graph G.

LEMMA 2-4. The rank of the circuit matrix B_a of a connected graph G of v vertices and e edges is no greater than $e - v + 1$; that is,

$$R(B_a) \leqq e - v + 1 \tag{2-20}$$

Proof: To prove this lemma, we shall resort to a theorem in matrix theory known as *Sylvester's law of nullity*,[4] which states that if $P = [p_{ij}]_{m\times n}$ and $Q = [q_{ij}]_{n\times r}$ such that $PQ = 0$, then

$$R(P) + R(Q) \leq n \tag{2-21}$$

By Equation (2-14) of Theorem 2-5, $A_a B_a^T = 0$. Thus, if we let $A_a = P$ and $B_a^T = Q$ so that $e = n$, we conclude from (2-21) that

$$R(A_a) + R(B_a^T) \leq e \tag{2-22}$$

But in Section 2-2, we showed that

$$R(A_a) = v - 1 \tag{2-23}$$

Hence, combining Equations (2-22) and (2-23) and rearranging the terms, we obtain

$$R(B_a^T) \leq e - v + 1$$

Finally, since for any matrix M, $R(M) = R(M^T)$, we conclude that

$$R(B_a) \leq e - v + 1 \tag{2-24}$$

The proof of the lemma is now complete.

Now, our last step is to show that the rank of B_a is exactly $e - v + 1$ for a nonoriented, connected graph G of v vertices and e edges.

THEOREM 2-6. For a nonoriented, connected graph G of v vertices and e edges, the rank of its circuit matrix B_a is given by

$$R(B_a) = e - v + 1 \tag{2-25}$$

Proof: To prove this theorem, we need only to observe that for expression (2-19) of Lemma 2-3 and expression (2-20) of Lemma 2-4 to hold simul-

[4]For a more detailed discussion and proof of this theorem, the reader is referred to Seshu and Reed [76], pp. 67–68.

taneously, the equality sign must be used in both expressions. Hence the theorem.

We next define the fundamental circuit matrix of a nonoriented, connected graph.

DEFINITION 2-5. Fundamental Circuit Matrix. A *fundamental circuit matrix with respect to a tree* of a nonoriented, connected graph G of v vertices and e edges is the matrix B_f of order $(e - v + 1) \times e$ with each row identified by a fundamental circuit (with respect to T) and each column by an edge such that

$$B_f = [b_{ij}]_{(e-v+1)\times e}$$

where

$$b_{ij} = 1 \text{ if edge } e_j \text{ is in fundamental circuit } c_i$$

$$b_{ij} = 0 \text{ if edge } e_j \text{ is not in } c_i$$

The next theorem is obvious.

THEOREM 2-7. The rank of a fundamental circuit matrix B_f with respect to a tree T of a given nonoriented, connected graph G of v vertices and e edges is given by

$$R(B_f) = e - v + 1 \tag{2-26}$$

Proof: This theorem follows directly from the fact that each row of B_f is identified by a distinct fundamental circuit that is independent from the others within the set of $e - v + 1$ fundamental circuits with respect to T. Hence the set of all of the $e - v + 1$ rows constitutes an independent set, and by a theorem in matrix theory, the rank of B_f is $e - v + 1$. This completes the proof.

With Theorems 2-6 and 2-7, our algorithm for obtaining a circuit matrix B_a from a connected graph G can be revised as stated in the following theorem:

THEOREM 2-8. Let G be a nonoriented, connected graph of v vertices and e edges. The set of all circuits of G may be obtained by the following steps:

Step 1: Choose a tree T of G and form B_f with the set of $e - v + 1$ fundamental circuits as its rows, and the e edges of G as its columns.

Step 2: Perform all possible ring sums among the rows of B_f to obtain B_1 with its rows including all of the $e - v + 1$ rows of B_f and the new rows generated by the ring-sum operations on the rows of B_f.

Step 3: Eliminate all of the redundant rows among the generated rows in B_1 (a row is redundant if it is an edge-disjoint union of some of the rows preceding it).

Step 4: Collecting all the remaining rows of B_1, we get the desired circuit matrix B_a.

Since each step in Theorem 2-8 has been discussed to a certain degree, the formal proof is left to the reader as an exercise (Problem 2-11).

EXAMPLE 2-9. Once more, let us consider G_1 of Figure 2-6. The graph is shown in Figure 2-8 with the tree T: $\{e_4, e_5, e_6\}$ indicated in heavy strokes.

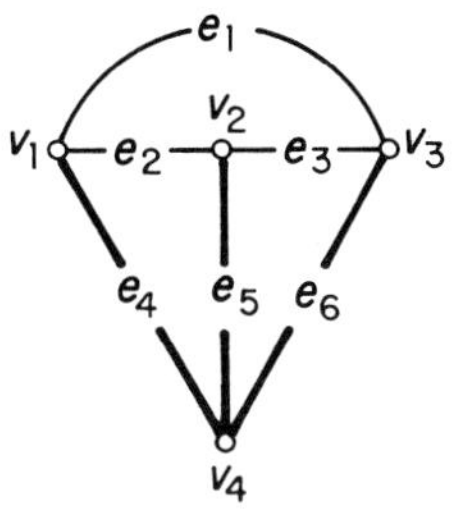

Figure 2-8 Graph G with tree T: $\{e_4,e_5,e_6\}$.

Step 1: The fundamental circuit matrix B_f with respect to T is

$$B_f = \begin{array}{c} \\ c_{e_1} \\ c_{e_2} \\ c_{e_3} \end{array} \begin{array}{c} \begin{array}{cccccc} e_1 & e_2 & e_3 & e_4 & e_5 & e_6 \end{array} \\ \begin{bmatrix} 1 & 0 & 0 & 1 & 0 & 1 \\ 0 & 1 & 0 & 1 & 1 & 0 \\ 0 & 0 & 1 & 0 & 1 & 1 \end{bmatrix} \end{array}$$

Step 2: Performing all possible ring sums among the three rows of B_f we have

$$B_1 = \begin{array}{r} \\ c_{e_1} \\ c_{e_2} \\ c_{e_3} \\ c_{e_1} \oplus c_{e_2} \\ c_{e_1} \oplus c_{e_3} \\ c_{e_2} \oplus c_{e_3} \\ c_{e_1} \oplus c_{e_2} \oplus c_{e_3} \end{array} \begin{array}{c} \begin{array}{cccccc} e_1 & e_2 & e_3 & e_4 & e_5 & e_6 \end{array} \\ \begin{bmatrix} 1 & 0 & 0 & 1 & 0 & 1 \\ 0 & 1 & 0 & 1 & 1 & 0 \\ 0 & 0 & 1 & 0 & 1 & 1 \\ 1 & 1 & 0 & 0 & 1 & 1 \\ 1 & 0 & 1 & 1 & 1 & 0 \\ 0 & 1 & 1 & 1 & 0 & 1 \\ 1 & 1 & 1 & 0 & 0 & 0 \end{bmatrix} \end{array} \begin{array}{l} \\ c_4 \\ c_1 \\ c_2 \\ c_6 \\ c_5 \\ c_7 \\ c_3 \end{array} \qquad (2\text{-}27)$$

Step 3: By inspection, we see that none of the generated rows in B_1 is a redundant row, and hence no elimination of rows in B_1 is necessary.

Step 4: Since we have eliminated no rows of B_1 in Step 3, we conclude that B_1 and B_{a_1} represent the same set of circuits.

By comparing the rows of B_{a_1} in Equation (2-12) with the rows of B_1 in (2-27), we see that

$$c_{e_1} = c_4 \qquad c_{e_1} \oplus c_{e_2} = c_6$$

$$c_{e_2} = c_1 \qquad c_{e_1} \oplus c_{e_3} = c_5$$

$$c_{e_3} = c_2 \qquad c_{e_2} \oplus c_{e_3} = c_7 \qquad c_{e_1} \oplus c_{e_2} \oplus c_{e_3} = c_3$$

as indicated in the matrix of (2-27).

Next, we shall give another example in which a more complicated graph is dealt with in order to better illustrate the usefulness of the algorithm for obtaining B_a.

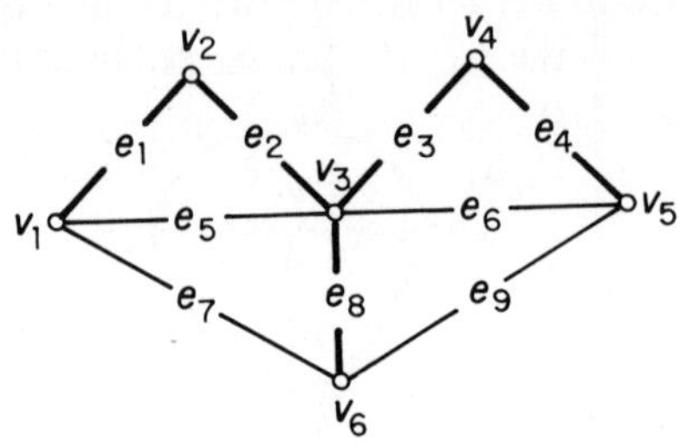

Figure 2-9 Graph G with tree T: $\{e_1, e_2, e_3, e_4, e_8\}$.

EXAMPLE 2-10. Consider the nonoriented, connected graph G shown in Figure 2-9 with tree T: $\{e_1, e_2, e_3, e_4, e_8\}$ indicated in heavy strokes. The fundamental circuit matrix with respect to T is

$$B_{f_1} = \begin{array}{r} \\ c_1(=c_{e_5}) \\ c_2(=c_{e_6}) \\ c_3(=c_{e_7}) \\ c_4(=c_{e_8}) \end{array} \begin{array}{c} \begin{array}{ccccccccc} e_1 & e_2 & e_3 & e_4 & e_5 & e_6 & e_7 & e_8 & e_9 \end{array} \\ \left[\begin{array}{ccccccccc} 1 & 1 & 0 & 0 & 1 & 0 & 0 & 0 & 0 \\ 0 & 0 & 1 & 1 & 0 & 1 & 0 & 0 & 0 \\ 1 & 1 & 0 & 0 & 0 & 0 & 1 & 1 & 0 \\ 0 & 0 & 1 & 1 & 0 & 0 & 0 & 1 & 1 \end{array}\right] \end{array}$$

The matrix B_1 is given by

$$B_1 = \begin{array}{r} \\ c_1 \\ c_2 \\ c_3 \\ c_4 \\ c_1 \oplus c_2 \\ c_1 \oplus c_3 \\ c_1 \oplus c_4 \\ c_2 \oplus c_3 \\ c_2 \oplus c_4 \\ c_3 \oplus c_4 \\ c_1 \oplus c_2 \oplus c_3 \\ c_1 \oplus c_2 \oplus c_4 \\ c_1 \oplus c_3 \oplus c_4 \\ c_2 \oplus c_3 \oplus c_4 \\ c_1 \oplus c_2 \oplus c_3 \oplus c_4 \end{array} \begin{array}{c} \begin{array}{ccccccccc} e_1 & e_2 & e_3 & e_4 & e_5 & e_6 & e_7 & e_8 & e_9 \end{array} \\ \left[\begin{array}{ccccccccc} 1 & 1 & 0 & 0 & 1 & 0 & 0 & 0 & 0 \\ 0 & 0 & 1 & 1 & 0 & 1 & 0 & 0 & 0 \\ 1 & 1 & 0 & 0 & 0 & 0 & 1 & 1 & 0 \\ 0 & 0 & 1 & 1 & 0 & 0 & 0 & 1 & 1 \\ 1 & 1 & 1 & 1 & 1 & 1 & 0 & 0 & 0 \\ 0 & 0 & 0 & 0 & 1 & 0 & 1 & 1 & 0 \\ 1 & 1 & 1 & 1 & 1 & 0 & 0 & 1 & 1 \\ 1 & 1 & 1 & 1 & 0 & 1 & 1 & 1 & 0 \\ 0 & 0 & 0 & 0 & 0 & 1 & 0 & 1 & 1 \\ 1 & 1 & 1 & 1 & 0 & 0 & 1 & 0 & 1 \\ 0 & 0 & 1 & 1 & 1 & 1 & 1 & 1 & 0 \\ 1 & 1 & 0 & 0 & 1 & 1 & 0 & 1 & 1 \\ 0 & 0 & 1 & 1 & 1 & 0 & 1 & 0 & 1 \\ 1 & 1 & 0 & 0 & 0 & 1 & 1 & 0 & 1 \\ 0 & 0 & 0 & 0 & 1 & 1 & 1 & 0 & 1 \end{array}\right] \end{array} \begin{array}{l} \\ c_1 \\ c_2 \\ c_3 \\ c_4 \\ \text{redundant} \\ c_5 \\ \text{redundant} \\ \text{redundant} \\ c_6 \\ c_7 \\ \text{redundant} \\ \text{redundant} \\ c_8 \\ c_9 \\ c_{10} \end{array}$$

Finally, removing the redundant rows as marked in B_1, we obtain the circuit matrix B_a

$$B_a = \begin{array}{c} \\ c_1 \\ c_2 \\ c_3 \\ c_4 \\ c_5 \\ c_6 \\ c_7 \\ c_8 \\ c_9 \\ c_{10} \end{array} \begin{array}{c} \begin{matrix} e_1 & e_2 & e_3 & e_4 & e_5 & e_6 & e_7 & e_8 & e_9 \end{matrix} \\ \begin{bmatrix} 1 & 1 & 0 & 0 & 1 & 0 & 0 & 0 & 0 \\ 0 & 0 & 1 & 1 & 0 & 1 & 0 & 0 & 0 \\ 1 & 1 & 0 & 0 & 0 & 0 & 1 & 1 & 0 \\ 0 & 0 & 1 & 1 & 0 & 0 & 0 & 1 & 1 \\ 0 & 0 & 0 & 0 & 1 & 0 & 1 & 1 & 0 \\ 0 & 0 & 0 & 0 & 0 & 1 & 0 & 1 & 1 \\ 1 & 1 & 1 & 1 & 0 & 0 & 1 & 0 & 1 \\ 0 & 0 & 1 & 1 & 1 & 0 & 1 & 0 & 1 \\ 1 & 1 & 0 & 0 & 0 & 1 & 1 & 0 & 1 \\ 0 & 0 & 0 & 0 & 1 & 1 & 1 & 0 & 1 \end{bmatrix} \end{array} \tag{2-28}$$

From this example, the reader should realize that the more complicated the graph, the more useful the algorithm becomes, because the chances of missing a circuit by inspection will greatly increase with the increase of the complexity of the graph.

DEFINITION 2-6. Circuit Submatrix. A *circuit submatrix* B is an $(e - v + 1) \times e$ submatrix of B_a of a nonoriented, connected graph G of v vertices and e edges such that

$$R(B) = e - v + 1 \tag{2-29}$$

In other words, B is obtained by choosing a set of $e - v + 1$ *independent* rows from B_a as its rows. Thus, all of the $e - v + 1$ rows of B correspond to a set of $e - v + 1$ independent circuit of G. Furthermore, since there can be more than one way of choosing a set of $e - v + 1$ independent circuits, the formation of B is not unique. However, since by definition each B must be of order $(e - v + 1) \times e$ and of rank $e - v + 1$, it is evident that all of the circuit submatrices are equivalent; and it is always possible to transform one circuit submatrix to another by (repeated) application of the ring-sum operation on the rows of the former.

EXAMPLE 2-11. For the graph G of Figure 2-9, the circuit matrix B_a is given in Equation (2-28). A circuit submatrix B_1 of G is

$$B_1 = \begin{array}{c} \\ c_1 \\ c_2 \\ c_5 \\ c_6 \end{array} \begin{array}{c} \begin{matrix} e_1 & e_2 & e_3 & e_4 & e_5 & e_6 & e_7 & e_8 & e_9 \end{matrix} \\ \begin{bmatrix} 1 & 1 & 0 & 0 & 1 & 0 & 0 & 0 & 0 \\ 0 & 0 & 1 & 1 & 0 & 1 & 0 & 0 & 0 \\ 0 & 0 & 0 & 0 & 1 & 0 & 1 & 1 & 0 \\ 0 & 0 & 0 & 0 & 0 & 1 & 0 & 1 & 1 \end{bmatrix} \end{array} \tag{2-30}$$

Another circuit submatrix B_2 is given by

$$B_2 = \begin{array}{c} \\ c_1 \\ c_3 \\ c_2 \\ c_4 \end{array}\begin{array}{c} \begin{array}{ccccccccc} e_1 & e_2 & e_3 & e_4 & e_5 & e_6 & e_7 & e_8 & e_9 \end{array} \\ \begin{bmatrix} 1 & 1 & 0 & 0 & 1 & 0 & 0 & 0 & 0 \\ 1 & 1 & 0 & 0 & 0 & 0 & 1 & 1 & 0 \\ 0 & 0 & 1 & 1 & 0 & 1 & 0 & 0 & 0 \\ 0 & 0 & 1 & 1 & 0 & 0 & 0 & 1 & 1 \end{bmatrix} \end{array} \tag{2-31}$$

Still another one, B_3, is

$$B_3 = \begin{array}{c} \\ c_2 \\ c_4 \\ c_8 \\ c_7 \end{array}\begin{array}{c} \begin{array}{ccccccccc} e_1 & e_2 & e_3 & e_4 & e_5 & e_6 & e_7 & e_8 & e_9 \end{array} \\ \begin{bmatrix} 0 & 0 & 1 & 1 & 0 & 1 & 0 & 0 & 0 \\ 0 & 0 & 1 & 1 & 0 & 0 & 0 & 1 & 1 \\ 0 & 0 & 1 & 1 & 1 & 0 & 1 & 0 & 1 \\ 1 & 1 & 1 & 1 & 0 & 0 & 1 & 0 & 1 \end{bmatrix} \end{array} \tag{2-32}$$

As indicated in each circuit submatrix, all three matrices are submatrices of B_a of Equation (2-28). In fact, B_1 is formed by taking rows 1, 2, 5, and 6 of B_a of (2-28); B_2 is obtained by choosing rows 1, 3, 2, and 4 of the same matrix; and B_3 is formed by taking rows 2, 4, 8, and 7 of B_a as its rows.

It is simple to show that all three circuit submatrices are of rank 4, which is equal to the number $e - v + 1$ where, in this case, $e = 9$, $v = 6$, so that $e - v + 1 = 9 - 6 + 1 = 4$, thus satisfying the condition given in Equation (2-29) under Definition 2-6.

It will be interesting for the reader to transform one circuit submatrix, say B_1, to another, say B_2, by the application of ring-sum operations on the rows of B_1.

Finally, it is appropriate to note that in most network problems, one is concerned only with the fundamental circuit matrix B_f (or B), such as in the case of writing loop equations. Thus the problem of obtaining the circuit matrix B_a of a given network is essentially a theoretical one, and the inclusion of such a discussion here is mainly for the sake of completeness. However, it should be pointed out also that by the application of computer implementation, the algorithm developed here for obtaining B_a can be used, for instance, to determine the planarity of various networks. The interested reader is referred to Dunn and Chan [27] for such applications.

2-4 THE CUTSET MATRIX

The third matrix that we will discuss is the cutset matrix, which also plays a very important role in network topology. As in the case of the incidence matrix and circuit matrix, it describes the relation between vertices and edges of a given graph. We shall define it formally as follows:

DEFINITION 2-7. Cutset Matrix. The *cutset matrix* of a nonoriented, connected graph G of v vertices and e edges is the matrix Q_a of order $k \times e$ with each row identified by a cutset and each column by an edge, such that

$$Q_a = [q_{ij}]_{k \times e}$$

where

$q_{ij} = 1$ if edge e_j is in cutset k_i

$q_{ij} = 0$ if e_j is not in cutset k_i

k is the total number of cutsets in G

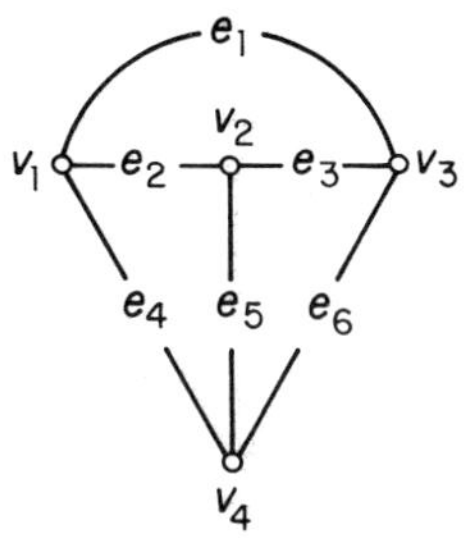

Figure 2-10 Graph G_1 of Figure 2-6.

EXAMPLE 2-12. Consider once again graph G_1 of Figure 2-6. The graph is repeated in Figure 2-10 and the cut-lines corresponding to the cutsets are shown in Figure 2-11. From Figure 2-11, we form the cutset matrix Q_a:

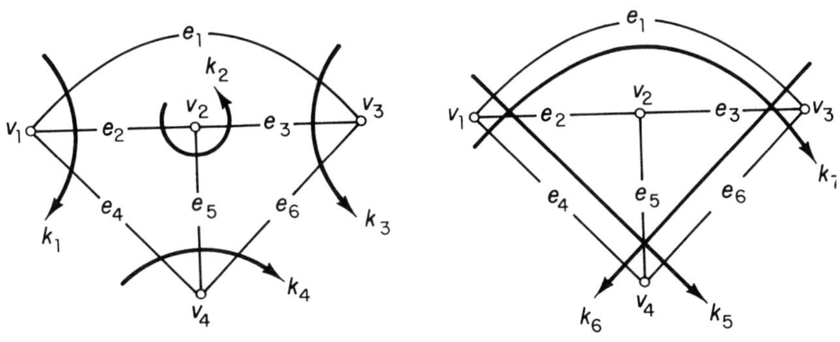

Figure 2-11 The cut-lines corresponding to the cutsets of G_1.

$$Q_a = \begin{matrix} & e_1 & e_2 & e_3 & e_4 & e_5 & e_6 \\ k_1 & 1 & 1 & 0 & 1 & 0 & 0 \\ k_2 & 0 & 1 & 1 & 0 & 1 & 0 \\ k_3 & 1 & 0 & 1 & 0 & 0 & 1 \\ k_4 & 0 & 0 & 0 & 1 & 1 & 1 \\ k_5 & 1 & 1 & 0 & 0 & 1 & 1 \\ k_6 & 1 & 0 & 1 & 1 & 1 & 0 \\ k_7 & 0 & 1 & 1 & 1 & 0 & 1 \end{matrix} \qquad (2\text{-}33)$$

Obviously, we are encountering the same kind of difficulty as we did with the circuit matrix B_a; that is, the total number of cutsets for a given graph is not readily determined by inspection, especially when the number of vertices v and the number of edges e of the graph are large. Thus, as in the case of a circuit matrix, we would like to, first, choose a set of (necessary and sufficient) independent cutsets from the graph, and then perform the ring-sum operations among these cutsets in order to generate the rest of the cutsets in the graph.

To accomplish this, first of all we need to know the rank of Q_a so that we will know how many independent cutsets will be *just* enough to produce the rest of the cutsets from the set of all cutsets corresponding to the rows of Q_a. But, before determining the rank of Q_a, we first define a fundamental cutset matrix and discuss some of the properties of this matrix.

DEFINITION 2-8. Fundamental Cutset Matrix. The *fundamental cutset matrix with respect to a tree* T of a nonoriented, connected graph G of v vertices and e edges is the matrix Q_f of order $(v - 1) \times e$ with each row identified by a fundamental cutset (with respect to T) and each column by an edge such that

$$Q_f = [q_{ij}]_{(v-1)\times e}$$

where

$$q_{ij} = 1 \text{ if edge } e_j \text{ is in the fundamental cutset } k_i$$

$$q_{ij} = 0 \text{ if edge } e_j \text{ is not in } k_i$$

Since, by definition, the set of $v - 1$ fundamental cutsets with respect to a tree constitutes an independent set, and since each fundamental cutset corresponds to a row of Q_f (with respect to the same tree, of course), the following theorem is self-evident.

THEOREM 2-9. The rank of Q_f of a connected, nonoriented graph G of v vertices and e edges is $v - 1$; that is,

$$R(Q_f) = v - 1 \tag{2-34}$$

Since Q_a is the cutset matrix including all possible cutsets of the graph, it is obvious that Q_f is a submatrix of Q_a, provided, of course, that the columns of both matrices are arranged in the same edge sequence. Thus we have the following corollary:

COROLLARY 2-1. The rank of the cutset matrix Q_a of a connected, nonoriented graph G of v vertices and e edges is no less than $v - 1$; that is,

$$R(Q_a) \geqq v - 1 \tag{2-35}$$

Using Sylvester's law of nullity and Theorem 2-6, we shall state and prove the next theorem.

THEOREM 2-10. The rank of the cutset matrix Q_a of a connected, non-oriented graph G of v vertices and e edges is exactly $v - 1$; that is,

$$R(Q_a) = v - 1$$

Proof: Recall that Sylvester's law of nullity states that if C and D are two matrices such that $C = [c_{ij}]_{m\times n}$, $D = [d_{ij}]_{n\times k}$, and $CD = 0$; then

$$R(C) + R(D) \leqq n \tag{2-36}$$

Now, let $Q_a = C$ and $B_a^T = D$. Using Theorem 1-6 and following the same pattern as in the procedure for the proof of Theorem 2-5, it can be easily shown that

$$Q_a B_a^T = 0 \tag{2-37}$$

so that all conditions in Sylvester's law of nullity are satisfied by Q_a and B_a^T.

Thus, from Equation (2-36) we write

$$R(Q_a) + R(B_a^T) \leqq e$$

or

$$R(Q_a) \leqq e - R(B_a^T) \tag{2-38}$$

But

$$R(B_a) = e - v + 1 \tag{2-39}$$

and

$$R(B_a^T) = R(B_a) \tag{2-40}$$

Hence, substituting Equations (2-39) and (2-40) into (2-38), we have

$$R(Q_a) \leqq e - (e - v + 1)$$

or

$$R(Q_a) \leqq v - 1 \tag{2-41}$$

Finally, for Equations (2-35) and (2-41) to hold true simultaneously, the equality sign must be used; that is,

$$R(Q_a) = v - 1$$

This completes the proof.

With the rank of Q_a known, the procedure for obtaining the set of all cutsets for a given graph G is clear. First, we obtain a set S_1 of $v - 1$ independent cutsets from G. Then we perform all possible ring sums among the $v - 1$ cutsets. Note that the ring sum of two cutsets is either another cutset or an edge-disjoint union of cutsets. (Why?) Hence a ring sum of some (or all) of the $v - 1$ cutsets in S_1 must be one of these two possibilities: (a) a (new) cutset not included in S_1; (b) a redundant term due to the fact that it is an edge-disjoint union of cutsets. Consequently, if we remove all of those redundant terms after the ring-sum operation, the remaining generated cutsets along with the $v - 1$ cutsets in S_1 constitute the set of all cutsets in G.

We now state this algorithm in the next theorem.

THEOREM 2-11. Let G be a connected, nonoriented graph of v vertices and e edges. The set of all cutsets of G (and hence Q_a) may be obtained by the following procedure:

Step 1: Choose a tree T of G and form Q_f with the set of $v - 1$ fundamental cutsets as its rows and the e edges of G as its columns.

Step 2: Perform all possible ring sums among the rows of Q_f to obtain Q_1 with its rows including all of the $v - 1$ rows of Q_f and the new rows generated by the ring-sum operation.

Step 3: Remove all of the redundant rows from Q_1.

Step 4: Collecting all of the remaining rows in Q_1, we obtain the desired cutset matrix Q_a.

The formal proof of Theorem 2-11 is left as an exercise (Problem 2-17). The following example will illustrate the ideas involved in the four steps stated in the theorem.

EXAMPLE 2-13. Refer to graph G_1 of Figure 2-8 in which a tree T: $\{e_4, e_5, e_6\}$ has been chosen. Graph G_1 is also the same graph as that of Figure 2-10 in Example 2-12.

Step 1: With respect to T: $\{e_4,e_5,e_6\}$, we form Q_f:

$$Q_f = \begin{matrix} \\ k_{e_4} \\ k_{e_5} \\ k_{e_6} \end{matrix} \begin{matrix} \begin{matrix} e_1 & e_2 & e_3 & e_4 & e_5 & e_6 \end{matrix} \\ \begin{bmatrix} 1 & 1 & 0 & 1 & 0 & 0 \\ 0 & 1 & 1 & 0 & 1 & 0 \\ 1 & 0 & 1 & 0 & 0 & 1 \end{bmatrix} \end{matrix} \begin{matrix} \\ = k_1 \\ = k_2 \\ = k_3 \end{matrix} \tag{2-42}$$

From the cutset matrix Q_a in Equation (2-33), we see that these are exactly the first three rows of Q_a.

Step 2: The ring sums of the three rows of Q_f are:

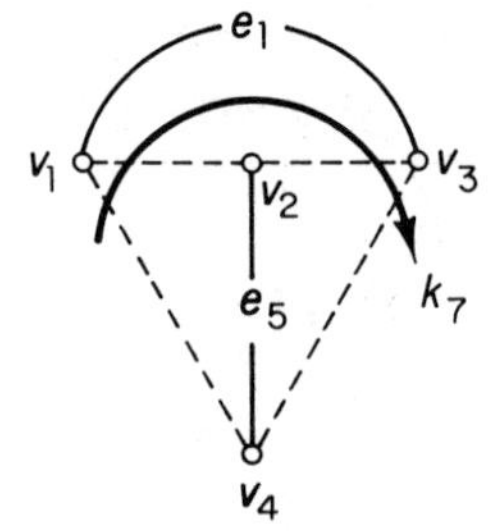

Figure 2-12 A nonconnected graph resulting from the removal of cutset k_7 from graph G_1 of Figure 2-8.

$$\begin{array}{r@{}cccccc@{}l} & e_1 & e_2 & e_3 & e_4 & e_5 & e_6 & \\ k_1 \oplus k_2[& 1 & 0 & 1 & 1 & 1 & 0 &] = k_6 \\ k_1 \oplus k_3[& 0 & 1 & 1 & 1 & 0 & 1 &] = k_7 \\ k_2 \oplus k_3[& 1 & 1 & 0 & 0 & 1 & 1 &] = k_5 \\ k_1 \oplus k_2 \oplus k_3[& 0 & 0 & 0 & 1 & 1 & 1 &] = k_4 \end{array} \tag{2-43}$$

Thus our resulting matrix Q_1 is

$$Q_1 = \begin{array}{c} \\ k_1 \\ k_2 \\ k_3 \\ k_6 \\ k_7 \\ k_5 \\ k_4 \end{array} \begin{array}{c} \begin{array}{cccccc} e_1 & e_2 & e_3 & e_4 & e_5 & e_6 \end{array} \\ \begin{bmatrix} 1 & 1 & 0 & 1 & 0 & 0 \\ 0 & 1 & 1 & 0 & 1 & 0 \\ 1 & 0 & 1 & 0 & 0 & 1 \\ 1 & 0 & 1 & 1 & 1 & 0 \\ 0 & 1 & 1 & 1 & 0 & 1 \\ 1 & 1 & 0 & 0 & 1 & 1 \\ 0 & 0 & 0 & 1 & 1 & 1 \end{bmatrix} \end{array} \tag{2-44}$$

Step 3: Comparing the generated rows of Q_1 in Equation (2-43) with the rows of Q_f in Equation (2-42), we see that none of the generated rows in Q_1 is a redundant row. Hence, no removal of rows from Q_1 is neccssary.

Step 4: Since no row of Q_1 has been removed in Step 3, we have $Q_1 \sim Q_a$. In fact, in comparison of the rows between Q_1 in Equation (2-44) and Q_a in (2-33), we see that there exists a one-to-one correspondence between the rows of Q_a and the rows of Q_1; that is, Q_1 can be made identical to Q_a upon the rearrangement of its rows as indicated in Equation (2-44).

Even from simple examples, such as Examples 2-12 and 2-13, we see that by inspection one probably would have missed cutset k_7: e_2, e_3, e_4, e_6, which is also a circuit. But, if all of the edges in k_7 are removed from G, an unconnected graph with two maximally connected subgraphs results, as shown in Figure 2-12. Thus, k_7 satisfies all of the conditions stated in Definition 1-14 and is, consequently, a cutset.

If the algorithm of Theorem 2-11 is employed, however, it will be impossible to omit any cutsets even if the graph is a complicated one (in the sense that both v and e are large numbers), since the procedure is systematic. Every step is well defined and involves only simple operations.

In order to better appreciate the third step in the procedure, in which elimination of redundant terms is carried out, the reader should try a more complicated graph; for example, the graph of Figure 2-9, and obtain Q_a by the algorithm of Theorem 2-11.

We might mention, also, that an easy way of checking the correctness of the row of Q_a is to form either one of the products, $Q_aB_a^T$ or $B_aQ_a^T$, and see whether or not it results in a zero matrix. However, this will only show that the rows of Q_a obtained are really cutsets (provided, of course, that B_a is correct), but it will not indicate whether the number of rows does represent the correct number of *all* cutsets for the graph. This check is used mainly against any error made in the ring-sum operation. If the product is not a zero matrix, an inspection of the graph and the row that causes a nonzero entry in the product will quickly show what the correct cutset should be. The correct number of rows of Q_a is guaranteed by the third and the fourth step in the procedure stated in Theorem 2-11. These remarks will become much clearer once the reader has worked a few problems similar to, but perhaps more complicated than, that of Example 2-13.

As in the case of a circuit submatrix B of the circuit matrix B_a, there also exists a cutset submatrix Q of the cutset matrix Q_a that plays a similar role among the cutsets as B does among the circuits. We shall define it as follows:

DEFINITION 2-9. Cutset Submatrix. A *cutset submatrix* Q is a submatrix of order $(v - 1) \times e$ of the cutset matrix Q_a of a connected, nonoriented graph G of v vertices and e edges such that

$$R(Q) = v - 1 \tag{2-45}$$

Definition 2-9 implies that Q is obtained from Q_a by selecting a set of $v - 1$ *independent* rows from Q_a as its rows. Similar to the argument given for the case of a circuit submatrix B, the formation of Q is not unique, and there can be more than one Q in Q_a. But if there is more than one Q in Q_a, all the Q's are equivalent and one can be transformed into another through the ring-sum operations on its rows.

EXAMPLE 2-14. Consider the graph G_1 of Figure 2-10. The cutset matrix Q_a is given in Equation (2-33). Since $v = 4$ for this graph, we need three independent rows from Q_a to form a cutset submatrix Q. Equation (2-46) is one of these cutset submatrices.

$$Q_1 = \begin{array}{c} \\ k_1 \\ k_2 \\ k_3 \end{array} \begin{array}{c} \begin{array}{cccccc} e_1 & e_2 & e_3 & e_4 & e_5 & e_6 \end{array} \\ \begin{bmatrix} 1 & 1 & 0 & 1 & 0 & 0 \\ 0 & 1 & 1 & 0 & 1 & 0 \\ 1 & 0 & 1 & 0 & 0 & 1 \end{bmatrix} \end{array} \tag{2-46}$$

Another one is given by

$$Q_2 = \begin{array}{c} \\ k_2 \\ k_3 \\ k_4 \end{array} \begin{array}{c} \begin{array}{cccccc} e_1 & e_2 & e_3 & e_4 & e_5 & e_6 \end{array} \\ \begin{bmatrix} 0 & 1 & 1 & 0 & 1 & 0 \\ 1 & 0 & 1 & 0 & 0 & 1 \\ 0 & 0 & 0 & 1 & 1 & 1 \end{bmatrix} \end{array} \tag{2-47}$$

Comparing the rows of Q_1 and Q_2 with the rows of Q_a in Equation (2-33), it is evident that both matrices Q_1 and Q_2 are submatrices of Q_a, as indicated in Equations (2-46) and (2-47), with the rows of Q_1 and Q_2 identified by the rows of Q_a. The fact that both Q_1 and Q_2 are of rank 3 can be seen simply by inspection of the nonsingular submatrices of order 3 contained in the two matrices.

Needless to say, there are still some more cutset submatrices contained in Q_a. Each of them represents a set of $v - 1$ independent cutsets in the graph.[5]

2-5 INTERRELATIONSHIPS BETWEEN MATRICES

In the preceding section we introduced the incidence matrix A_a, the circuit matrix B_a, and the cutset matrix Q_a of a connected, nonoriented graph G. We also discussed some of the properties of these three matrices and the implicit interrelationship between them. More specifically, we showed that

$$A_a B_a^T = 0$$

or

$$B_a A_a^T = 0$$

Similarly,

$$Q_a B_a^T = 0$$

or

$$B_a Q_a^T = 0$$

In this section we shall derive the interrelationship between the matrices A, B_f, and Q_f. In order to accomplish this, we shall first state a few theorems that serve as stepping stones leading to the desired results.

THEOREM 2-12. Let A be an incidence submatrix of a connected, nonoriented graph G of v vertices. A submatrix A_1 of order $v - 1$ of A is nonsingular if and only if the $v - 1$ columns of A_1 correspond to the set of $v - 1$ branches of a tree T of G.

Proof: Let A_1 be a submatrix of order $v - 1$ of A such that its $v - 1$ columns correspond to the $v - 1$ branches of a tree T. Then A_1 is an incidence submatrix of the tree T. But a tree is a connected graph of v vertices and $v - 1$ edges, and an incidence submatrix of T must be of order $(v - 1) \times (v - 1)$ and of rank $v - 1$. Hence A_1 is nonsingular.

Next, let A_1 be a nonsingular submatrix of A and of order $v - 1$. Since it is a submatrix of the incidence matrix A_a, the rows of A_1 represent $v - 1$ vertices of G and the columns of A_1 correspond to $v - 1$ edges of G. Now, consider the subgraph G_1 formed by the $v - 1$ edges that correspond to the

[5]The comments made about B_a, B_f, and B at the end of Section 2-3 also apply here about Q_a, Q_f, and Q in writing node equations.

columns of A_1. This subgraph has $v - 1$ edges and v vertices, and one of its incidence submatrices is A_1. This implies that G_1 must be connected; if it were not, its incidence matrix (or submatrix) would be of rank $v_1 - p_1$, where v_1 is the number of vertices and $p_1(> 1)$ the number of maximally connected subgraphs of G_1 (Theorem 2-3). Since, in this case, $p_1 = 1$, then G_1 is a connected subgraph of G. However, to connect all of the v vertices with only $v - 1$ edges is possible only if G_1 is a tree. (See the discussion on the four properties of a tree, Section 1-5, and note that certain two or any three of the four properties will serve as a necessary and sufficient condition for a tree; in this case, the two properties included are the connectivity and the right number, namely, $v - 1$, of edges.) Hence the theorem.

In fact, a theorem similar to Theorem 2-12 may be stated about the cutset submatrix Q. This is given as the next theorem.

THEOREM 2-13. Let Q be a cutset submatrix of a connected, nonoriented graph G of v vertices. The columns of a nonsingular submatrix Q_1 of order $v - 1$ of Q correspond to the set of all branches of some tree T of G.

Proof: This theorem can be proved by first showing that an incidence submatrix A can be transformed into a cutset submatrix Q through the application of the ring-sum operations on its rows, and vice versa. This will be left to the reader (Problem 2-32). After this has been accomplished, it should be evident that the nonsingular submatrix Q_1 of Q will be transformed into a nonsingular submatrix A_1 of A when Q is transformed into A through the ring-sum operations on its rows. Since only row operations have been performed, the edge order has not been changed by such operations. This statement implies that the columns of Q_1 and the columns of A_1 correspond to the same set of $v - 1$ edges in G. But, by Theorem 2-12, we know that the columns of A_1 correspond to the set of $v - 1$ branches of some tree. Hence the theorem.

We now state the next theorem about the circuit submatrix B, which is similar to Theorems 2-12 and 2-13 about A and Q, respectively.

THEOREM 2-14. Let B be a circuit submatrix of a connected, nonoriented graph G of v vertices and e edges. A submatrix B_1 of order $e - v + 1$ of B is nonsingular if and only if the $e - v + 1$ columns of B_1 correspond to the set of $e - v + 1$ chords with respect to some tree T of G.

The proof of this theorem is left as a problem (Problem 2-22).

With the results of Theorems 2-12, 2-13, and 2-14, we are in a position to state and prove the following theorem.

THEOREM 2-15. Let A be an incidence submatrix, let B_f and Q_f be a fundamental circuit matrix and a fundamental cutset matrix, respectively, with respect to some tree T of a connected, nonoriented graph G of v vertices and e edges. Furthermore, let the columns of the three matrices be arranged in the same edge order with the first $e - v + 1$ edges corresponding to chords and the rest of the edges corresponding to branches of the same tree T so that the matrices can be partitioned as

$$A = [A_{11} \quad A_{12}] \tag{2-48}$$

$$B_f = [U_N \quad B_{f_{12}}] \tag{2-49}$$

and

$$Q_f = [Q_{f_{11}} \quad U_R] \tag{2-50}$$

where $N = e - v + 1$ and $R = v - 1$.
Then the following interrelationships result:

$$Q_{f_{11}} = B^T_{f_{12}} = A^{-1}_{12} \cdot A_{11} \tag{2-51}$$

and

$$Q_f = [B^T_{f_{12}} \quad U_R] = A^{-1}_{12} \cdot A \tag{2-52}$$

Proof: From the assumptions made in the theorem, we see that the columns of A_{12} in Equation (2-48) correspond to the set of $v - 1$ branches of T. Thus, A_{12} is nonsingular and A^{-1}_{12} exists. Furthermore, $A \cdot B^T_f = 0$ in light of Equation (2-14); and, hence, taking the transpose of Equation (2-49) and premultiplying it by (2-50) we get

$$A \cdot B^T_f = [A_{11} \quad A_{12}] \cdot \begin{bmatrix} U^T_N \\ B^T_{f_{12}} \end{bmatrix} = 0$$

or

$$A_{11} \cdot U^T_N + A_{12} \cdot B^T_{f_{12}} = 0 \tag{2-53}$$

But $U^T_N = U_N$ and $A_{11} \cdot U_N = A_{11}$. Thus Equation (2-53) becomes

$$A_{11} + A_{12} \cdot B^T_{f_{12}} = 0 \tag{2-54}$$

Next, recall that the matrices we are dealing with all belong to the set of matrices with only 0's and 1's as elements, and the operations of addition (ring sum) and multiplication are governed by the addition and multiplication tables defined in Figure 2-3. Hence, for any two matrices, M and N, the equation $M + N = 0$ implies that $M = N$. (Why?) We conclude, therefore, that from Equation (2-54),

$$A_{12} \cdot B^T_{f_{12}} = A_{11} \tag{2-55}$$

Premultiplying Equation (2-55) by A^{-1}_{12} we get

$$B^T_{f_{12}} = A^{-1}_{12} \cdot A_{11} \tag{2-56}$$

Taking the transpose of Equation (2-56), we obtain

$$(B_{f12}^T)^T = (A_{12}^{-1} \cdot A_{11})^T$$

or

$$B_{f12} = A_{11}^T \cdot (A_{12}^{-1})^T \tag{2-57}$$

Similarly, starting from

$$Q_f \cdot B_f^T = [Q_{f11} \quad U_R] \cdot \begin{bmatrix} U_N^T \\ B_{f12}^T \end{bmatrix} = 0$$

and, carrying out the few obvious steps, we arrive at the conclusion that

$$Q_{f11} = B_{f12}^T \tag{2-58}$$

Finally, combining Equations (2-50), (2-56), (2-57), and (2-58), we obtain the desired results of Equations (2-51) and (2-52). The proof is now complete.

EXAMPLE 2-15. Consider graph G of Figure 2-8 with the tree T: $\{e_4, e_5, e_6\}$. The fundamental circuit matrix B_f for T with edges arranged in the order of chords and branches as specified in Theorem 2-15 is

$$B_f = \begin{matrix} \\ c_{e_1} \\ c_{e_2} \\ c_{e_3} \end{matrix} \begin{matrix} \begin{matrix} e_1 & e_2 & e_3 & e_4 & e_5 & e_6 \end{matrix} \\ \left[\begin{array}{ccc|ccc} 1 & 0 & 0 & 1 & 0 & 1 \\ 0 & 1 & 0 & 1 & 1 & 0 \\ 0 & 0 & 1 & 0 & 1 & 1 \end{array}\right] \end{matrix} = [U_3 \quad B_{f12}] \tag{2-59}$$

Similarly, the fundamental cutset matrix Q_f for T with the same edge order is

$$Q_f = \begin{matrix} \\ k_{e_4} \\ k_{e_5} \\ k_{e_6} \end{matrix} \begin{matrix} \begin{matrix} e_1 & e_2 & e_3 & e_4 & e_5 & e_6 \end{matrix} \\ \left[\begin{array}{ccc|ccc} 1 & 1 & 0 & 1 & 0 & 0 \\ 0 & 1 & 1 & 0 & 1 & 0 \\ 1 & 0 & 1 & 0 & 0 & 1 \end{array}\right] \end{matrix} = [Q_{f11} \quad U_3] \tag{2-60}$$

Here both unit matrices in Equations (2-59) and (2-60) are of the same order because $N = e - v + 1 = 6 - 4 + 1 = 3$ and $R = v - 1 = 4 - 1 = 3$, so that $U_N = U_R = U_3$.

Next, the incidence submatrix A with vertex v_1 taken as reference and with the columns arranged in the same edge order is

$$A = \begin{matrix} \\ v_2 \\ v_3 \\ v_4 \end{matrix} \begin{matrix} \begin{matrix} e_1 & e_2 & e_3 & e_4 & e_5 & e_6 \end{matrix} \\ \left[\begin{array}{ccc|ccc} 0 & 1 & 1 & 0 & 1 & 0 \\ 1 & 0 & 1 & 0 & 0 & 1 \\ 0 & 0 & 0 & 1 & 1 & 1 \end{array}\right] \end{matrix} = [A_{11} \quad A_{12}] \tag{2-61}$$

where

$$A_{12} = \begin{bmatrix} 0 & 1 & 0 \\ 0 & 0 & 1 \\ 1 & 1 & 1 \end{bmatrix}$$

so that

$$A_{12}^{-1} = \begin{bmatrix} 1 & 1 & 1 \\ 1 & 0 & 0 \\ 0 & 1 & 0 \end{bmatrix}$$

keeping in mind that the "negative" of an element is the element itself for matrices with only 0's and 1's as elements.

It is now a simple matter to verify the correctness of Equations (2-51) and (2-52) by making the appropriate substitutions as indicated in the two expressions. The task is again left to the reader as an exercise.

It is worth noting that with the interrelationships (2-51) and (2-52) between A, B_f, and Q_f, stated in Theorem 2-15, once matrix A is known, the two matrices, B_f and Q_f, can easily be determined by means of the two equations (see, also, Problem 2-32).

Finally, we shall define the following two terms before discussing other matrices of a graph.

DEFINITION 2-10. Rank of a Graph. The *rank of a graph* is defined as the rank of its incidence matrix.

DEFINITION 2-11. Nullity of a Graph. The *nullity of a graph* is defined as the rank of its circuit matrix.

Thus, if G is a connected graph of v vertices and e edges, the rank of G is $v - 1$ and the nullity of G is $e - v + 1$, and we write

$$R(G) = v - 1 \tag{2-62}$$

and

$$N(G) = e - v + 1 \tag{2-63}$$

We should point out that the symbol G in both (2-62) and (2-63) stands for a graph, not a matrix. Also, it should be obvious to the reader that if G is an unconnected graph with v vertices, e edges, and p maximally connected subgraphs, then we get

$$R(G) = v - p$$

and

$$N(G) = e - v + p$$

It is worth noting that the "nullity" of a graph is also known as the "connectivity," the "cyclomatic number," and the "first Betti number" in other literature of linear graph theory.

2-6 OTHER MATRICES OF A GRAPH

So far, we have introduced the following matrices: The incidence matrix A_a, an incidence submatrix A, the circuit matrix B_a, the fundamental circuit matrix B_f (with respect to some tree), a circuit submatrix B, the cutset matrix Q_a, the fundamental cutset matrix Q_f (with respect to some tree), and a cutset submatrix Q. There are still more matrices of a graph that have not been introduced here, but those we have covered are the ones most commonly known in network topology. Their properties have been widely applied to the analysis and synthesis of electrical networks, especially in the area of topological analysis of both passive and active networks.

Some of the graph matrices that have not yet been mentioned are, to name a few, the path matrix of a graph (with respect to all possible paths between two prescribed vertices), which is very useful in the study of switching networks; the transition matrix, which has applications in connection with sequential machines; and the relation matrix, which is associated with communication networks. Some of these matrices will be defined and studied in some detail in later chapters.

2-7 THE ORIENTED GRAPH

In the preceding sections we discussed the properties of the incidence matrix, the circuit matrix, and the cutset matrix of a nonoriented graph. We also derived the interrelationships between these matrices. In this section we shall consider the oriented graph and make the natural extension of the results that were obtained for the nonoriented graph. Since orientations will be assigned to all of the edges, circuits, and cutsets in an oriented graph, it becomes necessary for us to redefine some of the terms that have been introduced for the nonoriented graph.

DEFINITION 2-12. Incidence Matrix of a Connected, Oriented Graph. The *incidence matrix of a connected, oriented graph* of v vertices and e edges is the matrix A_a of order $v \times e$ with each row identified by a vertex and each column by an edge, such that

$$A_a = [a_{ij}]_{v\times e}$$

where

$a_{ij} = 1$ if edge e_j is incident with vertex v_i and is oriented *away from* v_i

$a_{ij} = -1$ if e_j is incident with v_i and is oriented *toward* v_i

$a_{ij} = 0$ if e_j is not incident with v_i

Note that in Definition 2-12 we used the same symbol A_a for the incidence matrix of an oriented graph, but now we have three values (instead of two in the case of a nonoriented graph) for the elements of A_a — namely, 1, -1, and 0 — because of the edge orientation. In a similar manner, we shall define the circuit matrix as follows:

DEFINITION 2-13. Circuit Matrix of a Connected, Oriented Graph. The *circuit matrix of a connected, oriented graph* G of v vertices and e edges is the matrix B_a of order $c \times e$ with each row identified by an oriented circuit and each column by an edge, such that to each circuit in G is assigned an orientation that is considered positive in the clockwise (or counterclockwise) direction, and that

$$B_a = [b_{ij}]_{c \times e}$$

where

$b_{ij} = 1$ if edge e_j is in circuit c_i and the orientation of e_j coincides with the orientation of c_i

$b_{ij} = -1$ if e_j is in c_i but the orientation of e_j is opposite to the orientation of c_i

$b_{ij} = 0$ if e_j is not in c_i

c is the total number of circuits in G

Likewise, we shall define the cutset matrix.

DEFINITION 2-14. Cutset Matrix of a Connected, Oriented Graph. The *cutset matrix of a connected oriented graph* G of v vertices and e edges is the matrix Q_a of order $k \times e$ with each row identified by an oriented cutset and each column by an edge such that to each cutset in G is assigned an orientation that is positive if the orientation is directed from left to right (or from right to left, if one so desires) across the cut-line, and that

$$Q_a = [q_{ij}]_{k \times e}$$

where

$q_{ij} = 1$ if edge e_j is in cutset k_i and the orientation of e_j coincides with the orientation of k_i

$q_{ij} = -1$ if e_j is in k_i but the orientation of e_j is opposite to the orientation of k_i

$q_{ij} = 0$ if e_j is not in k_i

k is the total number of cutsets in G

With A_a, B_a, and Q_a defined for a connected, oriented graph G, the other graph submatrices A, B, Q, B_f, and Q_f (with respect to some tree T) may be similarly defined. We shall leave this task to the reader as a problem.

It should be emphasized that the assignment of an orientation to a circuit is completely arbitrary; that is, a circuit may have a positive orientation in the clockwise direction, whereas another circuit of the same graph may have a positive orientation in the counterclockwise direction. A similar remark can be made about the orientation of a cutset; that is, a cutset may have a positive orientation from one side to the other across the cut-line.

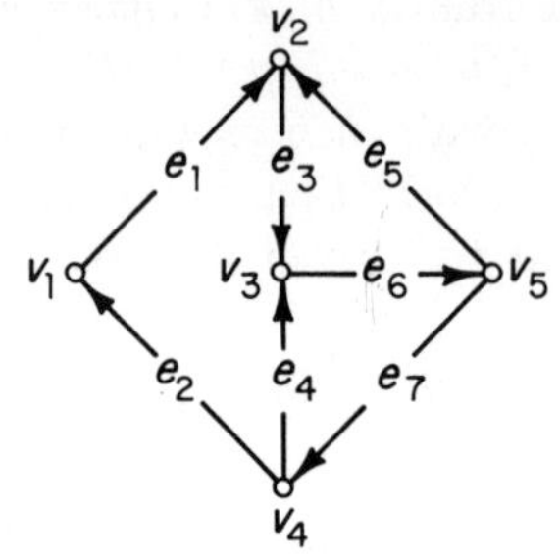

Figure 2-13 An oriented graph G.

EXAMPLE 2-16. Consider the oriented graph G of Figure 2-13. The incidence matrix A_a of G is

$$A_a = \begin{array}{c} \\ v_1 \\ v_2 \\ v_3 \\ v_4 \\ v_5 \end{array} \begin{array}{c} \begin{array}{ccccccc} e_1 & e_2 & e_3 & e_4 & e_5 & e_6 & e_7 \end{array} \\ \begin{bmatrix} 1 & -1 & 0 & 0 & 0 & 0 & 0 \\ -1 & 0 & 1 & 0 & -1 & 0 & 0 \\ 0 & 0 & -1 & -1 & 0 & 1 & 0 \\ 0 & 1 & 0 & 1 & 0 & 0 & -1 \\ 0 & 0 & 0 & 0 & 1 & -1 & 1 \end{bmatrix} \end{array} \qquad (2\text{-}64)$$

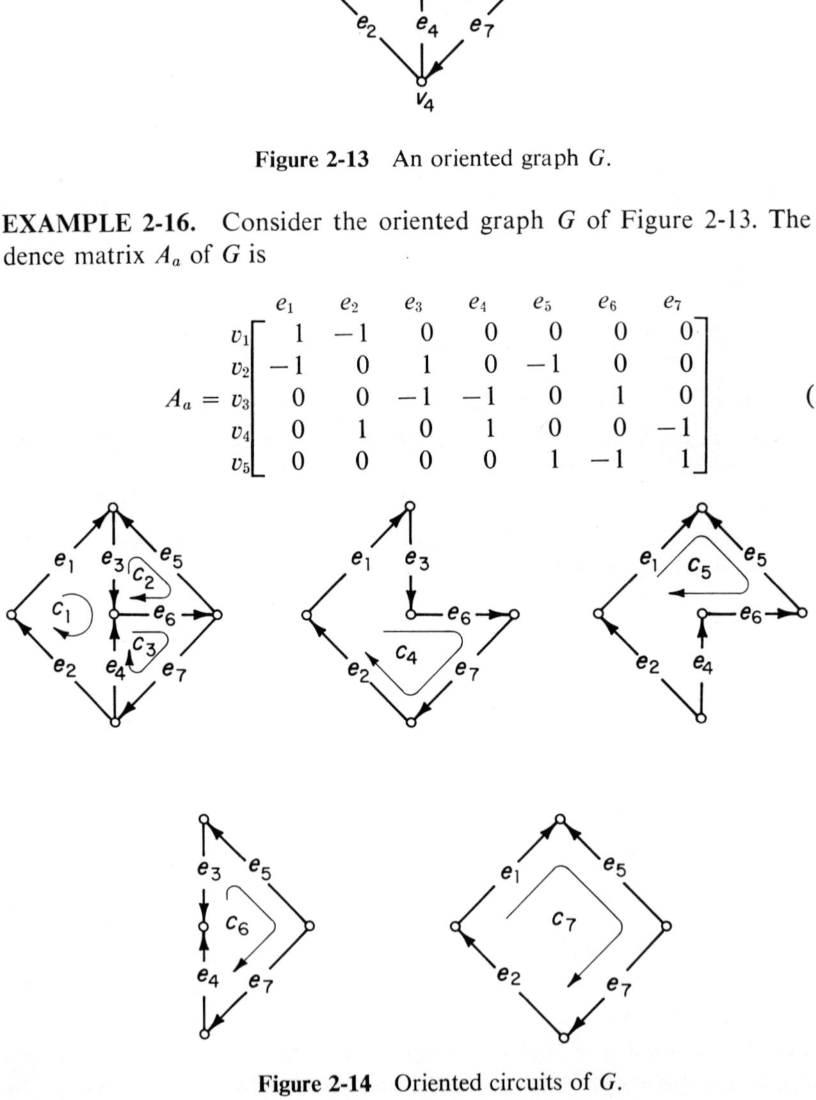

Figure 2-14 Oriented circuits of G.

Before the circuit matrix B_a can be formed, it is necessary to assign an orientation to each circuit of G, as indicated in Figure 2-14.

The circuit matrix B_a is given by

$$B_a = \begin{array}{c} \\ c_1 \\ c_2 \\ c_3 \\ c_4 \\ c_5 \\ c_6 \\ c_7 \end{array} \begin{array}{c} \begin{array}{ccccccc} e_1 & e_2 & e_3 & e_4 & e_5 & e_6 & e_7 \end{array} \\ \begin{bmatrix} 1 & 1 & 1 & -1 & 0 & 0 & 0 \\ 0 & 0 & -1 & 0 & -1 & -1 & 0 \\ 0 & 0 & 0 & 1 & 0 & 1 & 1 \\ 1 & 1 & 1 & 0 & 0 & 1 & 1 \\ 1 & 1 & 0 & -1 & -1 & -1 & 0 \\ 0 & 0 & -1 & 1 & -1 & 0 & 1 \\ 1 & 1 & 0 & 0 & -1 & 0 & 1 \end{bmatrix} \end{array} \tag{2-65}$$

Next, we assign the cutset orientations, as shown in Figure 2-15. The cutset matrix Q_a now is

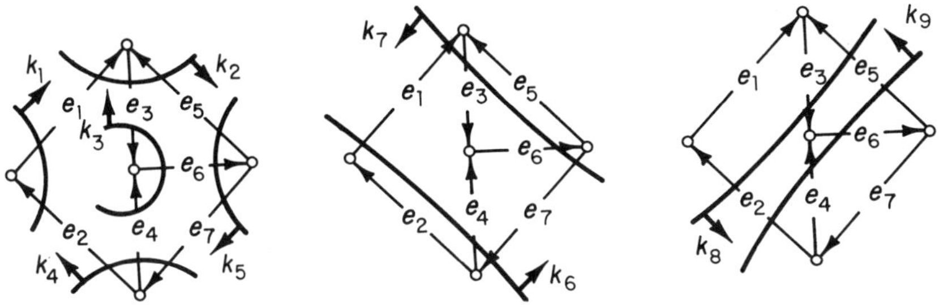

Figure 2-15 Oriented cutsets of G.

$$Q_a = \begin{array}{c} \\ k_1 \\ k_2 \\ k_3 \\ k_4 \\ k_5 \\ k_6 \\ k_7 \\ k_8 \\ k_9 \end{array} \begin{array}{c} \begin{array}{ccccccc} e_1 & e_2 & e_3 & e_4 & e_5 & e_6 & e_7 \end{array} \\ \begin{bmatrix} 1 & -1 & 0 & 0 & 0 & 0 & 0 \\ -1 & 0 & 1 & 0 & -1 & 0 & 0 \\ 0 & 0 & -1 & -1 & 0 & 1 & 0 \\ 0 & 1 & 0 & 1 & 0 & 0 & -1 \\ 0 & 0 & 0 & 0 & 1 & -1 & 1 \\ 1 & 0 & 0 & 1 & 0 & 0 & -1 \\ -1 & 0 & 1 & 0 & 0 & -1 & 1 \\ 0 & -1 & 1 & 0 & -1 & 0 & 0 \\ 0 & 1 & 0 & 1 & 1 & -1 & 0 \end{bmatrix} \end{array} \tag{2-66}$$

From Equation (2-64) we see that there are exactly two non-zero elements in each column of A_a, a "+1" and a "−1," so that the sum of all rows of A_a is still a row of zeros, keeping in mind that $1 + (-1) = 0$. Thus, treating the entries of A_a as real integers and following the same methods of proof as in the corresponding theorems (Lemmas 2-1 and 2-2 and Theorem 2-2) for the nonoriented graph, it is easy to show that

$$R(A_a) \leqq v - 1 \tag{2-67}$$

$$R(A_a) \geqq v - 1 \tag{2-68}$$

and, finally, that

$$R(A_a) = v - 1$$

Stating this last equation formally, we have the following theorem:

THEOREM 2-16. The rank of the incidence matrix A_a of a connected, oriented graph G of v vertices and e edges is given by

$$R(A_a) = v - 1 \tag{2-69}$$

Since the proof of Theorem 2-16 is almost identical to that of Theorem 2-2 with the proofs of Lemmas 2-1 and 2-2 as stepping stones, we shall not repeat it here.

In the remainder of this section, we shall state the following theorems without proof, and ask the reader to furnish each statement with a formal proof since all of these proofs are similar to their counterparts for the non-oriented graph.

THEOREM 2-17. For an unconnected, oriented graph G with v vertices and k maximally connected subgraphs, the rank of the incidence matrix A_a is given by (see Theorem 2-3)

$$R(A_a) = v - k$$

THEOREM 2-18. Let A_a and B_a be the incidence matrix and the circuit matrix, respectively, of a connected, oriented graph G and let the columns of the two matrices be arranged in the same edge order. Then (see Theorem 2-5)

$$A_a \cdot B_a^T = 0 \tag{2-70}$$

or

$$B_a \cdot A_a^T = 0 \tag{2-71}$$

We might remark here that in order to prove Equation (2-70) or (2-71), for example, we need only to observe that for any vertex v_i and any circuit c_j in G there are two cases to consider:

(a) Vertex v_i is in circuit c_j.

(b) Vertex v_i is not in circuit c_j.

In the second case, the product of the ith row of A_a and the jth column of the *transpose* of B_a is obviously zero, since v_i and c_j have no common edges and the product of the ith row of A_a and the jth column of B_a^T must take the form

$$0 \cdot 0 + 0 \cdot 0 + \cdots + 0 \cdot 0 = 0$$

However, in the first case, where v_i is one of the vertices in c_j, there are exactly two edges in c_j that are incident with v_i. Thus, if we consider all of the possible combinations ($4 \times 2 = 8$) among the edge orientations of the two common edges (four possible combinations) and the circuit orientations (two possible combinations), we shall see in Figure 2-16 that each of such combinations

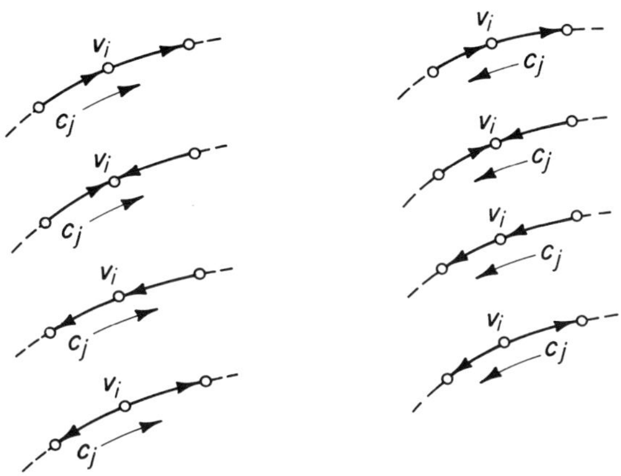

Figure 2-16 The eight combinations of the four edge orientations and the two circuit orientations in case (b).

will lead to a product of the ith row of A_a and the jth column of B_a^T of the following form:

$$\underbrace{0\cdot 0 + \cdots + (1)(-1) + 0\cdot 0 + \cdots + (1)(1) + 0\cdot 0 + \cdots + 0\cdot 0}_{e \text{ terms}} = 0 \qquad (2\text{-}72)$$

so that always one of the e terms in Equation (2-72) is $(1)(-1) = -1$, another term is $(1)(1) = 1$, and all the rest of the remaining $e - 2$ terms are zero.

EXAMPLE 2-17. Consider graph G of Figure 2-13. The three matrices A_a, B_a and Q_a are given in Equations (2-64), (2-65), and (2-66), respectively. The products $A_aB_a^T$ and $Q_aB_a^T$ are given by

$$A_aB_a^T = \begin{array}{c} \\ v_1 \\ v_2 \\ v_3 \\ v_4 \\ v_5 \end{array}\begin{array}{c} \begin{array}{ccccccc} e_1 & e_2 & e_3 & e_4 & e_5 & e_6 & e_7 \end{array} \\ \begin{bmatrix} 1 & -1 & 0 & 0 & 0 & 0 & 0 \\ -1 & 0 & 1 & 0 & -1 & 0 & 0 \\ 0 & 0 & -1 & -1 & 0 & 1 & 0 \\ 0 & 1 & 0 & 1 & 0 & 0 & -1 \\ 0 & 0 & 0 & 0 & 1 & -1 & 1 \end{bmatrix} \end{array}$$

$$\times\;\begin{array}{c} \\ e_1 \\ e_2 \\ e_3 \\ e_4 \\ e_5 \\ e_6 \\ e_7 \end{array}\!\!\begin{array}{c} \begin{array}{ccccccc} c_1 & c_2 & c_3 & c_4 & c_5 & c_6 & c_7 \end{array} \\ \begin{bmatrix} 1 & 0 & 0 & 1 & 1 & 0 & 1 \\ 1 & 0 & 0 & 1 & 1 & 0 & 1 \\ 1 & -1 & 0 & 1 & 0 & -1 & 0 \\ -1 & 0 & 1 & 0 & -1 & 1 & 0 \\ 0 & -1 & 0 & 0 & -1 & -1 & -1 \\ 0 & -1 & 1 & 1 & -1 & 0 & 0 \\ 0 & 0 & 1 & 1 & 0 & 1 & 1 \end{bmatrix} \end{array} = [0]_{5\times 7}$$

and

$$Q_aB_a^T = \begin{array}{c} \\ k_1 \\ k_2 \\ k_3 \\ k_4 \\ k_5 \\ k_6 \\ k_7 \\ k_8 \\ k_9 \end{array}\!\!\begin{array}{c} \begin{array}{ccccccc} e_1 & e_2 & e_3 & e_4 & e_5 & e_6 & e_7 \end{array} \\ \begin{bmatrix} 1 & -1 & 0 & 0 & 0 & 0 & 0 \\ -1 & 0 & 1 & 0 & -1 & 0 & 0 \\ 0 & 0 & -1 & -1 & 0 & 1 & 0 \\ 0 & 1 & 0 & 1 & 0 & 0 & -1 \\ 0 & 0 & 0 & 0 & 1 & -1 & 1 \\ 1 & 0 & 0 & 1 & 0 & 0 & -1 \\ -1 & 0 & 1 & 0 & 0 & -1 & 1 \\ 0 & -1 & 1 & 0 & -1 & 0 & 0 \\ 0 & 1 & 0 & 1 & 1 & -1 & 0 \end{bmatrix} \end{array}$$

$$\times\;\begin{array}{c} \\ e_1 \\ e_2 \\ e_3 \\ e_4 \\ e_5 \\ e_6 \\ e_7 \end{array}\!\!\begin{array}{c} \begin{array}{ccccccc} c_1 & c_2 & c_3 & c_4 & c_5 & c_6 & c_7 \end{array} \\ \begin{bmatrix} 1 & 0 & 0 & 1 & 1 & 0 & 1 \\ 1 & 0 & 0 & 1 & 1 & 0 & 1 \\ 1 & -1 & 0 & 1 & 0 & -1 & 0 \\ -1 & 0 & 1 & 0 & -1 & 1 & 0 \\ 0 & -1 & 0 & 0 & -1 & -1 & -1 \\ 0 & -1 & 1 & 1 & -1 & 0 & 0 \\ 0 & 0 & 1 & 1 & 0 & 1 & 1 \end{bmatrix} \end{array} = [0]_{9\times 7}$$

The two equations, $B_aA_a^T = 0$ and $B_aQ_a^T = 0$, are also satisfied by the three matrices since they are simply the transposition of $A_aB_a^T$ and $Q_aB_a^T$, respectively.

THEOREM 2-19. For a connected, oriented graph G of v vertices and e edges, the rank of its circuit matrix B_a is given by (see Theorem 2-6)

$$R(B_a) = e - v + 1 \tag{2-73}$$

THEOREM 2-20. The rank of a fundamental circuit matrix B_f with respect to a tree T of a connected, oriented graph G of v vertices and e edges is given by (see Theorem 2-7)

$$R(B_f) = e - v + 1 \tag{2-74}$$

A theorem similar to Theorem 2-8 may be stated for the connected, oriented graph with minor changes in some of the steps in the procedure for obvious reasons. The statement and its proof will be left as a problem (see Problem 2-26).

THEOREM 2-21. Let A be an incidence submatrix of a connected, oriented graph G of v vertices. A submatrix A_1 of order $v - 1$ of A is nonsingular if and only if the $v - 1$ columns of A_1 correspond to the $v - 1$ branches of some tree of G (see Theorem 2-12).

THEOREM 2-22. Let Q be a cutset submatrix of a connected, oriented graph G of v vertices. The columns of a nonsingular submatrix Q_1 of order $v - 1$ of Q corresponds to the $v - 1$ branches of some tree of G (see Theorem 2-13).

THEOREM 2-23. Let B be a circuit submatrix of a connected, oriented graph G of v vertices and e edges. A submatrix B_1 of order $e - v + 1$ of B is nonsingular if and only if the $e - v + 1$ columns of B_1 correspond to the $e - v + 1$ chords with respect to a tree T of G (see Theorem 2-14).

THEOREM 2-24. Let A be an incidence submatrix, let B_f and Q_f be a fundamental circuit matrix and a fundamental cutset matrix, respectively, with respect to some tree T of a connected, oriented graph of v vertices and e edges. Furthermore, let the columns of the three matrices be arranged in the same edge order, with the first $e - v + 1$ edges corresponding to chords and the rest of the edges corresponding to branches of the same tree T, so that the matrices are partitioned as

$$A = [A_{11} \quad A_{12}] \tag{2-75}$$

$$B_f = [U_N \quad B_{f_{12}}] \tag{2-76}$$

and

$$Q_f = [Q_{f_{11}} \quad U_R] \tag{2-77}$$

where $N = e - v + 1$ and $R = v - 1$. Then,

$$Q_{f_{11}} = -B^T_{f_{12}} = A^{-1}_{12} \cdot A_{11} \tag{2-78}$$

and

$$Q_f = [-B^T_{f_{12}} \quad U_R] = A^{-1}_{12} \cdot A \tag{2-79}$$

(See Theorem 2-15.)

In order to express B_f in the form of Equation (2-76) and Q_f in the form of (2-77), it is necessary that the circuit orientations coincide with the orientations of the corresponding chords and the cutset orientations coincide with the orientations of the corresponding branches so that U_N and U_R will result as indicated in Equations (2-76) and (2-77), respectively.

We should point out also that in the case of the oriented graph, the equation $Q_{f_{11}} + B^T_{f_{12}} = 0$ implies that $Q_{f_{11}} = -B^T_{f_{12}}$ since the negative of a matrix of the three-element kind (namely, 0, 1, and -1) is well defined.

EXAMPLE 2-18. Let us consider again graph G of Figure 2-13. We shall choose the tree T: $\{e_1, e_2, e_6, e_7\}$. The fundamental circuit matrix B_f with respect to T is

$$B_f = \begin{matrix} \\ c_{f_1} \\ c_{f_2} \\ c_{f_3} \end{matrix}\!\!\begin{matrix} \begin{matrix} e_3 & e_4 & e_5 & e_1 & e_2 & e_6 & e_7 \end{matrix} \\ \left[\begin{array}{ccc|cccc} 1 & 0 & 0 & 1 & 1 & 1 & 1 \\ 0 & 1 & 0 & 0 & 0 & 1 & 1 \\ 0 & 0 & 1 & -1 & -1 & 0 & -1 \end{array}\right] \end{matrix} \begin{matrix} \\ c_{f_1} = c_{e_3} \\ c_{f_2} = c_{e_4} \\ c_{f_3} = c_{e_5} \end{matrix} = [U_3 \quad B_{f_{12}}] \qquad (2\text{-}80)$$

and the fundamental cutset matrix Q_f with respect to the same tree T is

$$Q_f = \begin{matrix} \\ k_{f_1} \\ k_{f_2} \\ k_{f_3} \\ k_{f_4} \end{matrix}\!\!\begin{matrix} \begin{matrix} e_3 & e_4 & e_5 & e_1 & e_2 & e_6 & e_7 \end{matrix} \\ \left[\begin{array}{ccc|cccc} -1 & 0 & 1 & 1 & 0 & 0 & 0 \\ -1 & 0 & 1 & 0 & 1 & 0 & 0 \\ -1 & -1 & 0 & 0 & 0 & 1 & 0 \\ -1 & -1 & 1 & 0 & 0 & 0 & 1 \end{array}\right] \end{matrix} \begin{matrix} \\ k_{f_1} = k_{e_1} \\ k_{f_2} = k_{e_2} \\ k_{f_3} = k_{e_6} \\ k_{f_4} = k_{e_7} \end{matrix} = [Q_{f_{11}} \quad U_4] \qquad (2\text{-}81)$$

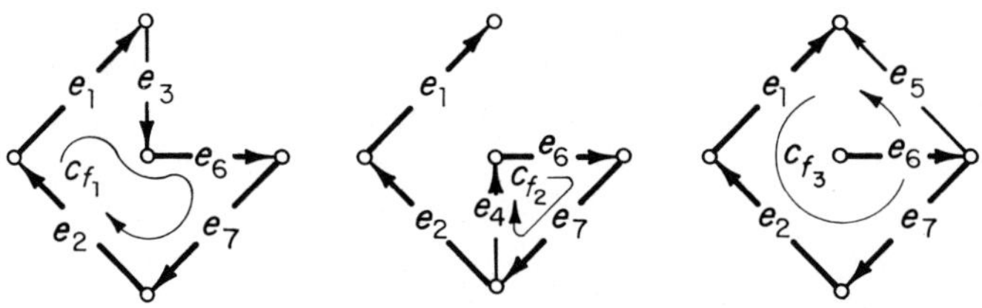

Figure 2-17 Oriented fundamental circuits with respect to T: $\{e_1, e_2, e_6, e_7\}$.

The oriented fundamental circuits are given in Figure 2-17 and the oriented fundamental cutsets in Figure 2-18.

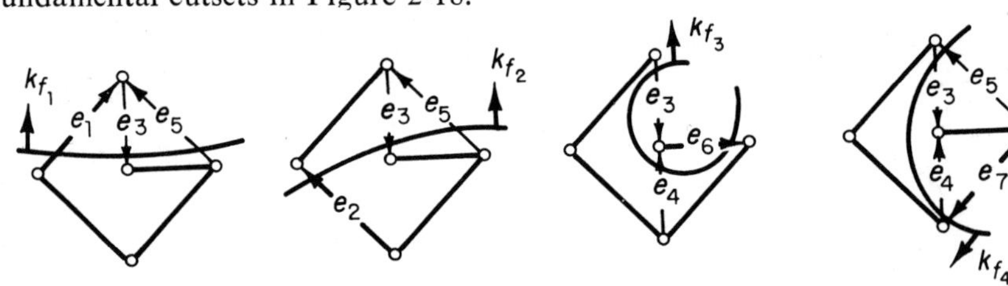

Figure 2-18 Oriented fundamental cutsets with respect to T: $\{e_1, e_2, e_6, e_7\}$.

From Equations (2-80) and (2-81) we see that

$$B_{f_{12}} = \begin{bmatrix} 1 & 1 & 1 & 1 \\ 0 & 0 & 1 & 1 \\ -1 & -1 & 0 & -1 \end{bmatrix} \quad \text{and} \quad Q_{f_{11}} = \begin{bmatrix} -1 & 0 & 1 \\ -1 & 0 & 1 \\ -1 & -1 & 0 \\ -1 & -1 & 1 \end{bmatrix}$$

so that clearly

$$Q_{f_{11}} = -B_{f_{12}}^T$$

Also, the incidence submatrix A with v_5 as the reference vertex and with the columns arranged in the same edge order as B_f and Q_f is

$$A = \begin{array}{c} \\ v_1 \\ v_2 \\ v_3 \\ v_4 \end{array} \begin{array}{c} \begin{array}{ccc:cccc} e_3 & e_4 & e_5 & e_1 & e_2 & e_6 & e_7 \end{array} \\ \left[\begin{array}{ccc:cccc} 0 & 0 & 0 & 1 & -1 & 0 & 0 \\ 1 & 0 & -1 & -1 & 0 & 0 & 0 \\ -1 & -1 & 0 & 0 & 0 & 1 & 0 \\ 0 & 1 & 0 & 0 & 1 & 0 & -1 \end{array}\right] \end{array} = [A_{11} \quad A_{12}] \tag{2-82}$$

where

$$A_{11} = \begin{bmatrix} 0 & 0 & 0 \\ 1 & 0 & -1 \\ -1 & -1 & 0 \\ 0 & 1 & 0 \end{bmatrix} \tag{2-83}$$

and

$$A_{12} = \begin{bmatrix} 1 & -1 & 0 & 0 \\ -1 & 0 & 0 & 0 \\ 0 & 0 & 1 & 0 \\ 0 & 1 & 0 & -1 \end{bmatrix} \tag{2-84}$$

so that

$$A_{12}^{-1} = \begin{bmatrix} 0 & -1 & 0 & 0 \\ -1 & -1 & 0 & 0 \\ 0 & 0 & 1 & 0 \\ -1 & -1 & 0 & -1 \end{bmatrix} \tag{2-85}$$

Hence, by Equations (2-82) and (2-85),

$$A_{12}^{-1} \cdot A = \begin{bmatrix} 0 & -1 & 0 & 0 \\ -1 & -1 & 0 & 0 \\ 0 & 0 & 1 & 0 \\ -1 & -1 & 0 & -1 \end{bmatrix} \cdot \begin{bmatrix} 0 & 0 & 0 & 1 & -1 & 0 & 0 \\ 1 & 0 & -1 & -1 & 0 & 0 & 0 \\ -1 & -1 & 0 & 0 & 0 & 1 & 0 \\ 0 & 1 & 0 & 0 & 1 & 0 & -1 \end{bmatrix}$$

$$= \begin{array}{c} \begin{array}{ccc:cccc} e_3 & e_4 & e_5 & e_1 & e_2 & e_6 & e_7 \end{array} \\ \left[\begin{array}{ccc:cccc} -1 & 0 & 1 & 1 & 0 & 0 & 0 \\ -1 & 0 & 1 & 0 & 1 & 0 & 0 \\ -1 & -1 & 0 & 0 & 0 & 1 & 0 \\ -1 & -1 & 1 & 0 & 0 & 0 & 1 \end{array}\right] \end{array}$$

which, when compared to Equation (2-81), is seen to be equal to Q_f as predicted by Equation (2-79) in Theorem 2-24.

The reader should also verify that by using submatrix A_{12}^{-1} of Equation (2-85) and submatrix A_{11} of (2-83), the product $A_{12}^{-1} \cdot A_{11}$ is indeed equal to $Q_{f_{11}}$, as predicted by (2-78).

2.8 SUMMARY

In this section, we shall conclude our discussion on the theory of graphs by listing the results of Chapters 1 and 2 in their logical order with all definitions stated first, and lemmas, theorems, and corollaries given subsequently. For the sake of compactness, some of the theorems are rewritten in abbreviated form, omitting the obvious assumptions.

(A) *DEFINITIONS*

- D 1-1 Linear graph (G) and its edges (e_j) and vertices (v_j)
- D 1-2 Degree of a vertex
- D 1-3 Subgraph, proper subgraph (both G_j), complement of a subgraph (G_j'), null graph ($\varnothing$)
- D 1-4 Path and path set (both p_j), terminals and interior vertices of a path.
- D 1-5 Circuit (loop) and circuit set (both c_j)
- D 1-6 Connected and unconnected graphs (both G)
- D 1-7 Series edges
- D 1-8 Parallel edges
- D 1-9 Separable and nonseparable graphs (both G), cut-vertex
- D 1-10 Disjoint graphs
- D 1-11 Oriented and nonoriented graphs (both G), edge-disjoint subgraphs (G_j)
- D 1-12 Tree (T_j) and branches
- D 1-13 Chord set (cotree) (T_j') and chord (link)
- D 1-14 Cutset and vertex cutset (both k_j), isolated vertex
- D 1-15 Incidence set
- D 1-16 Union ($\cup$)
- D 1-17 Intersection ($\cap$)
- D 1-18 Ring sum ($\oplus$)
- D 1-19 Difference ($-$)
- D 1-20 Fundamental circuit (f-circuit, c_{f_j})
- D 1-21 Fundamental cutset (f-cutset, k_{f_j})
- D 1-22 Congruence (1-isomorphism, or, simply, isomorphism) ($\equiv$)
- D 1-23 Maximally connected subgraph (G_m)
- D 1-24 Equivalence (2-isomorphism) ($\sim$)

(a) *For a connected, nonoriented graph of v vertices and e edges:*

- D 2-1 Incidence matrix (A_a)
- D 2-2 Incidence submatrix (A)
- D 2-3 Forest
- D 2-4 Circuit matrix (B_a)
- D 2-5 Fundamental circuit matrix (B_f)
- D 2-6 Circuit submatrix (B)
- D 2-7 Cutset matrix (Q_a)
- D 2-8 Fundamental cutset matrix (Q_f)

D 2-9 Cutset submatrix (Q)
D 2-10 Rank of a graph [$R(G)$]
D 2-11 Nullity of a graph [$N(G)$]

(b) *For a connected, oriented graph of v vertices and e edges.*

D 2-12 Incidence matrix (A_a)
D 2-13 Circuit matrix (B_a)
D 2-14 Cutset matrix (Q_a)
Other submatrices are similarly defined.

(B) *LEMMAS, THEOREMS, AND COROLLARIES*

(a) *For a connected, nonoriented graph G of v vertices and e edges:*

T 1-1 There exists at least one tree in G.

T 1-2 A tree of G contains $v - 1$ edges.

T 1-3 A graph G_t is a tree if and only if there exists exactly one path between any two vertices of G_t.
Four of the properties of a tree are:
(a) It is connected.
(b) It contains all of the v vertices.
(c) It contains exactly $v - 1$ edges.
(d) It contains no circuits.

T 1-4 If G_t is a subgraph containing $v - 1$ edges and no circuits, it is a tree.

T 1-5 For a given tree T of G there are exactly $v - 1$ branches and $e - v + 1$ chords with respect to T.

T 1-6 The number of common edges between a cutset and a circuit in G is even.

T 1-7 The ring sum of two circuits in G is either a circuit or a edge-disjoint union of circuits.

T 2-1 A matrix is the incidence matrix if and only if all of its elements are 1 or 0 and there are exactly two 1's in each column.

L 2-1 There are, at most, $v - 1$ independent rows in A_a.

L 2-2 There are at least $v - 1$ independent rows in A_a.

T 2-2 $R(A_a) = v - 1$.

T 2-4 Algorithm for obtaining B_a of G from a set of $R(B_a)$ independent circuits.

T 2-5 $A_a B_a^T = 0$ or $B_a A_a^T = 0$.

L 2-3 $R(B_a) \geqq e - v + 1$.

L 2-4 $R(B_a) \leqq e - v + 1$.

T 2-6 $R(B_a) = e - v + 1$.

T 2-7 $R(B_f) = e - v + 1$.

T 2-8 Algorithm for obtaining B_a of G from a B_f.

T 2-9 $R(Q_f) = v - 1$.

C 2-1 $R(Q_a) \geqq v - 1$.

T 2-10 $R(Q_a) = v - 1$.

T 2-11 Algorithm for obtaining Q_a from a Q_f.

T 2-12 A submatrix A_1 of order $v-1$ of A is nonsingular if the $v-1$ columns of A_1 correspond to the $v-1$ branches of a tree T of G.

T 2-13 A submatrix Q_1 of order $v-1$ of Q is non singular if the $v-1$ columns of Q_1 correspond to the $v-1$ branches of some tree T of G.

T 2-14 A submatrix B_1 of order $e-v+1$ of B is nonsingular if the $e-v+1$ columns of B_1 correspond to the $e-v+1$ chords with respect to some tree T of G.

T 2-15 Let $A = [A_{11} \quad A_{12}]$, $B_f = [U_N \quad B_{f12}]$, $Q_f = [Q_{f11} \quad U_R]$ with columns arranged in the same edge order. Then $Q_{f11} = B^T_{f12} = A^{-1}_{12} \cdot A_{11}$ and $Q_f = [B^T_{f12} \quad U_R] = A^{-1}_{12} \cdot A$.

(b) *For a connected, oriented graph G of v vertices and e edges:*

T 2-16 $R(A_a) = v - 1$.

T 2-18 $A_a \cdot B^T_a = 0$ or $B_a \cdot A^T_a = 0$.

T 2-19 $R(B_a) = e - v + 1$.

T 2-20 $R(B_f) = e - v + 1$.

T 2-21 Same statement as T 2-12.

T 2-22 Same statement as T 2-13.

T 2-23 Same statement as T 2-14.

T 2-24 $Q_{f11} = -B^T_{f12} = A^{-1}_{12} \cdot A_{11}$ and $Q_f = [B^T_{f12} \quad U_R] = A^{-1}_{12} \cdot A$. (See Theorem 2-15 for the assumptions made about A, B_f, Q_f and their related submatrices.

(c) *For a nonconnected graph G with v vertices and p maximally connected subgraphs* (*G* may be either oriented or nonoriented): T 2-3 and T 2-17 $R(A_a) = v - p$. Similarly, $R(B_a) = e - v + p$.

PROBLEMS

2-1 Write the incidence matrix A_a of each of the graphs shown in Figure P2-1.

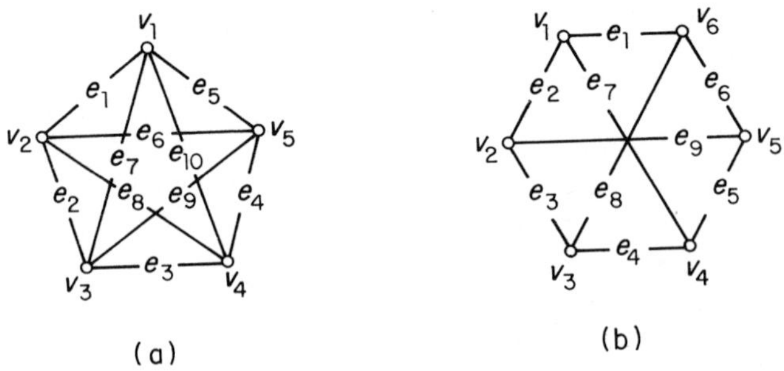

Figure P2-1

2-2 Find the incidence matrix A_a and, hence, the graph from each of the following incidence submatrices:

(a)

$$A_1 = \begin{bmatrix} 1 & 0 & 0 & 1 & 0 & 0 & 1 \\ 0 & 1 & 0 & 0 & 1 & 0 & 0 \\ 1 & 0 & 0 & 1 & 1 & 1 & 0 \\ 0 & 0 & 1 & 0 & 0 & 1 & 0 \end{bmatrix}$$

(b)

$$A_2 = \begin{bmatrix} 0 & 1 & 0 & 1 & 0 & 1 & 0 & 0 \\ 1 & 0 & 1 & 0 & 0 & 0 & 1 & 0 \\ 1 & 0 & 0 & 0 & 0 & 0 & 0 & 1 \\ 0 & 0 & 0 & 1 & 0 & 0 & 1 & 0 \\ 0 & 1 & 0 & 0 & 1 & 0 & 0 & 0 \end{bmatrix}$$

2-3 Prove the validity of each of the six properties, (a) through (f), of the incidence matrix A_a of a graph G as stated in Section 2-2.

2-4 Prove that there are at most $v - 1$ independent rows in the incidence matrix A_a of a connected graph G (Lemma 2-1).

2-5 Determine the rank of each of the following incidence submatrices by reducing it to normal form; then obtain the graph from the given incidence submatrix; and, finally, verify that $R(A) = v - 1$.

(a)

$$A_1 = \begin{bmatrix} 1 & 0 & 0 & 1 & 1 & 0 \\ 0 & 1 & 1 & 0 & 1 & 0 \\ 0 & 1 & 0 & 1 & 0 & 1 \\ 1 & 0 & 1 & 0 & 0 & 1 \end{bmatrix}$$

(b)

$$A_2 = \begin{bmatrix} 0 & 1 & 0 & 0 & 1 & 0 & 1 & 0 \\ 1 & 0 & 0 & 1 & 1 & 0 & 0 & 1 \\ 0 & 1 & 1 & 0 & 0 & 0 & 0 & 1 \\ 1 & 0 & 1 & 0 & 0 & 1 & 0 & 0 \\ 0 & 0 & 0 & 1 & 0 & 1 & 1 & 0 \end{bmatrix}$$

2-6 Find the incidence matrix A_a of the unconnected graph shown in Figure P2-6 and determine $R(A_a)$, the rank of A_a. Does your result verify Equation (2-8) in Theorem 2-3?

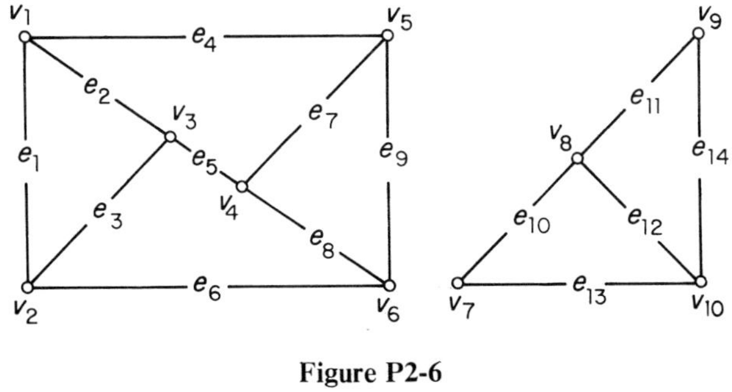

Figure P2-6

2-7 Find the circuit matrix B_a of each of the graphs shown in Figure P2-1 by identifying all possible circuits of the graph by inspection.

2-8 Determine the rank of each of the circuit matrices B_a obtained in Problem 2-7. Does your result verify Equation (2-25) in Theorem 2-6 for each B_a?

2-9 For the graph shown in Figure P2-1(a), find
(a) The fundamental circuit matrix B_f for the tree T: $\{e_6, e_7, e_8, e_9\}$.
(b) $R(B_f)$, the rank of B_f obtained in (a).

2-10 Repeat Problem 2-9 for the graph of Figure P2-1(b) with the tree T: $\{e_1, e_2, e_3, e_7, e_9\}$.

2-11 Prove Theorem 2-8.

2-12 Find the circuit matrix B_a of each of the graphs in Figure P2-1 by the application of Theorem 2-8 and compare each B_a with the corresponding B_a obtained in Problem 2-7.

2-13 Construct a graph G such that a circuit submatrix B of the circuit matrix B_a of G *cannot* be a fundamental circuit matrix B_f; that is, give an example of B which does not correspond to any tree of G. (Note that if B does correspond to some tree T of G, then $B = B_f$ with respect to that tree T.)

2-14 Find the cutset matrix Q_a of each of the graphs shown in Figure P2-1 by identifying all the cutsets of the graph by inspection.

2-15 For the graph shown in Figure P2-1(a), find
(a) The fundamental cutset matrix Q_f for the tree T: $\{e_6, e_7, e_8, e_9\}$.
(b) $R(Q_f)$, the rank of Q_f obtained in (a).

2-16 Repeat Problem 2-15 for the graph of Figure P2-1(b) with the tree T: $\{e_1, e_2, e_3, e_7, e_9\}$.

2-17 Prove Theorem 2-11.

2-18 Find the cutset matrix Q_a of each of the graphs of Figure P2-1 by the application of Theorem 2-11 and compare your results with those obtained in Problem 2-14.

2-19 Determine the rank of each of the cutset matrices Q_a obtained in Problem 2-18 and verify the validity of Theorem 2-10 in each case.

2-20 By applying the definition of linear dependence, show that
(a) The set of all rows of *any* fundamental circuit matrix B_f of a connected graph is an independent set.
(b) The set of all rows of *any* fundamental cutset matrix Q_f of a connected graph is an independent set.

2-21 For each of the graphs shown in Figure P2-1, verify the following statements: (a) $A_a B_a^T = 0$. (b) $Q_a B_a^T = 0$.

2-22 Prove Theorem 2-14.

2-23 Find the fundamental cutset matrix Q_f of a connected graph G with respect to a given tree T if the fundamental circuit matrix B_f with respect to T is

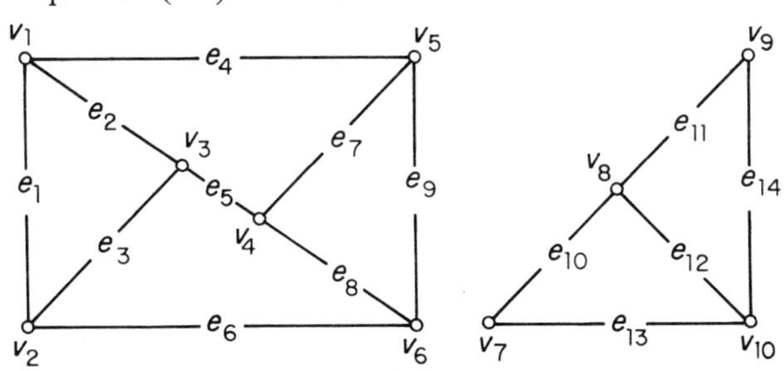

$$B_f = \begin{bmatrix} 1 & 1 & 0 & 0 & 0 & 0 & 0 & 0 \\ 1 & 0 & 0 & 1 & 0 & 1 & 0 & 1 \\ 0 & 0 & 0 & 0 & 1 & 1 & 0 & 1 \\ 1 & 0 & 1 & 0 & 0 & 1 & 1 & 1 \end{bmatrix}$$

2-24 Define (a) the incidence submatrix A, (b) the circuit submatrix B, (c) the cutset submatrix Q, (d) the fundamental circuit matrix B_f (with respect to some tree T), and (e) the fundamental cutset matrix Q_f (with respect to some tree T) of a connected, oriented graph G in the same way they are defined for a connected, nonoriented graph (Definitions 2-12, 2-13, and 2-14).

2-25 Find the incidence matrix A_a, the circuit matrix B_a, and the cutset matrix Q_a of the oriented graph depicted in Figure P2-25.

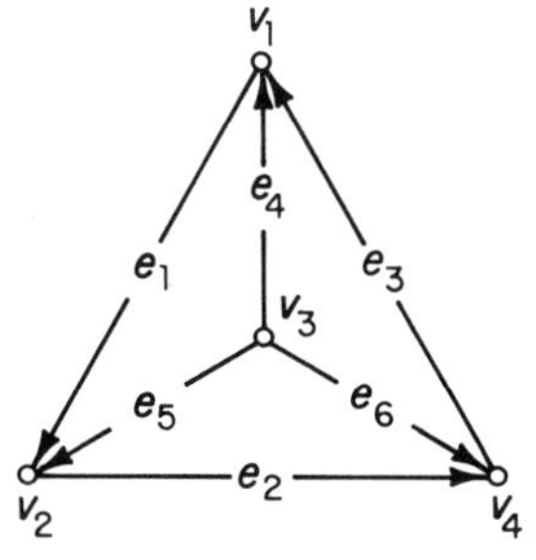

Figure P2-25

2-26 State and prove the theorem for an oriented graph G similar to Theorem 2-8.

2-27 Prove that the value of *any* nonzero major of an incidence submatrix A of an oriented graph G is either $+1$ or -1.

2-28 Given the circuit matrix B_a of some connected oriented graph G.

$$B_a = \begin{bmatrix} 1 & -1 & 0 & 0 & 1 & 0 \\ 0 & 1 & -1 & 0 & 0 & 0 \\ 0 & 0 & 1 & 1 & 0 & 1 \\ 1 & 0 & -1 & 0 & 1 & 0 \\ 0 & 1 & 0 & 1 & 0 & 1 \\ 1 & 0 & 0 & 1 & 1 & 1 \end{bmatrix}$$

Find the graph G. Use any *systematic* scheme.

2-29 A connected graph has a fundamental circuit matrix as follows:

$$B_f = \begin{bmatrix} a & b & c & d & e & f & g \\ 1 & 0 & 0 & 1 & 1 & 0 & 0 \\ 0 & 1 & 0 & 0 & 1 & 1 & 0 \\ 0 & 0 & 1 & 0 & 0 & 1 & 1 \end{bmatrix}$$

Without constructing the graph, determine whether each of the following statements is true:
(1) Edges *bcef* form a tree.
(2) Edges *acef* form a tree.
(3) Edges *abeg* form a cutset.
(4) Edges *bfc* form a cutset.
(5) Edges *defg* form a cutset.
(6) Edges *abcdg* form a circuit.
(7) Edges *acdegf* form a circuit.
Show your work.

2-30 For the oriented graph shown in Figure P2-30:

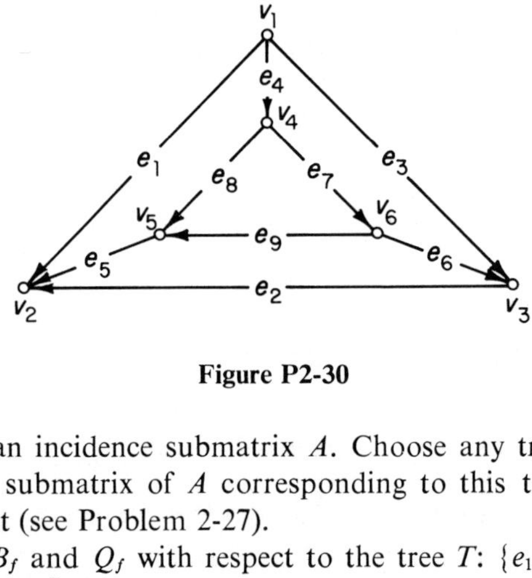

Figure P2-30

(a) Write an incidence submatrix A. Choose any tree and verify that the square submatrix of A corresponding to this tree has a nonzero determinant (see Problem 2-27).
(b) Write B_f and Q_f with respect to the tree T: $\{e_1, e_5, e_6, e_7, e_8\}$ and show that $B_f Q_f^T = 0$. What assumptions, if any, do you need to make for the above equation to be valid?
(c) Write a B matrix using the "meshes" (windowpanes) of the graph shown in Figure P2-30. Can this matrix be the fundamental circuit matrix with respect to some tree? Why? Does this matrix possess the property that each column has at most two nonzero elements, a $(+1)$ and a (-1)? Show that the rows of this matrix form an independent set.

2-31 Given a battery E, a lamp R, and eight switches labeled a, b, c, d, e, f, g, h as shown in Figure P2-31. Determine the connection of these components such that the following four conditions are met:

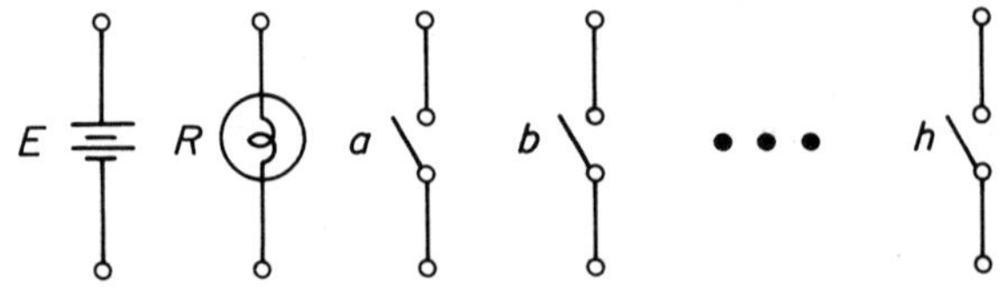

Figure P2-31

(1) With switches b, d, f open, the lamp will go on if and only if the switches a, c, e, g, and h are closed.
(2) With switches a, c, and e open, the lamp will go on if and only if switches b, d, f, g, and h are closed.
(3) With switches c, d, and g open, the lamp will go on if and only if switches a, b, e, f, and h are closed.
(4) With switches a, f, and h open, the lamp will go on if and only if switches b, c, d, e, and g are closed.
Show all your work in detail.

2-32 Prove that an incidence submatrix A and a cutset submatrix Q of a given nonoriented, connected graph G are related by an elementary matrix D such that $A = DQ$. This suggests that A can be obtained from Q by performing the ring-sum operations on the rows of Q.

2-33 Given that B_a is the circuit matrix of a nonoriented, connected graph G, as follows:

$$B_a = \begin{array}{c} \begin{matrix} a & b & c & d & e & f & g \end{matrix} \\ \begin{bmatrix} 1 & 0 & 1 & 0 & 1 & 0 & 0 \\ 0 & 1 & 0 & 1 & 1 & 0 & 0 \\ 0 & 0 & 1 & 1 & 0 & 1 & 0 \\ 0 & 0 & 0 & 0 & 0 & 1 & 1 \\ 1 & 1 & 0 & 0 & 0 & 0 & 1 \\ 0 & 0 & 1 & 1 & 0 & 0 & 1 \\ 1 & 1 & 0 & 0 & 0 & 1 & 0 \\ 0 & 1 & 1 & 0 & 1 & 1 & 0 \\ 1 & 0 & 0 & 1 & 1 & 0 & 1 \\ 0 & 1 & 1 & 0 & 1 & 0 & 1 \\ 1 & 0 & 0 & 1 & 1 & 1 & 0 \end{bmatrix} \end{array}$$

Obtain G by the following steps:
(a) Form a fundamental circuit matrix B_f: $B_f = [U_N \quad B_{f12}]$.
(b) Form the fundamental cutset matrix Q_f from B_f: $Q_f = [B_{f12}^T \quad U_R]$.
(c) Obtain the incidence matrix A (and, hence, A_a) from Q_f: $A = DQ_f$ (see Problem 2-32) where D is an elementary matrix.
(d) Draw the graph G identifying all the edges.

CHAPTER 3

DUALITY

3-1 INTRODUCTION

In an introductory network analysis course, one is often told that the number of independent mesh (loop) equations necessary to completely describe a given network is exactly equal to the number of regions or "windows" in the network. More precisely, the mesh equation obtained by summing up algebraically the voltages across the network elements around each "windowpane" represents an independent Kirchhoff voltage-law (KVL) equation; and, hence, for an m-windowpane network, we need exactly m independent mesh equations.

Also, one learns that for an n-node network, the number of independent node equations we need to write is $n - 1$. That is, after choosing one of the n nodes as the reference node, we can write a unique Kirchhoff current-law (KCL) equation for each of the $n - 1$ nodes by summing up algebraically the currents at that node.

Now, let us pose the following questions:

(1) Why does each window correspond to an independent KVL equation? If this question is answered satisfactorily, the next question probably is:

(2) Why does the total number of windows correspond exactly to the total number of independent KVL equations which are necessary (and sufficient) to completely describe the given network?

(3) In the case of the KCL equations, why do we need exactly $n - 1$ equations for an n-node network?

Actually, when we speak of an m-mesh or m-loop network, we automatically assume that the network is a *planar network* — a network that can be drawn in a plane such that no two branches (or edges when we refer to its graph) will "cross" or intersect at a point that is not a node (or vertex when the graph is referred to). However, there are many networks that do not possess such a property; that is, there are *nonplanar networks*. Two of these networks are shown in Figure 3-1. This, then, raises the following important question:

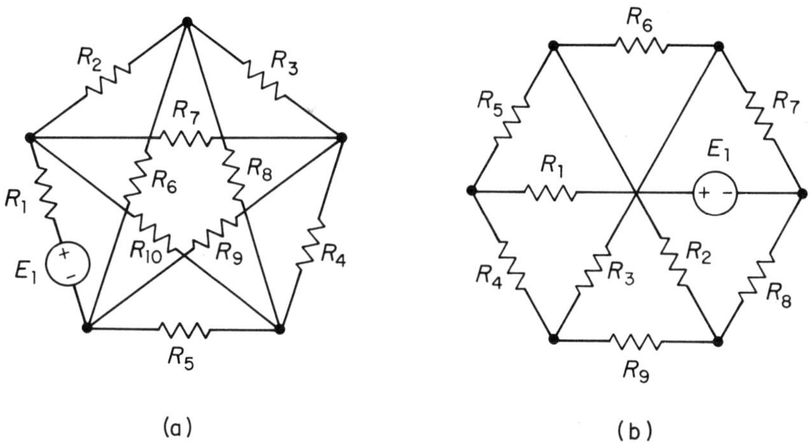

Figure 3-1 Two nonplanar networks.

(4) If the given network is a nonplanar one, the windowpane scheme will not apply. Then, how many independent loop equations do we need, and how do we choose the loops that will guarantee us that the corresponding KVL equations are all independent?

These questions, as well as some others the reader might have in mind already, can be answered quite readily after we study the topics of planarity and duality in this chapter. The answers to these questions will be given in Section 4-7.

In the next section, we shall discuss planar graphs and some of their properties.

3-2 PLANAR GRAPHS

DEFINITION 3-1. Planar and Nonplanar Graphs. A graph is *planar* if it can be drawn in a plane such that no two edges intersect at a point which is not a vertex. All other graphs are *nonplanar*.

EXAMPLE 3-1. The graphs G_{k_1} and G_{k_2} of the networks of Figure 3-1 are shown in Figure 3-2(a) and (b), respectively. To show that the graph of Figure

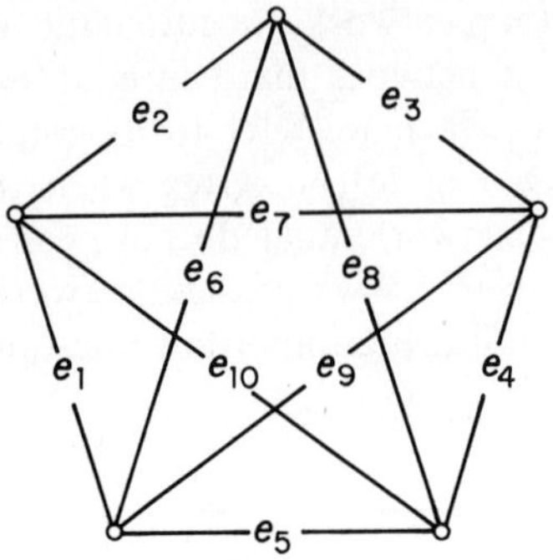

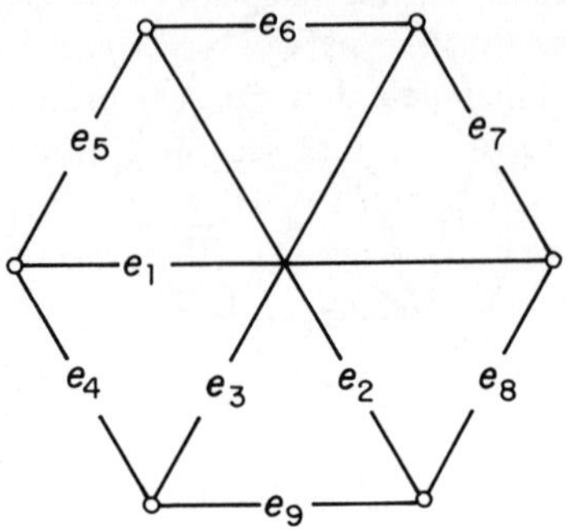

Figure 3-2 Two basic Kuratowski nonplanar graphs: (a) G_{k_1} and (b) G_{k_2}.

3-2(a) is nonplanar, for example, we can redraw some of the edges of the graph so as to eliminate as many "crossings" as we can. One of such redrawings is shown in Figure 3-3.

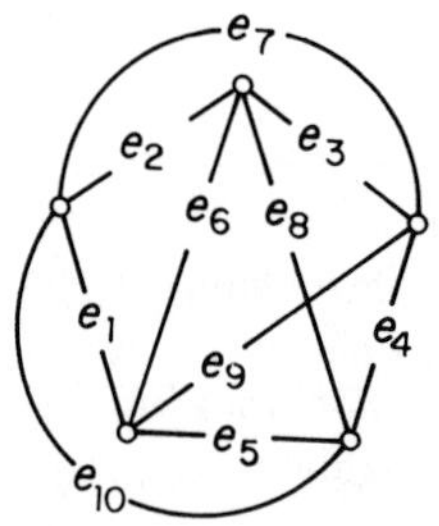

Figure 3-3 A redrawing of the graph of Figure 3-2.

From Figure 3-3 we see that four of the five crossings that appear in the graph G_{k_1} of Figure 3-2(a) have been eliminated by the redrawing of edges e_7 and e_{10}. However, the crossing between edges e_8 and e_9 cannot be avoided by redrawing either e_8 or e_9 (or both) without creating another crossing elsewhere; that is, if we redraw edge e_8 we will create a crossing between edges e_7 and e_8 as shown in Figure 3-4(a). On the other hand, if we rearrange edge e_9, another crossing will likewise be created between edges e_9 and e_{10}, as indicated in Figure 3-4(b). The reader may easily verify that no matter how the edges are redrawn, it is impossible to map this 5-vertex graph onto a plane without at least one crossing between some two edges. Thus, by Definition 3-1, this is a nonplanar graph.

Similarly, the reader may show that the graph G_{k_2} of the network of Figure 3-1(b) is also nonplanar because this graph cannot be redrawn on a plane avoiding all the crossings between edges. In fact, the graphs of Figure 3-2(a)

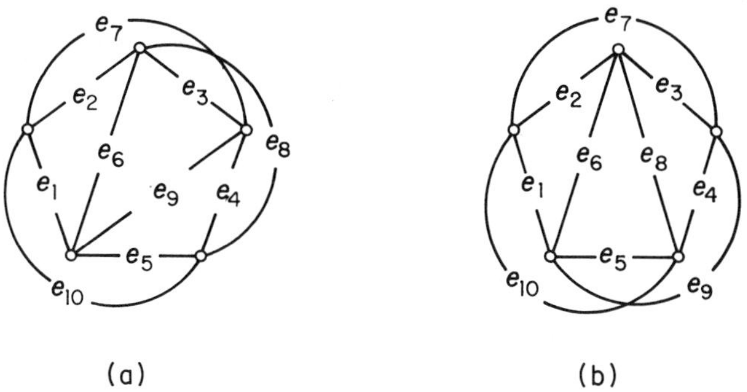

Figure 3-4 Two redrawings of the graph of Figure 3-2.

and (b) are the two *basic* nonplanar graphs in the sense that all other nonplanar graphs contain either one or both of these two nonplanar graphs as subgraphs. This is stated in the following theorem.

THEOREM 3-1. A graph is planar if and only if it contains as its subgraphs neither of the two basic nonplanar graphs of Figure 3-2, provided that each edge e_j in the two graphs is taken to be either a single edge or a series connection of edges.

Theorem 3-1 is the result of C. Kuratowski, and, for this reason, is referred to as *Kuratowski's theorem*. The proof of this theorem is beyond the scope of this text and, therefore, will not be discussed here. The interested reader, however, may refer to Kuratowski's original work for a complete proof.[1]

When a graph is planar, it can be drawn either on a plane or on a sphere such that no two edges will intersect at a point that is not a vertex. This means that a planar graph can always be mapped onto a plane from a sphere and vice versa. To show this, we shall establish that there exists a one-to-one correspondence between a point on the plane and a point on the sphere.

Let us consider a sphere S which lies on a plane P with the point of contact denoted by O. Next, we draw a line 1 through O such that 1 is perpendicular to P and intersects the surface of S at a second point N. Then the line segment $\overline{ON}$ is the diameter of S and points O and N may be considered as the south and north pole, respectively, of S.

Now, take any point X on P. A line $\overline{NX}$ can always be constructed so that $\overline{NX}$ intersects the surface of S at a point X'.[2] On the other hand, if we take any point X' on S, a line $\overline{NX'}$ can always be drawn such that the extension of $\overline{NX'}$ intersects P at a point X. Thus there exists a one-to-one correspondence between a point X' on S and a point X on P except when $X' = N$, in which case point X on P is not defined. Such a one-to-one correspondence between

[1]C. Kuratowski [50]. Refer to C. Berge [3].

[2]If X' happens to be the point O, then X is also the point O so that both X and X' coincide.

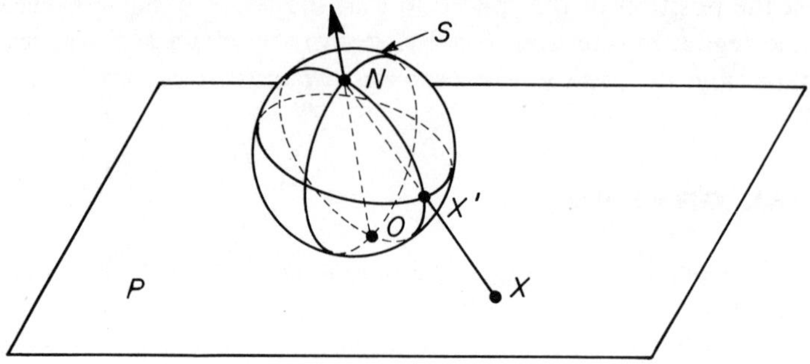

Figure 3-5 A one-to-one correspondence between a point X' on S and a point X on P.

points on P and S is shown in Figure 3-5. Consequently, if we wish to map a planar graph from a plane onto a sphere, the mapping technique just described may always be employed. Conversely, to map a planar graph from a sphere onto a plane, the same mapping technique may be used in the reverse order, provided that point N (the north pole) of the sphere is not a point in the graph.

When a planar graph is mapped onto a plane or a sphere, it divides the entire surface into a number of regions. For example, a 2-mesh graph divides the plane into three regions; a 3-mesh graph divides the plane into four regions; and, in general, an m-mesh graph divides the plane into $m + 1$ regions; m of these regions are bounded by the edges inside the graph and one region outside the graph (see Figure 3-6).

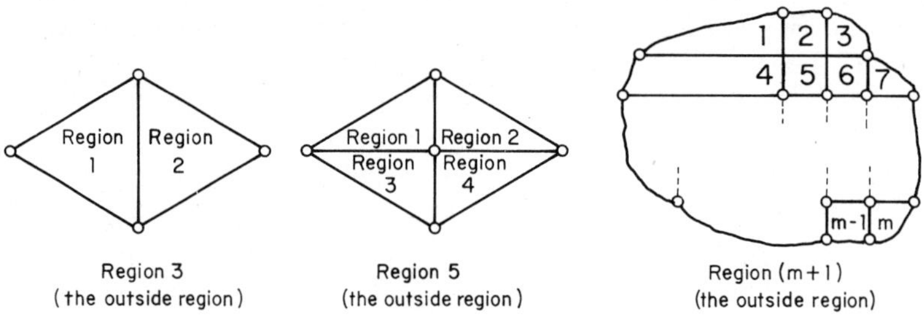

Figure 3-6 Division of a plane into regions by a planar graph.

As mentioned in the introduction of this chapter, the regions inside the graph bounded by the edges are often referred to as "windows" or, simply, "inside regions." The region outside the graph is called the "outside region."

It is interesting to note that any of the windows of a planar graph can be remapped onto a plane so that the window becomes the outside region. The process for such a remapping may be described as follows. First, map the graph onto a sphere employing the mapping technique as described earlier.

Then adjust the position of the sphere so that the north pole N of the sphere lies inside the region in question. Finally, remap the graph of the sphere back onto the plane and the region now becomes the outside region of the plane.

3-3 DUAL GRAPHS

In network theory, we study dual networks because of the obvious parallelism existing between the formulations of loop and node systems of equations. In order to better understand the application of the theory of duality to electrical networks, we shall first define a dual of a graph, and then discuss the properties of dual graphs before our discussion on dual networks in the next chapter.

DEFINITION 3-2. Dual Graph. A graph G_1 has a *dual* G_2 if (a) there exists a one-to-one correspondence between the edges of G_1 and the edges of G_2, and (b)

$$R(G'_{2_i}) = R(G_2) - N(G_{1_i}) \tag{3-1}$$

where G_{1_i} is *any* subgraph of G_1, and G'_{2_i} is the *complement* of the subgraph G_{2_i} of G_2 corresponding to G_{1_i}. The symbols $R(G)$ and $N(G)$, in general, stand for the rank and the nullity, respectively, of the graph G in question.

From this definition, it is evident that condition (b) is quite abstract, and it is not surprising that the reader finds it difficult to understand at the first attempt. However, after one or two illustrative examples, the idea involved will become much clearer.

EXAMPLE 3-2. Consider the graph G_1 of Figure 3-7(a) and a dual graph G_2 of Figure 3-7(b).

To verify that G_2 is the dual of G_1, we first show that condition (a) of Definition 3-2 is satisfied by the two graphs. This can be done most easily by super-

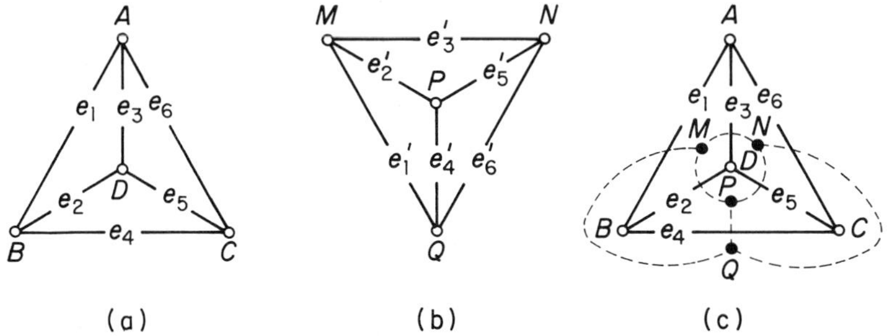

Figure 3-7 (a) Graph G_1. (b) Dual graph G_2. (c) G_1 and G_2.

imposing G_2 upon G_1 as shown in Figure 3-7(c), which clearly reveals the one-to-one correspondence between the edges of G_1 and the edges of G_2. In fact, each edge e_j of G_1 is crossed by the corresponding edge e_j' of G_2.

To illustrate that condition (b) of Definition 3-2 is also satisfied by the two graphs, let G_{1_i} of G_1 be the subgraph containing edges e_1, e_3, and e_6. Then the complement G_{2_i}' of G_{2_i} corresponding to G_{1_i} is the subgraph of G_2 containing edges e_2', e_4', and e_5'. These two subgraphs are shown in Figure 3-8(a) and (b), respectively.

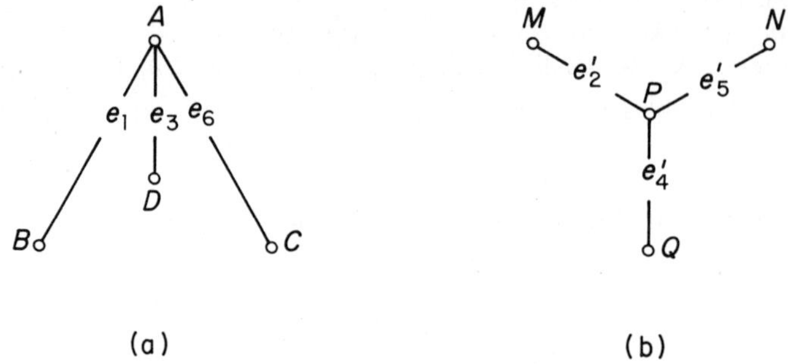

Figure 3-8 (a) Subgraph G_{1_i}. (b) Subgraph G_{2_i}'.

From Figures 3-7 and 3-8, we see that

$$R(G_{2_i}') = v_{2_i}' - 1 = 4 - 1 = 3$$

$$R(G_2) = v_2 - 1 = 4 - 1 = 3$$

$$N(G_{1_i}) = e_{1_i} - v_{1_i} + 1 = 3 - 4 + 1 = 0$$

so that Equation (3-1) is satisfied.

EXAMPLE 3-3. As another example, consider again G_1 and G_2 of Figure 3-7 but let G_{1_j} be the subgraph of G_1 containing edges e_3 and e_4 so that G_{2_j}' of G_2 is the one containing edges e_1', e_2', e_5', and e_6'. These two subgraphs are shown in Figure 3-9(a) and (b), respectively.

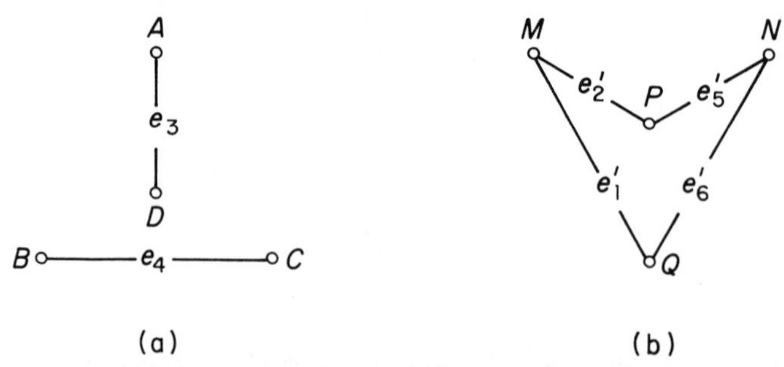

Figure 3-9 (a) Subgraph G_{1_j}. (b) Subgraph G_{2_j}'.

From Figures 3-7 and 3-9,

$$R(G'_{2_j}) = v'_{2_j} - 1 = 4 - 1 = 3$$

$$R(G_2) = v_2 - 1 = 4 - 1 = 3$$

$$N(G_{1_j}) = e_{1_j} - v_{1_j} + p_{1_j} = 2 - 4 + 2 = 0$$

Note that in the third equation above, p_{1_j} denotes the number of maximally connected subgraphs of G_{1_j}, which is a nonconnected subgraph of G_1.

Again, Equation (3-1) is satisfied with the subscript i replaced by j.

The appearance of the two graphs G_1 and G_2 in Figure 3-7(c) is similar to the so-called *dot method* for constructing a dual of a planar network. We shall discuss this method in some detail after our study of the following three theorems.

THEOREM 3-2. Let G_2 be a dual of G_1. Then

$$R(G_1) = N(G_2) \tag{3-2}$$

and

$$N(G_1) = R(G_2) \tag{3-3}$$

Proof: Let G_{1_i} of Equation (3-1) be the entire graph G_1. Then G_{2_i} is the entire graph G_2, and G'_{2_i} is a null graph. Thus Equation (3-1) becomes

$$0 = R(G_2) - N(G_1)$$

or

$$N(G_1) = R(G_2)$$

which is identical to Equation (3-3). Equation (3-2) may be derived from (3-3) as follows:

First, note that for any connected graph G,

$$N(G) + R(G) = (e - v + 1) + (v - 1) = e \tag{3-4}$$

where e is the total number of edges of G, and v is the total number of vertices of G.

Next, we note that since G_2 is the dual of G_1,

$$e_1 = e_2 \tag{3-5}$$

where e_1 is the total number of edges of G_1, and e_2 is the total number of edges of G_2.

Now, in light of Equation (3-4), we write

$$N(G_1) + R(G_1) = e_1 \tag{3-6}$$

and

$$N(G_2) + R(G_2) = e_2 \tag{3-7}$$

From Equations (3-5), (3-6), and (3-7),

$$N(G_1) + R(G_1) = N(G_2) + R(G_2)$$

or, rearranging the terms,

$$N(G_1) - R(G_2) = N(G_2) - R(G_1) \tag{3-8}$$

But, according to Equation (3-3),

$$N(G_1) - R(G_2) = 0$$

so that

$$N(G_2) - R(G_1) = 0$$

which implies that

$$R(G_1) = N(G_2)$$

This completes the proof.

THEOREM 3-3. If G_2 is a dual of G_1, then G_1 is a dual of G_2 (so that the two graphs are *dual graphs*).

Proof: Since G_2 is a dual of G_1, we have, according to Equation (3-1),

$$R(G'_{2_i}) = R(G_2) - N(G_{1_i})$$

or

$$N(G_{1_i}) = R(G_2) - R(G'_{2_i}) \tag{3-9}$$

To show that G_1 is a dual of G_2, we are required to show that

$$R(G'_{1_i}) = R(G_1) - N(G_{2_i})$$

or

$$R(G_{1_i}) = R(G_1) - N(G'_{2_i}) \tag{3-10}$$

since G_{1_i} can be *any* subgraph of G_1.

To derive Equation (3-10), we note that, from Theorem 3-2,

$$R(G_2) = N(G_1) \tag{3-11}$$

Also, since $e'_{1_i} = e'_{2_i}$, we write

$$e_{1_i} + e'_{2_i} = e_{1_i} + e'_{1_i} = e_1 \tag{3-12}$$

But, according to Equations (3-4) and (3-9),

$$\begin{aligned} R(G_{1_i}) &= e_{1_i} - N(G_{1_i}) \\ &= e_{1_i} - \{R(G_2) - R(G'_{2_i})\} \end{aligned}$$

Or, by Equation (3-11),

$$\begin{aligned} R(G_{1_i}) &= e_{1_i} - R(G_2) + R(G'_{2_i}) \\ &= e_{1_i} - N(G_1) + \{e'_{2_i} - N(G'_{2_i})\} \\ &= (e_{1_i} + e'_{2_i}) - N(G_1) - N(G'_{2_i}) \end{aligned}$$

which, due to Equation (3-12), becomes

$$\begin{aligned} R(G_{1_i}) &= e_1 - N(G_1) - N(G'_{2_i}) \\ &= \{e_1 - N(G_1)\} - N(G'_{2_i}) \end{aligned}$$

Finally, we write

$$R(G_{1_i}) = R(G_1) - N(G'_{2_i})$$

Hence the theorem.

The next theorem, which is the most significant one on duality, will now be stated without proof. We include the statement simply because of its importance, but we omit the proof since a formal proof is quite complicated and is beyond the scope of this book.[3]

THEOREM 3-4. A necessary and sufficient condition for a graph to have a dual is that the graph be planar.

Theorem 3-4 is very important because it guarantees the existence of a dual if the given graph is planar.

3-4 THE DOT METHOD FOR CONSTRUCTING A DUAL

Although Definition 3-2 is an accurate algebraic definition of duality, it renders very little help in constructing a dual G_2 of a given planar graph G_1 or in examining whether or not G_2 is a dual of G_1 when both of them are given. This is due to the fact that in order to apply Definition 3-2, one will have to examine and see whether Equation (3-1) of condition (b) is satisfied for *each possible* G_{1_i} of G_1; hence, one must exhaust all such possibilities before he can decide whether G_2 is a dual of G_1.

On the other hand, the dot method for constructing a dual from a given planar graph is very simple to apply. We shall first state the procedure to be followed in applying this method, then illustrate the ideas involved by means of an example; and, finally, explain briefly how the method is related to Definition 3-2.

To construct a dual by means of the dot method,

(a) Place a dot in each region (including the outside region) of the given planar graph G_1 with e edges. Thus, if G_1 contains m windows, it will require exactly $m + 1$ dots.

[3]For a formal proof, the interested reader may refer to Seshu and Reed [76], pp. 44–50.

(b) Connect the $m+1$ dots by the line segments such that, for each edge e_j between two windows, there exists a line segment e_j' which crosses e_j and is connected between the two dots corresponding to the two windows. Hence, if G_1 contains e edges, exactly e line segments will be required.

The resultant planar graph G_2 with the $m+1$ dots as vertices and the e line segments as edges is a dual of G_1.

EXAMPLE 3-4. Consider the graph G of Figure 3-10(a). At first sight, we might think that G is nonplanar; but, if edges e_7 and e_8 are redrawn to avoid the three crossings, it is seen to be a planar graph as shown in Figure 3-10(b). We therefore see that G has six regions — five windows and one outside region.

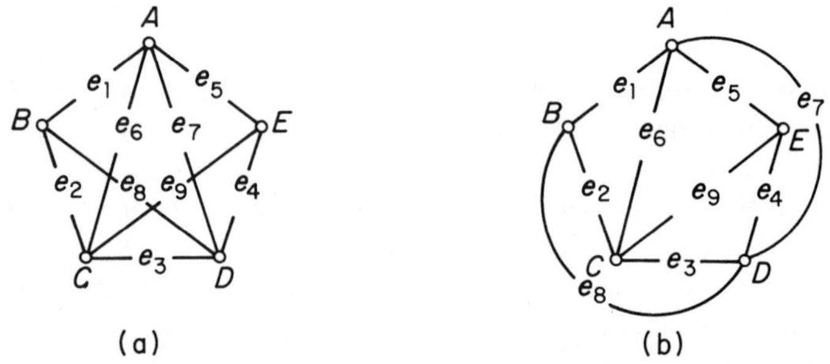

Figure 3-10 (a) A planar graph G. (b) A redrawing of G.

After completing steps (a) and (b) in the procedure, we have "crossed" each edge e_j of G_1 by a line segment e_j' which is connected between two dots corresponding to some two windows of G_1. This result is shown in Figure 3-11(a) with the two graphs "crossing" one another.

Finally, "extracting" from Figure 3-11(a) the new graph G_2, consisting of the $m+1$ dots as vertices and the e line segments as edges, we have the desired dual graph as shown in Figure 3-11(b).

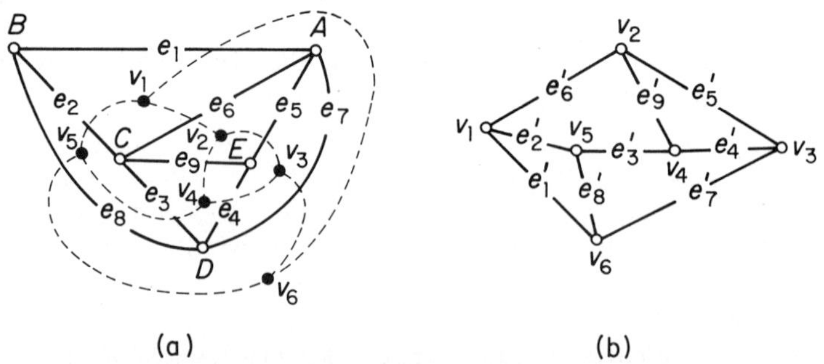

Figure 3-11 (a) G_1 and G_2 crossing each other. (b) Dual graph G_2.

To see how the dot method is related to Definition 3-2, we note that the one-to-one correspondence requirement between the edges of one graph and the edges of the other, as stated in condition (a) of Definition 3-2, is evidently satisfied by the graph resulting from the application of this method, because, for each e_j in the given graph, there exists an edge e_j' in the resultant graph to match it. To see how condition (b) of Definition 3-2 is related to the dot method, we observe that, according to Theorem 3-2, this condition implies that

$$R(G_1) = N(G_2) \tag{3-13}$$

and

$$N(G_1) = R(G_2) \tag{3-14}$$

Also, from Figure 3-11(a), we see that a cutset in G_1 corresponds to a unique circuit in G_2 (and vice versa); and a circuit in G_1 corresponds to a unique cutset in G_2 (and vice versa). This means that there exists a one-to-one correspondence between the set of all possible cutsets in G_1 and the set of all possible circuits in G_2 (and vice versa). Thus, the conditions stated in Equations (3-13) and (3-14) are satisfied by the resultant graph of the dot method.

3-5 INTERRELATIONSHIPS BETWEEN THE MATRICES OF DUAL GRAPHS

From our discussion in the preceding section, it should be evident that the cutset matrix Q_{a_1} of a planar graph G_1 becomes the circuit matrix B_{a_2} of its dual graph G_2 (and vice versa). The reason is that any cutset $k_{1_j} = \{e_1, e_2, \cdots, e_n\}$ of G_1 ($n \leqq e$) becomes a circuit $C'_{2_j} = \{e_1', e_2', \cdots, e_n'\}$ of G_2 (and vice versa). Similarly, the circuit matrix B_{a_1} of G_1 becomes the cutset matrix Q_{a_2} of G_2 (and vice versa). Thus we write

$$Q_{a_1} \leftrightarrow B_{a_2} \tag{3-15}$$

and

$$B_{a_1} \leftrightarrow Q_{a_2} \tag{3-16}$$

Furthermore, the circuit submatrix B_1 whose $e - v + 1$ rows are identified by the circuits corresponding to the $e - v + 1$ (why this number?) windows of G_1 becomes the incidence submatrix A_2 of G_2 with the omitted row identified by that vertex of G_2, which corresponds to the outside region of G_1 (and vice versa). Hence,

$$B_1 \leftrightarrow A_2 \tag{3-17}$$

$$A_1 \leftrightarrow B_2 \tag{3-18}$$

with the definitions of A_1 and B_2 omitted as obvious since they are similar to those of A_2 and B_1.

It is now evident that one of the reasons for studying dual graphs is that whatever information we obtain about a planar graph G_1, we automatically have exactly the same amount of information about a dual G_2 of G_1.

3-6 REMARKS

In the preceding sections we have discussed planar graphs versus nonplanar graphs, dual graphs, and the interrelationships between various matrices of dual graphs. In the next chapter we shall make use of the theory of duality developed here in connection with dual networks after our formal discussion on loop and node systems of network equations.

When we deal with two dual graphs, we always refer to one graph as being *a* dual, instead of *the* dual, of the other. The reader naturally might wonder whether or not a planar graph can have more than one dual. The answer to this question is affirmative and the easiest way to convince the reader of this is, perhaps, by means of a simple example.

EXAMPLE 3-5. Let G_1 be the separable, planar graph of Figure 3-12(a). A dual G_2 of G_1 is shown in Figure 3-12(b).

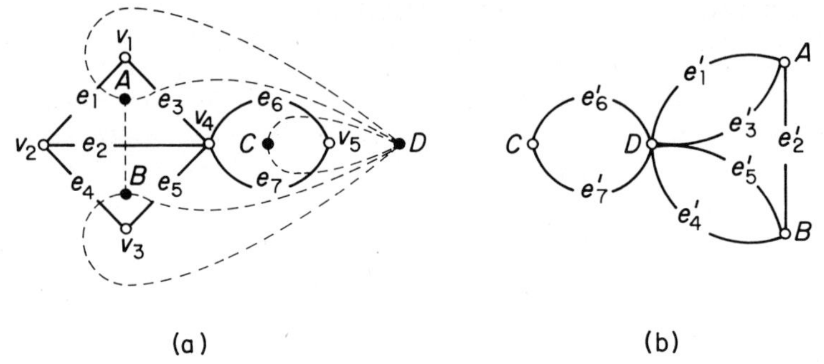

Figure 3-12 Dual graphs (a) G_1 and (b) G_2.

Now, let us consider another pair of separable, planar graphs G_3 and G_4, as shown in Figure 3-13(a) and (b), respectively. From Figures 3-12 and 3-13, we note the following observations:

(a) Graphs G_1 and G_3 are equivalent (or 2-isomorphic — see Definition 1-24), since they can be made identical by the application of operation (a) of Definition 1-24 to both graphs. That is, both graphs will be congruent if each of them is separated into two components by splitting the cut-vertex (v_4 of G_1, and v_1 of G_3).

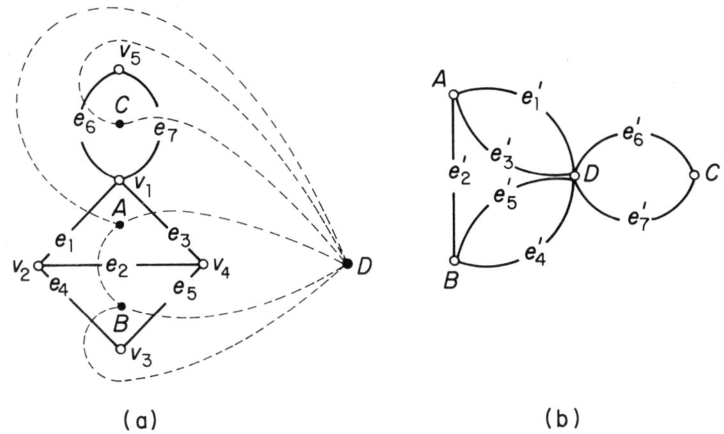

Figure 3-13 Dual graphs (a) G_3 and (b) G_4.

(b) Graphs G_2 and G_4 are congruent; that is, they are the same graph (see Definition 1-22).

Observation (b) points out the fact that two planar graphs may have the same dual. Since both G_1 and G_3 have G_2 (or G_4) as a dual, we conclude that *a planar graph may have more than one dual*, in light of Theorem 3-3.

Observation (a) suggests the conclusion that the duals of a planar graph are unique to within 2-isomorphism; that is, *the duals of a planar graph are equivalent graphs.*

PROBLEMS

3-1 Show that each of Kuratowski's basic nonplanar graphs can be drawn on the surface of a doughnut without any edges crossing each other except at the vertices.

3-2 Determine whether each of the graphs shown in Figure P3-2 is planar. If the graph is planar, redraw the graph to eliminate all crossings except at the vertices. If the graph is nonplanar, show the existence of a subgraph that is one of Kuratowski's basic nonplanar graphs (with some edges possibly replaced by a series connection of edges).

3-3 Verify that the two graphs, G and G', depicted in Figure P3-3 are dual graphs to each other by showing that the two conditions of Definition 3-2 are satisfied.

3-4 Find a dual of the graph given in Figure P3-4 by applying the dot method.

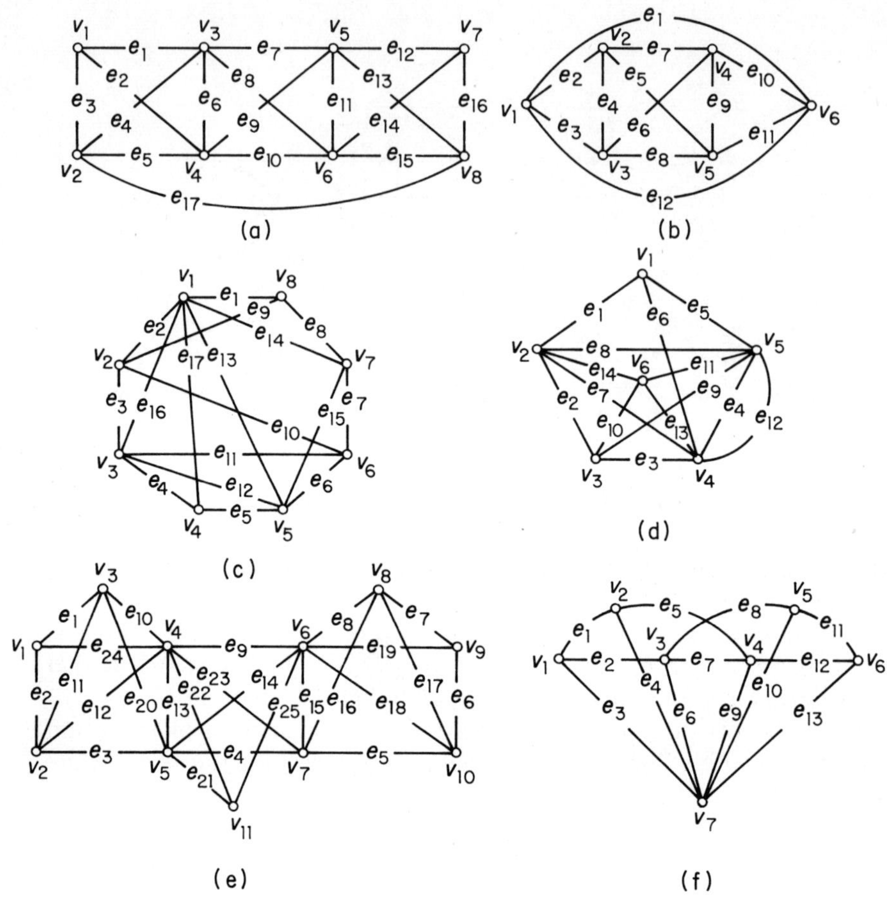

Figure P3-2

3-5 Determine whether each of the graphs of Figure P3-5 is planar. If the graph is planar, find a dual of this graph.

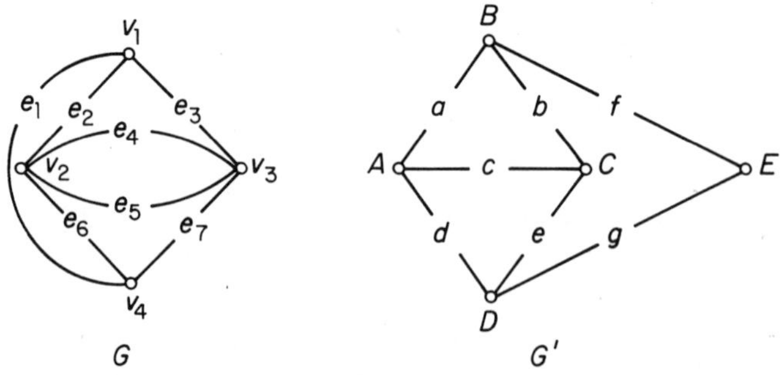

Figure P3-3

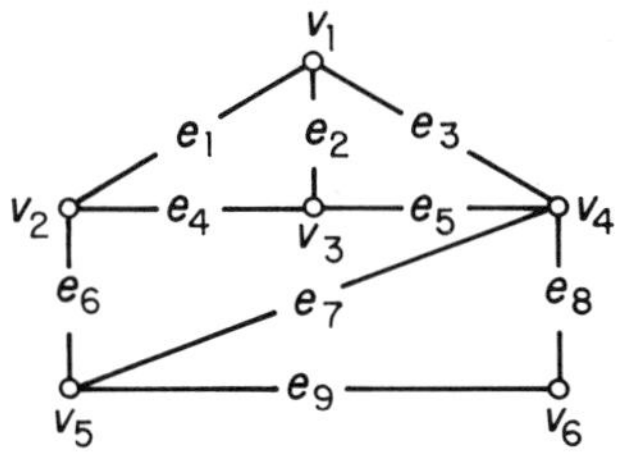

Figure P3-4

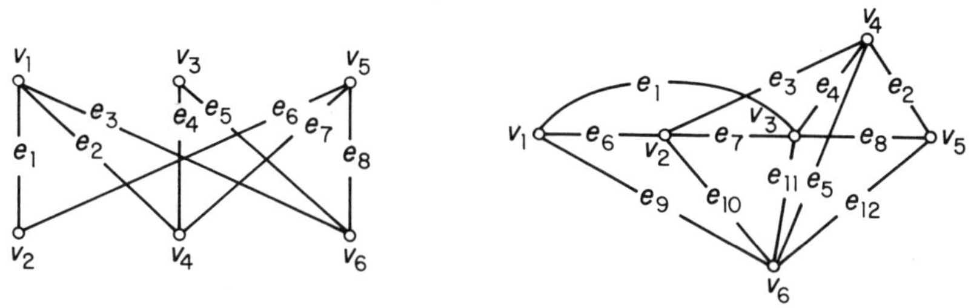

(a) (b)

Figure P3-5

3-6 Find the cutset matrix Q_a of the graph of Figure P3-6 by
(a) Working with the given graph directly.
(b) Obtaining a dual of the given graph first and then finding the circuit matrix B_a of the dual.
Compare the two matrices obtained in (a) and (b) and draw a conclusion.

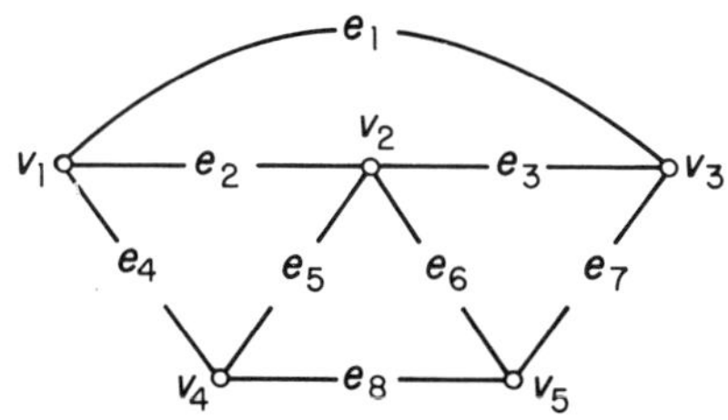

Figure P3-6

3-7 Construct a planar graph G such that some two duals of G, say G' and G'', are not congruent.

3-8 Prove that if G' and G'' are two duals of a planar graph G, then G' and G'' are equivalent (2-isomorphic) graphs. (Refer to Problem 3-6 and Definition 1-22.)

3-9 Prove that when a planar graph G of e edges and v vertices is mapped onto a plane, the total number of regions into which the plane is divided by the e edges of G is equal to $e - v + 2$.

3-10 Construct a graph G and its dual G' from the given incidence matrix A_a.

$$A_a = \begin{array}{c} \begin{matrix} a & b & c & d & e & f \end{matrix} \\ \begin{bmatrix} 1 & 0 & 1 & 0 & 1 & 0 \\ 1 & 1 & 0 & 0 & 0 & 1 \\ 0 & 1 & 0 & 1 & 1 & 0 \\ 0 & 0 & 1 & 1 & 0 & 1 \end{bmatrix} \end{array}$$

3-11 Verify that graphs G and G' obtained in Problem 3-10 are duals by showing that they satisfy conditions (a) and (b) of Definition 3-2.

3-12 Construct a graph G and its dual G' from the given circuit matrix B_a.

$$B_a = \begin{array}{c} \begin{matrix} a & b & c & d & e & f \end{matrix} \\ \begin{bmatrix} 1 & 1 & 0 & 1 & 0 & 1 \\ 0 & 0 & 0 & 0 & 1 & 1 \\ 0 & 0 & 1 & 1 & 0 & 1 \\ 1 & 1 & 1 & 0 & 0 & 0 \\ 0 & 0 & 1 & 1 & 1 & 0 \\ 1 & 1 & 0 & 1 & 1 & 0 \end{bmatrix} \end{array}$$

3-13 Refer to the matrix B_a in Problem 3-12. Suggest an algorithm for constructing the dual G' from B_a without first having to obtain G from B_a. Give your derivation of this algorithm in detail.

3-14 Answer each of the four questions posed in Section 3-1 as fully as possible.

3-15 The two graphs, G_1 and G_2, of Figure P3-15 are obviously congruent graphs. Show that they are also dual graphs.

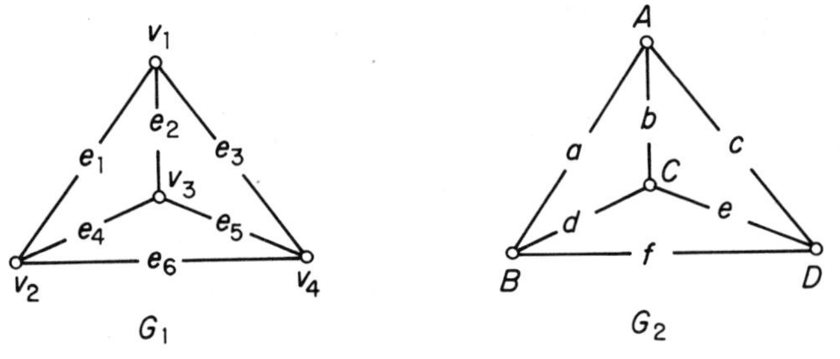

Figure P3-15

3-16 Give at least three examples (in addition to that in Problem 3-15) of dual graphs that are also congruent graphs.

3-17 Refer to the two graphs, G_1 and G_2, of Figure P3-15. Demonstrate the relationships of Equations (3-15) and (3-16) by writing the circuit matrix B_{a_1} of G_1 and show that it corresponds to the cutset matrix Q_{a_2} of G_2 in the sense that B_{a_1} can be made equal to Q_{a_2} by the rearrangement of its rows and columns.

3-18 Refer to the graph G_1 of Figure P3-15.
(a) Show that G_1 can be redrawn as G_K such that the region bounded by edges e_1, e_2, and e_4 become the "outside" region of G_K.
(b) Prove that a planar graph G can be redrawn as another planar graph G_K such that any "inside" region of G becomes the outside region of G_K.

CHAPTER 4

MATRIX REPRESENTATION OF NETWORK EQUATIONS

4-1 INTRODUCTION

The main objective of this chapter is to develop the various systems of network equations — namely, the loop equations, node equations, and the state equations — in matrix form using the concepts of the incidence matrix, the circuit matrix, the cutset matrix, and the interrelationships between these matrices. One of the advantages of representing the network equations in matrix notation is that such a notation is clear and concise in form (provided, of course, that each symbol used is well defined), and it is extremely convenient to manipulate the terms involved in the theoretical development of the loop and node systems of network equations. Furthermore, the matrix forms of such systems are best suited for digital computational purposes, especially when the given network is a complex one.

After the loop analysis, the node analysis, and the state equations have been developed, we shall then make use of the concept of duality to show the striking parallelism between the development of the loop-analysis and node-analysis systems, thus introducing the idea of dual networks. Finally, the last section will be devoted to the summarization of results for future reference.

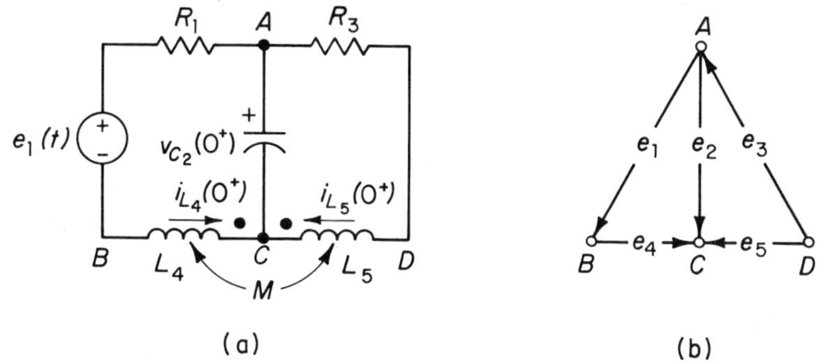

Figure 4-1 (a) A planar network N. (b) A graph G_1 of N.

4-2 PRIMARY SYSTEMS OF EQUATIONS

Consider the network N of Figure 4-1(a). An oriented graph G_1 of N is shown in Figure 4-1(b). Note that the orientations of the edges in the graph are purely arbitrary, and that the elements $e_1(t)$ and R_1 in the network are treated as single branches. Such a choice of branches is again arbitrary. For example, there is no reason why the network of Figure 4-1(a) cannot have a graph G_2 as shown in Figure 4-2, which is obtained by considering $e_1(t)$, R_1,

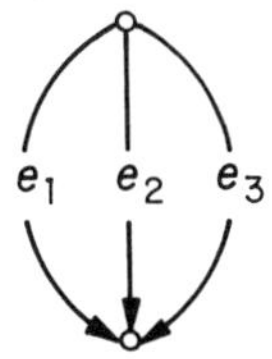

Figure 4-2 Another graph G_2 of the network N_1 of Figure 4-1(a).

and L_4 in N as one branch (corresponding to edge e_1 in the graph), C_2 as another (corresponding to edge e_2), and R_3 and L_5 as the third branch (corresponding to edge e_3).

With the graph G_1 of Figure 4-1(b) chosen as the graph of N, we can write the Kirchhoff voltage-law (KVL) and the Kirchhoff current-law (KCL) equations as shown in Equations (4-1) and (4-2), respectively.

$$\left\{\begin{array}{rrrrrl} -v_1 & +v_2 & & -v_4 & & = 0 \\ & -v_2 & -v_3 & & +v_5 & = 0 \\ -v_1 & & -v_3 & -v_4 & +v_5 & = 0 \end{array}\right. \tag{4-1}$$

$$\left\{\begin{array}{rrrrrl} i_1 & +i_2 & -i_3 & & & = 0 \\ -i_1 & & & +i_4 & & = 0 \\ & -i_2 & & -i_4 & -i_5 & = 0 \\ & & i_3 & & +i_5 & = 0 \end{array}\right. \tag{4-2}$$

Rewriting Equations (4-1) and (4-2) in matrix form, we get Equations (4-3) and (4-4), respectively.

$$\begin{array}{c} \\ c_1 \\ c_2 \\ c_3 \end{array}\begin{array}{c} \begin{array}{ccccc} e_1 & e_2 & e_3 & e_4 & e_5 \end{array} \\ \begin{bmatrix} -1 & 1 & 0 & -1 & 0 \\ 0 & -1 & -1 & 0 & 1 \\ -1 & 0 & -1 & -1 & 1 \end{bmatrix} \end{array}\begin{bmatrix} v_1(t) \\ v_2(t) \\ v_3(t) \\ v_4(t) \\ v_5(t) \end{bmatrix} = 0 \tag{4-3}$$

$$\begin{array}{c} \\ A \\ B \\ C \\ D \end{array}\begin{array}{c} \begin{array}{ccccc} e_1 & e_2 & e_3 & e_4 & e_5 \end{array} \\ \begin{bmatrix} 1 & 1 & -1 & 0 & 0 \\ -1 & 0 & 0 & 1 & 0 \\ 0 & -1 & 0 & -1 & -1 \\ 0 & 0 & 1 & 0 & 1 \end{bmatrix} \end{array}\begin{bmatrix} i_1(t) \\ i_2(t) \\ i_3(t) \\ i_4(t) \\ i_5(t) \end{bmatrix} = 0 \tag{4-4}$$

If we examine the two systems of equations in (4-3) and (4-4), we see that the coefficient matrix of Equation (4-3) is the circuit matrix B_a of the graph G_1 of Figure 4-1(b), and that the coefficient matrix of Equation (4-4) is the incidence matrix A_a of G_1. Written in compact matrix notation, Equations (4-3) and (4-4) become (4-5) and (4-6), respectively:

$$B_a V_e(t) = 0 \tag{4-5}$$

$$A_a I_e(t) = 0 \tag{4-6}$$

where $V_e(t)$ and $I_e(t)$ are the voltage vector and the current vector, respectively, for the five edges of G_1.

Since the rank of B_a is $e - v + 1$, or $5 - 4 + 1 = 2$, and the rank of A_a is $v - 1$, or $4 - 1 = 3$, where e is the number of edges and v is the number of vertices of G_1, we see that only two of the three equations in (4-3) and three of the four equations in (4-4) are needed to represent the network N fully. This is because the other equations are the dependent ones and they do not give additional information about the network. Hence, we write

$$BV_e(t) = 0 \tag{4-7}$$

and

$$AI_e(t) = 0 \tag{4-8}$$

where B is a circuit submatrix and A is an incidence submatrix of G_1, as defined in Chapter 2.

Now, let us consider the number of unknowns in the network N. Since there are five edges in G_1, and for each edge there is both a voltage and a current to be considered, there are ten unknowns in N—namely, the five edge voltages and the five edge currents. Or, in general, for a network with e branches, there are $2e$ unknowns.[1]

From Equations (4-7) and (4-8), we note that there are only $(e - v + 1) + (v - 1) = e$ equations, but that there are $2e$ unknowns. Furthermore, from our first network analysis course, we learned that for a linear physical system with n unknowns, we need to write exactly n independent equations in order to determine the n unknowns uniquely. Thus, we need to have e more equations for N.

Before we try to find the other e equations, let us first examine the e equations we already have on hand. The $e - v + 1$ equations in (4-7) are the independent KVL equations, each of which is in the form

$$\sum_k v_k = 0 \tag{4-9}$$

where v_k is the voltage across the kth branch of the network N. Thus, Equation (4-9) represents the algebraic sum of all the branch voltages around a circuit of N.

The $v - 1$ equations in (4-8) are the independent KCL equations, each of which is in the form

$$\sum_j i_j = 0 \tag{4-10}$$

where i_j is the current in the jth branch of N so that Equation (4-10) is the algebraic sum of all the branch currents at a node of N.

With these $e - v + 1$ KVL equations and $v - 1$ KCL equations written, it is now obvious what the other e equations will be. They are the e branch voltage-current-relationship (VCR) equations, each of which expresses the relationship between the voltage and the current of a branch of N.

To see what form these e VCR equations will take, let us consider a branch, say the kth branch in a network, as shown in Figure 4-3. The VCR equation

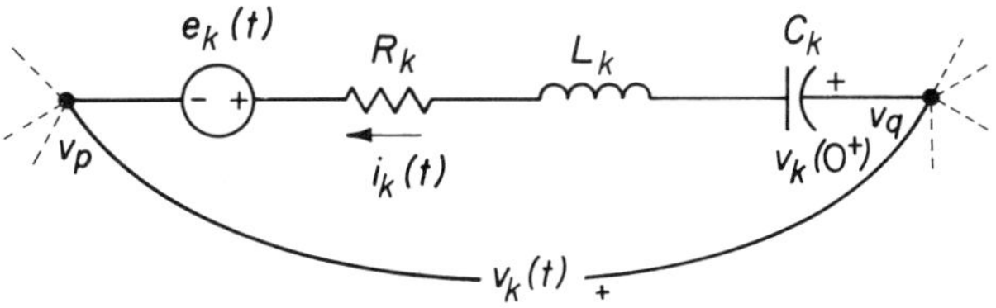

Figure 4-3 A typical branch of a network.

[1]This statement is valid for any network N. Note that, in this example, the edge currents, i_1 and i_4, are equal, and the edge currents, i_3 and i_5, are equal in magnitude but opposite in sign, so that, with equations $i_1 = i_4$ and $i_3 = -i_5$ being considered, the number of unknowns may be reduced. This amounts to considering e_1 and e_4 together as one edge and e_3 and e_5 together as another edge. Such reduction of unknowns, however, will not make our general statement invalid.

for this branch is

$$v_k(t) = e_k(t) + R_k i_k(t) + L_k \frac{d}{dt} i_k(t) + \sum_{\substack{j=1 \\ j \neq k}}^{e} M_{jk} \frac{d}{dt} i_j(t) + \frac{1}{C_k}\int_0^t i_k(t)\,dt + v_{c_k}(0^+) \tag{4-11}$$

where M_{jk} is the mutual inductance between L_k in branch k and L_j in branch j, and $i_j(t)$ is the current in branch j.

For the network N of Figure 4-1(a), the five VCR equations are:

$$\begin{aligned} v_1(t) &= e_1(t) + R_1 i_1(t) \\ v_2(t) &= \frac{1}{C_2}\int_0^t i_2(t)\,dt + v_{c_2}(0^+) \\ v_3(t) &= R_3 i_3(t) \\ v_4(t) &= L_4 \frac{d}{dt} i_4(t) + M_{54}\frac{d}{dt} i_5(t) \\ v_5(t) &= L_5 \frac{d}{dt} i_5(t) + M_{45}\frac{d}{dt} i_4(t) \end{aligned} \tag{4-12}$$

When written in matrix form the system of Equation (4-12) becomes

$$\begin{bmatrix} v_1(t) \\ v_2(t) \\ v_3(t) \\ v_4(t) \\ v_5(t) \end{bmatrix} = \begin{bmatrix} e_1(t) \\ 0 \\ 0 \\ 0 \\ 0 \end{bmatrix} + \begin{bmatrix} R_1 & 0 & 0 & 0 & 0 \\ 0 & 0 & 0 & 0 & 0 \\ 0 & 0 & R_3 & 0 & 0 \\ 0 & 0 & 0 & 0 & 0 \\ 0 & 0 & 0 & 0 & 0 \end{bmatrix} \begin{bmatrix} i_1(t) \\ i_2(t) \\ i_3(t) \\ i_4(t) \\ i_5(t) \end{bmatrix} + \begin{bmatrix} 0 & 0 & 0 & 0 & 0 \\ 0 & 0 & 0 & 0 & 0 \\ 0 & 0 & 0 & 0 & 0 \\ 0 & 0 & 0 & L_4 & M_{54} \\ 0 & 0 & 0 & M_{45} & L_5 \end{bmatrix} \frac{d}{dt} \begin{bmatrix} i_1(t) \\ i_2(t) \\ i_3(t) \\ i_4(t) \\ i_5(t) \end{bmatrix} + \begin{bmatrix} 0 & 0 & 0 & 0 & 0 \\ 0 & 1/C_2 & 0 & 0 & 0 \\ 0 & 0 & 0 & 0 & 0 \\ 0 & 0 & 0 & 0 & 0 \\ 0 & 0 & 0 & 0 & 0 \end{bmatrix} \int_0^t \begin{bmatrix} i_1(t) \\ i_2(t) \\ i_3(t) \\ i_4(t) \\ i_5(t) \end{bmatrix} dt + \begin{bmatrix} 0 \\ v_{c_2}(0^+) \\ 0 \\ 0 \\ 0 \end{bmatrix} \tag{4-13}$$

The fact that Equation (4-13) is equivalent to (4-12) might not be easily recognized at first sight, but it can be readily verified by carrying out the matrix operations in Equation (4-13) and showing that the two systems of equations are identical.

Thus, in general, the VCR matrix equation for a network of e branches may be written in the form

$$V_e(t) = E_e(t) + R_eI_e(t) + L_e\frac{d}{dt}I_e(t) + D_e\int_0^t I_e(t)\,dt + V_e(0^+) \quad (4\text{-}14)$$

where

$$V_e(t) = \begin{bmatrix} v_1(t) \\ v_2(t) \\ \cdot \\ \cdot \\ \cdot \\ v_e(t) \end{bmatrix} \qquad I_e(t) = \begin{bmatrix} i_1(t) \\ i_2(t) \\ \cdot \\ \cdot \\ \cdot \\ i_e(t) \end{bmatrix} \qquad E_e(t) = \begin{bmatrix} e_1(t) \\ e_2(t) \\ \cdot \\ \cdot \\ \cdot \\ e_e(t) \end{bmatrix}$$

$$V_e(0^+) = \begin{bmatrix} V_{c_1}(0^+) \\ V_{c_2}(0^+) \\ \cdot \\ \cdot \\ \cdot \\ V_{c_e}(0^+) \end{bmatrix} \qquad R_e = \begin{bmatrix} R_1 & & & 0 \\ & R_2 & & \\ & & \ddots & \\ 0 & & & R_e \end{bmatrix}$$

$$L_e = \begin{bmatrix} L_1 & M_{12} & \cdots & M_{1e} \\ M_{21} & L_2 & \cdots & M_{2e} \\ & \cdots & & \\ & \cdots & & \\ M_{e1} & M_{e2} & & L_e \end{bmatrix} \qquad D_e = \begin{bmatrix} 1/C_1 & & & 0 \\ & 1/C_2 & & \\ & & \ddots & \\ 0 & & & 1/C_e \end{bmatrix}$$

More specifically, $V_e(t)$ is the e-vector of branch voltages such that the voltage $v_j(t)$ across the jth branch is the jth element of the vector; $I_e(t)$ is the e-vector of branch currents such that the current $i_j(t)$ in the jth branch is the jth element of the vector; $E_e(t)$ is the e-vector of voltage sources such that its jth element is equal to the voltage source $e_j(t)$ in branch j and is equal to zero if there is no voltage source in that branch; $V_e(0^+)$ is the e-vector of capacitor initial voltages such that the jth element of the vector is equal to the capacitor initial voltage $V_{c_j}(0^+)$ and is equal to zero if there is no capacitor in the jth branch or if the initial condition of the capacitor in that branch is zero; R_e is a diagonal matrix of order e such that the resistance R_j in the jth branch is the (j,j) element of R_e; L_e is a symmetric matrix of order e such that the self-inductance L_j in the jth branch is the (j,j) element and the mutual inductance M_{ij} (which is equal to M_{ji} for all i and j) between branch i and branch j is the (i,j) element of L_e; and, finally, D_e is the diagonal matrix of order e such that the *inverse*, $1/C$, of the capacitance C_j in branch j is the (j,j) element of the matrix.

The three matrix equations — (4-7), (4-8), and (4-14) — are the three postulates we shall use in the development of the loop-analysis and node-analysis systems of equations. These three matrix equations constitute the *primary* systems of network equations in the loop analysis and the node analysis.

We should emphasize at this point that the three postulates we have just introduced are matrix equations in the time domain; that is, all the variables in the three matrix equations are functions of time t.

If we now take the Laplace transform of these equations, we get

$$BV_e(s) = 0 \tag{4-15}$$

$$AI_e(s) = 0 \tag{4-16}$$

and

$$V_e(s) = E_e(s) + R_eI_e(s) + L_e\{sI_e(s) - I_e(0^+)\} + D_e\left\{\frac{1}{s}I_e(s)\right\} + \frac{1}{s}V_e(0^+) \tag{4-17}$$

keeping in mind that, from the Laplace transform theory, for any function $f(t)$ which is Laplace transformable with its Laplace transform $\mathcal{L}[f(t)]$ denoted by $F(s)$, we have the following theorems:

$$\mathcal{L}\left[\frac{d}{dt}f(t)\right] = sF(s) - f(0^+)$$

$$\mathcal{L}\left[\int_0^t f(t)\,dt\right] = \frac{1}{s}F(s)$$

and

$$\mathcal{L}[V_e(0^+)] = \frac{1}{s}V_e(0^+)$$

where $V_e(0^+)$ is a constant vector and

$$f(0^+) \triangleq \lim_{t\to 0^+} f(t)$$

is the limit of $f(t)$ as $t \to 0$ from the right-hand side (that is, from a positive value).

Thus it is obvious that in Equation (4-17), $I_e(0^+)$ is the e-vector of initial currents in the inductors with a nonzero entry $i_j(0^+)$ in the jth position (j = 1, 2, . . . , e) *only if* the jth edge of the graph represents an inductor with an initial current $i_j(0^+)$, which is positive if it is in the same direction as the edge orientation and negative otherwise. Similarly, $V_e(0^+)$ is the e-vector of initial voltages in the capacitors with a nonzero entry $v_k(0^+)$ in the $(k,1)$ position (k = 1, 2, . . . , e) *only if* the kth edge of the graph contains a capacitor with an initial voltage $v_k(0^+)$, which is positive if its polarity agrees with the edge orientation and negative otherwise.

Equation (4-17) may be further simplified as

$$V_e(s) = E_e(s) + \left\{R_e + sL_e + \frac{1}{s}D_e\right\}I_e(s) + \frac{1}{s}V_e(0^+) - L_eI_e(0^+)$$

which may be written in a more compact form as

$$V_e(s) = E_e(s) + Z_e(s)I_e(s) + \frac{1}{s} V_e(0^+) - L_e I_e(0^+) \tag{4-18}$$

where

$$Z_e(s) \triangleq R_e + sL_e + \frac{1}{s} D_e$$

Equations (4-15), (4-16), and (4-18) are the Laplace transforms of the three postulates and are called the primary systems of transformed equations.

EXAMPLE 4-1. Consider the network N of Figure 4-4(a) and its (oriented) graph G as shown in Figure 4-4(b). The orientations of the edges and of the circuits in G are completely arbitrary.

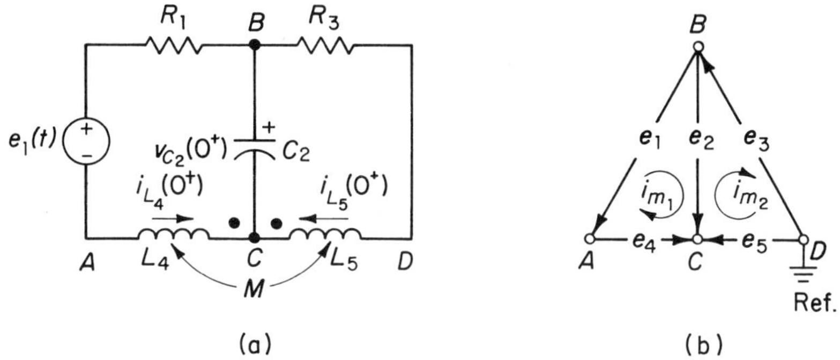

Figure 4-4 (a) A network N. (b) Its graph G.

The KVL matrix equation, $BV_e(t) = 0$, is

$$\begin{matrix} 1 & 2 & 3 & 4 & 5 \end{matrix}$$
$$\begin{bmatrix} -1 & 1 & 0 & -1 & 0 \\ 0 & -1 & -1 & 0 & 1 \end{bmatrix} \cdot \begin{bmatrix} v_1(t) \\ v_2(t) \\ v_3(t) \\ v_4(t) \\ v_5(t) \end{bmatrix} = 0 \tag{4-19}$$

and the KCL matrix equation, $AI_e(t) = 0$, is

$$\begin{matrix} & 1 & 2 & 3 & 4 & 5 \end{matrix}$$
$$\begin{matrix} A \\ B \\ C \end{matrix} \begin{bmatrix} -1 & 0 & 0 & 1 & 0 \\ 1 & 1 & -1 & 0 & 0 \\ 0 & -1 & 0 & -1 & -1 \end{bmatrix} \cdot \begin{bmatrix} i_1(t) \\ i_2(t) \\ i_3(t) \\ i_4(t) \\ i_5(t) \end{bmatrix} = 0 \tag{4-20}$$

The VCR matrix equation may be obtained by first writing the VCR equation for each branch and then combining the five equations into a matrix equation. Thus,

$$
\begin{aligned}
v_1(t) &= e_1(t) + R_1 i_1(t) \\
v_2(t) &= \frac{1}{C_2}\int_0^t i_2(t)\,dt + V_{C_2}(0^+) \\
v_3(t) &= R_3 i_3(t) \\
v_4(t) &= L_4 \frac{d}{dt} i_4(t) + M \frac{d}{dt} i_5(t) \\
v_5(t) &= M \frac{d}{dt} i_4(t) + L_5 \frac{d}{dt} i_5(t)
\end{aligned}
\tag{4-21}
$$

Putting Equation (4-21) in matrix form,

$$
V_e(t) = E_e(t) + R_e I_e(t) + L_e \frac{d}{dt} I_e(t) + D_e \int_0^t I_e(t)\,dt + V_e(0^+) \tag{4-22}
$$

we have

$$
\begin{bmatrix} v_1 \\ v_2 \\ v_3 \\ v_4 \\ v_5 \end{bmatrix} = \begin{bmatrix} e_1 \\ 0 \\ 0 \\ 0 \\ 0 \end{bmatrix} + \begin{bmatrix} R_1 & 0 & 0 & 0 & 0 \\ 0 & 0 & 0 & 0 & 0 \\ 0 & 0 & R_3 & 0 & 0 \\ 0 & 0 & 0 & 0 & 0 \\ 0 & 0 & 0 & 0 & 0 \end{bmatrix} \begin{bmatrix} i_1 \\ i_2 \\ i_3 \\ i_4 \\ i_5 \end{bmatrix} + \begin{bmatrix} 0 & 0 & 0 & 0 & 0 \\ 0 & 0 & 0 & 0 & 0 \\ 0 & 0 & 0 & 0 & 0 \\ 0 & 0 & 0 & L_4 & M \\ 0 & 0 & 0 & M & L_5 \end{bmatrix} \frac{d}{dt} \begin{bmatrix} i_1 \\ i_2 \\ i_3 \\ i_4 \\ i_5 \end{bmatrix}
$$

$$
+ \begin{bmatrix} 0 & 0 & 0 & 0 & 0 \\ 0 & 1/C_2 & 0 & 0 & 0 \\ 0 & 0 & 0 & 0 & 0 \\ 0 & 0 & 0 & 0 & 0 \\ 0 & 0 & 0 & 0 & 0 \end{bmatrix} \int_0^t \begin{bmatrix} i_1 \\ i_2 \\ i_3 \\ i_4 \\ i_5 \end{bmatrix} dt + \begin{bmatrix} 0 \\ V_{C_2}(0^+) \\ 0 \\ 0 \\ 0 \end{bmatrix} \tag{4-23}
$$

where all the v_j's and i_j's and the voltage source e_1 are functions of time t, and the differentiation and integration of the current vector signify the corresponding operations in the elements of the current vector.

Sometimes it is convenient to define the operators, p and $1/p$, as

$$
p \triangleq \frac{d}{dt} \quad \text{and} \quad \frac{1}{p} \triangleq \int_0^t dt \tag{4-24}
$$

which, when operated on any function $f(t)$ will yield

$$
pf(t) \equiv \frac{d}{dt} f(t)
$$

and

$$
\frac{1}{p} f(t) \equiv \int_0^t f(t)\,dt
$$

so that $Z_e(p)$ may be defined as

$$Z_e(p) \triangleq R_e + pL_e + \frac{1}{p} D_e \tag{4-25}$$

Then Equation (4-23) may be rewritten as

$$\begin{bmatrix} v_1 \\ v_2 \\ v_3 \\ v_4 \\ v_5 \end{bmatrix} = \begin{bmatrix} e_1 \\ 0 \\ 0 \\ 0 \\ 0 \end{bmatrix} + \begin{bmatrix} R_1 & 0 & 0 & 0 & 0 \\ 0 & 1/C_2p & 0 & 0 & 0 \\ 0 & 0 & R_3 & 0 & 0 \\ 0 & 0 & 0 & L_4p & Mp \\ 0 & 0 & 0 & Mp & L_5p \end{bmatrix} \begin{bmatrix} i_1 \\ i_2 \\ i_3 \\ i_4 \\ i_5 \end{bmatrix} + \begin{bmatrix} 0 \\ V_{C_2}(0^+) \\ 0 \\ 0 \\ 0 \end{bmatrix} \tag{4-26}$$

or, in shorthand matrix notation,

$$V_e(t) = E_e(t) + Z_e(p)\, I_e(t) + V_e(0^+) \tag{4-27}$$

We should emphasize that $Z_e(p)$, as defined in Equation (4-25), is *not* an impedance function but a function of the operator p, because up to now in this example we have been dealing with equations in the time domain; whereas, impedances are functions of the complex frequency variable s.

To obtain the transformed KVL, KCL, and VCR matrix equations, we simply take the Laplace transform of Equations (4-19), (4-20) and (4-23). Thus, for the KVL equation, $BV_e(s) = 0$, we get

$$\begin{bmatrix} -1 & 1 & 0 & -1 & 0 \\ 0 & -1 & -1 & 0 & 1 \end{bmatrix} \begin{bmatrix} V_1(s) \\ V_2(s) \\ V_3(s) \\ V_4(s) \\ V_5(s) \end{bmatrix} = 0 \tag{4-28}$$

for the KCL equation, $AI_e(s) = 0$, we get

$$\begin{bmatrix} -1 & 0 & 0 & 1 & 0 \\ 1 & 1 & -1 & 0 & 0 \\ 0 & -1 & 0 & -1 & -1 \end{bmatrix} \begin{bmatrix} I_1(s) \\ I_2(s) \\ I_3(s) \\ I_4(s) \\ I_5(s) \end{bmatrix} = 0 \tag{4-29}$$

and, for the VCR equation,

$$V_e(s) = E_e(s) + Z_e(s)I_e(s) + \frac{1}{s} V_e(0^+) - L_eI_e(0^+)$$

we get

$$\begin{bmatrix} V_1(s) \\ V_2(s) \\ V_3(s) \\ V_4(s) \\ V_5(s) \end{bmatrix} = \begin{bmatrix} E_1(s) \\ 0 \\ 0 \\ 0 \\ 0 \end{bmatrix} + \begin{bmatrix} R_1 & 0 & 0 & 0 & 0 \\ 0 & 1/sC_2 & 0 & 0 & 0 \\ 0 & 0 & R_3 & 0 & 0 \\ 0 & 0 & 0 & sL_4 & sM \\ 0 & 0 & 0 & sM & sL_5 \end{bmatrix} \begin{bmatrix} I_1(s) \\ I_2(s) \\ I_3(s) \\ I_4(s) \\ I_5(s) \end{bmatrix}$$

$$+\begin{bmatrix} 0 \\ \frac{1}{s}V_{C_2}(0^+) \\ 0 \\ 0 \\ 0 \end{bmatrix} - \begin{bmatrix} 0 & 0 & 0 & 0 & 0 \\ 0 & 0 & 0 & 0 & 0 \\ 0 & 0 & 0 & 0 & 0 \\ 0 & 0 & 0 & L_4 & M \\ 0 & 0 & 0 & M & L_5 \end{bmatrix} \begin{bmatrix} 0 \\ 0 \\ 0 \\ I_{L_4}(0^+) \\ I_{L_5}(0^+) \end{bmatrix} \tag{4-30}$$

4-3 SECONDARY SYSTEM OF EQUATIONS

In Section 4-2, we introduced the three postulates — the KVL, the KCL, and the VCR matrix equations — in both the time domain and the frequency domain. The three matrix equations are called the *primary systems* of equations because together they constitute a necessary and sufficient set of equations for the complete solution of a given network.

Note that, for any connected network N with e branches and v nodes, we have $(e - v + 1)$ KVL equations, $(v - 1)$ KCL equations, and e VCR equations. But, the number of unknowns for N is equal to $2e$ (e branch voltages and e branch currents). If we sum up the number of equations, we have

$$(e - v + 1) + (v - 1) + e = 2e$$

which is exactly equal to the number of unknowns.

Furthermore, from our discussions on the rank of B and A in Chapter 2 and the way a VCR equation is formed, it is evident that all of the $2e$ equations are independent ones. Therefore, there exists a unique solution for the $2e$ unknowns and, consequently, we conclude that the KVL, the KCL, and the VCR matrix equations together form a necessary and sufficient set of network equations for the complete solution of N as previously asserted.

In the discussions that follow, we shall derive the so-called *secondary matrix equations*, which serve as auxiliary equations in the formulation of the loop-analysis and the node-analysis systems.

In Chapter 2 we learned that, for a connected graph of e edges and v vertices, if A and B are the incidence submatrix and the circuit submatrix, respectively, with their corresponding columns arranged in the same order, we have

$$AB^T = 0 \tag{4-31}$$

or

$$BA^T = 0 \tag{4-32}$$

Also, in Section 4-2, we have

$$AI_e(t) = 0 \tag{4-8}$$

and

$$BV_e(t) = 0 \tag{4-7}$$

Comparing Equation (4-7) with (4-32), it seems reasonable to expect a linear relationship between the two quantities $V_e(t)$ and A^T, namely,

$$V_e(t) = KA^T \tag{4-33}$$

where K denotes a matrix operator or transformation of some kind that transforms A^T into the vector $V_e(t)$. But, if we examine the orders of the two matrices, we see at once that Equation (4-33) is incorrect because $V_e(t)$ is of order $e \times 1$ and A^T is of order $e \times (v - 1)$, and it is impossible to find matrix K of any order so that the product KA^T be of order $e \times 1$ as demanded by Equation (4-33). However, if we write

$$V_e(t) = A^TK \tag{4-34}$$

and let K be a matrix of order $(v - 1) \times 1$, we see that the order of the product A^TK is indeed $e \times 1$, which is the order of $V_e(t)$. Thus matrix A^T is conformable to K for multiplication and the product A^TK in Equation (4-34) exists.

Then, our next question: What kind of unit shall we attach to this column matrix K?

The answer to this question is quite simple. We note that $V_e(t)$ has a unit of volts, being a vector of branch voltages. Furthermore, we observe that A^T is a matrix of pure constants and, hence, is dimensionless. It requires, therefore, that K be a vector of voltages of some kind such that the product A^TK also has a unit of volts. Thus we rewrite Equation (4-34) as

$$\begin{bmatrix} v_1(t) \\ v_2(t) \\ \cdot \\ \cdot \\ \cdot \\ v_e(t) \end{bmatrix}_{e\times 1} = \begin{bmatrix} \\ \\ A^T \\ \\ \\ \end{bmatrix}_{e\times R} \begin{bmatrix} v_{n_1}(t) \\ v_{n_2}(t) \\ \cdot \\ \cdot \\ \cdot \\ v_{n_R}(t) \end{bmatrix}_{R\times 1} \tag{4-35}$$

where $R = v - 1$ is the rank of G; or, in compact form,

$$V_e(t) = A^TV_n(t) \tag{4-36}$$

As we shall show in our next example, the elements in the matrix $V_n(t)$ correspond to the $v - 1$ node-to-datum voltages with v_{n_j} being the voltage between node j and the reference node with the plus polarity mark at node j. For this reason, $V_n(t)$ is called the vector of node voltages and Equation (4-36) is referred to as the *node-transformation* matrix equation.

Following a development parallel to the one discussed above, one can derive a similar expression for the *mesh-transformation* matrix equation

$$\begin{bmatrix} i_1(t) \\ i_2(t) \\ \cdot \\ \cdot \\ \cdot \\ i_e(t) \end{bmatrix}_{e\times 1} = \begin{bmatrix} \\ \\ B^T \\ \\ \\ \end{bmatrix}_{e\times N} \cdot \begin{bmatrix} i_{m_1}(t) \\ i_{m_2}(t) \\ \cdot \\ \cdot \\ \cdot \\ i_{m_N}(t) \end{bmatrix}_{N\times 1} \tag{4-37}$$

where $N = e - v + 1$ is the nullity of G; or, in compact form,

$$I_e(t) = B^T I_m(t) \tag{4-38}$$

where $I_m(t)$ is the vector of loop currents of order $(e - v + 1) \times 1$. The derivation of Equation (4-37) is obvious and is left to the reader as a problem.

EXAMPLE 4-2. Let us consider the network N of Figure 4-1(a). From Example 4-1, the matrices B and A are known so that the transposes B^T and A^T are:

$$B^T = \begin{bmatrix} -1 & 0 \\ 1 & -1 \\ 0 & -1 \\ -1 & 0 \\ 0 & 1 \end{bmatrix} \quad \text{and} \quad A^T = \begin{bmatrix} -1 & 1 & 0 \\ 0 & 1 & -1 \\ 0 & -1 & 0 \\ 1 & 0 & -1 \\ 0 & 0 & -1 \end{bmatrix}$$

Thus the node transformation, $V_e(t) = A^T V_n(t)$, is

$$\begin{bmatrix} v_1(t) \\ v_2(t) \\ v_3(t) \\ v_4(t) \\ v_5(t) \end{bmatrix} = \begin{bmatrix} -1 & 1 & 0 \\ 0 & 1 & -1 \\ 0 & -1 & 0 \\ 1 & 0 & -1 \\ 0 & 0 & -1 \end{bmatrix} \begin{bmatrix} v_A(t) \\ \\ v_B(t) \\ \\ v_C(t) \end{bmatrix} \tag{4-39}$$

and the mesh transformation, $I_e(t) = B^T I_m(t)$, is

$$\begin{bmatrix} i_1(t) \\ i_2(t) \\ i_3(t) \\ i_4(t) \\ i_5(t) \end{bmatrix} = \begin{bmatrix} -1 & 0 \\ 1 & -1 \\ 0 & -1 \\ -1 & 0 \\ 0 & 1 \end{bmatrix} \begin{bmatrix} \\ i_{m_1}(t) \\ \\ i_{m_2}(t) \\ \\ \end{bmatrix} \tag{4-40}$$

The network is repeated in Figure 4-5(a) and its graph G is shown in Figure 4-5(b) with various quantities defined in the graph. If the reader carries out the matrix multiplications in Equations (4-39) and (4-40), he will immediately realize that Equation (4-39) corresponds to the set of equations that express the branch voltages in terms of the node-pair voltages with node D chosen

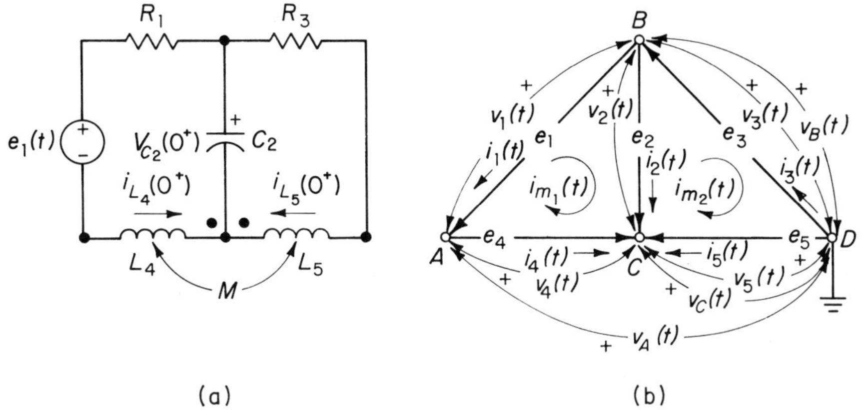

Figure 4-5 (a) The network N. (b) Its graph G of Figure 4-1.

as the reference node. Similarly, Equation (4-40) corresponds to the set of equations that express the branch currents in terms of the mesh (or loop) currents [see Figure 4-5(b)].

In conclusion, Equations (4-7), (4-8), and (4-14) together constitute the primary systems of matrix equations, and Equations (4-36) and (4-38) form the secondary systems of matrix equations. The transforms of these five matrix equations provide the bases for the formulation of the loop-analysis and the node-analysis matrix equations.

4-4 THE LOOP EQUATIONS

Consider the following matrix equations:

$$BV_e(t) = 0 \tag{4-41}$$

$$V_e(t) = E_e(t) + R_eI_e(t) + L_e \frac{d}{dt} I_e(t) + D_e \int_0^t I_e(t)\, dt + V_e(0^+) \tag{4-42}$$

where B equals the circuit submatrix of order $(e - v + 1) \times e$

$$V_e(t) = \begin{bmatrix} V_1(t) \\ V_2(t) \\ \cdot \\ \cdot \\ \cdot \\ V_e(t) \end{bmatrix} \qquad I_e(t) = \begin{bmatrix} i_1(t) \\ i_2(t) \\ \cdot \\ \cdot \\ \cdot \\ i_e(t) \end{bmatrix}$$

$E_e(t)$ = column matrix of order $e \times 1$ with a zero entry if the corresponding edge does not have a source, and a nonzero entry if the corresponding edge does have a source, in which case the entry has a value equal to the source value

$V_e(0^+)$ = column matrix of order $e \times 1$ with a zero entry if the corresponding edge is not a capacitor, or with a nonzero entry if the corresponding edge is a capacitor, in which case the entry has a value equal to the initial capacitor voltage (which may be zero)

R_e = a square matrix of order e with zeros for all off-diagonal entries and $r_{ii} = R_i$ which is the resistance of edge i (it is, therefore, a diagonal matrix)

D_e = a square matrix of order e with zeros for all off-diagonal entries and $d_{ii} = 1/C_i$ where C_i is the capacitance of edge i (and, hence, a diagonal matrix also)

L_e = a square matrix of order e with the (i,i) element equal to L_i: $1_{ii} = L_i$, which is the inductance of edge i, and $1_{ij} = M_{ij}$, which is the mutual inductance between edges i and j.

Now, taking the Laplace transform of Equations (4-41) and (4-42), we get

$$BV_e(s) = 0 \tag{4-43}$$

and

$$V_e(s) = E_e(s) + R_eI_e(s) + L_e\{sI_e(s) - I_e(0^+)\} + D_e\frac{1}{s}I_e(s) + \frac{1}{s}V_e(0^+)$$

or

$$V_e(s) = E_e(s) + \left\{R_e + sL_e + \frac{1}{s}D_e\right\}I_e(s) + \frac{1}{s}V_e(0^+) - L_eI_e(0^+)$$

which may be written in a more compact form,

$$V_e(s) = E_e(s) + Z_e(s)I_e(s) + \frac{1}{s}V_e(0^+) - L_eI_e(0^+) \tag{4-44}$$

where

$$Z_e(s) \triangleq R_e + sL_e + \frac{1}{s}D_e$$

Substituting Equation (4-44) into (4-43), we get

$$BV_e(s) = B\left\{E_e(s) + Z_e(s)I_e(s) + \frac{1}{s}V_e(0^+) - L_eI_e(0^+)\right\} = 0$$

Rearranging the terms, we have

$$BZ_e(s)I_e(s) = -B\left\{E_e(s) + \frac{1}{s}V_e(0^+) - LI_e(0^+)\right\} \tag{4-45}$$

Making use of the mesh-transformation matrix equation,

$$I_e(s) = B^TI_m(s)$$

Equation (4-45) may be written as

$$BZ_e(s)B^T I_m(s) = -B\left\{E_e(s) + \frac{1}{s}V_e(0^+) - L_e I_e(0^+)\right\}$$

or

$$Z_m(s)I_m(s) = -B\left\{E_e(s) + \frac{1}{s}V_e(0^+) - L_e I_e(0^+)\right\} \tag{4-46}$$

where $Z_m(s) \triangleq BZ_e(s)B^T$ is called the *loop impedance matrix*. Equation (4-46) is known as the loop-analysis equation, which gives a system of $e - v + 1$ independent loop equations, and may be written in a more concise form as

$$Z_m(s)I_m(s) = E_m(s) + \frac{1}{s}E_c(0^+) + E_\lambda(0^+) \tag{4-47}$$

where

$E_m(s) = -BE_e(s)$, the voltage rises around the loops
$E_c(0^+) = -BV_e(0^+)$, the loop-voltage rises due to initial capacitance changes
$E_\lambda(0^+) = +BL_e I_e(0^+)$, the loop-voltage rise due to initial inductance currents

when the loops are traced in the same direction as the assigned positive directions of the loop currents.

It is worthwhile to take a little time to investigate the unit of each of the terms in Equation (4-47). We note that the transform $V(s)$ of any voltage $v(t)$ has the unit of volt-seconds. Similarly, the transform $I(s)$ of any current $i(t)$ has the unit of ampere-seconds. Consequently, the impedance $Z(s)$, which is the ratio of $V(s)$ to $I(s)$, must have a unit of volt-seconds/ampere-seconds, or ohms. Thus, on the left-hand side of Equation (4-47), $Z_m(s)I_m(s)$ has the unit of

$$\text{ohms} \times (\text{ampere} - \text{seconds}) = \frac{\text{volts}}{\text{ampere}}(\text{ampere} - \text{seconds}) = \text{volt-seconds}$$

On the right-hand side of Equation (4-47), the first term

$$E_m(s) = -BE_e(s)$$

is evidently in volt-seconds since B is dimensionless. Next, consider the second term on the right-hand side,

$$\frac{1}{s}E_c(0^+) = -\frac{1}{s}BV_e(0^+)$$

The quantity $V_e(0^+)$ is the column matrix of initial capacitance voltages in the branches and, hence, has the unit of volts. But the variable s has the unit of seconds^{-1} and B is dimensionless. Therefore, $(1/s)E_c(0^+)$ is in volt-seconds, and $E_c(0^+)$ has the same unit as $V_e(0^+)$ except that it represents the algebraic sum of the initial capacitance voltages around the loops. Finally, we represent the third term as

$$E_\lambda(0^+) \equiv BL_eI_e(0^+)$$

which has a unit of

$$\frac{\text{volt-seconds}}{\text{ampere}} \times \text{amperes} = \text{volt-seconds}$$

noting again that B is dimensionless. Thus we conclude that Equation (4-47) is dimensionally correct.

The subscripts c and λ and the symbol 0^+ in $E_c(0^+)$ and $E_\lambda(0^+)$ serve to remind us of the fact that $E_c(0^+)$ is due to the initial voltages in the capacitors, and $E_\lambda(0^+)$ is due to the initial currents in the inductors in the loops.

We shall next outline in steps the procedure for deriving the loop-analysis matrix equation for a given network.

(a) Write the KVL equation:

$$BV_e(s) = 0$$

(b) Write the VCR equation:

$$V_e(s) = E_e(s) + Z_e(s)I_e(s) + \frac{1}{s}V_e(0^+) - L_eI_e(0^+)$$

(c) From (b), write:

$$BZ_e(s)I_e(s) = -B\left\{E_e(s) + \frac{1}{s}V_e(0^+) - L_eI_e(0^+)\right.$$

(d) Defining

$$Z_m(s) \equiv BZ_e(s)B^T$$

and noting that

$$I_e(s) = B^TI_m(s)$$

write, from (c),

$$Z_m(s)I_m(s) = -B\left\{E_e(s) + \frac{1}{s}V_e(0^+) - L_eI_e(0^+)\right\}$$

or

$$Z_m(s)I_m(s) = E_m(s) + \frac{1}{s}E_c(0^+) + E_\lambda(0^+)$$

EXAMPLE 4-3. Refer again to the network N of Figure 4-1(a). To obtain the transformed loop-analysis matrix equation, we write, as the first step,

(a) $BV_e(s) = 0$

$$\begin{bmatrix} -1 & 1 & 0 & -1 & 0 \\ 0 & -1 & -1 & 0 & 1 \end{bmatrix} \begin{bmatrix} V_1(s) \\ V_2(s) \\ V_3(s) \\ V_4(s) \\ V_5(s) \end{bmatrix} = 0 \tag{4-48}$$

(b) Next we write the transformed VCR matrix equation,

$$V_e(s) = E_e(s) + Z_e(s)I_e(s) + \frac{1}{s}V_e(0^+) - L_eI_e(0^+)$$

$$\underbrace{\begin{bmatrix} V_1(s) \\ V_2(s) \\ V_3(s) \\ V_4(s) \\ V_5(s) \end{bmatrix}}_{V_e(s)} = \underbrace{\begin{bmatrix} E_1(s) \\ 0 \\ 0 \\ 0 \\ 0 \end{bmatrix}}_{E_e(s)} + \underbrace{\begin{bmatrix} R_1 & 0 & 0 & 0 & 0 \\ 0 & 1/sC_2 & 0 & 0 & 0 \\ 0 & 0 & R_3 & 0 & 0 \\ 0 & 0 & 0 & sL_4 & sM \\ 0 & 0 & 0 & sM & sL_5 \end{bmatrix}}_{Z_e(s)} \underbrace{\begin{bmatrix} I_1(s) \\ I_2(s) \\ I_3(s) \\ I_4(s) \\ I_5(s) \end{bmatrix}}_{I_e(s)}$$

$$+ \underbrace{\begin{bmatrix} 0 \\ \frac{1}{s}V_{C_2}(0^+) \\ 0 \\ 0 \\ 0 \end{bmatrix}}_{\frac{1}{s}V_e(0)} - \underbrace{\begin{bmatrix} 0 & 0 & 0 & 0 & 0 \\ 0 & 0 & 0 & 0 & 0 \\ 0 & 0 & 0 & 0 & 0 \\ 0 & 0 & 0 & L_4 & M \\ 0 & 0 & 0 & M & L_5 \end{bmatrix}}_{L_e} \underbrace{\begin{bmatrix} 0 \\ 0 \\ 0 \\ I_{L_4}(0^+) \\ I_{L_5}(0^+) \end{bmatrix}}_{I_e(0^+)} \tag{4-49}$$

(c) Substituting Equation (4-49) into (4-48) and rearranging, we get

$$BZ_e(s)I_e(s) = -B\left\{E_e(s) + \frac{1}{s}V_e(0^+) - L_eI_e(0^+)\right\}$$

$$\underbrace{\begin{bmatrix} -1 & 1 & 0 & -1 & 0 \\ 0 & -1 & -1 & 0 & 1 \end{bmatrix}}_{B} \underbrace{\begin{bmatrix} R_1 & 0 & 0 & 0 & 0 \\ 0 & 1/sC_2 & 0 & 0 & 0 \\ 0 & 0 & R_3 & 0 & 0 \\ 0 & 0 & 0 & sL_4 & sM \\ 0 & 0 & 0 & sM & sL_5 \end{bmatrix}}_{Z_e(s)} \underbrace{\begin{bmatrix} I_1(s) \\ I_2(s) \\ I_3(s) \\ I_4(s) \\ I_5(s) \end{bmatrix}}_{I_e(s)}$$

$$= \underbrace{\begin{bmatrix} 1 & -1 & 0 & 1 & 0 \\ 0 & 1 & 1 & 0 & -1 \end{bmatrix}}_{-B} \left\{ \underbrace{\begin{bmatrix} E_1(s) \\ 0 \\ 0 \\ 0 \\ 0 \end{bmatrix}}_{E_e(s)} + \underbrace{\begin{bmatrix} 0 \\ \frac{1}{s}V_{C_2}(0) \\ 0 \\ 0 \\ 0 \end{bmatrix}}_{\frac{1}{s}V_e(0^+)} \right.$$

$$
- \underbrace{\begin{bmatrix} 0 \\ 0 \\ 0 \\ L_4 I_{L_4}(0^+) + M I_{L_5}(0^+) \\ L_5 I_{L_5}(0^+) + M I_{L_4}(0^+) \end{bmatrix}}_{L_e I_e(0^+)} \qquad (4\text{-}50)
$$

(d) Noting that

$$Z_m(s) = B Z_e(s) B^T$$

$$
Z_m(s) = \begin{bmatrix} -1 & 1 & 0 & -1 & 0 \\ 0 & -1 & -1 & 0 & 1 \end{bmatrix}
\times \begin{bmatrix} R_1 & 0 & 0 & 0 & 0 \\ 0 & 1/sC_2 & 0 & 0 & 0 \\ 0 & 0 & R_3 & 0 & 0 \\ 0 & 0 & 0 & sL_4 & sM \\ 0 & 0 & 0 & sM & sL_5 \end{bmatrix}
\begin{bmatrix} -1 & 0 \\ 1 & -1 \\ 0 & -1 \\ -1 & 0 \\ 0 & 1 \end{bmatrix}
$$

$$
= \begin{bmatrix} -R_1 & \dfrac{1}{sC_2} & 0 & -sL_4 & -sM \\ 0 & -\dfrac{1}{sC_2} & -R_3 & sM & sL_5 \end{bmatrix}
\begin{bmatrix} -1 & 0 \\ 1 & -1 \\ 0 & -1 \\ -1 & 0 \\ 0 & 1 \end{bmatrix}
$$

or

$$
Z_m(s) = \begin{bmatrix} \left(R_1 + sL_4 + \dfrac{1}{sC_2}\right) & -\left(\dfrac{1}{sC_2} + sM\right) \\ -\left(\dfrac{1}{sC_2} + sM\right) & \left(R_3 + sL_5 + \dfrac{1}{sC_2}\right) \end{bmatrix}
$$

We have the transformed loop-analysis matrix equation:

$$
\begin{aligned}
Z_m(s) I_m(s) &= -B \left\{ E_e(s) + \frac{1}{s} V_e(0^+) - L_e I_e(0^+) \right\} \\
&= E_m(s) + \frac{1}{s} E_c(0^+) + E_\lambda(0^+)
\end{aligned}
$$

$$
\underbrace{\begin{bmatrix} \left(R_1 + sL_4 + \dfrac{1}{sC_2}\right) & -\left(\dfrac{1}{sC_2} + sM\right) \\ -\left(\dfrac{1}{sC_2} + sM\right) & \left(R_3 + sL_5 + \dfrac{1}{sC_2}\right) \end{bmatrix}}_{Z_m(s)}
\underbrace{\begin{bmatrix} I_{m_1}(s) \\ I_{m_2}(s) \end{bmatrix}}_{I_m(s)}
$$

$$
= \underbrace{\begin{bmatrix} 1 & -1 & 0 & 1 & 0 \\ 0 & 1 & 1 & 0 & -1 \end{bmatrix}}_{-B}
$$

$$
\times \underbrace{\begin{bmatrix} E_1(s) \\ \frac{1}{s}V_{C_2}(0^+) \\ 0 \\ -L_4 I_{L_4}(0^+) - M I_{L_5}(0^+) \\ -M I_{L_4}(0^+) - L_5 I_{L_5}(0^+) \end{bmatrix}}_{\left\{E_e(s) + \frac{1}{s}V_e(0^+) - L_e I_e(0^+)\right\}}
$$

$$
= \underbrace{\begin{bmatrix} E_1(s) - \frac{1}{s}V_{C_2}(0^+) - L_4 I_{L_4}(0^+) - M I_{L_5}(0^+) \\ \frac{1}{s}V_{C_2}(0^+) + M I_{L_4}(0^+) + L_5 I_{L_5}(0^+) \end{bmatrix}}_{\left\{E_m(s) + \frac{1}{s}E_c(0^+) + E_\lambda(0^+)\right\}}
$$

which yields the following two loop equations:

$$
\left(R_1 + sL_4 + \frac{1}{sC_2}\right)I_{m_1} - \left(\frac{1}{sC_2} + sM\right)I_{m_2} = E_1(s) - \frac{1}{s}V_{C_2}(0^+) - \{L_4 I_{L_4}(0^+) + M I_{L_5}(0^+)\} \tag{4-51}
$$

$$
-\left(\frac{1}{sC_2} + sM\right)I_{m_1} + \left(R_3 + sL_5 + \frac{1}{sC_2}\right)I_{m_2} = \frac{1}{s}V_{C_2}(0^+) + M I_{L_4}(0^+) + L_5 I_{L_5}(0^+) \tag{4-52}
$$

The reader should check these equations, obtaining them by using the conventional method of loop analysis, which was taught in the first network analysis course.

It should be emphasized that the importance of the development leading to the formation of the loop-analysis matrix equation (4-48) lies in the fact that it gives a rigorous mathematical basis for developing the theory of the loop analysis, and it answers the questions concerning the number of loop equations that are necessary and sufficient for the complete solution of a given network. It also explains how the loop equations (which correspond to the so-called windowpanes in the case of a planar network) are formulated. Once the development of the theory is fully understood, it will be foolish to obtain the loop equations by actually going through all the steps listed in the foregoing procedure. In fact, the reader should be able to obtain these loop equations even by inspection, provided he has acquired sufficient experience after working a number of problems.

4-5 THE NODE EQUATIONS

As the parallel to the loop equations, consider the KCL matrix equation

$$AI_e(t) = 0$$

which has the Laplace transform

$$AI_e(s) = 0 \tag{4-53}$$

Also consider in the Laplace transform of the VCR matrix equation

$$V_e(s) = E_e(s) + Z_e(s)I_e(s) + \frac{1}{s}V_e(0^+) - L_eI_e(0^+) \tag{4-54}$$

Our next step is to express $I_e(s)$ explicitly by solving Equation (4-54) for $I_e(s)$. Therefore, we write

$$Z_e(s)I_e(s) = V_e(s) - E_e(s) + L_eI_e(0^+) - \frac{1}{s}V_e(0^+) \tag{4-55}$$

and see that $I_e(s)$ may be obtained by premultiplying Equation (4-55) through by $Z_e^{-1}(s)$ so that

$$I_e(s) = Z_e(s)^{-1}\left\{V_e(s) - E_e(s) + L_eI_e(0^+) - \frac{1}{s}V_e(0^+)\right\} \tag{4-56}$$

provided $Z_e(s)$ is nonsingular [so that $Z^{-1}(s)$ exists]. However, if $Z_e(s)$ is singular [corresponding to the case where some edges are purely voltage or current sources so that there will be one or more rows of zeros in $Z_e(s)$], then Equation (4-56) may not be used to obtain $I_e(s)$. Some other method must be used to overcome this difficulty. But, for the time being, let us assume that $Z_e(s)$ is nonsingular and proceed to find the node equations. We can then come back to the case where $Z_e^{-1}(s)$ does not exist. Substituting Equation (4-56) into (4-53) and rearranging the terms, we get

$$AZ_e^{-1}(s)V_e(s) = AZ_e^{-1}(s)\left\{E_e(s) - L_eI_e(0^+) + \frac{1}{s}V_e(0^+)\right\} \tag{4-57}$$

Now, substituting the transformed node transformation matrix equation (4-36),

$$V_e(s) = A^TV_n(s)$$

into Equation (4-57), we obtain

$$AZ_e^{-1}(s)A^TV_n(s) = AZ_e^{-1}(s)\left\{E_e(s) - L_eI_e(0^+) + \frac{1}{s}V_e(0^+)\right\}$$

or

$$Y_n(s)V_n(s) = AZ_e^{-1}(s)\left\{E_e(s) - L_eI_e(0^+) + \frac{1}{s}V_e(0^+)\right\} \tag{4-58}$$

where

$$Y_n(s) \equiv AZ_e^{-1}(s)A^T$$

is known as the *node admittance matrix.*

Equation (4-58) is the matrix equation for the system of node equations and may be written in a more compact form as

$$Y_n(s)V_n(s) = I_n(s) + \frac{1}{s} I_L(0^+) + Q_c(0^+) \tag{4-59}$$

[as compared to Equation (4-47)] where

$$I_n(s) = AZ_e^{-1}(s)\,E_e(s)$$

the current source at the nodes with positive signs to denote currents flowing toward the nodes;

$$\frac{1}{s} I_L(0^+) = -AZ_e^{-1}(s)L_eI_e(0^+)$$

with $I_L(0^+)$ being the current sources due to initial currents flowing in the inductors with the same sign convention; and

$$Q_c(0^+) = \frac{1}{s} AZ_e^{-1}(s)V_e(0^+)$$

the initial charges across the capacitors, with positive signs denoting the capacitors' initial voltages and with positive polarity marks toward the nodes.

Before discussing the case where $Z_e^{-1}(s)$ does not exist, it may be worthwhile to explain briefly why the second and the third terms on the right-hand side of Equation (4-59) are expressed in such form.

If we consider the term $Z_e^{-1}(s)L_eI_e(0^+)$, we see that the $Z_e^{-1}(s)$ is an admittance function and L_e is an inductance function; thus sL_e will be an impedance, and the product $Z_e^{-1}(s)sL_e$ will be dimensionless. Therefore, we may express this term as

$$-AZ_e^{-1}(s)L_eI_e(0^+) = -AZ_e^{-1}(s)sL_e\frac{I_e(0^+)}{s} = \frac{1}{s} K_1I_e(0^+) \qquad K_1 = -AZ_e^{-1}(s)sL_e$$

where K_1 is a matrix of pure constants. Finally,

$$-AZ_e^{-1}(s)L_eI_e(0^+) = \frac{1}{s} I_L(0^+)$$

where

$$I_L(0^+) = K_1I_e(0^+)$$

In a similar manner, consider the term

$$\frac{1}{s} AZ_e^{-1}(s)V_c(0^+)$$

which may be written as

$$\frac{1}{s}AZ_e^{-1}(s)D_eQ(0^+) = AZ_e^{-1}(s)\frac{1}{s}\,D_eQ(0^+) = K_2Q(0^+)$$

where

$$K_2 = AZ_e^{-1}(s)\frac{1}{s}D_e$$

is a $(v - 1) \times e$ matrix of constants and is dimensionless. Therefore, we write

$$\frac{1}{s}AZ_e^{-1}(s)V_c(0^+) = Q_c(0^+)$$

where

$$Q_c(0^+) = K_2Q(0^+)$$

Now let us come back to the case where $Z_e^{-1}(s)$ does not exist. This corresponds to a network in which some edges are pure current or pure voltage sources. Then we may arrange the $Z_e(s)$ matrix so that those edges of the generators will give rise to the first k rows and k columns of zeros in the new $Z_e(s)$. Thus, we write

$$\left[\begin{array}{c} v_1 \\ v_2 \\ \vdots \\ v_k \\ \hline v_{k+1} \\ v_{k+2} \\ \vdots \\ v_e \end{array}\right] = \left[\begin{array}{c} E_{g1} \\ E_{g2} \\ \vdots \\ E_{gk} \\ \hline E_{k+1} \\ E_{k+2} \\ \vdots \\ E_e \end{array}\right] + \begin{array}{c} k \text{ rows} \\ \\ e-k \text{ rows} \end{array} \overset{\substack{k \quad e-k \\ \text{cols. cols.}}}{\left[\begin{array}{c|c} 0 & 0 \\ \hline 0 & Z_{22} \end{array}\right]} \left[\begin{array}{c} I_1 \\ I_2 \\ \vdots \\ I_k \\ \hline I_{k+1} \\ I_{k+2} \\ \vdots \\ I_e \end{array}\right] + \frac{1}{s}\left[\begin{array}{c} 0 \\ 0 \\ \vdots \\ 0 \\ \hline v_{k+1}(0^+) \\ v_{k+2}(0^+) \\ \vdots \\ v_e(0^+) \end{array}\right] - \left[\begin{array}{c|c} 0 & 0 \\ \hline 0 & L_{22} \end{array}\right] \left[\begin{array}{c} 0 \\ 0 \\ \vdots \\ 0 \\ \hline I_{k+1}(0^+) \\ I_{k+2}(0^+) \\ \vdots \\ I_e(0^+) \end{array}\right] \tag{4-60}$$

This matrix equation may be broken down into two matrix equations:

$$\begin{bmatrix} v_1 \\ v_2 \\ \cdot \\ \cdot \\ \cdot \\ v_k \end{bmatrix} = \begin{bmatrix} E_{g_1} \\ E_{g_2} \\ \cdot \\ \cdot \\ \cdot \\ E_{gk} \end{bmatrix} \tag{4-61}$$

and

$$\begin{bmatrix} v_{k+1} \\ v_{k+2} \\ \cdot \\ \cdot \\ \cdot \\ v_e \end{bmatrix} = \begin{bmatrix} E_{k+1} \\ E_{k+2} \\ \cdot \\ \cdot \\ \cdot \\ E_e \end{bmatrix} + \begin{bmatrix} \\ Z_{22} \\ \\ \end{bmatrix} \begin{bmatrix} I_{k+1} \\ I_{k+2} \\ \cdot \\ \cdot \\ \cdot \\ I_e \end{bmatrix} + \frac{1}{s} \begin{bmatrix} v_{k+1}(0^+) \\ v_{k+2}(0^+) \\ \cdot \\ \cdot \\ \cdot \\ v_e(0^+) \end{bmatrix} - \begin{bmatrix} \\ L_{22} \\ \\ \end{bmatrix} \begin{bmatrix} I_{k+1}(0^+) \\ I_{k+2}(0^+) \\ \cdot \\ \cdot \\ \cdot \\ I_e(0^+) \end{bmatrix} \tag{4-62}$$

or simply

$$V_k(s) = E_k(s) \tag{4-63}$$

and

$$V_{e'}(s) = E_{e'}(s) + Z_{e'}(s)I_{e'}(s) + \frac{1}{s}V_{e'}(0^+) - L_{e'}I_{e'}(0^+) \tag{4-64}$$

where

$$Z_{e'}(s) \equiv Z_{22}(s) \quad \text{and} \quad e' = e - k$$

From Equation (4-64) we may solve for $I_{e'}(s)$ just as we did in Equation (4-56):

$$I_{e'}(s) = Z_{e'}^{-1}(s)\left\{V_{e'}(s) - E_{e'}(s) + L_{e'}I_{e'}(0^+) - \frac{1}{s}V_{e'}(0^+)\right\} \tag{4-65}$$

Note that Equation (4-65) may be obtained simply by substituting the subcript e by e' in (4-56).

Next we shall introduce a transformation which, when applied to the network, will result in a new network with no branches containing voltage generators alone. That is, after such a transformation, each of the voltage generators in the resultant network will be in series with some passive elements (R, L, or C) if there are any voltage generators in the original network.

Consider a branch, say the kth branch, of a network which contains a voltage generator $E_k(s)$ only, as depicted in Figure 4-6(a), where only the pertinent portion of the network N has been shown. If we coalesce the endpoints A and B of the kth branch by "pushing" the generator $E_k(s)$ through node A into the other branches that are connected to A so that each of the other branches contains $E_k(s)$ in series with the element impedance, the resultant network N_a is shown in Figure 4-6(b). Similarly, if we push $E_k(s)$ through node B, another resultant network N_b will be obtained, as shown in

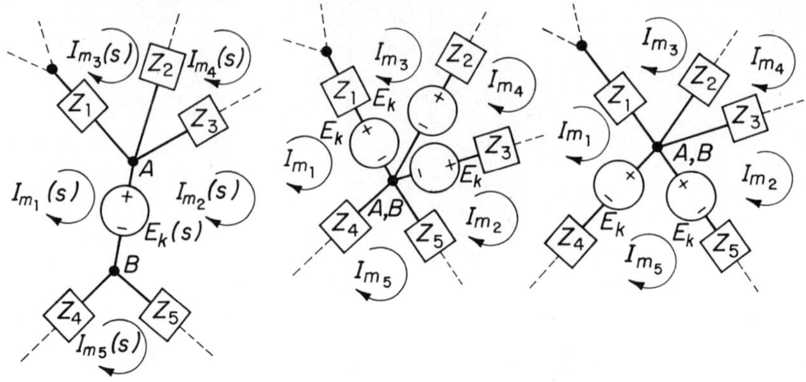

(a) (b) (c)

Figure 4-6 (a) The network N and its equivalent networks (b) N_a and (c) N_b.

Figure 4-6(c). Note the polarity mark of $E_k(s)$ in each branch in the new networks N_a and N_b.

It is evident that both networks, N_a and N_b, are equivalents of the original network N as far as the loop equations are concerned. This can be readily verified by writing the loop equations for each network and then comparing the corresponding equations.

The transformation described is known as the *Blakesley transformation*, or sometimes referred to as the *e-shift operation.*

Now let us come back to our discussion of the node analysis. We assume that all the generators in the first k edges are current generators. If any one of them is a voltage generator, the Blakesley transformation (e-shift operation) may be applied to it so that the difficulty will be overcome, since the voltage generators that appear in the resulting network after the transformation will be each in series with an impedance. Therefore, no generality will be lost by making such an assumption. With this understanding, we express $I_e(s)$ as follows:

$$\begin{bmatrix} I_k(s) \\ I_{e'}(s) \end{bmatrix} = \begin{bmatrix} 0 & 0 \\ 0 & Z_{e'}^{-1}(s) \end{bmatrix} \left\{ \begin{bmatrix} V_k(s) \\ V_{e'}(s) \end{bmatrix} - \begin{bmatrix} 0 \\ E_{e'}(s) \end{bmatrix} + \begin{bmatrix} 0 & 0 \\ 0 & L_{e'} \end{bmatrix} \begin{bmatrix} 0 \\ I_{e'}(0+) \end{bmatrix} - \frac{1}{s} \begin{bmatrix} 0 \\ V_{e'}(0+) \end{bmatrix} \right\} + \begin{bmatrix} I_{gk}(s) \\ 0 \end{bmatrix} \tag{4-66}$$

or, in a more compact form,

$$I_e(s) = Y_e(s) \left\{ V_e(s) - E_e(s) + L_e I_e(0+) - \frac{1}{s} V_e(0+) \right\} + I_g(s) \tag{4-67}$$

where

$$Y_e(s) = \begin{bmatrix} 0 & 0 \\ 0 & Z_{e'}^{-1}(s) \end{bmatrix} = \begin{bmatrix} 0 & 0 \\ 0 & Z_{22}^{-1}(s) \end{bmatrix}$$

and

$$I_g(s) = \begin{bmatrix} I_{g1}(s) \\ I_{g2}(s) \\ \cdot \\ \cdot \\ \cdot \\ I_{gk}(s) \\ 0 \\ 0 \\ \cdot \\ \cdot \\ \cdot \\ 0 \end{bmatrix}$$

and $V_e(s)$, $I_e(s)$, $E_e(s)$, L_e, $I_e(0^+)$, $V_e(0^+)$ are as defined in Equation (4-44).

It should be noted that Equations (4-66) and (4-67) are identical, since (4-66) may very well be written as

$$\begin{bmatrix} I_1 \\ I_2 \\ \cdot \\ \cdot \\ \cdot \\ I_e \end{bmatrix} = \begin{bmatrix} 0 & 0 \\ & \\ 0 & Z_{22}^{-1}(s) \end{bmatrix} \left\{ \begin{bmatrix} v_1 \\ v_2 \\ \cdot \\ \cdot \\ \cdot \\ v_e \end{bmatrix} - \begin{bmatrix} E_1 \\ E_2 \\ \cdot \\ \cdot \\ \cdot \\ E_e \end{bmatrix} + \begin{bmatrix} 0 & 0 \\ & \\ 0 & L_{22} \end{bmatrix} \begin{bmatrix} I_1(0^+) \\ I_2(0^+) \\ \cdot \\ \cdot \\ \cdot \\ I_e(0^+) \end{bmatrix} - \frac{1}{s} \begin{bmatrix} v_1(0^+) \\ v_2(0^+) \\ \cdot \\ \cdot \\ \cdot \\ v_e(0^+) \end{bmatrix} \right\} + \begin{bmatrix} I_{g_1} \\ I_{g_2} \\ \cdot \\ \cdot \\ \cdot \\ I_{g_k} \\ 0 \\ 0 \\ \cdot \\ \cdot \\ \cdot \\ 0 \end{bmatrix} \tag{4-68}$$

When Equation (4-66) is expanded by carrying out the multiplication on the right-hand side, we see, for example, that

$$\begin{bmatrix} 0 & 0 \\ 0 & Z_{22}^{-1}(s) \end{bmatrix} \begin{bmatrix} 0 \\ E_{e'}(s) \end{bmatrix} = \begin{bmatrix} 0 & 0 \\ 0 & Z_{22}^{-1}(s) \end{bmatrix} \begin{bmatrix} E_k(s) \\ E_{e'}(s) \end{bmatrix} \equiv Y_e(s)E_e(s)$$

The presence of the zero submatrices makes it clear that the replacement of the zero submatrix by $E_k(s)$ in $E_e(s)$ will not alter the result. Now, substituting Equation (4-67) into (4-53) and rearranging the terms, we get

$$AY_e(s)V_e(s) = -AI_g(s) + AY_e(s)\left\{E_e(s) + \frac{1}{s}V_e(0^+) - L_eI_e(0^+)\right\} \tag{4-69}$$

Making use of the transformed node-transformation matrix equation,

$$V_e(s) = A^T V_n(s)$$

(4-69) may be rewritten in the form

$$AY_e(s)A^TV_n(s) = -AI_g(s) + AY_e(s)E_e(s) + \frac{1}{s}AY_e(s)V_e(0^+) - AY_e(s)L_eI_e(0^+)$$

or simply,

$$Y_n^*(s)V_n(s) = I_g^*(s) + I_n^*(s) + Q_c^*(0^+) + \frac{1}{s}I_L^*(0^+) \tag{4-70}$$

where

$$Y_n^*(s) = AY_e(s)A^T$$

$$I_g^*(s) = -AI_g(s)$$

$$I_n^*(s) = AY_e(s)E_e(s)$$

$$Q_c^*(0^+) = \frac{1}{s}AY_e(s)V_e(0^+)$$

and

$$\frac{1}{s}I_L^*(0^+) = -AY_e(s)L_eI_e(0^+)$$

The reader should compare Equation (4-70) to (4-59), term by term, so that the significance of each term may be appreciated.

As in the case of loop analysis, we shall outline the procedure for writing the node-analysis matrix equation for a given network as follows:

Case (A): $\mathbf{Z}_e(s)$ is nonsingular.

(a) Write the KCL equation: $AI_e(s) = 0$

(b) Write the VCR matrix equation:

$$V_e(s) = E_e(s) + Z_e(s)I_e(s) + \frac{1}{s}V_e(0^+) - L_eI_e(0^+)$$

(c) From (b), write

$$Z_e(s)I_e(s) = V_e(s) - E_e(s) + L_eI_e(0^+) - \frac{1}{s}V_e(0^+)$$

and then solve for $I_e(s)$:

$$I_e(s) = Z_e^{-1}(s)\left\{V_e(s) - E_e(s) + L_eI_e(0^+) - \frac{1}{s}V_e(0^+)\right\}$$

(d) Substituting $I_e(s)$ into $AI_e(s) = 0$ and rearranging the terms, we get

$$AZ_e^{-1}(s)V_e(s) = AZ_e^{-1}(s)\left\{E_e(s) + \frac{1}{s}V_e(0^+) - L_eI_e(0^+)\right\}$$

(e) Define

$$Y_n(s) = AZ_e^{-1}(s)A^T$$

and write

$$Y_n(s)V_n(s) = AZ_e^{-1}(s)\left\{E_e(s) + \frac{1}{s}V_e(0^+) - L_eI_e(0^+)\right\}$$

or

$$Y_n(s)V_n(s) = I_n(s) + Q_c(0^+) + \frac{1}{s}I_L(0^+)$$

Case (B): *$Z_e(s)$ is singular.*

(a) Apply the Blakesley transformation to the original network N so that the resultant equivalent, N^*, has no branches containing voltage generators without series impedances. If N contains no voltage generators as single branches to begin with, this step will be omitted.

(b) Relabel the branches so that the first k branches correspond to those branches that contain current generators as single branches (k being the total number of such branches).

(c) Write the KCL equation for N^*: $AI_e(s) = 0$

(d) Write the VCR equation for N^*:

$$V_e(s) = E_e(s) + Z_e(s)I_e(s) + \frac{1}{s}V_e(0^+) - L_eI_e(0^+)$$

(e) In the VCR equation, solve for $I_{e'}(s)$:

$$I_{e'}(s) = Z_{e'}^{-1}(s)\left\{V_{e'}(s) - E_{e'}(s) + L_{e'}I_{e'}(0^+) - \frac{1}{s}V_{e'}(0^+)\right\}$$

(f) Write

$$I_e(s) = Y_e(s)\left\{V_e(s) - E_e(s) + L_eI_e(0^+) - \frac{1}{s}V_e(0^+)\right\} + I_g(s)$$

or, when written in a more detailed form,

$$\begin{bmatrix} I_k(s) \\ I_{e'}(s) \end{bmatrix} = \begin{bmatrix} 0 & 0 \\ 0 & Z_{e'}^{-1}(s) \end{bmatrix}\left\{\begin{bmatrix} V_k(s) \\ V_{e'}(s) \end{bmatrix} - \begin{bmatrix} 0 \\ E_{e'}(s) \end{bmatrix} + \begin{bmatrix} 0 & 0 \\ 0 & L_{e'} \end{bmatrix}\begin{bmatrix} 0 \\ I_{e'}(0^+) \end{bmatrix} - \frac{1}{s}\begin{bmatrix} 0 \\ V_{e'}(0^+) \end{bmatrix}\right\} + \begin{bmatrix} I_{gk}(s) \\ 0 \end{bmatrix}$$

where

$$Y_e(s) = \begin{bmatrix} 0 & 0 \\ 0 & Z_{e'}^{-1}(s) \end{bmatrix}$$

(g) Substituting $I_e(s)$ into $AI_e(s) = 0$ and rearranging the terms, we get

$$AY_e(s)V_e(s) = -AI_g(s) + AY_e(s)\left\{E_e(s) + \frac{1}{s}V_e(0^+) - L_eI_e(0^+)\right\}$$

(h) Define: $Y_n(s) \triangleq AY_e(s)A^T$
and write

$$Y_n(s)V_n(s) = -AI_g(s) + AY_e(s)E_e(s) + \frac{1}{s}AY_e(s)V_e(0^+) - AY_e(s)L_eI_e(0^+)$$

or

$$Y_n(s)V_n(s) = I_g^*(s) + I_n^*(s) + Q_c^*(0^+) + \frac{1}{s}I_L^*(0^+)$$

EXAMPLE 4-4. Once again, consider the network N of Figure 4-1(a). Let us obtain the node-analysis matrix equation following the steps outlined in the procedure. From Example 4-3, we find that

$$Z_e(s) = \left[\begin{array}{ccc|cc} R_1 & 0 & 0 & 0 & 0 \\ 0 & 1/sC_2 & 0 & 0 & 0 \\ 0 & 0 & R_3 & 0 & 0 \\ \hline 0 & 0 & 0 & sL_4 & sM \\ 0 & 0 & 0 & sM & sL_5 \end{array}\right] = \left[\begin{array}{c|c} Z_{11} & 0 \\ \hline 0 & Z_{22} \end{array}\right]$$

Taking the inverse, we get

$$Z_e^{-1}(s) = \left[\begin{array}{c|c} Z_{11}^{-1} & 0 \\ \hline 0 & Z_{22}^{-1} \end{array}\right] = \left[\begin{array}{ccc|cc} \frac{1}{R_1} & 0 & 0 & 0 & 0 \\ 0 & sC_2 & 0 & 0 & 0 \\ 0 & 0 & \frac{1}{R_3} & 0 & 0 \\ \hline 0 & 0 & 0 & \frac{L_5}{s\Delta} & -\frac{M}{s\Delta} \\ 0 & 0 & 0 & -\frac{M}{s\Delta} & \frac{L_4}{s\Delta} \end{array}\right] \tag{4-71}$$

where

$$\Delta = L_4L_5 - M^2$$

Thus, $Z_e(s)$ is nonsingular, and we shall follow the steps for Case (A).

(a) Write $AI_e(s) = 0$:

$$\begin{array}{c} \\ A \\ B \\ C \end{array}\begin{array}{c} \begin{array}{ccccc} e_1 & e_2 & e_3 & e_4 & e_5 \end{array} \\ \begin{bmatrix} -1 & 0 & 0 & 1 & 0 \\ 1 & 1 & -1 & 0 & 0 \\ 0 & -1 & 0 & -1 & -1 \end{bmatrix} \end{array} \begin{bmatrix} I_1(s) \\ I_2(s) \\ I_3(s) \\ I_4(s) \\ I_5(s) \end{bmatrix} = 0$$

(b) Write $V_e(s) = E_e(s) + Z_e(s)I_e(s) + \frac{1}{s}V_e(0^+) - L_eI_e(0^+)$:

$$\begin{bmatrix} V_1(s) \\ V_2(s) \\ V_3(s) \\ V_4(s) \\ V_5(s) \end{bmatrix} = \begin{bmatrix} E_1(s) \\ 0 \\ 0 \\ 0 \\ 0 \end{bmatrix} + \begin{bmatrix} R_1 & 0 & 0 & 0 & 0 \\ 0 & 1/sC_2 & 0 & 0 & 0 \\ 0 & 0 & R_3 & 0 & 0 \\ 0 & 0 & 0 & sL_4 & sM \\ 0 & 0 & 0 & sM & sL_5 \end{bmatrix} \begin{bmatrix} I_1(s) \\ I_2(s) \\ I_3(s) \\ I_4(s) \\ I_5(s) \end{bmatrix}$$

$$+ \frac{1}{s}\begin{bmatrix} 0 \\ V_{C2}(0^+) \\ 0 \\ 0 \\ 0 \end{bmatrix} - \begin{bmatrix} 0 & 0 & 0 & 0 & 0 \\ 0 & 0 & 0 & 0 & 0 \\ 0 & 0 & 0 & 0 & 0 \\ 0 & 0 & 0 & L_4 & M \\ 0 & 0 & 0 & M & L_5 \end{bmatrix} \begin{bmatrix} 0 \\ 0 \\ 0 \\ i_{L4}(0^+) \\ i_{L5}(0^+) \end{bmatrix}$$

(c) Write $Z_e(s)I_e(s) = V_e(s) - \left\{E_e(s) - L_eI_e(0^+) + \frac{1}{s} V_e(0^+)\right\}$:

$$\begin{bmatrix} R_1 & 0 & 0 & 0 & 0 \\ 0 & 1/sC_2 & 0 & 0 & 0 \\ 0 & 0 & R_3 & 0 & 0 \\ 0 & 0 & 0 & sL_4 & sM \\ 0 & 0 & 0 & sM & sL_5 \end{bmatrix} \begin{bmatrix} I_1(s) \\ I_2(s) \\ I_3(s) \\ I_4(s) \\ I_5(s) \end{bmatrix} = \begin{bmatrix} V_1(s) \\ V_2(s) \\ V_3(s) \\ V_4(s) \\ V_5(s) \end{bmatrix} - \left\{ \begin{bmatrix} E_1(s) \\ 0 \\ 0 \\ 0 \\ 0 \end{bmatrix} \right.$$

$$\left. - \begin{bmatrix} 0 & 0 & 0 & 0 & 0 \\ 0 & 0 & 0 & 0 & 0 \\ 0 & 0 & 0 & 0 & 0 \\ 0 & 0 & 0 & L_4 & M \\ 0 & 0 & 0 & M & L_5 \end{bmatrix} \begin{bmatrix} 0 \\ 0 \\ 0 \\ i_{L4}(0^+) \\ i_{L5}(0^+) \end{bmatrix} + \frac{1}{s}\begin{bmatrix} 0 \\ V_{C2}(0^+) \\ 0 \\ 0 \\ 0 \end{bmatrix} \right\}$$

which may be simplified as

$$\begin{bmatrix} R_1 & 0 & 0 & 0 & 0 \\ 0 & 1/sC_2 & 0 & 0 & 0 \\ 0 & 0 & R_3 & 0 & 0 \\ 0 & 0 & 0 & sL_4 & sM \\ 0 & 0 & 0 & sM & sL_5 \end{bmatrix} \begin{bmatrix} I_1(s) \\ I_2(s) \\ I_3(s) \\ I_4(s) \\ I_5(s) \end{bmatrix} = \begin{bmatrix} V_1(s) \\ V_2(s) \\ V_3(s) \\ V_4(s) \\ V_5(s) \end{bmatrix} + \begin{bmatrix} -E_1(s) \\ -\frac{1}{s} V_{C2}(0^+) \\ 0 \\ L_4 i_{L4}(0^+) + M i_{L5}(0^+) \\ M i_{L4}(0^+) + L_5 i_{L5}(0^+) \end{bmatrix}$$

With $Z_e^{-1}(s)$ calculated as shown in Equation (4-71), we write

$$I_e(s) = Z_e^{-1}(s)V_e(s) - Z_e^{-1}(s)\left\{E_e(s) + \frac{1}{s}V_e(0^+) - L_eI_e(0^+)\right\}$$

$$\begin{bmatrix} I_1(s) \\ I_2(s) \\ I_3(s) \\ I_4(s) \\ I_5(s) \end{bmatrix} = \begin{bmatrix} \frac{1}{R_1} & 0 & 0 & 0 & 0 \\ 0 & sC_2 & 0 & 0 & 0 \\ 0 & 0 & \frac{1}{R_3} & 0 & 0 \\ 0 & 0 & 0 & \frac{L_5}{s\Delta} & \frac{-M}{s\Delta} \\ 0 & 0 & 0 & \frac{-M}{s\Delta} & \frac{L_4}{s\Delta} \end{bmatrix} \begin{bmatrix} V_1(s) \\ V_2(s) \\ V_3(s) \\ V_4(s) \\ V_5(s) \end{bmatrix}$$

$$+\begin{bmatrix} \frac{1}{R_1} & 0 & 0 & 0 & 0 \\ 0 & sC_2 & 0 & 0 & 0 \\ 0 & 0 & \frac{1}{R_3} & 0 & 0 \\ 0 & 0 & 0 & \frac{L_5}{s\Delta} & \frac{-M}{s\Delta} \\ 0 & 0 & 0 & \frac{-M}{s\Delta} & \frac{L_4}{s\Delta} \end{bmatrix} \begin{bmatrix} -E_1(s) \\ -\frac{1}{s}V_{C_2}(0^+) \\ 0 \\ L_4 i_{L_4}(0^+) + M i_{L_5}(0^+) \\ M i_{L_4}(0^+) + L_5 i_{L_5}(0^+) \end{bmatrix}$$

where, as defined in Equation (4-71),

$$\Delta = L_4 L_5 - M^2$$

(d) Write $AZ_e^{-1}(s)V_e(s) = AZ_e^{-1}(s)\ \{E_e(s) + \frac{1}{s}V_e(0^+) - L_e I_e(0^+)\}$:

$$\begin{bmatrix} -1 & 0 & 0 & 1 & 0 \\ 1 & 1 & -1 & 0 & 0 \\ 0 & -1 & 0 & -1 & -1 \end{bmatrix} \begin{bmatrix} \frac{1}{R_1} & 0 & 0 & 0 & 0 \\ 0 & sC_2 & 0 & 0 & 0 \\ 0 & 0 & \frac{1}{R_3} & 0 & 0 \\ 0 & 0 & 0 & \frac{L_5}{s\Delta} & \frac{-M}{s\Delta} \\ 0 & 0 & 0 & \frac{-M}{s\Delta} & \frac{L_4}{s\Delta} \end{bmatrix} \begin{bmatrix} V_1(s) \\ V_2(s) \\ V_3(s) \\ V_4(s) \\ V_5(s) \end{bmatrix}$$

$$= -\begin{bmatrix} -1 & 0 & 0 & 1 & 0 \\ 1 & 1 & -1 & 0 & 0 \\ 0 & -1 & 0 & -1 & -1 \end{bmatrix} \begin{bmatrix} \frac{1}{R_1} & 0 & 0 & 0 & 0 \\ 0 & sC_2 & 0 & 0 & 0 \\ 0 & 0 & \frac{1}{R_3} & 0 & 0 \\ 0 & 0 & 0 & \frac{L_5}{s\Delta} & \frac{-M}{s\Delta} \\ 0 & 0 & 0 & \frac{-M}{s\Delta} & \frac{L_4}{s\Delta} \end{bmatrix}$$

$$\times \begin{bmatrix} -E_1(s) \\ -\frac{1}{s}V_{C_2}(0^+) \\ 0 \\ L_4 i_{L_4}(0^+) + M i_{L_5}(0^+) \\ M i_{L_4}(0^+) + L_5 i_{L_5}(0^+) \end{bmatrix}$$

(e) Write $Y_n(s) = AZ_e^{-1}(s)\ A^T$:

$$Y_n(s) = \begin{bmatrix} -1 & 0 & 0 & 1 & 0 \\ 1 & 1 & -1 & 0 & 0 \\ 0 & -1 & 0 & -1 & -1 \end{bmatrix}$$

$$\times \begin{bmatrix} \frac{1}{R_1} & 0 & 0 & 0 & \\ 0 & sC_2 & 0 & 0 & 0 \\ 0 & 0 & \frac{1}{R_3} & 0 & 0 \\ 0 & 0 & 0 & \frac{L_5}{s\Delta} & \frac{-M}{s\Delta} \\ 0 & 0 & 0 & \frac{-M}{s\Delta} & \frac{L_4}{s\Delta} \end{bmatrix} \begin{bmatrix} -1 & 1 & 0 \\ 0 & 1 & -1 \\ 0 & -1 & 0 \\ 1 & 0 & -1 \\ 0 & 0 & -1 \end{bmatrix}$$

$$= \begin{bmatrix} -\frac{1}{R_1} & 0 & 0 & \frac{L_5}{s\Delta} & \frac{-M}{s\Delta} \\ \frac{1}{R_1} & sC_2 & -\frac{1}{R_3} & 0 & 0 \\ 0 & -sC_2 & 0 & \frac{M-L_5}{s\Delta} & \frac{M-L_4}{s\Delta} \end{bmatrix} \begin{bmatrix} -1 & 1 & 0 \\ 0 & 1 & -1 \\ 0 & -1 & 0 \\ 1 & 0 & -1 \\ 0 & 0 & -1 \end{bmatrix}$$

or

$$Y_n(s) = \begin{bmatrix} \frac{1}{R_1} + \frac{L_5}{s\Delta} & -\frac{1}{R_1} & \frac{M-L_5}{s\Delta} \\ -\frac{1}{R_1} & \frac{1}{R_1} + sC_2 + \frac{1}{R_3} & -sC_2 \\ \frac{M-L_5}{s\Delta} & -sC_2 & sC_2 + \frac{L_4 + L_5 - 2M}{s\Delta} \end{bmatrix}$$

so that the equation

$$Y_n(s)V_n(s) = AZ_e^{-1}(s)\left\{E_e(s) + \frac{1}{s}V_e(0^+) - L_eI_e(0^+)\right\}$$

becomes

$$\underbrace{\begin{bmatrix} \frac{1}{R_1} + \frac{L_5}{s\Delta} & -\frac{1}{R_1} & \frac{M-L_5}{s\Delta} \\ -\frac{1}{R_1} & \frac{1}{R_1} + sC_2 + \frac{1}{R_3} & -sC_2 \\ \frac{M-L_5}{s\Delta} & -sC_2 & sC_2 + \frac{L_4 + L_5 - 2M}{s\Delta} \end{bmatrix}}_{Y_n(s)} \underbrace{\begin{bmatrix} V_A(s) \\ V_B(s) \\ V_C(s) \end{bmatrix}}_{V_n(s)}$$

$$= \underbrace{\begin{bmatrix} -\frac{1}{R_1} & 0 & 0 & \frac{L_5}{s\Delta} & \frac{-M}{s\Delta} \\ \frac{1}{R_1} & sC_2 & -\frac{1}{R_3} & 0 & 0 \\ 0 & -sC_2 & 0 & \frac{M-L_5}{s\Delta} & \frac{M-L_4}{s\Delta} \end{bmatrix}}_{AZ_e^{-1}(s)} \underbrace{\begin{bmatrix} E_1(s) \\ \frac{1}{s}V_{C_2}(0^+) \\ 0 \\ -L_4 i_{L_4}(0^+) - M i_{L_5}(0^+) \\ -M i_{L_4}(0^+) - L_5 i_{L_5}(0^+) \end{bmatrix}}_{\left\{E_e(s) + \frac{1}{s}V_e(0^+) - L_eI_e(0^+)\right\}}$$

or

$$\begin{bmatrix} \left(\frac{1}{R_1}+\frac{L_5}{s\Delta}\right)V_A(s)-\frac{1}{R_1}V_B(s)+\left(\frac{M-L_5}{s\Delta}\right)V_C(s) \\ -\frac{1}{R_1}V_A(s)+\left(\frac{1}{R_1}+sC_2+\frac{1}{R_3}\right)V_B(s)-sC_2V_C(s) \\ \left(\frac{M-L_5}{s\Delta}\right)V_A(s)-sC_2V_B(s)+\left(sC_2+\frac{L_4+L_5-2M}{s\Delta}\right)V_C(s) \end{bmatrix}$$

$$= \begin{bmatrix} -\frac{E_1(s)}{R_1}-\frac{L_5}{s\Delta}\{L_4i_{L_4}(0^+)+Mi_{L_5}(0^+)\}+\frac{M}{s\Delta}\{Mi_{L_4}(0^+)+L_5i_{L_5}(0^+)\} \\ \frac{E_1}{R_1}+C_2V_{C_2}(0^+) \\ -C_2V_{C_2}(0^+)-\frac{M-L_5}{s\Delta}\{L_4i_{L_4}(0^+)+Mi_{L_5}(0^+)\}-\frac{M-L_4}{s\Delta} \\ \{Mi_{L_4}(0^+)+L_5i_{L_5}(0^+)\} \end{bmatrix}$$

EXAMPLE 4-5. As another example of formulating the node-analysis matrix equation, let us consider the network N of Figure 4-7. By inspection, we see that $Z_e(s)$ of N is singular and, hence, Case (B) of the procedure applies.

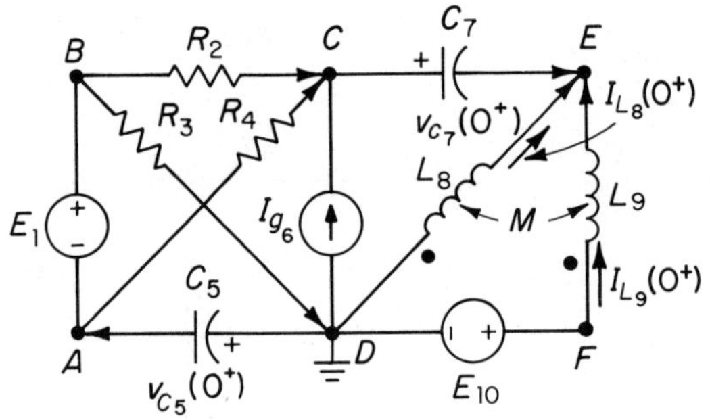

Figure 4-7 A network N with voltage sources alone as branches.

As the result of the application of the Blakesley transformation (e-shift operation), we obtain the equivalent N^* as shown in Figure 4-8(a). The graph of N^* is shown in Figure 4-8(b). Next, we obtain the matrices A, $Z_{22}(s) = Z_{e'}(s)$, $Z_{22}^{-1}(s)$, $I_g(s)$, $E_e(s)$, $V_e(0^+)$, $I_e(0^+)$, and

$$Y_e(s) = \begin{bmatrix} 0 & 0 \\ 0 & Z_{22}^{-1} \end{bmatrix}$$

as follows:

$$A = \begin{matrix} \\ A, B \\ C \\ E \end{matrix} \begin{matrix} \begin{matrix} 6 & \;\;2 & \;\;3 & \;\;4 & \;\;5 & \;\;7 & \;\;8 & \;\;9 \end{matrix} \\ \begin{bmatrix} 0 & 1 & 1 & 1 & -1 & 0 & 0 & 0 \\ -1 & -1 & 0 & -1 & 0 & 1 & 0 & 0 \\ 0 & 0 & 0 & 0 & 0 & -1 & -1 & -1 \end{bmatrix} \end{matrix}$$

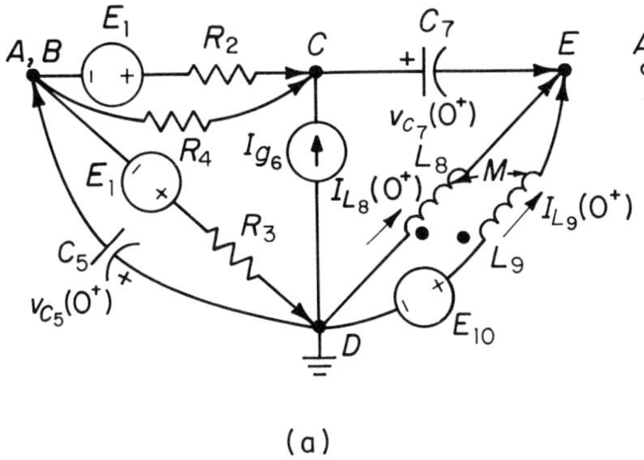

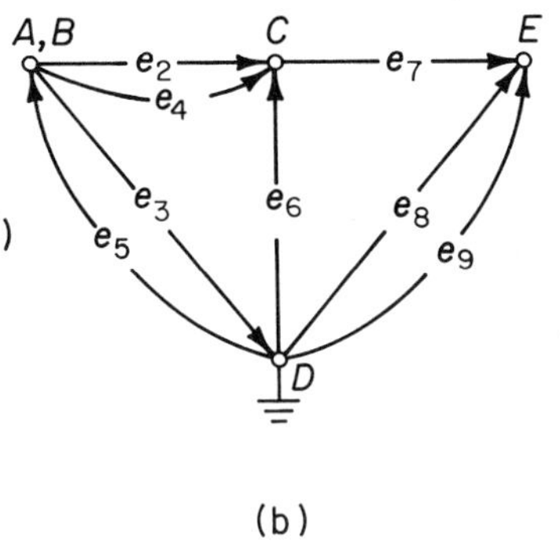

Figure 4-8 (a) The equivalent N^* of N. (b) The graph of N^*.

$$
Z_{22}(s) = \begin{array}{c} \\ 2 \\ 3 \\ 4 \\ 5 \\ 7 \\ 8 \\ 9 \end{array}
\begin{array}{c}
\begin{array}{ccccc|cc} 2 & 3 & 4 & 5 & 7 & 8 & 9 \end{array} \\
\left[\begin{array}{ccccc|cc}
R_2 & 0 & 0 & 0 & 0 & 0 & 0 \\
0 & R_3 & 0 & 0 & 0 & 0 & 0 \\
0 & 0 & R_4 & 0 & 0 & 0 & 0 \\
0 & 0 & 0 & 1/sC_5 & 0 & 0 & 0 \\
0 & 0 & 0 & 0 & 1/sC_7 & 0 & 0 \\
\hline
0 & 0 & 0 & 0 & 0 & sL_8 & sM \\
0 & 0 & 0 & 0 & 0 & sM & sL_9
\end{array}\right]
\end{array}
$$

$$
Z_{22}^{-1}(s) = \begin{bmatrix}
1/R_2 & 0 & 0 & 0 & 0 & 0 & 0 \\
0 & 1/R_3 & 0 & 0 & 0 & 0 & 0 \\
0 & 0 & 1/R_4 & 0 & 0 & 0 & 0 \\
0 & 0 & 0 & sC_5 & 0 & 0 & 0 \\
0 & 0 & 0 & 0 & sC_7 & 0 & 0 \\
0 & 0 & 0 & 0 & 0 & a & b \\
0 & 0 & 0 & 0 & 0 & c & d
\end{bmatrix}
$$

where

$$
a = \frac{L_9}{s(L_8L_9 - M^2)} \qquad b = \frac{-M}{s(L_8L_9 - M^2)} \qquad c = b \qquad d = \frac{L_8}{s(L_8L_9 - M^2)}
$$

$$
I_g(s) = \begin{bmatrix} I_{g_6} \\ 0 \\ 0 \\ 0 \\ 0 \\ 0 \\ 0 \\ 0 \end{bmatrix}\begin{array}{c} 6 \\ 2 \\ 3 \\ 4 \\ 5 \\ 7 \\ 8 \\ 9 \end{array} \qquad E_e(s) = \begin{bmatrix} 0 \\ -E_1 \\ -E_1 \\ 0 \\ 0 \\ 0 \\ 0 \\ -E_{10} \end{bmatrix}
$$

$$V_e(0^+) = \begin{bmatrix} 0 \\ 0 \\ 0 \\ 0 \\ v_{C_5}(0^+) \\ v_{C_7}(0^+) \\ 0 \\ 0 \end{bmatrix} \qquad I_e(0^+) = \begin{bmatrix} 0 \\ 0 \\ 0 \\ 0 \\ 0 \\ 0 \\ i_{L_8}(0^+) \\ i_{L_9}(0^+) \end{bmatrix}$$

and

$$Y_e(s) = \left[\begin{array}{c|c} 0 & 0 \\ \hline 0 & Z_{22}^{-1} \end{array}\right] = \left[\begin{array}{c|ccccccc} 0 & 0 & 0 & 0 & 0 & 0 & 0 & 0 \\ \hline 0 & 1/R_2 & 0 & 0 & 0 & 0 & 0 & 0 \\ 0 & 0 & 1/R_3 & 0 & 0 & 0 & 0 & 0 \\ 0 & 0 & 0 & 1/R_4 & 0 & 0 & 0 & 0 \\ 0 & 0 & 0 & 0 & sC_5 & 0 & 0 & 0 \\ 0 & 0 & 0 & 0 & 0 & sC_7 & 0 & 0 \\ 0 & 0 & 0 & 0 & 0 & 0 & a & b \\ 0 & 0 & 0 & 0 & 0 & 0 & c & d \end{array}\right]$$

Now,

$$Y_n^*(s) = A\{Y_e(s)A^T\} = \begin{bmatrix} 0 & 1 & 1 & 1 & -1 & 0 & 0 & 0 \\ -1 & -1 & 0 & -1 & 0 & 1 & 0 & 0 \\ 0 & 0 & 0 & 0 & 0 & -1 & -1 & -1 \end{bmatrix}$$

$$\times \begin{bmatrix} 0 & 0 & 0 \\ 1/R_2 & -1/R_2 & 0 \\ 1/R_3 & 0 & 0 \\ 1/R_4 & -1/R_4 & 0 \\ -sC_5 & 0 & 0 \\ 0 & sC_7 & -sC_7 \\ 0 & 0 & -(a+b) \\ 0 & 0 & -(c+d) \end{bmatrix}$$

or

$$Y_n^*(s) = \left[\begin{array}{c|c|c} \frac{1}{R_2} + \frac{1}{R_3} + \frac{1}{R_4} + sC_5 & -\frac{1}{R_2} - \frac{1}{R_4} & 0 \\ \hline -\frac{1}{R_2} - \frac{1}{R_4} & \frac{1}{R_2} + \frac{1}{R_4} + sC_7 & -sC_7 \\ \hline 0 & -sC_7 & sC_7 + a + b + c + d \end{array}\right]$$

Also,

$$V_n^*(s) = \begin{bmatrix} V_{AB} \\ V_C \\ V_E \end{bmatrix}$$

$$-AI_g(s) = -\begin{bmatrix} 0 & 1 & 1 & 1 & -1 & 0 & 0 & 0 \\ -1 & -1 & 0 & -1 & 0 & 1 & 0 & 0 \\ 0 & 0 & 0 & 0 & 0 & -1 & -1 & -1 \end{bmatrix} \begin{bmatrix} I_{g6} \\ 0 \\ 0 \\ 0 \\ 0 \\ 0 \\ 0 \\ 0 \end{bmatrix} = \begin{bmatrix} 0 \\ I_{g6} \\ 0 \end{bmatrix}$$

$$AY_e(s)E_e(s) = \begin{bmatrix} 0 & 1 & 1 & 1 & -1 & 0 & 0 & 0 \\ -1 & -1 & 0 & -1 & 0 & 1 & 0 & 0 \\ 0 & 0 & 0 & 0 & 0 & -1 & -1 & -1 \end{bmatrix}$$

$$\times \begin{bmatrix} 0 & 0 & 0 & 0 & 0 & 0 & 0 & 0 \\ 0 & 1/R_2 & 0 & 0 & 0 & 0 & 0 & 0 \\ 0 & 0 & 1/R_3 & 0 & 0 & 0 & 0 & 0 \\ 0 & 0 & 0 & 1/R_4 & 0 & 0 & 0 & 0 \\ 0 & 0 & 0 & 0 & sC_5 & 0 & 0 & 0 \\ 0 & 0 & 0 & 0 & 0 & sC_7 & 0 & 0 \\ 0 & 0 & 0 & 0 & 0 & 0 & a & b \\ 0 & 0 & 0 & 0 & 0 & 0 & c & d \end{bmatrix} \begin{bmatrix} 0 \\ -E_1 \\ -E_1 \\ 0 \\ 0 \\ 0 \\ 0 \\ -E_{10} \end{bmatrix}$$

$$= \begin{bmatrix} -\dfrac{E_1}{R_2} - \dfrac{E_1}{R_3} \\ \dfrac{E_1}{R_2} \\ E_{10}(b + d) \end{bmatrix}$$

$$\frac{1}{s} AY_e v_e(0^+) = AY_e \frac{1}{s} v_e(0^+) = \begin{bmatrix} 0 & 1 & 1 & 1 & -1 & 0 & 0 & 0 \\ -1 & -1 & 0 & -1 & 0 & 1 & 0 & 0 \\ 0 & 0 & 0 & 0 & 0 & -1 & -1 & -1 \end{bmatrix}$$

$$\times \begin{bmatrix} 0 & 0 & 0 & 0 & 0 & 0 & 0 & 0 \\ 0 & 1/R_2 & 0 & 0 & 0 & 0 & 0 & 0 \\ 0 & 0 & 1/R_3 & 0 & 0 & 0 & 0 & 0 \\ 0 & 0 & 0 & 1/R_4 & 0 & 0 & 0 & 0 \\ 0 & 0 & 0 & 0 & sC_5 & 0 & 0 & 0 \\ 0 & 0 & 0 & 0 & 0 & sC_7 & 0 & 0 \\ 0 & 0 & 0 & 0 & 0 & 0 & a & b \\ 0 & 0 & 0 & 0 & 0 & 0 & c & d \end{bmatrix}$$

$$\times \begin{bmatrix} 0 \\ 0 \\ 0 \\ 0 \\ \dfrac{v_{C_5}(0^+)}{s} \\ \dfrac{v_{C_7}(0^+)}{s} \\ 0 \\ 0 \end{bmatrix} = \begin{bmatrix} -C_5 v_{C_5}(0^+) \\ C_7 v_{C_7}(0^+) \\ -C_7 v_{C_7}(0^+) \end{bmatrix}$$

$$-AY_e(s)L_eI_e(0^+) = -\begin{bmatrix} 0 & 1 & 1 & 1 & -1 & 0 & 0 & 0 \\ -1 & -1 & 0 & -1 & 0 & 1 & 0 & 0 \\ 0 & 0 & 0 & 0 & 0 & -1 & -1 & -1 \end{bmatrix}$$

$$\times \begin{bmatrix} 0 & 0 & 0 & 0 & 0 & 0 & 0 & 0 \\ 0 & 1/R_2 & 0 & 0 & 0 & 0 & 0 & 0 \\ 0 & 0 & 1/R_3 & 0 & 0 & 0 & 0 & 0 \\ 0 & 0 & 0 & 1/R_4 & 0 & 0 & 0 & 0 \\ 0 & 0 & 0 & 0 & sC_5 & 0 & 0 & 0 \\ 0 & 0 & 0 & 0 & 0 & sC_7 & 0 & 0 \\ 0 & 0 & 0 & 0 & 0 & 0 & a & b \\ 0 & 0 & 0 & 0 & 0 & 0 & c & d \end{bmatrix}$$

$$\times \begin{bmatrix} 0 & 0 & 0 & 0 & 0 & 0 & 0 & 0 \\ 0 & 0 & 0 & 0 & 0 & 0 & 0 & 0 \\ 0 & 0 & 0 & 0 & 0 & 0 & 0 & 0 \\ 0 & 0 & 0 & 0 & 0 & 0 & 0 & 0 \\ 0 & 0 & 0 & 0 & 0 & 0 & 0 & 0 \\ 0 & 0 & 0 & 0 & 0 & 0 & 0 & 0 \\ 0 & 0 & 0 & 0 & 0 & 0 & L_8 & M \\ 0 & 0 & 0 & 0 & 0 & 0 & M & L_9 \end{bmatrix} \begin{bmatrix} 0 \\ 0 \\ 0 \\ 0 \\ 0 \\ 0 \\ i_{L_8}(0^+) \\ i_{L_9}(0^+) \end{bmatrix}$$

or

$$-AY_e(s)L_eI_e(0^+)$$

$$= \begin{bmatrix} 0 \\ 0 \\ \{L_8 i_{L_8}(0^+) + M i_{L_9}(0^+)\}\{a + c\} + \{M i_{L_8}(0^+) + L_9 i_{L_9}(0^+)\}\{b + d\} \end{bmatrix}$$

Finally, the equation

$$\begin{aligned} Y_n^*(s)V_n^*(s) &= AY_e(s)A^TV_n(s) \\ &= -AI_g(s) + AY_e(s)E_e(s) + \frac{1}{s}AY_e(s)V_e(0^+) - AY_e(s)L_eI_e(0^+) \end{aligned}$$

becomes

$$\left[\begin{array}{c|c|c} \frac{1}{R_2}+\frac{1}{R_3}+\frac{1}{R_4}+sC_5 & -\frac{1}{R_2}-\frac{1}{R_4} & 0 \\ \hline -\frac{1}{R_2}-\frac{1}{R_4} & \frac{1}{R_2}+\frac{1}{R_4}+sC_7 & -sC_7 \\ \hline 0 & -sC_7 & sC_7+\frac{L_8+L_9-2M}{s(L_8L_9-M^2)} \end{array}\right]\begin{bmatrix} V_{AB} \\ V_C \\ V_E \end{bmatrix}$$

$$= \begin{bmatrix} -\frac{E_1}{R_2}-\frac{E_1}{R_3}-C_5v_{C_5}(0^+) \\ I_{g6}+\frac{E_1}{R_2}+C_7v_{C_7}(0^+) \\ \frac{E_{10}(L_8-M)}{(L_8L_9-M^2)s}-C_7v_{C_7}(0^+)+\frac{\{L_8i_{L_8}(0^+)+Mi_{L_9}(0^+)\}(L_9-M)}{s(L_8L_9-M^2)} \\ +\frac{\{L_9i_{L_9}(0^+)+Mi_{L_8}(0^+)\}(L_8-M)}{s(L_8L_9-M^2)} \end{bmatrix}$$

If we assume zero initial currents in the inductors, the system reduces to

$$\begin{bmatrix} \frac{1}{R_2}+\frac{1}{R_3}+\frac{1}{R_4}+sC_5 & -\frac{1}{R_2}-\frac{1}{R_4} & 0 \\ -\frac{1}{R_2}-\frac{1}{R_4} & \frac{1}{R_2}+\frac{1}{R_4}+sC_7 & -sC_7 \\ 0 & -sC_7 & sC_7+\frac{L_8+L_9-2M}{s(L_8L_9-M^2)} \end{bmatrix}\begin{bmatrix} V_{AB} \\ V_C \\ V_E \end{bmatrix}$$

$$= \begin{bmatrix} -\frac{E_1}{R_2}-\frac{E_1}{R_3}-C_5v_{C_5}(0^+) \\ I_{g6}+\frac{E_1}{R_2}+C_7v_{C_7}(0^+) \\ \frac{E_{10}(L_8-M)}{s(L_8L_9-M^2)}-C_7v_{C_7}(0^+) \end{bmatrix}$$

If the initial currents are not zero, the matrix equation may still be simplified to

$$\begin{bmatrix} \frac{1}{R_2}+\frac{1}{R_3}+\frac{1}{R_4}+sC_5 & -\left(\frac{1}{R_2}+\frac{1}{R_4}\right) & 0 \\ -\left(\frac{1}{R_2}+\frac{1}{R_4}\right) & \frac{1}{R_2}+\frac{1}{R_4}+sC_7 & -sC_7 \\ 0 & -sC_7 & sC_7+\frac{L_8+L_9-2M}{s(L_8L_9-M^2)} \end{bmatrix}\begin{bmatrix} V_{AB} \\ V_C \\ V_E \end{bmatrix}$$

$$= \begin{bmatrix} -E_1\left(\frac{R_2+R_3}{R_2R_3}\right)-C_5v_{C_5}(0^+) \\ I_{g6}+\frac{E_1}{R_2}+C_7v_{C_7}(0^+) \\ \frac{E_{10}(L_8-M)}{s(L_8L_9-M^2)}-C_7v_{C_7}(0^+)+\frac{1}{s}\{i_{L_8}(0^+)+i_{L_9}(0^+)\} \end{bmatrix}$$

4-6 THE STATE EQUATIONS

Thus far, we have discussed the more familiar loop-analysis and node-analysis representations of an electrical network. In this section, we shall present a brief introduction to the so-called "state-variable" characterization of a linear system. More precisely, we shall show how the topological properties may be applied to formulate an algorithm for obtaining the state equations for an electrical network. Since our primary objective is to illustrate some of the topological applications in connection with the state-variable approach, we shall attempt no more than a brief informal discussion of the state-variable concept in network analysis. We shall further limit ourselves to the formulation of the *state equations* of an RLC network without going into the methods of solution. We should point out, however, that state-variable techniques are a very powerful tool in system analysis, especially in handling complex engineering problems not amenable to classical methods, since the state-variable characterization of a system is most suitable for computer solutions, and, perhaps more importantly, it is the only effective way to analyze time-varying and nonlinear networks. On the other hand, the theory behind the state-variable approach to the solution of a system is by no means new from a mathematical point of view. It is merely a new application of certain existing mathematical techniques. More precisely, it is an application of matrix and vector methods to manipulate a large number of variables called state variables in a given physical system.

The concept of "state" is perhaps originally from classical mechanics where, in a physical system, particles are generally characterized by their positions, say x_j, and velocities, which are the time rate of change, $\dot{x}_j = (d/dt)x_j$, of the position variables x_j. Thus, variables x_j and $\dot{x}_j$ together describe the complete *state* of the system, and, consequently, they are called the *state variables* of the system. In ordinary language, the state of a system is actually the minimal amount of information necessary at any time to characterize completely any possible future behavior of the system.[2]

One of the main differences between the state-variable technique and the conventional methods in network analysis lies in the fact that the state-variable approach emphasizes the "dynamic" character of the given system, whereas the conventional methods, such as the loop analysis and node analysis, do not. To illustrate this point, let us take a passive network N, for example. The procedure for the formulation of the loop (node) equations in the loop (node) analysis is always the same regardless of the types of elements contained in N. That is, no matter whether N is an all-resistance network or an RL, RC, LC, or RLC network, the procedure for selecting the variables and formulating the network equations in a conventional method remains un-

[2]This simple description of the state of a system is taken from a paper by Kuh and Rohrer [48].

altered. On the other hand, by the state-variable approach, quantities such as the voltage across a capacitor and the current in an inductor, which represent the stored energy and controls the *dynamic* behavior of the network, play an important role in the formulation of the state equations, as will be seen later in the examples.

When the set of ordinary differential equations which describes a dynamic physical system is expressed in the form

$$\dot{x}_j = f_j(x_1, x_2, \ldots, x_n, h_1, h_2, \ldots, h_m) \qquad (j = 1, 2, \ldots, n) \tag{4-72}$$

where $x_1, x_2, \ldots, x_n$ are unknown variables (outputs) and $h_1, h_2, \ldots, h_m$ are prescribed functions of time t (inputs), the set of equations is said to be in the *normal* (canonical, or standard) form. If the system is linear, Equation (4-72) can be written in the matrix form,

$$\dot{X} = AX + BU \tag{4-73}$$

where the n-vector X is called the *state vector*, U is an m-vector, called the *input vector*, A is an $n \times n$ matrix called the *system matrix*, B is an $n \times m$ matrix called the *distribution matrix*, and $\dot{X}$ is the time derivative of X, which is defined as

$$\dot{X} = \frac{d}{dt}X = \frac{d}{dt}\begin{bmatrix} x_1 \\ x_2 \\ \cdot \\ \cdot \\ \cdot \\ x_n \end{bmatrix} = \begin{bmatrix} \dfrac{dx_1}{dt} \\ \dfrac{dx_2}{dt} \\ \cdot \\ \cdot \\ \cdot \\ \dfrac{dx_n}{dt} \end{bmatrix} \tag{4-74}$$

In the state-variable approach, the variables $x_1, x_2, \ldots, x_n$ are called the state variables of the given linear system, and Equation (4-73) is the matrix state equation of the system. When a set of values is assigned to the state variables in X at a given time, say t_0, a particular state vector results which represents a particular state of the system (at time t_0).

We shall illustrate the ideas introduced by means of the following examples.

EXAMPLE 4-6. Consider the *RLC* network N shown in Figure 4-9. Since there are two energy storage elements in N, a natural choice of the state variables is the following:

$$\begin{aligned} x_1 &= i_2 \\ x_2 &= v_4 \end{aligned} \tag{4-75}$$

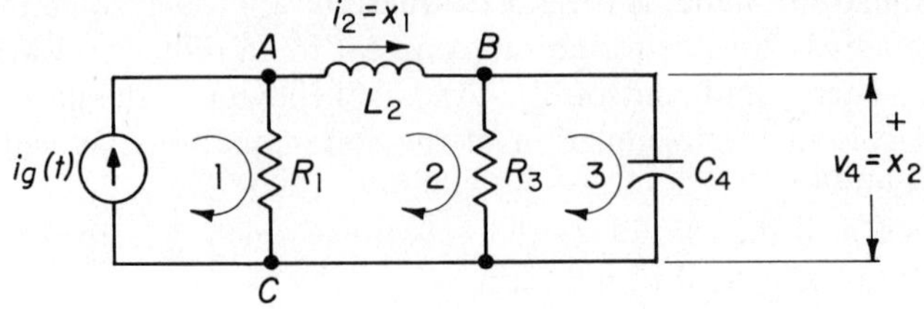

Figure 4-9 The *RLC* network *N* for Example 4-6.

Thus, we write

$$\text{Around loop 2:}\quad L_2\dot{x}_1 + x_2 - R_1(i_g - x_1) = 0 \tag{4-76a}$$

$$\text{At node } B\text{:}\quad C_4\dot{x}_2 + \frac{x_2}{R_3} - x_1 = 0 \tag{4-76b}$$

which can be rewritten as

$$\begin{aligned} \dot{x}_1 &= -\frac{R_1}{L_2}x_1 - \frac{1}{L_2}x_2 + \frac{R_1}{L_2}i_g \\ \dot{x}_2 &= \frac{1}{C_4}x_1 - \frac{1}{R_3C_4}x_2 \end{aligned} \tag{4-77}$$

Expressing Equation (4-77) in matrix form, we have

$$\begin{bmatrix} \dot{x}_1 \\ \dot{x}_2 \end{bmatrix} = \begin{bmatrix} -\dfrac{R_1}{L_2} & -\dfrac{1}{L_2} \\ \dfrac{1}{C_4} & -\dfrac{1}{R_3C_4} \end{bmatrix} \begin{bmatrix} x_1 \\ x_2 \end{bmatrix} + \begin{bmatrix} \dfrac{R_1}{L_2} \\ 0 \end{bmatrix} [i_g] \tag{4-78}$$

Comparing Equation (4-78) with (4-73), we see that the state-variable vector X and its time derivative $\dot{X}$ are

$$X = \begin{bmatrix} x_1 \\ x_2 \end{bmatrix} \quad \text{and} \quad \dot{X} = \begin{bmatrix} \dot{x}_1 \\ \dot{x}_2 \end{bmatrix}$$

respectively; the system matrix A is given by

$$A = \begin{bmatrix} -\dfrac{R_1}{L_2} & -\dfrac{1}{L_2} \\ \dfrac{1}{C_4} & -\dfrac{1}{R_3C_4} \end{bmatrix}$$

the distribution matrix is

$$B = \begin{bmatrix} \dfrac{R_1}{L_2} \\ 0 \end{bmatrix}$$

and the input vector U is

$$U = [i_g]$$

which, as in this case, is a scalar for a single source.

Note that Equation (4-76a) is the KVL equation

$$\sum_k v_k = 0$$

around loop 2, and (4-76b) is the KCL equation

$$\sum_j i_j = 0$$

at node B.

EXAMPLE 4-7. Let us obtain the matrix state equation of the RLC network N shown in Figure 4-10. Following the procedure illustrated in Example 4-6, we let

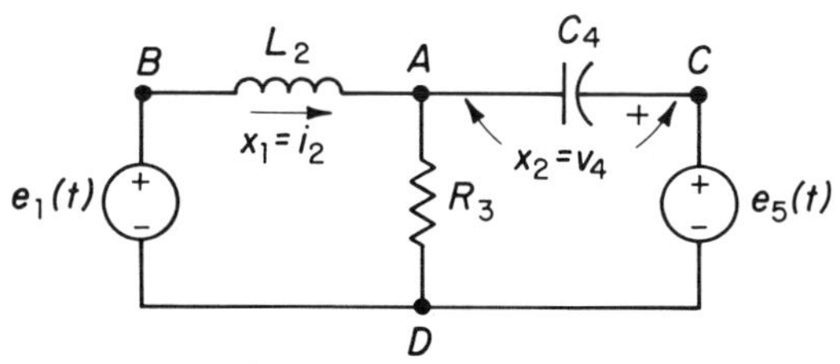

Figure 4-10 The RLC network N for Example 4-7.

$$x_1 = i_2 \qquad x_2 = v_4$$

and write the KVL equation around the outside loop consisting of e_1, L_2, C_4, and e_5:

$$L_2\dot{x}_1 - x_2 + e_5 - e_1 = 0 \tag{4-79}$$

Next we write the KCL equation at node A:

$$\frac{e_5 - x_2}{R_3} - x_1 - C_4\dot{x}_2 = 0 \tag{4-80}$$

Rewriting Equations (4-79) and (4-80) in normal form, we have

$$\dot{x}_1 = 0 \cdot x_1 + \frac{1}{L_2}x_2 + \frac{1}{L_2}e_1 - \frac{1}{L_2}e_5$$

$$\dot{x}_2 = -\frac{1}{C_4}x_1 - \frac{1}{R_3C_4}x_2 + 0 \cdot e_1 + \frac{1}{R_3C_4}e_5$$

which can be written in matrix form as

$$\begin{bmatrix} \dot{x}_1 \\ \dot{x}_2 \end{bmatrix} = \begin{bmatrix} 0 & \dfrac{1}{L_2} \\ -\dfrac{1}{C_4} & -\dfrac{1}{R_3C_4} \end{bmatrix} \begin{bmatrix} x_1 \\ x_2 \end{bmatrix} + \begin{bmatrix} \dfrac{1}{L_2} & \dfrac{-1}{L_2} \\ 0 & \dfrac{1}{R_3C_4} \end{bmatrix} \begin{bmatrix} e_1 \\ e_5 \end{bmatrix} \tag{4-81}$$

Upon comparison of Equations (4-81) and (4-73), we immediately obtain

$$X = \begin{bmatrix} x_1 \\ x_2 \end{bmatrix} \qquad \dot{X} = \begin{bmatrix} \dot{x}_1 \\ \dot{x}_2 \end{bmatrix} \qquad A = \begin{bmatrix} 0 & \dfrac{1}{L_2} \\ -\dfrac{1}{C_4} & -\dfrac{1}{R_3 C_4} \end{bmatrix}$$

$$B = \begin{bmatrix} \dfrac{1}{L_2} & -\dfrac{1}{L_2} \\ 0 & \dfrac{1}{R_3 C_4} \end{bmatrix} \qquad U = \begin{bmatrix} e_1 \\ e_5 \end{bmatrix}$$

From the two simple examples just given, we see how the matrix state equation (4-73) can be obtained from a given physical system. We shall devote the remainder of this section to the application of network topology in the formulation of state equations.

From the preceding discussion of the state-variable approach and from the two examples, we see that the quantities that control the dynamic character of a general *RLC* network are the capacitor voltages and the inductor currents. In formulating the state equations, the first task is to determine the minimum number α of (independent) state variables. This number is evidently the (minimum) number of quantities required to characterize the state of the system and is called the *order* of the system, or the *order of complexity*[3] of a network in network theory. For an *RLC* network without circuits consisting of capacitors alone and cutsets consisting of inductors alone, α is simply equal to the total number N_{lc} of inductors and capacitors contained in the network; that is,

$$\alpha = N_{lc} \tag{4-82}$$

However, for a general *RLC* network, allowing the possibility of the presence of all-capacitor circuits and all-inductor cutsets, the order of complexity α (which is the number of independent state variables to be used in formulating the state of equations) is given by[3]

$$\alpha = N_{lc} - C_c - K_l \tag{4-83}$$

where C_c is the total number of all-capacitor (with possibly voltage sources) circuits and K_l is the total number of all-inductor (with possibly current sources) cutsets.

Since our main concern here is to show how network topology may be applied to formulate the state equations of a given network, it will serve our purpose to consider the simpler case where no all-capacitor (and voltage sources) circuits and all-inductor (and current sources) cutsets are present in the network. For an *RLC* network under the aforementioned assumptions,

[3]Refer to Definition 10-1 and Theorem 10-1.

the state variables are simply the set of all capacitor voltages and inductor currents. To formulate the state equations, we shall first select a tree T such that T includes all of the (known) voltage sources and none of the (known) current sources. It can be shown that such a tree always exists for an RLC network which satisfies the aforementioned constraints.[4] The purpose of selecting such a tree is that it will enable us to express the chord voltages in terms of the voltage sources and the (tree-) branch voltages; and the branch currents in terms of the current sources and the chord currents from the fundamental circuit equations and fundamental cutset equations for the tree T. Thus, we can eliminate the undesired voltages and currents (which are the voltages and currents other than the state variables and the known sources) by substituting the circuit and cutset equations into the voltage-current-relationship equations for the passive elements. Finally, with all the undesired variables eliminated, the resultant equations can be arranged in the normal form [Equation (4-73)], which is the desired matrix state equation for the given network.

We shall summarize the procedure for the formulation of the state equations for an RLC network N without mutual inductances, all-capacitor (and/or voltage source), all-inductor (and/or current source) loops and cutsets as follows:

(1) Obtain from N the directed graph G.

(2) Select a tree T such that T contains

- (a) All of the voltage sources
- (b) None of the current sources
- (c) A maximum number of capacitors
- (d) A minimum number of inductors.

(3) Write the VCR equations for the passive elements and arrange these equations in two groups:

- (a) Branch capacitors and chord inductors are in one group.
- (b) All other passive elements are in the second group.

(4) Write the fundamental circuit and cutset equations and arrange them according to the grouping of elements in Step (3) (for easy substitution in the next elimination process).

(5) Eliminate all the undesired variables (which are all variables other than the state variables and the sources) by substituting the equations in Step (4) into those in Step (3).

(6) Rearrange the terms and express the resultant equations in standard form [Equation (4-73)], which is the matrix state equation for the given network N.

EXAMPLE 4-8. Consider the network N shown in Figure 4-9. We shall obtain the matrix state equation using the algorithm just discussed.

[4]See Seshu and Reed [76].

Step 1: The oriented graph G of N is shown in Figure 4-11.

Step 2: The tree T is chosen to contain R_1 and C_4 as shown in heavy strokes in Figure 4-11.

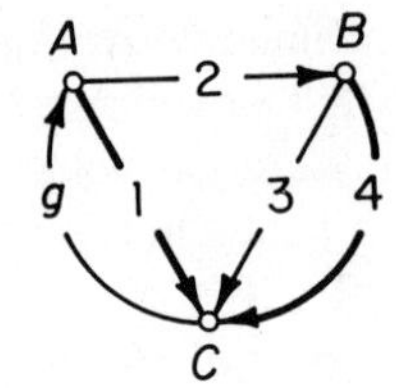

Figure 4-11 The graph G of network N for Figure 4-9.

Step 3: The VCR equations for the passive elements of N are:

(a)
$$\begin{Bmatrix} i_2 = \dfrac{1}{L_2}\displaystyle\int v_2\,dt \\ v_4 = \dfrac{1}{C_4}\displaystyle\int i_4\,dt \end{Bmatrix} \Rightarrow \begin{Bmatrix} \dot{i}_2 = \dfrac{1}{L_2}v_2 \\ \dot{v}_4 = \dfrac{1}{C_4}i_4 \end{Bmatrix}$$

or

$$\begin{bmatrix} \dot{i}_2 \\ \dot{v}_4 \end{bmatrix} = \begin{bmatrix} \dfrac{1}{L_2} & 0 \\ 0 & \dfrac{1}{C_4} \end{bmatrix} \begin{bmatrix} v_2 \\ i_4 \end{bmatrix} \tag{4-84}$$

(b)
$$\begin{cases} v_1 = R_1 i_1 \\ v_3 = R_3 i_3 \quad \text{or} \quad i_3 = \dfrac{1}{R_3}v_3 \end{cases}$$

Step 4: The Kirchhoff voltage-law (KVL) equations for the fundamental circuits and the Kirchhoff current-law (KCL) equations for the fundamental cutsets of T are:

$$\text{KVL}\begin{Bmatrix} -v_1 + v_2 + v_4 = 0 \\ -v_3 + v_4 = 0 \end{Bmatrix} \Rightarrow \begin{cases} v_2 = v_1 - v_4 \\ v_3 = v_4 \end{cases}$$

$$\text{KCL}\begin{Bmatrix} -i_g + i_1 + i_2 = 0 \\ -i_2 + i_3 + i_4 = 0 \end{Bmatrix} \Rightarrow \begin{cases} i_1 = i_g - i_2 \\ i_4 = i_2 - i_3 \end{cases}$$

Step 5: Using the method of elimination by substitution, v_2 and i_4 may be written from the equations in Step 3 and Step 4 as

$$\begin{cases} v_2 = v_1 - v_4 = R_1 i_1 - v_4 = R_1(i_g - i_2) - v_4 = R_1 i_g - R_1 i_2 - v_4 \\ i_4 = i_2 - i_3 = i_2 - \dfrac{1}{R_3}v_3 = i_2 - \dfrac{1}{R_3}v_4 \end{cases}$$

or

$$\begin{bmatrix} v_2 \\ i_4 \end{bmatrix} = \begin{bmatrix} -R_1 & -1 \\ 1 & -\dfrac{1}{R_3} \end{bmatrix} \begin{bmatrix} i_2 \\ v_4 \end{bmatrix} + \begin{bmatrix} R_1 \\ 0 \end{bmatrix} [i_g] \tag{4-85}$$

Step 6: Finally, substituting Equation (4-85) into (4-84) and simplifying, we have

$$\begin{bmatrix} \dot{i}_2 \\ \dot{v}_4 \end{bmatrix} = \begin{bmatrix} -\dfrac{R_1}{L_2} & -\dfrac{1}{L_2} \\ \dfrac{1}{C_4} & -\dfrac{1}{R_3C_4} \end{bmatrix} \begin{bmatrix} i_2 \\ v_4 \end{bmatrix} + \begin{bmatrix} \dfrac{R_1}{L_2} \\ 0 \end{bmatrix} [i_g] \tag{4-86}$$

which is the matrix state equation for the given network N. Note that Equation (4-86) is already in standard form,

$$\dot{X} = AX + BU$$

where

$$X = \begin{bmatrix} x_1 \\ x_2 \end{bmatrix} = \begin{bmatrix} i_2 \\ v_4 \end{bmatrix} \qquad \text{so that } \dot{X} = \begin{bmatrix} \dot{x}_1 \\ \dot{x}_2 \end{bmatrix} = \begin{bmatrix} \dot{i}_2 \\ \dot{v}_4 \end{bmatrix}$$

$$A = \begin{bmatrix} -\dfrac{R_1}{L_2} & -\dfrac{1}{L_2} \\ \dfrac{1}{C_4} & -\dfrac{1}{R_3C_4} \end{bmatrix} \qquad B = \begin{bmatrix} \dfrac{R_1}{L_2} \\ 0 \end{bmatrix}$$

and
$U = [i_g]$, a scalar (in case of a single source) in this example.

EXAMPLE 4-9. Consider again the network N of Figure 4-10. Let us follow the steps outlined in the procedure to obtain the matrix state equation.

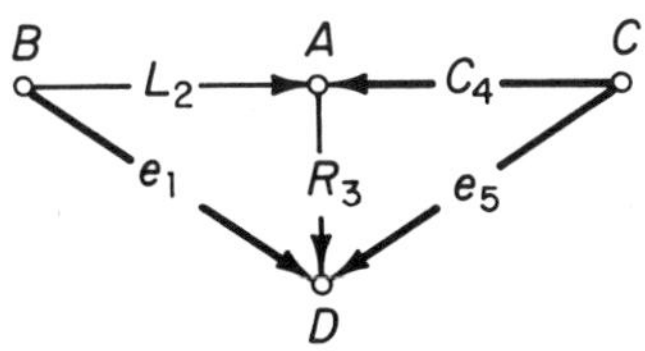

Figure 4-12 The directed graph G of the network in Example 4-7.

Step 1: The directed graph G of N is shown in Figure 4-12.
Step 2: The tree T selected is the one consisting of e_1, e_5, and c_4.
Step 3: The VCR equations are written and grouped as follows:

(a)

$$\begin{bmatrix} \dot{v}_4 \\ \dot{i}_2 \end{bmatrix} = \begin{bmatrix} \dfrac{1}{C_4} & 0 \\ 0 & \dfrac{1}{L_2} \end{bmatrix} \begin{bmatrix} i_4 \\ v_2 \end{bmatrix} \tag{4-87}$$

(b)
$$\begin{bmatrix} i_3 \\ 0 \end{bmatrix} = \begin{bmatrix} 0 & \frac{1}{R_3} \\ 0 & 0 \end{bmatrix} \begin{bmatrix} 0 \\ v_3 \end{bmatrix} \tag{4-88}$$

Step 4: The fundamental circuit and cutset equations are written according to the grouping of elements in Step 3 (for easy substitution in the next step).

(a)
$$\begin{bmatrix} i_4 \\ v_2 \end{bmatrix} = \begin{bmatrix} 0 & -1 \\ 1 & 0 \end{bmatrix} \begin{bmatrix} v_4 \\ i_2 \end{bmatrix} + \begin{bmatrix} 1 & 0 \\ 0 & 0 \end{bmatrix} \begin{bmatrix} i_3 \\ 0 \end{bmatrix} + \begin{bmatrix} 0 & 0 \\ 1 & -1 \end{bmatrix} \begin{bmatrix} e_1 \\ e_5 \end{bmatrix} \tag{4-89}$$

(b)
$$\begin{bmatrix} 0 \\ v_3 \end{bmatrix} = \begin{bmatrix} 0 & 0 \\ -1 & 0 \end{bmatrix} \begin{bmatrix} v_4 \\ i_2 \end{bmatrix} + \begin{bmatrix} 0 & 0 \\ 0 & 1 \end{bmatrix} \begin{bmatrix} e_1 \\ e_5 \end{bmatrix} \tag{4-90}$$

Step 5: Substituting Equation (4-89) into (4-87), we find

$$\begin{bmatrix} \dot{v}_4 \\ \dot{i}_2 \end{bmatrix} = \begin{bmatrix} 0 & -\frac{1}{C_4} \\ \frac{1}{L_2} & 0 \end{bmatrix} \begin{bmatrix} v_4 \\ i_2 \end{bmatrix} + \begin{bmatrix} \frac{1}{C_4} & 0 \\ 0 & 0 \end{bmatrix} \begin{bmatrix} i_3 \\ 0 \end{bmatrix} + \begin{bmatrix} 0 & 0 \\ \frac{1}{L_2} & -\frac{1}{L_2} \end{bmatrix} \begin{bmatrix} e_1 \\ e_5 \end{bmatrix} \tag{4-91}$$

Next, substituting Equation (4-88) and then (4-90) into (4-91), we have

$$\begin{bmatrix} \dot{v}_4 \\ \dot{i}_2 \end{bmatrix} = \begin{bmatrix} 0 & -\frac{1}{C_4} \\ \frac{1}{L_2} & 0 \end{bmatrix} \begin{bmatrix} v_4 \\ i_2 \end{bmatrix} + \begin{bmatrix} -\frac{1}{R_3C_4} & 0 \\ 0 & 0 \end{bmatrix} \begin{bmatrix} v_4 \\ i_2 \end{bmatrix} + \begin{bmatrix} 0 & \frac{1}{R_3C_4} \\ 0 & 0 \end{bmatrix} \begin{bmatrix} e_1 \\ e_5 \end{bmatrix}$$
$$+ \begin{bmatrix} 0 & 0 \\ \frac{1}{L_2} & -\frac{1}{L_2} \end{bmatrix} \begin{bmatrix} e_1 \\ e_5 \end{bmatrix} \tag{4-92}$$

Step 6: Finally, simplifying the right-hand side of Equation (4-92), we obtain the desired result:

$$\begin{bmatrix} \dot{v}_4 \\ \dot{i}_2 \end{bmatrix} = \begin{bmatrix} -\frac{1}{R_3C_4} & -\frac{1}{C_4} \\ \frac{1}{L_2} & 0 \end{bmatrix} \begin{bmatrix} v_4 \\ i_2 \end{bmatrix} + \begin{bmatrix} 0 & \frac{1}{R_3C_4} \\ \frac{1}{L_2} & \frac{-1}{L_2} \end{bmatrix} \begin{bmatrix} e_1 \\ e_5 \end{bmatrix}$$

which gives the same two state equations as those in Equation (4-81) with $v_4 = x_2$ and $i_2 = x_1$ as defined in Equation (4-75).

Obviously, the algorithm for formulating the matrix state equation of a given network of the system described here is not always the best method to use as we have seen from the foregoing examples. In fact, for a simple network such as those used in the examples, it is much easier to obtain the state equations by inspection or by mixed use of the loop equations and node equations. This point becomes evident by comparing the procedure used in Example 4-6 with that in Example 4-8, and the method used in Example 4-7 with the algorithm used in Example 4-9.

The usefulness of the algorithm, then, lies in the fact that the method lends itself readily to computer programming so that the algorithm can be used as a part of the network analysis program.

After the matrix state equation is obtained for a given network, the state vector X can be solved using any method.[5] Once the state vector is known, the next step is to express the output variables, $y_1, y_2, \ldots, y_p$, in matrix form,

$$Y = CX + DU \tag{4-93}$$

where Y is the p-vector, the *output vector*, containing all of the output variables $(y_1, y_2, \ldots, y_p)$; C and D are the matrices of constant entries of order $p \times n$ and $p \times m$, respectively, relating the output vector Y to the state vector X and the input vector U. Finally, by substituting the known vector X into Equation (4-93), the Y vector is determined and the final answer (the values of $y_1, y_2, \ldots, y_p$) obtained.

We should point out that the procedure for obtaining the state equations outlined above applies also to a general RLC network, allowing the possibility of the presence of all-inductor (and/or current-source) cutsets and/or all-capacitor (and/or voltage-source) loops. It can even be extended, with additional topological considerations, to include networks with mutual inductances as well as other multiterminal elements. The interested reader is referred to literature elsewhere.[6]

In conclusion, it is important to note that the state-variable approach is very useful for time-domain analysis, and is especially suited for computer calculation. There have been considerable significant developments in this area in recent years; however, the coverage of such topics is beyond the scope of this text.[6]

4.7 DUAL NETWORKS

We defined duality and discussed the properties of dual graphs in Chapter 3. We also observed a striking parallelism between the developments of the loop-analysis and the node-analysis systems of equations in the preceding sections of this chapter. In this section, we shall make use of the theory of duality in the analysis of electrical networks.

DEFINITION 4-1. Dual Networks. A network N_1 has a *dual* N_2 if (a) There exists a one-to-one correspondence between the branches of N_1 and the branches of N_2.

(b) The *v-i* equations for the branches of N_1 become the *i-v* equations for the corresponding branches of N_2 upon the interchange of symbols. For

[5]See Kuh and Rohrer [48] and Zadeh and Desoer [85], for example.

[6]The interested reader may refer to Kuh and Rohrer [48], Pottle [69, 70], Koenig and Tokad [47], Grigsby and Blackwell [37], LePage [51], and Branin [5, 6], to name a few.

any branch j of N_1 and the corresponding branch j' of N_2, we have the following correspondence:

$$v_j \leftrightarrow i_{j'}$$
$$i_j \leftrightarrow v_{j'}$$
$$R_j \leftrightarrow G_{j'}$$
$$L_j \leftrightarrow C_{j'}$$
$$C_j \leftrightarrow L_{j'}$$

where v stands for voltage, i for current, R for resistance, G for conductance, L for inductance, and C for capacitance.

(c) The KCL equations of N_1 become the KVL equations of N_2, and the KVL equations of N_1 become the KCL equations of N_2 upon the interchange of symbols as described in (b).

Similarly, we might note that the charge q in the network N_1 corresponds to the flux linkages λ in the dual network N_2, and vice versa. Since there exists no duality to the quality of mutual inductance M, we state without proof the following theorem, based on Theorem 3-4, which concerns the existence of a dual graph.

THEOREM 4-1. A network N_1 has a dual network N_2 if and only if
(a) The graph of N_1 is planar, and
(b) N_1 contains no mutual inductances.

EXAMPLE 4-10. Let us consider the network N of Figure 4-13. We see that N is actually a planar network, although it might not appear to be one

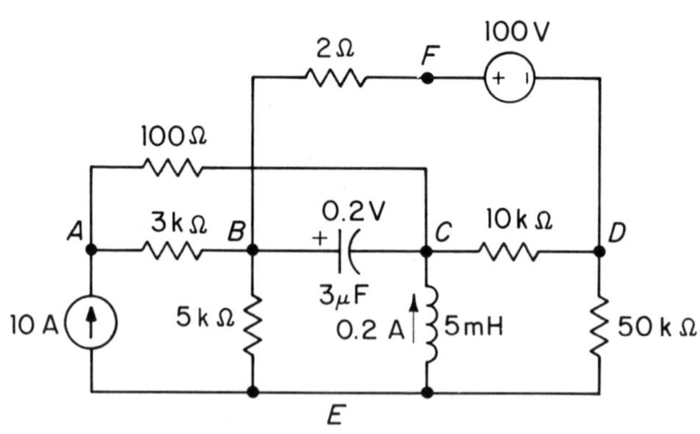

Figure 4-13 Network N.

at first sight. This fact becomes evident when N is redrawn as depicted in Figure 4-14(a). The graph of the redrawn network is shown in Figure 4-14(b). To obtain the dual N' of N, first we employ the dot method to find the dual

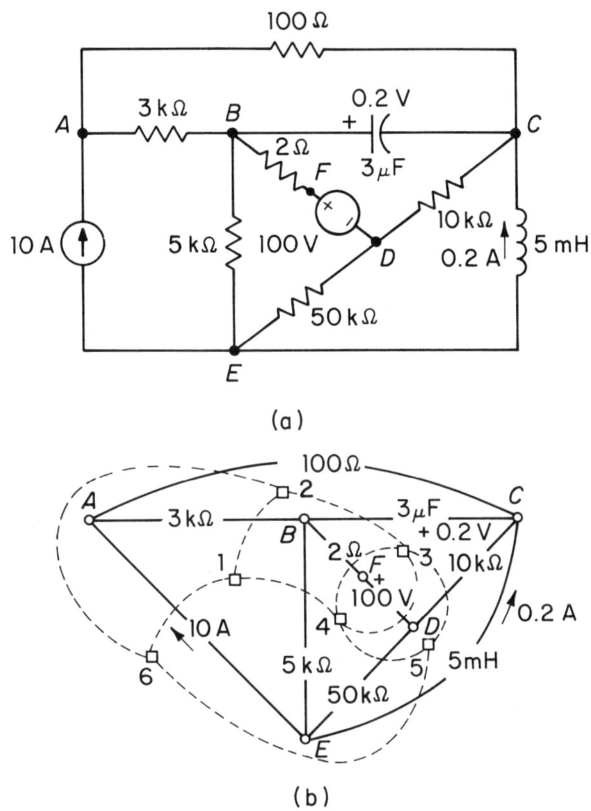

Figure 4-14 (a) The redrawn network of N. (b) The graph of (a).

of its graph, and then change the symbols and, hence, the units in each branch in the dual graph, retaining the numerical value in that branch. The resultant dual network N' is shown in Figure 4-15.

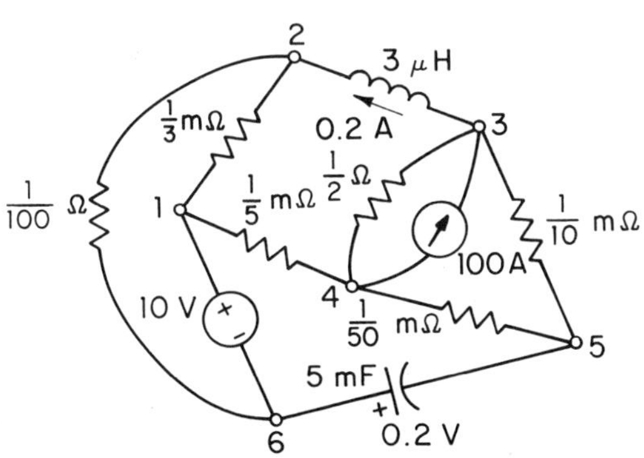

Figure 4-15 The dual network N' of the network of Figure 4-14(a).

It should be noted that the plus voltage-polarity marks for the voltages (voltage sources as well as initial capacitor voltages) and the positive directions for the currents (current sources and initial inductor currents) in the dual network are obtained by observing the correspondence between the loops of N and the nodes of N'. For example, if we trace in N the loop corresponding to node 2 in N' in the clockwise direction (which is assumed to be positive), we see that a voltage rise (which is a negative quantity) of 0.2 volt is encountered as the initial voltage across the capacitor. Thus, in N', the inductor corresponding to the capacitor in N has a negative initial current of 0.2 ampere with respect to node 2; that is, the initial inductor current flows *toward* the node since a positive current flows away from the node according to our convention.

We should emphasize that in determining the voltage-polarity marks and current directions in the dual, we always follow the convention that the "+" mark is placed at the tail of the current arrow even though the edge represents a source. We should also remark that when a switch is involved in obtaining duality in the dot method, it can be treated as a network element. The corresponding element in the dual network is again a switch, noting that the *opening* of a switch in the given network corresponds to the *closing* of the corresponding switch in the dual. The reasoning behind it is quite obvious and is thus left to the reader.

We are now ready to answer the four questions that we posed at the beginning of Chapter 3. To answer the first question: Why does each window correspond to a unique independent KVL equation?, we observe that each windowpane corresponds to a loop (KVL) equation in the given (planar) network N. This loop (KVL) equation, in turn, corresponds to a node (KCL) equation in the dual network N'. Thus, if m denotes the total number of windows in N, then the m KVL equations written for the m loops in N correspond to the m independent (KCL) equations for the m $(= v' - 1)$ nodes of N', where v' is the total number of nodes of N'. That all the m (KCL) equations are independent follows directly from the fact that the rank of the incidence matrix, $A_{N'}$, of N', is $v' - 1$ $(= m)$.

To answer the second question (Why does the total number of windows correspond exactly to the total number of independent KVL equations that completely describe the given network?) we simply need to prove that m, the total number of windows, is exactly equal to $e - v + 1$, where e is the total number of branches and v is the total number of nodes of N.

If we let B_N be a circuit submatrix of N, and $Q_{N'}$ be a cutset submatrix of N', then, from our previous discussion on dual graphs, we know that

$$B_N \leftrightarrow Q_{N'}$$

so that

$$R(B_N) = R(Q_{N'})$$

or

$$e - v + 1 = v' - 1 \tag{4-94}$$

But

$$m = v' - 1 \tag{4-95}$$

Hence, from Equations (4-94) and (4-95), we find that

$$m = e - v + 1 \tag{4-96}$$

The third question in Section 3-1 (Why do we need exactly $n - 1$ KCL equations for an n-node network?) can be answered readily by observing the fact that the n KCL equations for the n nodes of the network N correspond to the rows of the incidence matrix, A_a of N, and that the rank of A_a is exactly $n - 1$. This means that there are exactly $n - 1$ independent KCL equations for N that, together, completely describe N.

Finally, to answer the fourth question (If N is nonplanar, how many independent loop equations do we need, and how do we choose the loops so that each of them will correspond to an independent KVL equation?) we simply choose an arbitrary tree from the graph of N and form the $e - v + 1$ independent fundamental circuits by adding to the tree one chord at a time (as discussed in Section 2-3). Since each of the $e - v + 1$ fundamental circuits gives rise to an independent KVL equation, we have a set of $e - v + 1$ independent KVL equations that completely describe N.

4-8 SUMMARY

We shall now summarize our discussions on the primary and secondary systems of equations and the developments of the loop and the node systems of equations in Table 4-1.

TABLE 4-1

A Summary of Developments of Loop System and Node System of Equations

(A)	Primary equations (postulates): KCL: $AI_e(s) = 0$ (1) KVL: $BV_e(s) = 0$ (2) VCR: $V_e(s) = Ze(s)I_e(s) + E_e(s) + \frac{1}{s}V_e(0^+) - L_eI_e(0^+)$ (3)
(B)	Secondary equations Mesh transformation: $I_e(s) = B^TI_m(s)$ (4) Node transformation: $V_e(s) = A^TV_n(s)$ (5)

TABLE 4-1 (Continued)

Step ↓ Loop Analysis	Step ↓ Node Analysis (assuming that Z_e^{-1} exists)
(a) From (2) and (3) obtain $BZ_eI_e = -B\left\{E_e + \frac{1}{s}V_e(0^+) - L_eI_e(0^+)\right\}$ (6)	(a) Solve for $I_e(s)$ in (3) $I_e = Z_e^{-1}\left\{V_e - E_e + L_eI_e(0^+) - \frac{1}{s}V_e(0^+)\right\}$ (9)
(b) Substitute (4) into (6) to get $Z_mI_m = -B\left\{E_e + \frac{1}{s}V_e(0^+) - L_eI_e(0^+)\right\}$ where $Z_m = BZ_eB^T$ (7)	(b) From (1) and (9) obtain $AZ_e^{-1}V_e = AZ_e^{-1}\left\{E_e + \frac{1}{s}V_e(0^+) - L_eI_e(0^+)\right\}$ (10)
(c) Define $E_m(s) = -BE_e(s)$ $E_c(0^+) = -BV_e(0^+)$ $E_\lambda(0^+) = BL_eI_e(0^+)$ and rewrite (7) as $Z_m(s)I_m(s) = E_m(s) + \frac{1}{s}E_c(0^+) + E_\lambda(0^+)$ (8)	(c) Substitute (5) into (10) to get $Y_nV_n = AZ_e^{-1}\left\{E_e + \frac{1}{s}V_e(0^+) - L_eI_e(0^+)\right\}$ (11) where $Y_n = AZ_e^{-1}A^T$
	(d) Define $I_n(s) = AZ_e^{-1}(s)E_e(s)$ $Q_c(s) = \frac{1}{s}AZ_e^{-1}(s)V_e(0^+)$ $\frac{1}{s}I_L(0^+) = -AZ_e^{-1}(s)L_eI_e(0^+)$ and rewrite (11) as $Y_n(s)V_n(s) = I_n(s) + Q_c(0^+) + \frac{1}{s}I_L(0^+)$ (12)

In Table 4-1, the outline for the development of the node-analysis matrix equation is obviously for the case where $Z_e^{-1}(s)$ exists. The case where $Z_e(s)$ is singular is not included in the table since the main objective of this table is merely to show the parallelism between the loop and the node analysis.

Next, the procedure for formulating the state equations for an *RLC* network without mutual inductances, all-capacitor voltage-source loops, and all-inductor current-source cutsets is summarized in Table 4-2.

In Table 4-2, we note once again that the state equations are formulated from the VCR, the KVL, and the KCL equations that constitute the necessary and sufficient set of equations for a given network as asserted earlier in this chapter.

TABLE 4-2

Procedure for Formulating State Equations of an *RLC* Network *N*

Step	*Operation*
(1)	Obtain from the given network N the directed graph G.
(2)	Select a tree T from G such that T contains (a) All of the voltage sources (b) None of the current sources (c) A maximum number of capacitors (d) A minimum number of inductors.
(3)	Write the v-i relationship (VCR) equations for the passive elements and arrange these equations in two groups: Group (a): Branch capacitors and chord inductors Group (b): All other passive elements.
(4)	Write the fundamental circuit (KVL) and the fundamental cutset (KCL) equation and arrange these equations according to the grouping of elements in Step (3) for each substitution in the next elimination process.
(5)	Eliminate all the undesired variables (which are all variables other than the state variables and the sources) by substituting the equations in Step (4) into those equations in Step (3).
(6)	Rearrange the terms and express the resultant equations in standard form [Equation (4-73)], which is the matrix state equation for the given network N.

Finally, we summarize the one-to-one correspondence of various quantities between a planar network N (without mutual inductances) and its dual network N' in Table 4-3.

TABLE 4-3

Correspondence of Quantities between N and N'

Network N	*Relationship*	*Network N'*
B_N (or KVL equations) Q_N (or KCL equations)	$\longleftrightarrow$ $\longleftrightarrow$	$Q_{N'}$ (or KCL equations) $B_{N'}$ (or KVL equations)
$R(B_N)$ $R(Q_N)$	$=$ $=$	$R(Q_{N'})$ $R(Q_{N'})$
For any branch j:		For the corresponding branch j':
v-i equation v_j i_j R_j L_j C_j q_j switch s (open at $t = 0$) (closed at $t = 0$)	$\longleftrightarrow$ $\longleftrightarrow$ $\longleftrightarrow$ $\longleftrightarrow$ $\longleftrightarrow$ $\longleftrightarrow$ $\longleftrightarrow$ $\longleftrightarrow$ $\longleftrightarrow$	i-v equation $i_{j'}$ $v_{j'}$ $G_{j'}$ $C_{j'}$ $L_{j'}$ $\lambda_{j'}$ switch s' (closed at $t = 0$) (open at $t = 0$)

PROBLEMS

4-1 For the network N shown in Figure P4-1, write the KVL and KCL equations in matrix form.

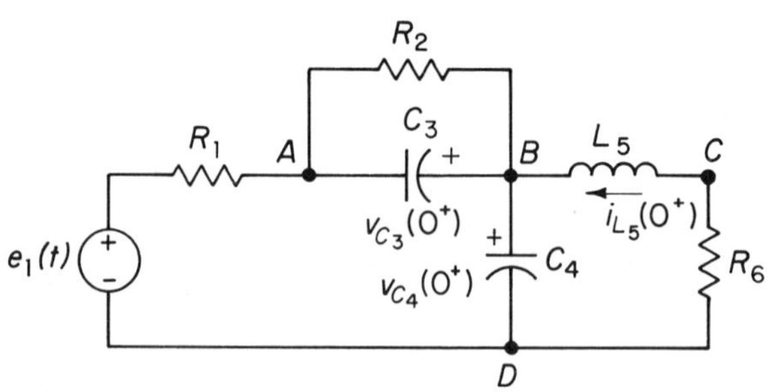

Figure P4-1

4-2 Refer to Figure P4-1. Write the VCR equations in matrix form and then write the transformed VCR matrix equation in s.

4-3 Derive the mesh-transformation matrix equation:

$$I_e(t) = B^T I_m(t)$$

where $I_e(t)$ is the vector of branch currents and $I_m(t)$ the vector of loop currents.

4-4 Obtain the loop-analysis matrix equation for the network depicted in Figure P4-4.

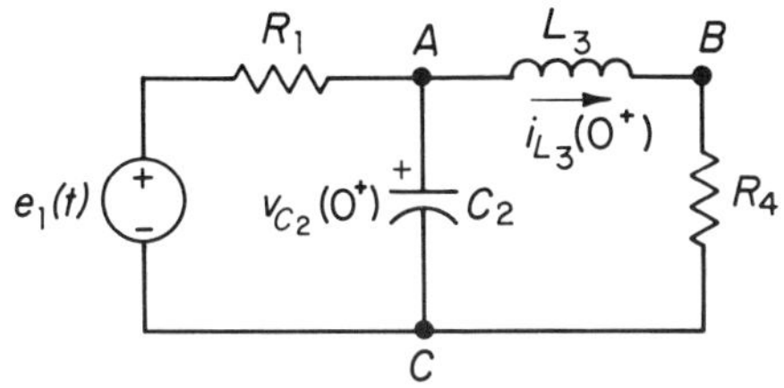

Figure P4-4

4-5 Find the loop-analysis matrix equation for the network N of Figure P4-1.

4-6 Obtain the loop-analysis matrix equation and solve for i_{m_3} for the network shown in Figure P4-6, assuming an ideal transformer with perfect coupling.

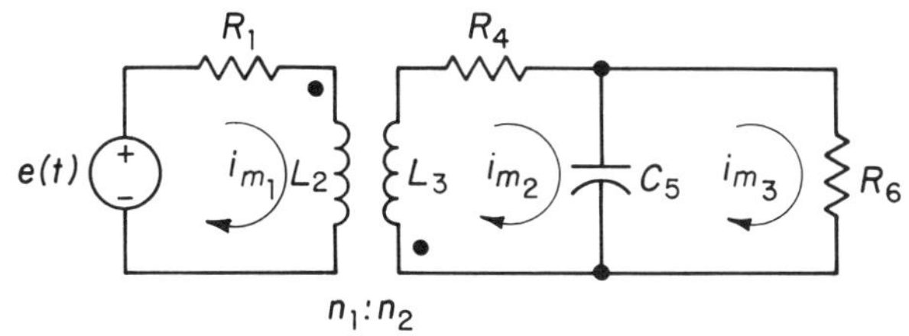

Figure P4-6

4-7 Find the node-analysis matrix equation for the network of Figure P4-4.

4-8 Refer to Figure P4-1. Find the node-analysis matrix equation for network N.

4-9 Find the node-analysis equation for the network shown in Figure P4-9.

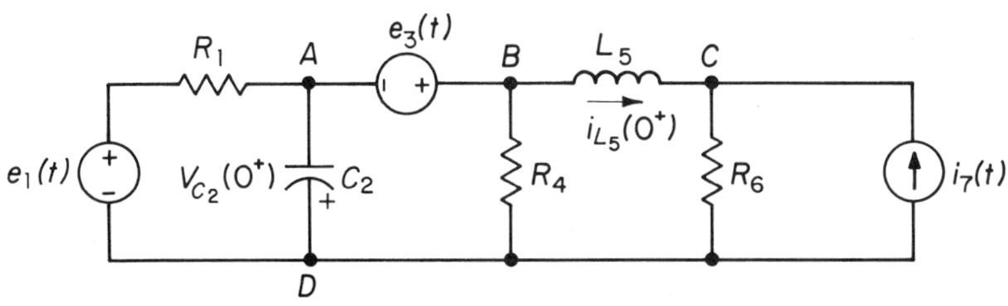

Figure P4-9

4-10 Find the node-analysis matrix equation for the network shown in Figure P4-10.

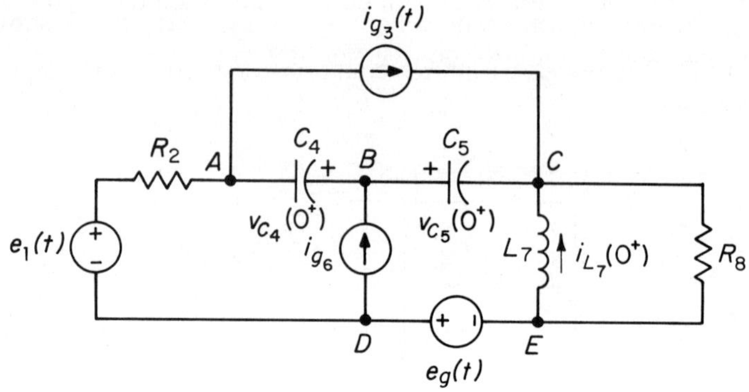

Figure P4-10

4-11 Obtain a dual network N' of the network N depicted in Figure P4-11.

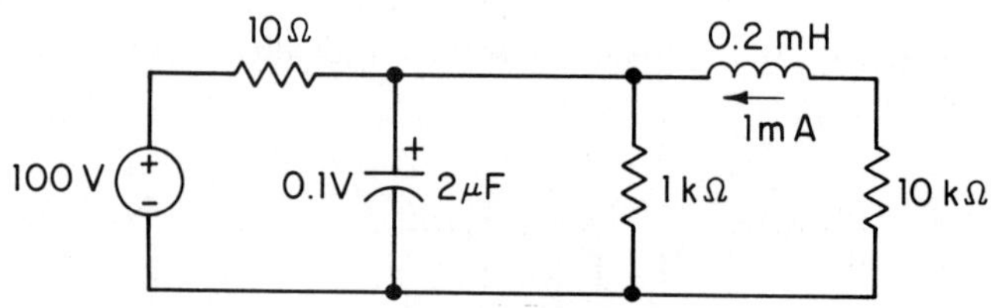

Figure P4-11

4-12 Find a dual network of the network shown in Figure P4-12.

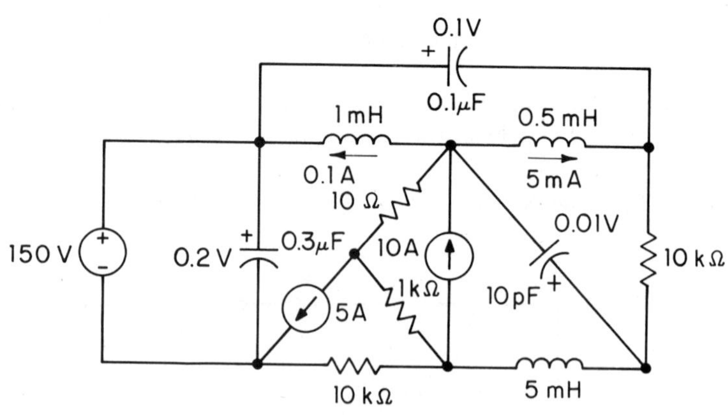

Figure P4-12

4-13 Show that any two of the following equations imply the third:

(a) $AI_e(t) = 0$. (b) $BV_e(t) = 0$. (c) $I_e^T(t)V_e(t) = 0$.

4-14 Obtain the loop-analysis matrix equation for the network N shown in Figure P4-14(a) by the procedure given in Table 4-1.

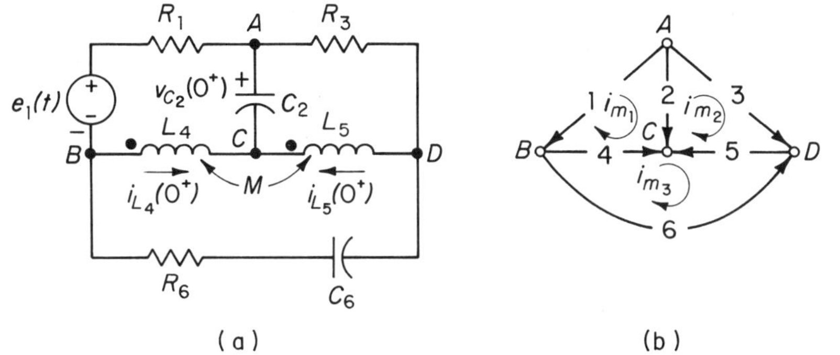

Figure P4-14

4-15 For the network N of Figure P4-15(a), obtain the node-analysis matrix equation following Case (B) of the procedure given in Section 4-5. Take node E as the reference node. Make necessary generator transformations (Blakesley transformations) so that after the transformations those edges containing generators alone contain current generators only.

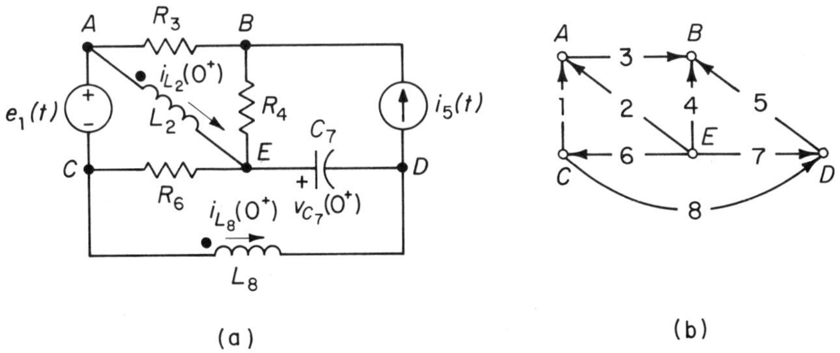

Figure P4-15

4-16 In the network of Figure P4-16, the arrows indicate combined current and voltage references (the voltage + is kept at the tail of each arrow). Select a tree, then express each chord voltage in terms of tree-branch voltages. Also, express each tree-branch current in terms of chord currents (for the same tree). Put the answers in the form of matrix products. How are the two "transformation matrices" related?

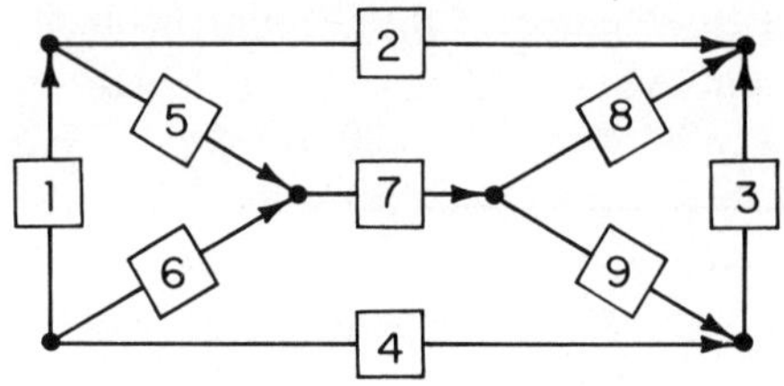

Figure P4-16

4-17 Based on Kirchhoff's current law and voltage law, and the properties of directed graphs, prove the following statement: For any network consisting of e branches whose voltages and currents satisfy Kirchhoff's laws,

$$\sum_{j=1}^{e} v_j(t) i_j(t) = 0$$

Remarks: In elementary circuit theory, this relationship is usually justified through the "conservation of power." This problem shows that the relationship is a direct consequence of Kirchhoff's laws and network topology.
Hint: Express $I_e(t)$ in terms of $I_m(t)$, and $V_e(t)$ in terms of $V_n(t)$, then calculate $V_e^T I_e$.

4-18 (a) In the network shown in Figure P4-18(a), R_n is a nonlinear resistance whose v-i relationship is given by

$$\frac{v_n}{i_n} = 1 + 0.5 i_n$$

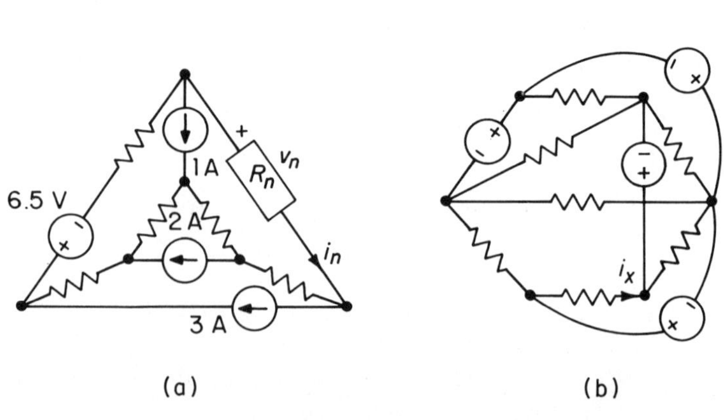

Figure P4-18

Others are 1-ohm resistors. Find i_n.

(b) In the network shown in Figure P4-18(b), each resistance is one ohm, and each voltage source is one volt. Find i_x.

Remarks: Any EE sophomore knows how to solve the two parts to this problem in principle. Your credit, therefore, lies in the simplicity of the solution. Above all, you should make a correct choice among different methods of attack, with the aim of reducing the number of equations to be solved simultaneously.

4-19 Let N be a connected network with e edges and v vertices.

(a) State *as many methods as you can* for choosing a set of $e - v + 1$ circuits that leads to $e - v + 1$ independent Kirchhoff's voltage equations.

(b) Let $S(a, \cdots)$ be a set of edges of N not containing any cutsets. State *as many methods as you can* for choosing $e - v + 1$ independent circuits such that each edge of S appears only once in the chosen circuits.

4-20 A connected network N has v nodes and e "composite" branches of the form shown in Figure P4-20, where z is a resistance, a capacitance, or an inductance (coupled to other inductances). At $t = 0$, the stored energy of the network is zero.

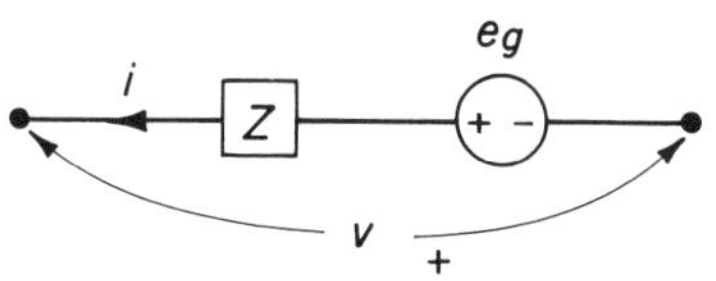

Figure P4-20

(a) Express the vector $I_e(s)$ explicitly in terms of the branch impedance matrix $Z_e(s)$, the voltage source vector $E_e(s)$, and a circuit submatrix B of the directed graph G associated with N.

(b) Discuss the conditions under which a unique solution of $I_e(s)$ exists.

4-21 Obtain the dual N' of the network N shown in Figure P4-21. In the dual network N', indicate numeral values, units, current directions and voltage polarities.

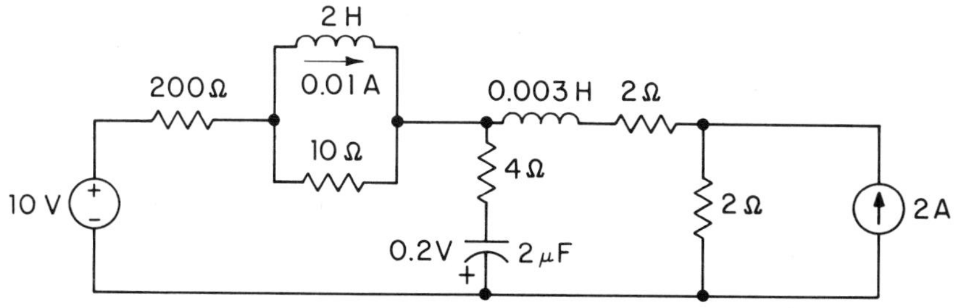

Figure P4-21

4-22 Repeat Problem 4-21 for each of the networks shown in Figure P4-22. Element values are in ohms, henrys, and farads. Diodes are ideal.

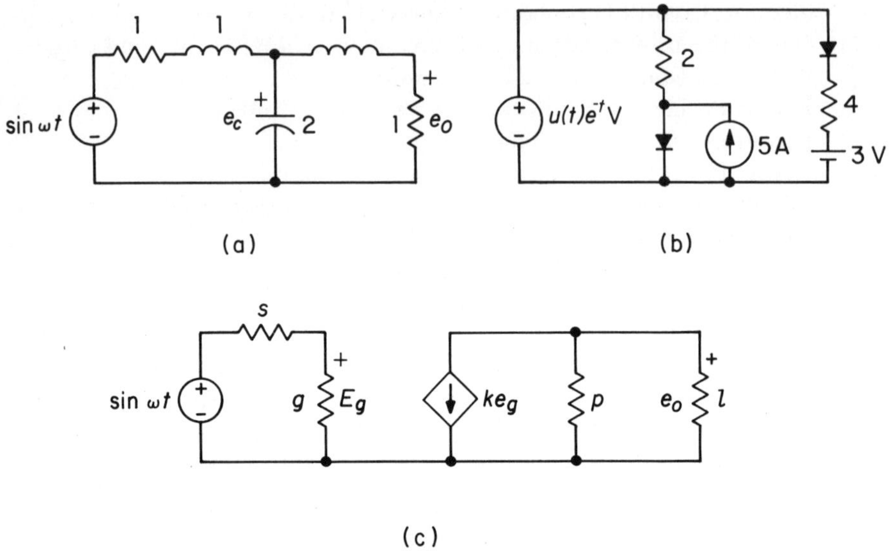

Figure P4-22

4-23 For each of the networks shown in Figure P4-23, using the concept of dual networks, construct a one-port network having a driving-point impedance $Z_b(s)$ related to $Z_a(s)$ by $Z_a(s)Z_b(s) = 4$ ohm^2.
Note: A one-port network or a one-terminal-pair network is a network with a pair of nodes specified as input terminals.

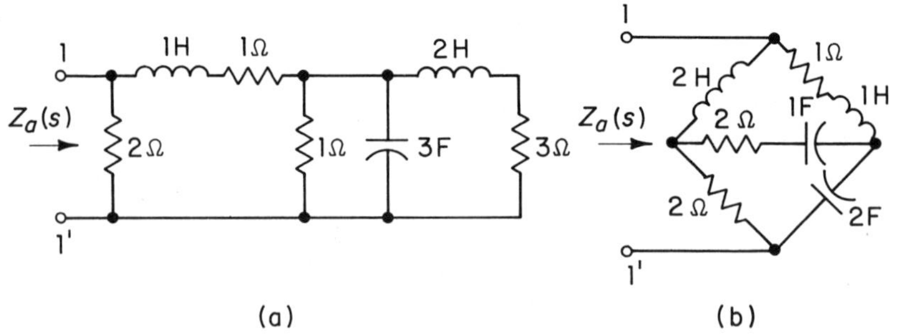

Figure P4-23

4-24 Present a scheme for writing
(a) The loop-analysis matrix equation [Equation (4-47)]

(b) The node-analysis matrix equation [Equation (4-59) or (4-70)] directly from a network.

Explain clearly in each case how each term in the equation may be obtained from the network by inspection.

4-25 Obtain the state equations for the network of Figure P4-1 and express them in standard form [Equation (4-73)].

4-26 Repeat Problem 4-25 for the network of Figure P4-4.

4-27 Repeat Problem 4-25 for the network of Figure P4-9.

4-28 Repeat Problem 4-25 for the network of Figure P4-10.

4-29 Modify the procedure for obtaining state equations outlined in Section 4-6 to include networks with all-capacitor and/or voltage-source loops, and/or all-inductor and/or current-source cutsets.

4-30 Apply the procedure obtained in Problem 4-29 to an *RLC* network with one all-capacitor and/or voltage-source loop, and one all-inductor and/or current-source cutset. (Make up your own network.)

4-31 Generalize your result obtained in Problem 4-29 to include networks with mutual inductances.

CHAPTER 5

TOPOLOGICAL ANALYSIS OF PASSIVE ONE-PORTS

5-1 INTRODUCTION

The development of the analysis of passive electrical networks using topological concepts may be dated back to 1847 when Kirchhoff[1] formulated his set of topological formulas in terms of resistances and the branch-current system of equations. In 1892, Maxwell[2] developed another set of topological formulas, which are the duals of those of Kirchhoff, based on the node equations. These two sets of formulas were supported mainly by heuristic reasoning and no formal proofs were then available.

In this chapter, we shall derive Maxwell's topological formulas for the passive *one-port*[3] without mutual inductances. The proofs of the theorems developed in this chapter, as well as the next, are based mainly on the work of Mayeda and Seshu [62], and we shall follow the terminology used in that paper as closely as possible.

Consider a network N with n independent nodes as shown in Figure 5-1. The node $1'$ is taken as the reference (datum) node. The voltages, $V_1, V_2, \ldots, V_n$, which are functions of s, are the transforms of the node-pair voltages (or

[1]For his original work, see G. Kirchhoff [46].

[2]For his original work, see J. C. Maxwell [58].

[3]A one-port or a one-terminal-pair network is a network with a pair of nodes specified as input terminals (Figure 5-2). In general, an n-port or an n-terminal-pair network is a network in which n pairs of nodes are of interest. Refer to Figures 6-1 and 6-3(a) for two illustrations of two-ports.

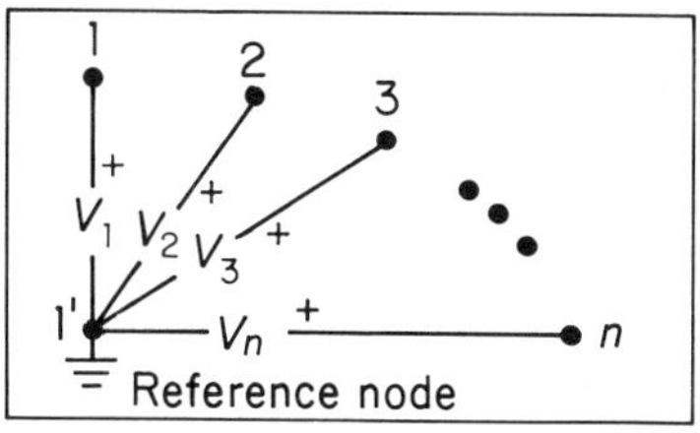

Figure 5-1 A network N with n independent nodes.

simply node voltages), $v_1, v_2, \ldots, v_n$, which are functions of t, between the n nodes and the reference node 1′ with the plus polarity marks at the n nodes. Using the node analysis method developed in Chapter 4, we obtain the n independent node equations:

$$\begin{aligned} \mathcal{Y}_{11}V_1 + \mathcal{Y}_{12}V_2 + \cdots + \mathcal{Y}_{1n}V_n &= I_1 \\ \mathcal{Y}_{21}V_1 + \mathcal{Y}_{22}V_2 + \cdots + \mathcal{Y}_{2n}V_n &= I_2 \\ &\cdot \\ &\cdot \\ &\cdot \\ \mathcal{Y}_{n1}V_1 + \mathcal{Y}_{n2}V_2 + \cdots + \mathcal{Y}_{nn}V_n &= I_n \end{aligned} \tag{5-1}$$

where, for any j ($j = 1, 2, \ldots, n$), $\mathcal{Y}_{jj}$ is the sum of the admittances of all the branches connected to node j; and, for any j and k ($j \neq k; j, k = 1, 2, \ldots, n$), $\mathcal{Y}_{jk}$ is the negative of the sum of the admittances of all the branches connected between nodes j and k. (Thus, all such branches are connected in parallel across the two nodes.) Furthermore, I_j is the sum of the transforms of the known current sources entering node j; and V_j is the transform of the voltage at node j with respect to the reference node 1′.

Written in matrix form, Equation (5-1) becomes

$$\begin{bmatrix} \mathcal{Y}_{11} & \mathcal{Y}_{12} & \cdots & \mathcal{Y}_{1n} \\ \mathcal{Y}_{21} & \mathcal{Y}_{22} & \cdots & \mathcal{Y}_{2n} \\ \cdot & & & \\ \cdot & & & \\ \cdot & & & \\ \mathcal{Y}_{n1} & \mathcal{Y}_{n2} & \cdots & \mathcal{Y}_{nn} \end{bmatrix} \begin{bmatrix} V_1 \\ V_2 \\ \cdot \\ \cdot \\ \cdot \\ V_n \end{bmatrix} = \begin{bmatrix} I_1 \\ I_2 \\ \cdot \\ \cdot \\ \cdot \\ I_n \end{bmatrix} \tag{5-2}$$

or, in abbreviated matrix notation,

$$Y_n V_n = I_n \tag{5-3}$$

where Y_n is the node admittance matrix, V_n the n-vector of the node voltage transforms, and I_n the n-vector of the transforms of the known current sources as expressed in Equation (5-2).

Solving for the node voltages upon the application of Cramer's rule, we find

$$V_j = \frac{\Delta_{1j}}{\Delta}I_1 + \frac{\Delta_{2j}}{\Delta}I_2 + \cdots + \frac{\Delta_{nj}}{\Delta}I_n \qquad (j = 1, 2, \cdots, n) \tag{5-4}$$

where Δ is the determinant of the node admittance matrix Y_n and Δ_{jk} is the (i,k) cofactor of Δ.

Now, let us assume that the network N is a relaxed passive one-port (so that the initial conditions are all zero) and drive it with a single current source I_1 at nodes 1 and 1′ as depicted in Figure 5-2.

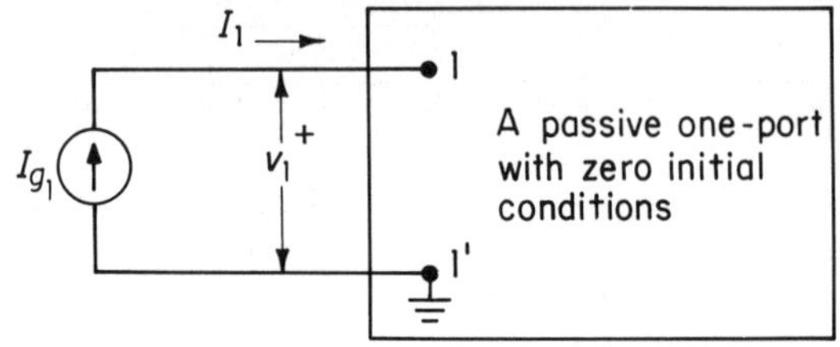

Figure 5-2 The network N driven by a single current source.

Thus, with the conditions

$$I_1 = I_{g1} \qquad \text{and} \qquad I_2 = I_3 = \cdots = I_n = 0$$

Equation (5-4) becomes

$$V_j = \frac{\Delta_{1j}}{\Delta} I_1 \qquad (j = 1, 2, \cdots, n) \tag{5-5}$$

from which, letting $j = 1$, we obtain

$$V_1 = \frac{\Delta_{11}}{\Delta} I_1 \tag{5-6}$$

Hence the driving-point impedance function, $Z_d(s)$, and its reciprocal, namely, the driving-point admittance function $Y_d(s)$, are readily obtained from Equation (5-6) as

$$Z_d(s) \triangleq \frac{V_1}{I_1} = \frac{\Delta_{11}}{\Delta} \tag{5-7}$$

and

$$Y_d(s) = \frac{1}{Z_d(s)} = \frac{\Delta}{\Delta_{11}} \tag{5-8}$$

respectively.

We see from Equations (5-7) and (5-8) that both $Z_d(s)$ and $Y_d(s)$ can be expressed in terms of the determinant of Y_n, Δ, and its (1,1) cofactor Δ_{11}.

It is our objective, in this chapter, to derive a set of topological formulas that relate quantities like Δ and Δ_{11} to the topological properties of the given network so that functions such as $Z_d(s)$ and $Y_d(s)$ may be determined topologically through the application of these formulas. But, before going into the discussion of topological formulas, we shall first give an example to illustrate the use of Equations (5-7) and (5-8) for the determination of $Z_d(s)$ and $Y_d(s)$.

EXAMPLE 5-1. Let it be required to obtain the driving-point impedance $Z_d(s)$ and its reciprocal $Y_d(s)$ of the passive one-port, as shown in Figure 5-3. The matrix equation for the three independent nodes, 1, 2, and 3, is

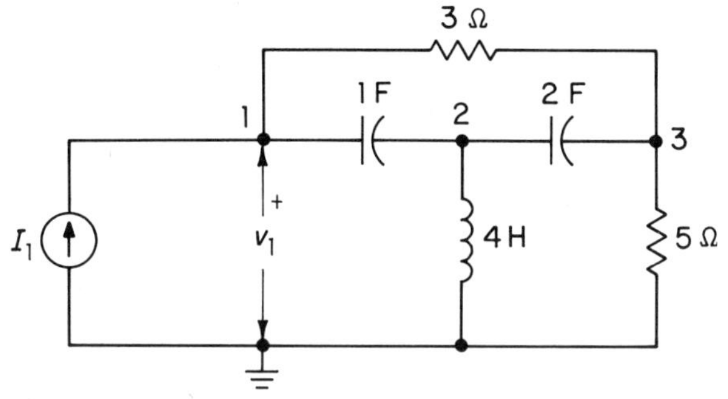

Figure 5-3 A passive one-port with zero initial conditions.

$$\begin{bmatrix} \dfrac{3s+1}{3} & -s & -\dfrac{1}{3} \\ -s & \dfrac{12s^2+1}{4s} & -2s \\ -\dfrac{1}{3} & -2s & \dfrac{30s+8}{15} \end{bmatrix} \begin{bmatrix} V_1 \\ V_2 \\ V_3 \end{bmatrix} = \begin{bmatrix} I_1 \\ 0 \\ 0 \end{bmatrix}$$

Thus,

$$\Delta = \begin{vmatrix} \dfrac{3s+1}{3} & -s & -\dfrac{1}{3} \\ -s & \dfrac{12s^2+1}{4s} & -2s \\ -\dfrac{1}{3} & -2s & \dfrac{30s+8}{15} \end{vmatrix}$$

$$= \left(\frac{3s+1}{3}\right)\left(\frac{12s^2+1}{4s}\right)\left(\frac{30s+8}{15}\right) + (-s)(-2s)\left(-\frac{1}{3}\right) + \left(-\frac{1}{3}\right)\left(-s\right)\left(-2s\right)$$
$$-\left(-\frac{1}{3}\right)\left(\frac{12s^2+1}{4s}\right)\left(-\frac{1}{3}\right) - (-2s)(-2s)\left(\frac{3s+1}{3}\right) - \left(\frac{30s+8}{15}\right)(-s)(-s)$$

or, after simplification,

$$\Delta = \frac{24s^3 + 42s^2 + 18s + 1}{60s} \tag{5-9}$$

Next, we evaluate the (1,1) cofactor of Δ.

$$\Delta_{11} = \begin{vmatrix} \dfrac{12s^2+1}{4s} & -2s \\ -2s & \dfrac{30s+8}{15} \end{vmatrix}$$

$$= \left(\frac{12s^2+1}{4s}\right)\left(\frac{30s+8}{15}\right) - (-2s)(-2s)$$

or

$$\Delta_{11} = \frac{120s^3 + 96s^2 + 30s + 8}{60s} \tag{5-10}$$

From Equations (5-7), (5-9), and (5-10) we find,

$$Z_d(s) = \frac{120s^3 + 96s^2 + 30s + 8}{24s^3 + 42s^2 + 18s + 1} \tag{5-11}$$

and

$$Y_d(s) = \frac{24s^3 + 42s^2 + 18s + 1}{120s^3 + 96s^2 + 30s + 8} \tag{5-12}$$

It should be pointed out that if loop analysis is employed instead of node analysis, two other expressions, which are the duals of Equations (5-7) (5-8), will be obtained for $Z_d(s)$ and $Y_d(s)$. They are

$$Z_d(s) = \frac{\Delta'}{\Delta'_{11}} \tag{5-13}$$

and

$$Y_d(s) = \frac{\Delta'_{11}}{\Delta'} \tag{5-14}$$

where Δ' is the determinant of the loop impedance matrix Δ'_{11} is the (1,1) cofactor of Δ'. The derivations of Equations (5-13) and (5-14) are exactly parallel to the developments of Equations (5-7) and (5-8), and, hence, are left to the reader as a problem.

Obviously, if Equations (5-13) and (5-14) were used instead of (5-7) and (5-8), the results would be identical to those given by (5-11) and (5-12).

5-2 TOPOLOGICAL FORMULA FOR Δ

In Chapter 4 we defined the node admittance matrix as

$$Y_n = AY_eA^T \tag{5-15}$$

where A is the incidence submatrix and Y_e, which may be called the branch admittance matrix, is the inverse of the branch impedance matrix Z_e provided, of course, that Z_e is nonsingular. Throughout this section, we shall assume that the network N under consideration is a passive one-port with zero initial conditions and no mutual inductances. Under such assumptions, Z_e is nonsingular so that $Y_e = Z_e^{-1}$ is a diagonal matrix[4] in the form

[4]It can be shown that under our present assumptions Y_e, which is the inverse of Z_e, always exists. This is included at the end of this chapter as Problem 5-4.

$$Y_e = \begin{bmatrix} y_1 & 0 & 0 & \cdots & 0 \\ 0 & y_2 & 0 & \cdots & 0 \\ \cdot & & & & \\ \cdot & & & & \\ \cdot & & & & \\ 0 & 0 & 0 & \cdots & y_e \end{bmatrix} \tag{5-16}$$

where y_i is the admittance of the ith branch and e is the total number of branches of N.

The determinant of Y_n is given by

$$\Delta = \det Y_n = \det \{AY_eA^T\} = \det \{(AY_e)A^T\} \tag{5-17}$$

In order to derive the topological formula for Δ, we shall need a theorem of matrix theory (the so-called Binet-Cauchy theorem) which is now stated without proof.[5]

THEOREM 5-1. If C and D are two matrices of orders $p \times q$ and $q \times p$, respectively, and if $p \leq q$, then

$$\det CD = \sum_j C_j D_j$$

where C_j, called a *major* of C, is a determinant of order p obtained from C by deleting some $q - p$ *columns;* and D_j, called the *corresponding major* of D to C_j, is a determinant of order p obtained from D by deleting those $q - p$ *rows* corresponding to the $q - p$ deleted columns of C. The summation is over the products of all such corresponding majors.

We now illustrate the ideas involved in this theorem by means of the following example.

EXAMPLE 5-2. Let C and D be the following matrices:

$$C = \begin{bmatrix} 1 & 2 & -1 \\ -2 & 3 & 1 \end{bmatrix} \qquad D = \begin{bmatrix} -1 & 1 \\ 3 & 2 \\ 2 & -1 \end{bmatrix}$$

Then

$$CD = \begin{bmatrix} 3 & 6 \\ 13 & 3 \end{bmatrix} \qquad \text{and} \qquad \det CD = -69$$

The majors of C are:

$$C_1 = \begin{vmatrix} 1 & 2 \\ -2 & 3 \end{vmatrix} = 7 \qquad C_2 = \begin{vmatrix} 1 & -1 \\ -2 & 1 \end{vmatrix} = -1 \qquad C_3 = \begin{vmatrix} 2 & -1 \\ 3 & 1 \end{vmatrix} = 5$$

[5]The proof of this theorem may be found in A. C. Aitken [1], p. 86, or in F. E. Hohn [43], pp. 64–65.

and the corresponding majors of D are, respectively,

$$D_1 = \begin{vmatrix} -1 & 1 \\ 3 & 2 \end{vmatrix} = -5 \qquad D_2 = \begin{vmatrix} -1 & 1 \\ 2 & -1 \end{vmatrix} = -1 \qquad D_3 = \begin{vmatrix} 3 & 2 \\ 2 & -1 \end{vmatrix} = -7$$

Thus, by Theorem 5-1, we find that

$$\begin{aligned} \det CD &= C_1D_1 + C_2D_2 + C_3D_3 \\ &= (7)(-5) + (-1)(-1) + (5)(-7) \\ &= -35 + 1 - 35 \\ &= -69 \end{aligned}$$

which agrees with the value of det CD obtained by first forming the product CD and then evaluating the determinant of the product, as was done earlier.

We are now ready to state and prove the following theorem:

THEOREM 5-2. Let N be a passive network without mutual inductances. The determinant Δ of the node admittance matrix Y_n is equal to the sum of all the tree admittance products of N, where a *tree admittance product*, $T^{(i)}(y)$, is defined to be the product of the admittances of all the branches of the tree $T^{(i)}$; that is,

$$\Delta = \det Y_n = \sum_i T^{(i)}(y) \tag{5-18}$$

Proof: To prove the theorem, we shall show that when Δ is expanded as a sum of the products, each nonzero term in the expansion of Δ corresponds to a tree-admittance product $T^{(i)}(y)$ in

$$\sum_i T^{(i)}(y)$$

We see from Equation (5-17) that

$$\Delta = \det \{(AY_e)\cdot A^T\}$$

Also, since Y_e is a diagonal matrix, we note that the matrices AY_e and A are identical in form with the same order and rank. More precisely, AY_e is obtained from A by multiplying the columns of A by their corresponding edge admittances. Thus, in light of Theorem 5-1,

$$\Delta = \det \{(AY_e)A^T\} = \sum \begin{pmatrix} \text{products of corresponding} \\ \text{majors of } AY_e \text{ and } A^T \end{pmatrix} \tag{5-19}$$

and there exists a one-to-one correspondence between the nonzero majors of AY_e and the nonzero majors of A. This can be easily seen by noting that if det A_i is a nonzero major of A, then the corresponding major of AY_e must also be nonzero since the edge admittances are always nonzero. Next, we recall that the columns of a nonzero major of A correspond to the branches of a tree of N (Theorem 2-13). The value of this nonzero major is either $+1$

or -1.[6] Thus, if we denote by det A_iY_{v-1} the major of AY_e, which corresponds to the nonzero major det A_i of A (Y_{v-1} is a diagonal matrix of order $v-1$ with the (j,j) element equal to y_{k_j}, the admittance value of the edge e_{k_j} identified by the jth column of A_i), the value of det A_iY_{v-1} is $\pm y_{k_1}y_{k_2}\cdots y_{k_{v-1}}$ because

$$\det AY_{v-1} = \det A \det Y_{v-1} = (\pm 1)(y_{k_1}y_{k_2}\cdots y_{k_{v-1}})$$

noting that

$$\det Y_{v-1} = \det \begin{bmatrix} y_{k_1} & 0 & 0 & \cdots & 0 \\ 0 & y_{k_2} & 0 & \cdots & 0 \\ \cdot & & & & \\ \cdot & & & & \\ \cdot & & & & \\ 0 & 0 & 0 & \cdots & y_{k_{v-1}} \end{bmatrix} = y_{k_1}y_{k_2}\cdots y_{k_{v-1}}$$

Furthermore, since, for any nonsingular matrix M, det M = det M^T, the major corresponding to the nonzero major, det AY_{v-1} of AY_e, is also nonzero and has a value of ± 1 ($+1$ if the sign of det AY_{v-1} is plus, and -1 if it has a minus sign). This establishes the one-to-one correspondence between a nonzero product of two corresponding nonzero majors of AY_e and A^T in Equation (5-19) and the tree admittance product, $y_{k_1}y_{k_2}\cdots y_{k_{v-1}}$, of some tree T_i of N with tree branches e_{k_1}, e_{k_2}, . . . , and $e_{k_{v-1}}$. Hence the theorem is true.

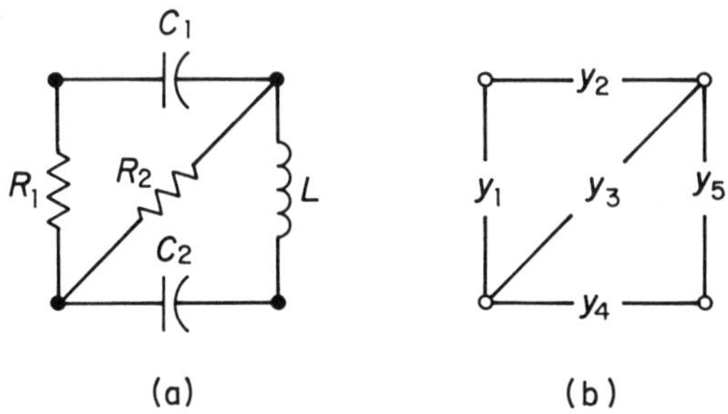

Figure 5-4 (a) A passive network N. (b) Its graph G.

EXAMPLE 5-3. Let us consider the network N of Figure 5-4(a). This graph G of N is shown in Figure 5-4(b) with the admittances of the edges specified as follows:

$$v_1 = \frac{1}{R_1} \qquad y_2 = sC_1 \qquad y_3 = \frac{1}{R_2} \qquad y_4 = sC_2 \qquad y_5 = \frac{1}{sL} \tag{5-20}$$

The node admittance matrix, Y_n of N, is given by

$$Y_n = \begin{bmatrix} y_1 + y_2 & -y_2 & 0 \\ -y_2 & y_2 + y_3 + y_5 & -y_5 \\ 0 & -y_5 & y_4 + y_5 \end{bmatrix}$$

[6]The proof of this statement has been listed as a problem in Chapter 2.

Thus, the determinant Δ of Y by straightforward expansion technique is

$$\Delta = \det Y_n = (y_1 + y_2)(y_2 + y_3 + y_5)(y_4 + y_5) - y_5^2(y_1 + y_2) - y_2^2(y_4 + y_5)$$

or, after cancellation,

$$\Delta = y_1y_2y_4 + y_1y_2y_5 + y_1y_3y_4 + y_1y_3y_5 + y_1y_4y_5 + y_2y_3y_4 + y_2y_3y_5 + y_2y_4y_5 \tag{5-21}$$

Substituting Equation (5-20) into (5-21) and after simplification, we find

$$\Delta = \left(\frac{C_1C_2}{R_1} + \frac{C_1C_2}{R_2}\right)s^2 + \left(\frac{C_2}{R_1R_2} + \frac{C_1C_2}{L}\right)s + \left(\frac{C_1}{R_1L} + \frac{C_2}{R_1L} + \frac{C_1}{R_2L}\right) + \left(\frac{1}{R_1R_2L}\right)\frac{1}{s} \tag{5-22}$$

Next, we shall list all the trees of N and form the tree admittance products in order to evaluate Δ topologically using Equation (5-18).

The set of all trees of G is shown in Figure 5-5. By inspection, it is evident that each tree admittance product of the trees in Figure 5-5 corresponds to a unique term in Equation (5-21). Thus, using Equation (5-18) we obviously will obtain the same result for Δ as that in Equation (5-22).

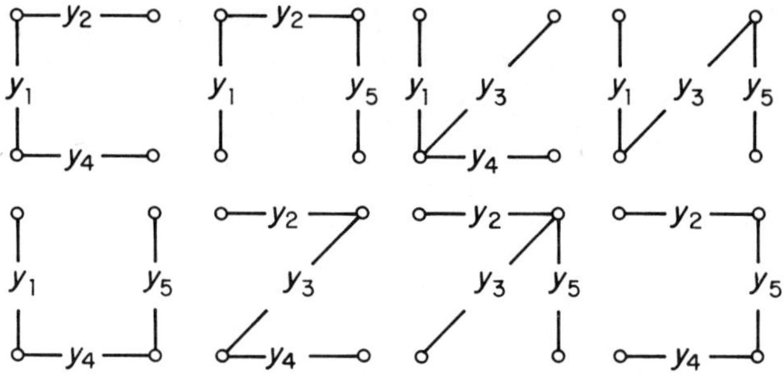

Figure 5-5 The complete set of trees of G.

Example 5-3 not only illustrates the application of the topological formula [Equation (5-18)] but also brings out one of the most significant properties common to this formula and to others which will be developed in the remainder of this chapter and in the next. This is:

In the application of Equation (5-18), no cancellation of terms is involved and, thus, the errors that might occur in the cancellation process are eliminated.

This is, indeed, a great advantage of the topological approach over the conventional method for the evaluation of the determinant Δ.

5-3 TOPOLOGICAL FORMULA FOR Δ_{11}

Before we develop a topological formula for Δ_{11}, the (1,1) cofactor of Δ, we shall define a "2-tree" and a "2-tree admittance product" of a connected graph (network).

DEFINITION 5-1. 2-tree. A *2-tree* T_2 of a connected graph G is a (proper) subgraph of a tree T of G obtained by removing *any* one of the $v - 1$ branches from T.

From this definition, we see that a 2-tree, T_2, of G contains exactly $v - 2$ edges and it includes all of the v vertices of G. Also, T_2 has two *disjoint*, circuitless subgraphs, each of which is connected and either (or both) of which may be an isolated vertex.

DEFINITION 5-2. 2-tree Admittance Product. A *2-tree admittance product* $W^{(i)}(y)$ is the product of the admittances of the $v - 2$ edges of a 2-tree, $T_2^{(i)}$.

Since a 2-tree has two separate parts with a set of k vertices contained in one part (k being a positive integer, $1 \leq k < v$) and the remaining $v - k$ vertices contained in the other, it is sometimes convenient to explicitly specify certain vertices as being separated from certain others in a given 2-tree. Thus we use the symbol

$$T_{2_{p_1 p_2 \cdots p_j, q_1 q_2 \cdots q_k}}$$

to mean that vertices $p_1, p_2, \ldots, p_j$ belong to one part and vertices $q_1, q_2, \ldots, q_k$ belong to the other part of the 2-tree T_2.

EXAMPLE 5-4. To illustrate the meanings of Definitions 5-1 and 5-2, let us use the same network N and its graph G of Figure 5-1 again. A 2-tree, $T_{2_{1,1'}}$, of G is shown in Figure 5-6(a). Of course, the same 2-tree of Figure

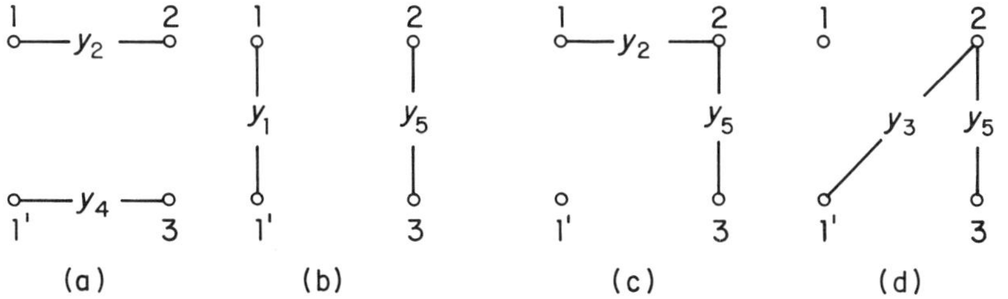

Figure 5-6 The various 2-trees of G of Figure 5-4(b).

5-6(a) may be denoted by $T_{2_{12,1'3}}$ since nodes 1 and 2 are in one part, and nodes 1′ and 3 are in the other part of the 2-tree. Similarly, a 2-tree, $T_{2_{11',23}}$, is illustrated in Figure 5-6(b); a 2-tree, $T_{2_{123,1'}}$, is shown in Figure 5-6(c); and a

2-tree, $T_{2_{1,1'23}}$, is depicted in Figure 5-6(d). The 2-tree admittance products of the 2-trees of Figure 5-6(a), (b), (c), and (d) are y_2y_4, y_1y_5, y_2y_5, and y_3y_5, respectively. Note that each of the 2-trees of Figure 5-6(c) and (d) contains an isolated vertex and we may conveniently define the "admittance" of an isolated vertex to be unity for the sake of completeness.

To evaluate the (j,j) cofactor Δ_{jj} of Δ for any positive integer j ($j \leq n$, where n is the order of Δ), we note that Δ_{jj} is obtained by deleting the jth row and the jth column from Δ. Thus,

$$\Delta_{jj} = \det A_{-j}Y_eA_{-j}^T \tag{5-23}$$

where A_{-j} is the submatrix of A by deleting the jth row from A, and A_{-j}^T is the transpose of A_{-j}. That Equation (5-23) is true can be seen by noting that

$$\Delta = \det \{(AY_e)A^T\} \tag{5-24}$$

and that any row of Δ can be deleted by deleting the same row from the product AY_e; and any column of Δ can be deleted by deleting the same column of A^T in Equation (5-24).

We shall next study the significance of Δ_{jj} with respect to the graph G of the given network N. We observe that Δ is the determinant of the node admittance matrix of N. What, then, does Δ_{jj} represent topologically?

The answer to the above question will be quite evident if one only observes the effect on G or N in obtaining Δ_{jj} from Δ by deleting its jth row and jth column. In deleting the jth row of Δ, we delete the jth row of the product AY_e, which amounts to deleting the jth row from A in the product, thus yielding $A_{-j}Y_e$. Similar argument holds for the deletion of the jth column from Δ and, consequently, Equation (5-23) results. But, A_{-j} satisfies all the requirements of an incidence submatrix (of some graph). In fact, it is the incidence submatrix of the graph N_{-j} obtained from N by coalescing the jth node with the reference node $1'$. Thus, Δ_{jj} is the determinant of the node admittance matrix of N_{-j}.

We observe that, in coalescing the jth node with the reference node $1'$, self-loops may be produced if there are any branches connecting across the two nodes. These self-loops will not affect Δ_{jj} in any manner and thus can be considered removed from the resultant network N_{-j}. Moreover, by virtue of Theorem 5-2, we may evaluate Δ_{jj} in terms of the tree admittance products of N_{-j}.

Finally, we note that a tree of N_{-j} is, in fact, a 2-tree of N. More precisely, any tree of N_{-j} is a 2-tree, $T_{2_{j,1'}}$, of N with node j in one part and the reference node $1'$ in the other. Consequently, in light of Theorem 5-2, Δ_{jj} may be evaluated topologically by summing all the 2-trees, $T_{2_{j,1'}}$, of N.

We now summarize this discussion in the next theorem.

THEOREM 5-3. Let Δ be the determinant of the node admittance matrix Y_n of a passive network N with $n + 1$ nodes and without mutual inductances. Also, let the reference node be denoted by $1'$. Then the (j,j) cofactor, Δ_{jj} of

Δ, is equal to the sum of all the 2-tree admittance products, $T_{2_{j,1'}}(y)$ of N, each of which contains node j in one part and node $1'$ in the other. That is,

$$\Delta_{jj} = \sum_k T_{2_{j,1'}}^{(k)}(y) \tag{5-25}$$

where the summation is taken over all the 2-tree admittance products of the form $T_{2_{j,1'}}(y)$.

EXAMPLE 5-5. Once again, consider network N and its graph G of Figure 5-4. From Example 5-3, we find

$$\Delta = \det Y_n = \begin{vmatrix} y_1 + y_2 & -y_2 & 0 \\ -y_2 & y_2 + y_3 + y_5 & -y_5 \\ 0 & -y_5 & y_4 + y_5 \end{vmatrix}$$

Thus,

$$\Delta_{11} = \begin{vmatrix} y_2 + y_3 + y_5 & -y_5 \\ -y_5 & y_4 + y_5 \end{vmatrix}$$

By direct expansion, we find

$$\Delta_{11} = (y_2 + y_3 + y_5)(y_4 + y_5) - y_5^2$$

or, after cancellation of terms,

$$\Delta_{11} = y_2y_4 + y_2y_5 + y_3y_4 + y_3y_5 + y_4y_5 \tag{5-26}$$

Now let us obtain Δ_{11} by applying Equation (5-25). For convenience, graph G is repeated in Figure 5-7(a). The graph G_{-1}, which is obtained by coalescing vertex 1 to the reference vertex $1'$, is shown in Figure 5-7(b). By inspection,

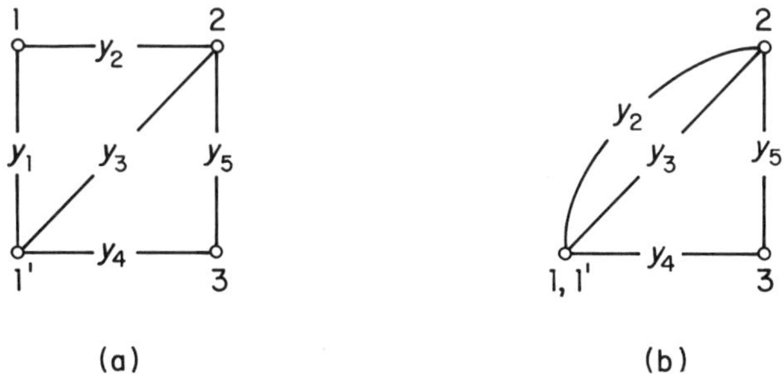

Figure 5-7 (a) Graph G. (b) Graph G_{-1}.

the tree admittance products of G_{-1} are y_2y_4, y_2y_5, y_3y_4, y_3y_5, and y_4y_5, which are exactly identical to those in Equation (5-26). These are, indeed, the 2-trees, $T_{2_{1,1'}}$, of G (with vertex 1 in one part and vertex $1'$ in the other) as shown in Figure 5-8. Note that in forming G_{-1} by shorting vertex 1 to vertex $1'$ of

G, edge y_2 becomes a self-loop and, hence, is removed from G_{-1} as shown in Figure 5-7(b). Each one of the 2-trees in Figure 5-8 is, indeed, a tree of G_{-1}.

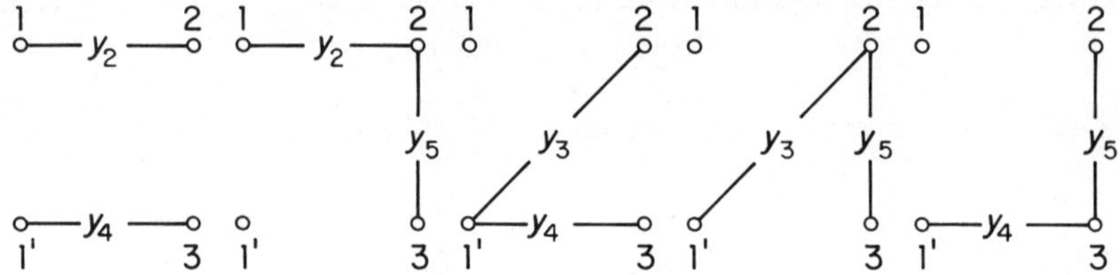

Figure 5-8 The complete set of 2-trees $T_{2_1,1'}$ of G.

EXAMPLE 5-6. As another example, let us evaluate Δ_{22} of the determinant Δ from Example 5-5. Thus,

$$\Delta_{22} = \begin{vmatrix} y_1 + y_2 & 0 \\ 0 & y_4 + y_5 \end{vmatrix} = (y_1 + y_2)(y_4 + y_5)$$

or

$$\Delta_{22} = y_1 y_4 + y_1 y_5 + y_2 y_4 + y_2 y_5$$

From Figure 5-7(a), we obtain G_{-2} by shorting vertex 2 to vertex 1′ as shown in Figure 5-9(a). The trees of G_{-2} are the 2-trees $T_{2_2,1'}$ of G of Figure 5-7(a). They are shown in Figure 5-9(b).

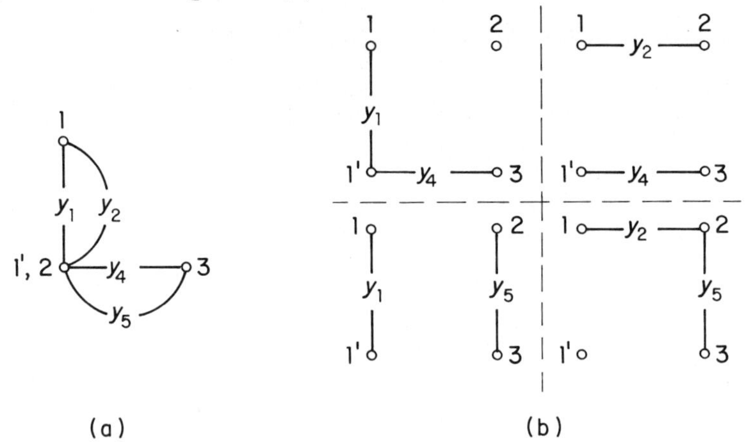

Figure 5-9 (a) Graph G_{-2}. (b) The complete set of 2-trees $T_{2_2,1'}$ of G of Figure 5-7(a).

5-4 TOPOLOGICAL FORMULAS FOR $Y_d(s)$ AND $Z_d(s)$

Combining our results from the preceding three sections — namely, Equations (5-7), (5-8), (5-18), and (5-25) — we readily obtain the topological formulas for the driving-point admittance $Y_d(s)$ and the driving-point impedance $Z_d(s)$ as stated in the following theorem:

THEOREM 5-4. With the same notation as in Theorem 5-3, the driving-point admittance $Y_d(s)$ and the driving-point impedance $Z_d(s)$ of a passive one-port containing no mutual indutances at terminals 1 and 1′ are given by

$$Y_d(s) = \frac{\Delta}{\Delta_{11}} = \frac{\sum_i T^{(i)}(y)}{\sum_k T^{(k)}_{2_{1,1'}}(y)} \tag{5-27}$$

and

$$Z_d(s) = \frac{\Delta_{11}}{\Delta} = \frac{\sum_k T^{(k)}_{2_{1,1'}}(y)}{\sum_i T^{(i)}(y)} \tag{5-28}$$

respectively.

For convenience, we may define the following shorthand notation:

$$V(Y) \triangleq \sum_i T^{(i)}(y)$$
$$= \text{the sum of all tree admittance products} \tag{5-29}$$

$$W_{j,r}(Y) \triangleq \sum_k T_{2_{j,r}}(y)$$
$$= \text{the sum of all 2-tree admittance products with node } j \text{ and the reference node } r \text{ contained in different parts} \tag{5-30}$$

Thus, Equations (5-27) and (5-28) may be rewritten as

$$Y_d(s) = \frac{V(Y)}{W_{1,1'}(Y)} \tag{5-31}$$

and

$$Z_d(s) = \frac{W_{1,1'}(Y)}{V(Y)} \tag{5-32}$$

respectively.

EXAMPLE 5-7. Let us obtain the driving-point impedance of the network N of Figure 5-4(a). From Examples 5-3 and 5-5, we find that

$$V(Y) = y_1y_2y_4 + y_1y_2y_5 + y_1y_3y_4 + y_1y_3y_5 + y_1y_4y_5 + y_2y_3y_4 + y_2y_3y_5 + y_2y_4y_5 \tag{5-33}$$

and

$$W_{1,1'}(Y) = y_2y_4 + y_2y_5 + y_3y_4 + y_3y_5 + y_4y_5 \tag{5-34}$$

But, from Equation (5-20),

$$y_1 = \frac{1}{R_1} = G_1 \qquad y_2 = sC_1 \qquad y_3 = \frac{1}{R_2} = G_2 \qquad y_4 = sC_2 \qquad y_5 = \frac{1}{sL} = \frac{\Gamma}{s} \tag{5-35}$$

Hence, substituting Equations (5-33), (5-34), and (5-35) into (5-31) and after simplification, we get

$$Y_d(s) = \frac{(C_1C_2G_1 + C_1C_2G_2)s^3 + (C_2G_1G_2 + C_1C_2\Gamma)s^2 + (C_1G_1\Gamma + C_1G_2\Gamma + C_2G_1\Gamma)s + G_1G_2\Gamma}{C_1C_2s^3 + C_2G_2s^2 + (C_1\Gamma + C_2\Gamma)s + G_2\Gamma} \tag{5-36}$$

Next we obtain the driving-point impedance $Z_d(s)$ by combining the branch impedances:

$$Z_d(s) = \frac{\left\{\dfrac{\left(sL + \dfrac{1}{sC_2}\right)R_2}{sL + \dfrac{1}{sC_2} + R_2} + \dfrac{1}{sC_1}\right\}R_1}{\dfrac{\left(sL + \dfrac{1}{sC_2}\right)R_2}{sL + \dfrac{1}{sC_2} + R_2} + \dfrac{1}{sC_1} + R_1}$$

With

$$Y_d(s) = \frac{1}{Z_d(s)} \qquad \Gamma = \frac{1}{L} \qquad G_1 = \frac{1}{R_1} \qquad G_2 = \frac{1}{R_2}$$

we find that

$$Y_d(s) = \frac{\dfrac{\left(\dfrac{s}{\Gamma} + \dfrac{1}{C_2s}\right)\dfrac{1}{G_2}}{\dfrac{s}{\Gamma} + \dfrac{1}{C_2s} + \dfrac{1}{G_2}} + \dfrac{1}{C_1s} + \dfrac{1}{G_1}}{\left\{\dfrac{\left(\dfrac{s}{\Gamma} + \dfrac{1}{C_2s}\right)\dfrac{1}{G_2}}{\dfrac{s}{\Gamma} + \dfrac{1}{C_2s} + \dfrac{1}{G_2}} + \dfrac{1}{C_1s}\right\}\dfrac{1}{G_1}}$$

After simplification, we get

$$Y_d(s) = \frac{\left(\dfrac{1}{G_2\Gamma} + \dfrac{1}{G_1\Gamma}\right)s + \left(\dfrac{1}{C_1\Gamma} + \dfrac{1}{G_1G_2}\right) + \left(\dfrac{1}{C_2G_2} + \dfrac{1}{C_2G_1} + \dfrac{1}{C_1G_2}\right)\dfrac{1}{s} + \left(\dfrac{1}{C_1C_2}\right)\dfrac{1}{s^2}}{\left(\dfrac{1}{G_1G_2\Gamma}\right)s + \left(\dfrac{1}{C_1G_1\Gamma}\right) + \left(\dfrac{1}{C_1G_1G_2} + \dfrac{1}{C_2G_1G_2}\right)\dfrac{1}{s} + \left(\dfrac{1}{C_1C_2G_1}\right)\dfrac{1}{s^2}} \tag{5-37}$$

Multiplying Equation (5-37) by $C_1C_2G_1G_2\Gamma s^2$, we get

$$Y_d(s) = \frac{(C_1C_2G_1 + C_1C_2G_2)s^3 + (C_2G_1G_2 + C_1C_2\Gamma)s^2 + (C_1G_1\Gamma + C_1G_2\Gamma + C_2G_1\Gamma)s + G_1G_2\Gamma}{C_1C_2s^3 + C_2G_2s^2 + (C_1\Gamma + C_2\Gamma)s + G_2\Gamma}$$

which is identical to the expression in Equation (5-36).

EXAMPLE 5-8. For the purpose of comparing results, let us find $Y_d(s)$ at the terminals 1 and 1′ of the network N of Figure 5-3 as used in Example 5-1. The graph of N is shown in Figure 5-10.

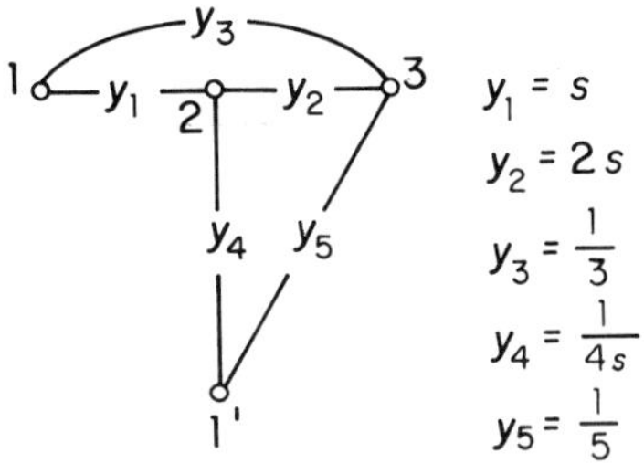

Figure 5-10 The graph G of the network N of Figure 5-3.

The set of all tree admittance products of N is:

$$\begin{aligned}
y_1y_2y_4 &= (s)(2s)\left(\frac{1}{4s}\right) = \frac{1}{2}s & y_1y_4y_5 &= (s)\left(\frac{1}{4s}\right)\left(\frac{1}{5}\right) = \frac{1}{20} \\
y_1y_2y_5 &= (s)(2s)\left(\frac{1}{5}\right) = \frac{2}{5}s^2 & y_2y_3y_4 &= (2s)\left(\frac{1}{3}\right)\left(\frac{1}{4s}\right) = \frac{1}{6} \\
y_1y_3y_4 &= (s)\left(\frac{1}{3}\right)\left(\frac{1}{4s}\right) = \frac{1}{12} & y_2y_3y_5 &= (2s)\left(\frac{1}{3}\right)\left(\frac{1}{5}\right) = \frac{2}{15}s \\
y_1y_3y_5 &= (s)\left(\frac{1}{3}\right)\left(\frac{1}{5}\right) = \frac{1}{15}s & y_3y_4y_5 &= \left(\frac{1}{3}\right)\left(\frac{1}{4s}\right)\left(\frac{1}{5}\right) = \frac{1}{60s}
\end{aligned} \tag{5-38}$$

Thus, from Equations (5-28) and (5-38),

$$\begin{aligned}
\Delta &= \frac{1}{2}s + \frac{2}{5}s^2 + \frac{1}{12} + \frac{1}{15}s + \frac{1}{20} + \frac{1}{6} + \frac{2}{15}s + \frac{1}{60s} \\
&= \frac{2}{5}s^2 + \left(\frac{1}{2} + \frac{1}{15} + \frac{2}{15}\right)s + \left(\frac{1}{12} + \frac{1}{20} + \frac{1}{6}\right) + \frac{1}{60s} \\
&= \frac{2}{5}s^2 + \frac{21}{30}s + \frac{18}{60} + \frac{1}{60s}
\end{aligned} \tag{5-39}$$

Next, short-circuiting vertex 1 to vertex 1′, we obtain the resultant graph G_{-1}, as shown in Figure 5-11.

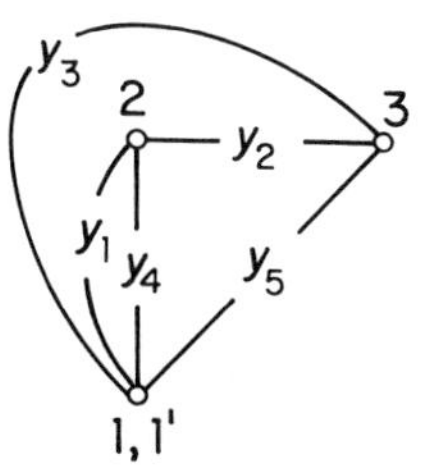

Figure 5-11 The resultant graph G_{-1} from short-circuiting vertices 1 and 1′ together.

The set of all 2-tree admittance products of N can be found by obtaining the set of all tree admittance products of G_{-1} as follows:

$$\begin{aligned} y_1y_2 &= 2s^2 & y_2y_4 &= \frac{1}{2} \\ y_1y_3 &= \frac{1}{3}s & y_2y_5 &= \frac{2}{5}s \\ y_1y_5 &= \frac{1}{5}s & y_3y_4 &= \frac{1}{12s} \\ y_2y_3 &= \frac{2}{3}s & y_4y_5 &= \frac{1}{20s} \end{aligned} \tag{5-40}$$

Substituting Equation (5-40) into (5-25), with $j = 1$, we find that

$$\begin{aligned} \Delta_{11} &= 2s^2 + \left(\frac{1}{3} + \frac{1}{5} + \frac{2}{3} + \frac{2}{5}\right)s + \frac{1}{2} + \left(\frac{1}{12} + \frac{1}{20}\right)\frac{1}{s} \\ &= 2s^2 + \frac{24}{15}s + \frac{1}{2} + \frac{8}{60s} \end{aligned} \tag{5-41}$$

Finally, from Equations (5-27), (5-39), and (5-41), we obtain

$$Y_d(s) = \frac{\Delta}{\Delta_{11}} = \frac{\dfrac{2}{5}s^2 + \dfrac{21}{30}s + \dfrac{18}{60} + \dfrac{1}{60s}}{2s^2 + \dfrac{24}{15}s + \dfrac{1}{2} + \dfrac{8}{60s}} = \frac{24s^3 + 42s^2 + 18s + 1}{120s^3 + 96s^2 + 30s + 8} \tag{5-42}$$

which is identical to Equation (5-12).

5-5 REMARKS

As we mentioned earlier in this chapter, one of the obvious advantages of topological formulas over the conventional methods for the evaluation of determinants and cofactors in obtaining the driving-point immittance (that is, impedance or admittance) functions is that the former avoids the usual cancellations inherent in the expansion of determinants and cofactors in the latter. Furthermore, topological analysis may be considered as a shortcut in evaluating network determinants and cofactors because there is no need to obtain the determinants and cofactors when topological formulas are used. The terms may be obtained directly from the network by inspection. This point has been adequately illustrated by the examples in Section 5-4. Thus, as another advantage, topological methods provide better physical insight to the problem since each term in the expression corresponds to a part of the system. [In the case of a passive one-port, each term in $Y_d(s)$ or $Z_d(s)$ corresponds either to a tree or to a 2-tree.]

However, the reader may have the following question in his mind: Since each tree (or 2-tree) is obtained from the network by inspection, it is quite possible that a tree may be omitted simply due to oversight, and thus how can one be sure that all the trees (or 2-trees) have been included? To overcome this difficulty, we state and prove the following important theorem:

THEOREM 5-5. The total number of trees of a connected network N (graph) is equal to the value of the determinant of the product of its incidence submatrix A and its transpose A^T; that is,

$$\text{Total number of trees} = \det (A \cdot A^T) \tag{5-43}$$

Proof: The proof of this important theorem is very simple. We recall that

$$\Delta = \det \{(AY_e)A^T\} = V(Y) \tag{5-44}$$

Now, if we replace every branch of N by a one-ohm resistor, then each branch has an admittance of unity and the branch admittance matrix becomes a unit matrix. Thus Equation (5-44) will be reduced to

$$\Delta = \det Y_n = \det (A \cdot A^T) = V(Y) \tag{5-45}$$

Next we note that under this assumption each term in $V(Y)$ must be unity since it is a tree admittance product, and each branch of the tree has a unit admittance value. Consequently, $V(Y)$ is equal to the total number of trees of N. This completes the proof.

Thus, it is a simple matter to determine the total number of trees of a given network, and Theorem 5-5 serves as a check against any omission of terms in evaluating Δ topologically.

As for the 2-trees of the type $T_{2_{1,1'}}$, Theorem 5-5 still applies because then the network of interest will be the modified one, N_{-1}, which is obtained by coalescing the two terminals, 1 and 1′, of N.

A systematic method of listing the trees and 2-trees may be helpful when topological formulas are applied. One of the methods[7] is described as follows:

Since the number of edges in a tree of a connected graph of e edges and v vertices is $v - 1$, one can determine the upper bound of the total number of trees T_t by $C(e, v - 1)$ where $C(e, v - 1)$ is the combination of e objects taken $v - 1$ at a time. Thus,

$$T_t \leqq C(e, v - 1) \tag{5-46}$$

But, from algebra,[8] we find that

$$C(e, v - 1) = \frac{e!}{(v - 1)!(e - v + 1)!} \tag{5-47}$$

[7]The method described here is the most elementary and is well suited for computer calculations. For other more sophisticated tree-listing methods, see G. J. Minty [64], Mayeda and Seshu [63], and W. K. Chen [21].

[8]Equation (5-47) can be found in most college algebra texts. See, for example, P. R. Rider [71], p. 252.

where, for any positive integer k,

$$k! \triangleq k(k-1)(k-2)\cdots 2\cdot 1$$

Thus, if we list all the $C(e, v-1)$ terms by forming all the possible combinations of e edges taken $v-1$ at a time, and subtract from these $C(e, v-1)$ terms those that contain one or more circuits, the result must be T_t:

$$T_t = C(e, v-1) - \left\{\begin{matrix}\text{all those combinations of } e \text{ edges taken} \\ v-1 \text{ at a time containing circuits}\end{matrix}\right\} \tag{5-48}$$

We now illustrate this idea by giving the following example:

EXAMPLE 5-9. Once again, consider the network N of Figure 5-3. The oriented graph G of N is shown in Figure 5-12. The orientations of the edges

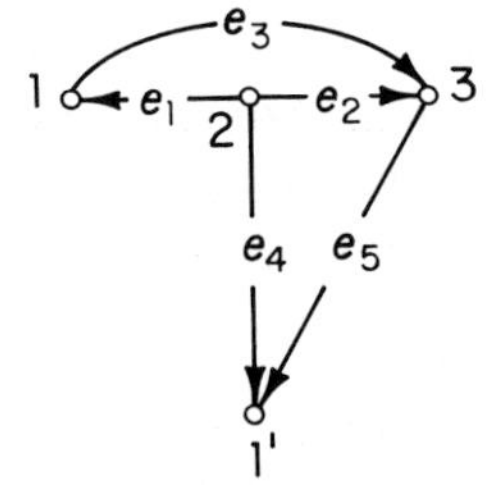

Figure 5-12 The oriented graph G of the network N of Figure 5-3.

of G are, of course, completely arbitrary. The incidence submatrix of G with vertex $1'$ as reference is

$$A = \begin{array}{c} \\ 1 \\ 2 \\ 3 \end{array}\begin{array}{c} \begin{matrix} e_1 & e_2 & e_3 & e_4 & e_5 \end{matrix} \\ \begin{bmatrix} -1 & 0 & 1 & 0 & 0 \\ 1 & 1 & 0 & 1 & 0 \\ 0 & -1 & -1 & 0 & 1 \end{bmatrix}\end{array}$$

Thus,

$$A \cdot A^T = \begin{bmatrix} -1 & 0 & 1 & 0 & 0 \\ 1 & 1 & 0 & 1 & 0 \\ 0 & -1 & -1 & 0 & 1 \end{bmatrix}\begin{bmatrix} -1 & 1 & 0 \\ 0 & 1 & -1 \\ 1 & 0 & -1 \\ 0 & 1 & 0 \\ 0 & 0 & 1 \end{bmatrix} = \begin{bmatrix} 2 & -1 & -1 \\ -1 & 3 & -1 \\ -1 & -1 & 3 \end{bmatrix}$$

which, indeed, *is the node admittance matrix of N with node 1′ as the reference node and every branch of N replaced by a 1-ohm resistor.*

By Theorem 5-5, we find that

$$T_t = \det(A \cdot A^T) = \det\begin{bmatrix} 2 & -1 & -1 \\ -1 & 3 & -1 \\ -1 & -1 & 3 \end{bmatrix}$$

Upon expansion and simplification, we get

$$T_t = 8$$

which indicates the correct number of terms in the expansion of Δ as seen in Equation (5-38).

Next we list the possible combinations of five branches taken three at a time. There are ten such possible combinations, since

$$C(5,3) = \frac{5!}{3!(5-3)!} = \frac{5 \cdot 4 \cdot 3 \cdot 2 \cdot 1}{3 \cdot 2 \cdot 1 \cdot 2 \cdot 1} = 10$$

Listing these combinations systematically, we have: 123, 124, 125, 134, 135, 145, 234, 235, 245, and 345. But, by inspection of the graph of Figure 5-12, we see that each of the combinations 123 and 245 contains a circuit and, therefore, they are not trees. Subtracting these two from the ten possible combinations, we thus obtain the set of all (eight) trees of G.

From Example 5-9, we see that by applying Theorem 5-5 and the procedure just described, one can be sure that no terms will be omitted in evaluating the network determinants and cofactors using the topological formulas derived in this chapter.

It is worth noting that the element values used in Example 5-1 are not meant to be practical. Needless to say, 1- and 2-farad capacitors and 4-henry inductors are not realistic components. These values are so chosen simply because we want to make our calculations easy to follow, while, at the same time, the ideas involved will be illustrated. We shall again choose these "non-realistic" element values in later chapters for the aforementioned reasons without further comment.

Finally, from the proof of Theorem 5-5 and Example 5-9, we see that the determinant, det $(A \cdot A^T)$, is equal to the determinant, det Y_{n1}, of the node admittance matrix Y_{n1} of a given passive network N with each element of N replaced by a 1-ohm resistor. This interesting observation gives rise to the following corollary:

COROLLARY 5-1. The total number T_t of trees of a connected passive network N is equal to the determinant, det Y_{n1}, of the node admittance matrix Y_{n1} of the network N_1, where N_1 is obtained from N by replacing every element of N by a 1-ohm resistor; that is,

$$T_t = \det Y_{n1} \tag{5-49}$$

Equation (5-49) is a very useful tool for determining the total number of trees in a given graph or network.

PROBLEMS

5-1 Using node analysis, determine the driving-point impedance $Z_d(s)$ and the driving-point admittance $Y_d(s)$ of the passive one-port shown in Figure P5-1.

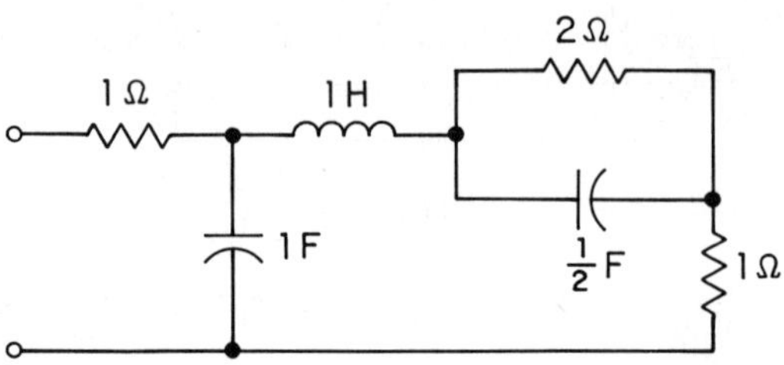

Figure P5-1

5-2 Derive the formulas for the driving-point impedance $Z_d(s)$ and the driving-point admittance $Y_d(s)$ of a passive one-port in terms of the determinant of the loop impedance matrix and its cofactors.

5-3 Using the formulas derived in Problem 5-2, determine the driving-point admittance $Y_d(s)$ of the passive one-port in Problem 5-1.

5-4 Show that the branch impedance matrix $Z_e(s)$ of a relaxed (zero initial conditions) passive one-port without mutual inductances is always nonsingular so that

$$Y_e(s) = Z_e^{-1}(s)$$

exists and is a diagonal matrix.

5-5 Evaluate the following determinants by the Binet-Cauchy theorem:

(a)
$$\det AB = \det \begin{bmatrix} 2 & 0 & -1 & 2 \\ 1 & 2 & 5 & 0 \\ 3 & 4 & 6 & -3 \end{bmatrix} \begin{bmatrix} 0 & 5 & -2 \\ 1 & 3 & 2 \\ -3 & 0 & 1 \\ 4 & 2 & -1 \end{bmatrix}$$

(b)
$$\det CD = \det \begin{bmatrix} 1 & 0 & 3 & 4 & 5 \\ -1 & 1 & 2 & 1 & 3 \\ 0 & 5 & -2 & 1 & 2 \end{bmatrix} \begin{bmatrix} 1 & 3 & 2 \\ 0 & -1 & 5 \\ 6 & 0 & 1 \\ 3 & 2 & -2 \\ 0 & 1 & 3 \end{bmatrix}$$

5-6 Evaluate the determinant Δ of the node admittance matrix of the network shown in Figure P5-6 by
(a) The determinant expansion technique.
(b) The topological approach.

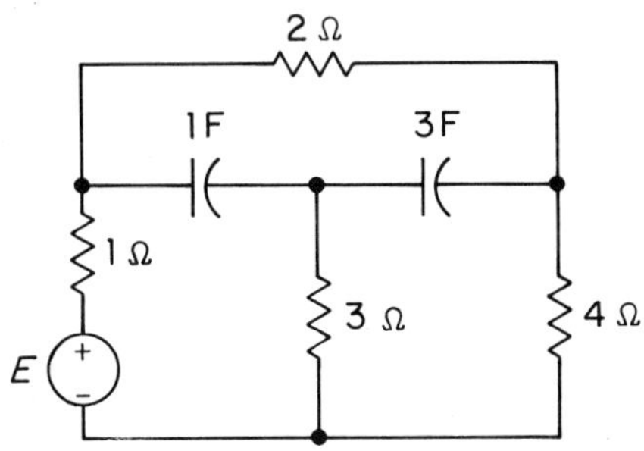

Figure P5-6

5-7 Evaluate the determinant Δ ($= \det Y_n$) of the network shown in Figure P5-7 topologically.

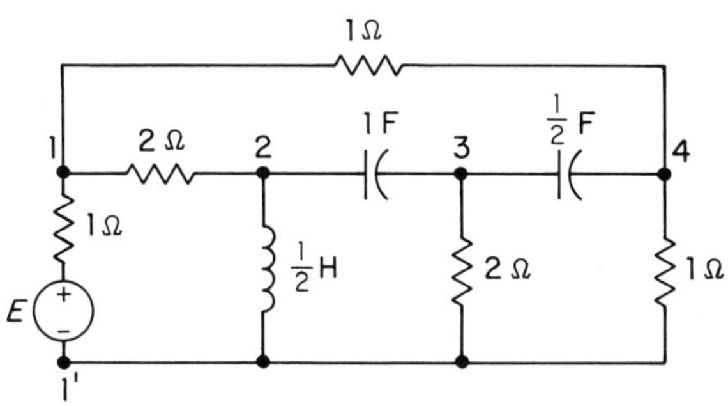

Figure P5-7

5-8 For the graph shown in Figure P5-8, find the sets of all 2-trees of the type:

(a) $T_{2_{1,1'}}$ (b) $T_{2_{1,2}}$ (c) $T_{2_{1,3}}$ (d) $T_{2_{13,2}}$

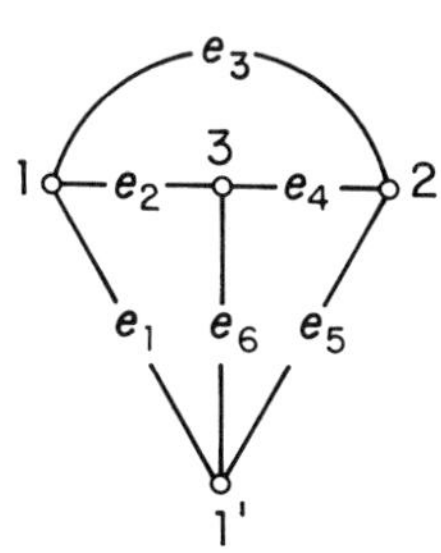

Figure P5-8

5-9 Evaluate Δ_{11} and Δ_{22} of Δ of the node admittance matrix of the network of Figure P5-7.

5-10 Using topological formulas, find the driving-point impedance $Z_d(s)$ at the terminals 1 and 1′ of the passive one-port shown in Figure P5-6.

5-11 Find, topologically, the driving-point admittance $Y_d(s)$ of the passive one-port shown in Figure P5-11.

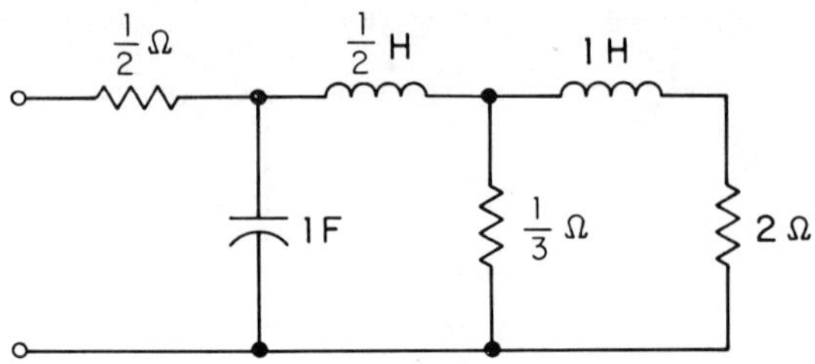

Figure P5-11

5-12 Determine the total number of trees of the network shown in Figure P5-12.

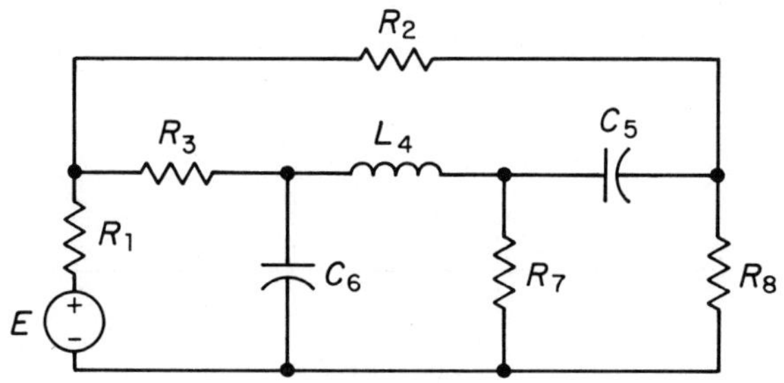

Figure P5-12

5-13 *Without performing any multiplication or addition*, prove that the driving-point impedance of the bridge network of Figure P5-13 is independent of s and equal to 1 ohm.

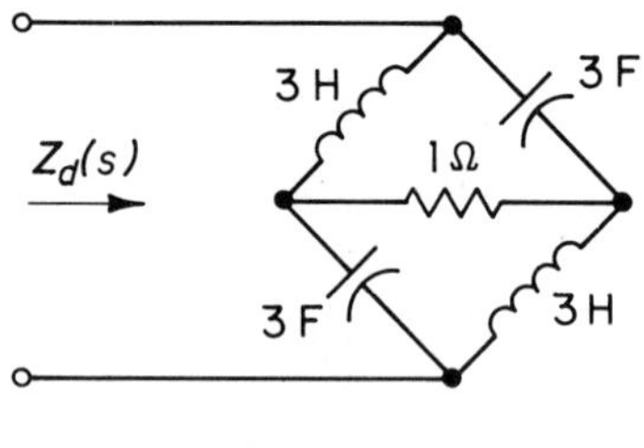

Figure P5-13

5-14 For each of the networks shown in Figure P5-14, calculate the driving-point impedance $Z_d(s)$, first by any conventional method that you prefer, and next by the use of topological formulas. It will be interesting to record the time required by each method and compare the results.

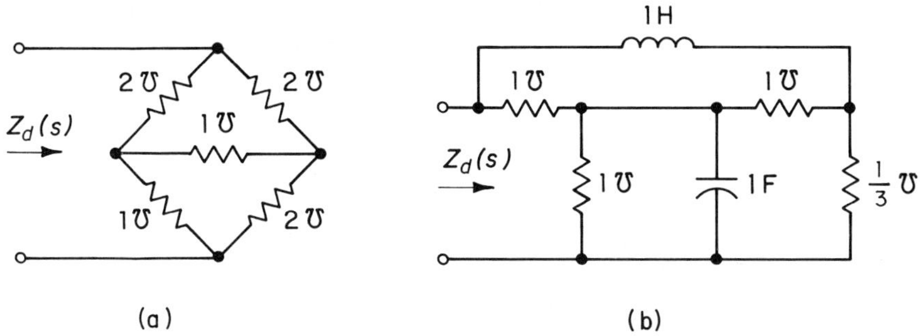

Figure P5-14

5-15 Using topological formulas, find the driving-point admittance $Y_d(s)$ for each of the networks shown in Figure P5-15. (Since negative resistances are present, you should not be surprised if the results look unusual.) Discuss some possible applications of these networks.

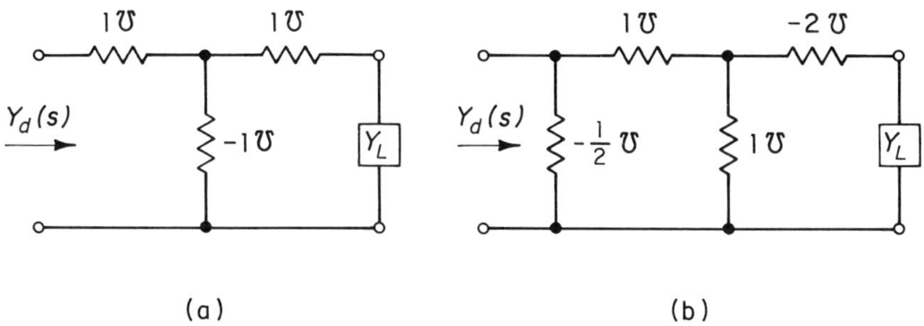

Figure P5-15

5-16 (a) In Figure P5-16(a), find the admittance measured across terminal pairs (1,2), (1,3), and (1,4) by the use of topological formulas.

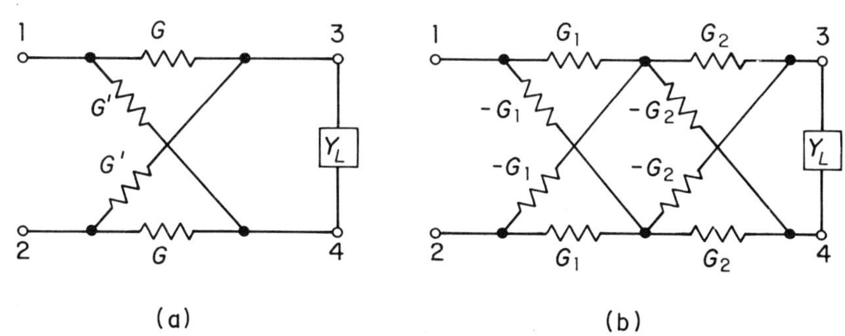

Figure P5-16

(b) If $G' = -G$, simplify the results of part (a)

(c) Using the results of part (b), find the admittances looking into terminal pairs (1,2) and (1,4) of Figure P5-16(b). Discuss the significance of the results.

5-17 In Section 5-2, we assume that no mutual inductances are present in the network of interest. What difficulties are avoided in making this assumption when deriving the topological formula for $Y_d(s)$?

5-18 List all the (a) advantages and (b) disadvantages of the topological methods over the conventional techniques for the analysis of passive networks.

5-19 Suggest a new topological approach to the analysis of passive one-ports that is different from and/or superior to the one presented in this chapter.

5-20 If mutual inductances are present in a given passive one-port, how would you modify the present formula to overcome this difficulty? (See, also, Problem 5-17.)

PASSIVE TWO-PORT TOPOLOGICAL ANALYSIS

CHAPTER 6

6-1 INTRODUCTION

In this chapter, we shall derive the topological formulas for various network functions of a passive *two-port* — that is, a passive two-terminal-pair network. We shall be dealing with passive two-port networks with zero initial conditions and without mutual inductances.

First, we shall define the various transfer functions — such as the open-circuit impedances, short-circuit admittances, and voltage-ratio transfer function — for the passive two-port, and derive the expressions for these functions in terms of the determinant Δ of the node admittance matrix and its cofactors. Then topological formulas will be developed for the transfer functions of the passive two-port without mutual inductances. Finally, a summary of results obtained in Chapter 5 and in the present one will be listed for future reference.

6-2 OPEN-CIRCUIT IMPEDANCES AND SHORT-CIRCUIT ADMITTANCES

(a) Open-circuit Impedance Functions

Starting from the node equations for a network with n independent nodes (refer to Figure 5-1), we have, from Equation (5-1),

$$
\begin{aligned}
y_{11}V_1 + y_{12}V_2 + \cdots + y_{1n}V_n &= I_1 \\
y_{21}V_1 + y_{22}V_2 + \cdots + y_{2n}V_n &= I_2 \\
&\vdots \\
y_{n1}V_1 + y_{n2}V_2 + \cdots + y_{nn}V_n &= I_n
\end{aligned} \tag{6-1}
$$

where the notation has been defined in Section 5-1. The general expression [Equation (5-4)] for the voltage V_j at node j with respect to the reference node $1'$ is repeated here for convenience.

$$V_j = \frac{\Delta_{1j}}{\Delta}I_1 + \frac{\Delta_{2j}}{\Delta}I_2 + \cdots + \frac{\Delta_{nj}}{\Delta}I_n \qquad (j = 1, 2, \cdots, n) \tag{6-2}$$

Now consider a passive two-port driven by two current generators, I_{g_1} and I_{g_2}, at ports 1 and 2, respectively, as shown in Figure 6-1. The following observations are noted:

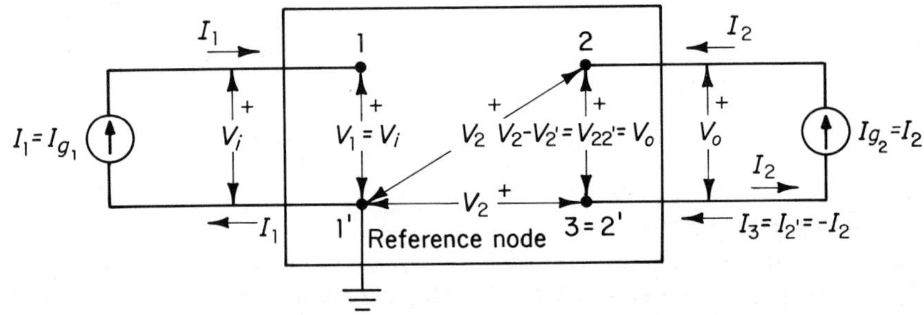

Figure 6-1 A passive two-port driven by two current generators.

(1) Since there are only two sources, I_{g_1} and I_{g_2}, we have

$$I_1 = I_{g_1} \qquad I_2 = I_{g_2} \qquad I_3 = I_{2'} = -I_2 \qquad I_4 = I_5 = \cdots = I_n = 0 \tag{6-3}$$

(2) The four external nodes are 1, 2, 2′, and 1′, with 1′ chosen as the reference node.

(3) With the notation used in the figure, we write

$$V_i = V_1$$

and (6-4)

$$V_o = V_{22'} = V_2 - V_{2'} = V_2 - V_3 \text{ (since node 3 equals node 2')}$$

Thus, from Equation (6-2), by letting $j = 1$, 2, and 3, we find that

$$
\begin{aligned}
V_1 &= \frac{\Delta_{11}}{\Delta}I_1 + \frac{\Delta_{21}}{\Delta}I_2 + \frac{\Delta_{31}}{\Delta}I_3 \\
V_2 &= \frac{\Delta_{12}}{\Delta}I_1 + \frac{\Delta_{22}}{\Delta}I_2 + \frac{\Delta_{32}}{\Delta}I_3 \\
V_3 &= \frac{\Delta_{13}}{\Delta}I_1 + \frac{\Delta_{23}}{\Delta}I_2 + \frac{\Delta_{33}}{\Delta}I_3
\end{aligned} \tag{6-5}
$$

Substituting Equation (6-3) into (6-5) and replacing the subscript 3 by 2′, we get

$$
\begin{aligned}
V_1 &= \frac{\Delta_{11}}{\Delta}I_1 + \left(\frac{\Delta_{21} - \Delta_{2'1}}{\Delta}\right)I_2 \\
V_2 &= \frac{\Delta_{12}}{\Delta}I_1 + \left(\frac{\Delta_{22} - \Delta_{2'2}}{\Delta}\right)I_2 \\
V_{2'} &= \frac{\Delta_{12'}}{\Delta}I_1 + \left(\frac{\Delta_{22'} - \Delta_{2'2'}}{\Delta}\right)I_2
\end{aligned}
\tag{6-6}
$$

Substituting Equation (6-4) into (6-6), we find that

$$
\begin{aligned}
V_i &= \left(\frac{\Delta_{11}}{\Delta}\right)I_1 + \left(\frac{\Delta_{21} - \Delta_{2'1}}{\Delta}\right)I_2 \\
V_o &= \left(\frac{\Delta_{12} - \Delta_{12'}}{\Delta}\right)I_1 + \left(\frac{\Delta_{22} + \Delta_{2'2'} - \Delta_{22'} - \Delta_{2'2}}{\Delta}\right)I_2
\end{aligned}
\tag{6-7}
$$

From Equation (6-7) we see that the input voltage V_i at port (1,1′) and the output voltage V_o at port (2,2′) may be expressed in terms of the currents I_1 and I_2 at the two ports. Thus we write

$$
V_i = z_{11}I_1 + z_{12}I_2 \qquad V_o = z_{21}I_1 + z_{22}I_2 \tag{6-8}
$$

and observe the physical significance of the coefficients z_{11}, z_{12}, z_{21}, and z_{22} in Equation (6-8).

Suppose that we open-circuit port (2,2′) by removing the current generator I_{g2} from the system [Figure 6-2(a)]. Then $I_2 = 0$, and Equation (6-8) becomes

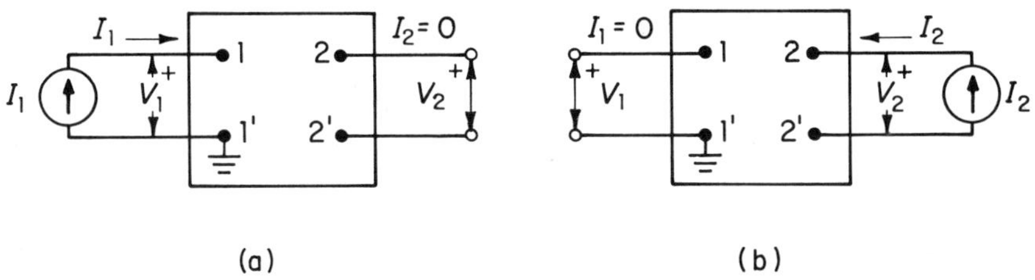

Figure 6-2 Arrangements in the two-port for determining the open-circuit impedance.

$$
V_i = z_{11}I_1 \qquad V_o = z_{21}I_1 \qquad (I_2 = 0) \tag{6-9}
$$

Solving Equation (6-9) for z_{11} and z_{21}, we find that

$$
z_{11} = \left.\frac{V_i}{I_1}\right|_{I_2=0} \tag{6-10}
$$

$$
z_{21} = \left.\frac{V_o}{I_1}\right|_{I_2=0} \tag{6-11}
$$

Next, if we open-circuit port (1,1′) by removing the current generator I_{g_1}

from the system [Figure 6-2(b)], making $I_1 = 0$, from Equation (6-8) we get the following:

$$V_i = z_{12}I_2 \qquad V_o = z_{22}I_2 \qquad (I_1 = 0) \tag{6-12}$$

Solving for z_{12} and z_{22} in Equation (6-12), we find that

$$z_{12} = \left.\frac{V_i}{I_2}\right|_{I_1=0} \tag{6-13}$$

$$z_{22} = \left.\frac{V_o}{I_2}\right|_{I_1=0} \tag{6-14}$$

Equations (6-10), (6-11), (6-13), and (6-14) together define the four network functions, z_{11}, z_{12}, z_{21}, and z_{22}, respectively, in Equation (6-8). These functions are called the *open-circuit impedance functions* of the passive two-port shown in Figure 6-1 for the obvious reason. More precisely, z_{11} and z_{22} are the open-circuit driving-point impedances at ports (1,1′) and (2,2′), respectively; and z_{12} and z_{21} are the open-circuit transfer impedances of the two-port.

Comparing Equation (6-8) with (6-7), we find immediately the following relationships:

$$\begin{aligned} z_{11} &= \frac{\Delta_{11}}{\Delta} & z_{21} &= \frac{\Delta_{12} - \Delta_{12'}}{\Delta} \\ z_{12} &= \frac{\Delta_{21} - \Delta_{2'1}}{\Delta} & z_{22} &= \frac{\Delta_{22} + \Delta_{2'2'} - \Delta_{2'2} - \Delta_{22'}}{\Delta} \end{aligned} \tag{6-15}$$

When the network is reciprocal (which is the case here), $\Delta_{ij} = \Delta_{ji}$, and Equation (6-15) may be written as

$$z_{11} = \frac{\Delta_{11}}{\Delta} \tag{6-16a}$$

$$z_{12} = z_{21} = \frac{\Delta_{12} - \Delta_{12'}}{\Delta} \tag{6-16b}$$

$$z_{22} = \frac{\Delta_{22} + \Delta_{2'2'} - 2\Delta_{22'}}{\Delta} \tag{6-16c}$$

which are the desired expressions for the open-circuit parameters.

EXAMPLE 6-1. Consider the two-port N as shown in Figure 6-3(a). The graph G of N is shown in Figure 6-3(b). The admittances of the six branches of N are

$$\begin{aligned} y_1 &= \frac{1}{R_1} = G_1 & y_3 &= \frac{1}{R_3} = G_3 & y_5 &= sC_5 \\ y_2 &= sC_2 & y_4 &= \frac{1}{R_4} = G_4 & y_6 &= \frac{1}{R_6} = G_6 \end{aligned} \tag{6-17}$$

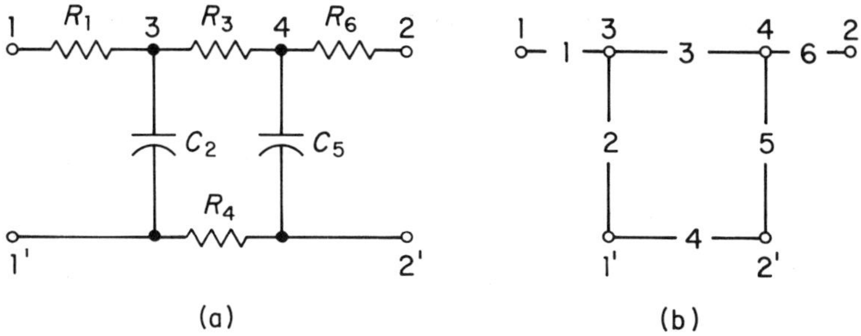

Figure 6-3 (a) A passive two-port N. (b) The graph G of N.

The determinant Δ of the node admittance matrix, Y_n of N, taking $1'$ as the reference node, is given by

$$\Delta = \begin{array}{c|ccccc|} & 1 & 2 & 2' & 3 & 4 \\ 1 & y_1 & 0 & 0 & -y_1 & 0 \\ 2 & 0 & y_6 & 0 & 0 & -y_6 \\ 2' & 0 & 0 & y_4 + y_5 & 0 & -y_5 \\ 3 & -y_1 & 0 & 0 & y_1 + y_2 + y_3 & -y_3 \\ 4 & 0 & -y_6 & -y_5 & -y_3 & y_3 + y_5 + y_6 \end{array} \tag{6-18a}$$

which, when expanded, yields

$$\begin{aligned} \Delta = {} & y_1^2y_3y_4y_6 + y_1^2y_4y_5y_6 + y_1^2y_4y_6^2 + y_1y_2y_3y_4y_6 + y_1y_2y_4y_5y_6 \\ & + y_1y_2y_4y_6^2 + y_1y_3^2y_4y_6 + y_1y_3y_4y_5y_6 + y_1y_2y_4y_6^2 + y_1^2y_3y_5y_6 \\ & + y_1^2y_5^2y_6 + y_1^2y_5y_6^2 + y_1y_2y_3y_5y_6 + y_1y_2y_5^2y_6 + y_1y_2y_5y_6^2 \\ & + y_1y_3^2y_5y_6 + y_1y_3y_5^2y_6 + y_1y_3y_5y_6^2 + y_1^2y_5^2y_6 + y_1^2y_4y_6^2 \\ & + y_1^2y_5y_6^2 - y_1y_3^2y_4y_6 - y_1y_3^2y_5y_6 - y_1^2y_5^2y_6 - y_1y_2y_5^2y_6 \\ & - y_1y_3y_5^2y_6 - y_1^2y_4y_6^2 - y_1y_2y_4y_6^2 - y_1y_3y_4y_6^2 - y_1^2y_5y_6^2 \\ & - y_1y_2y_5y_6^2 - y_1y_3y_5y_6^2 - y_1^2y_3y_4y_6 - y_1^2y_4y_5y_6 - y_1^2y_4y_6^2 \\ & - y_1^2y_3y_5y_6 - y_1^2y_5^2y_6 - y_1^2y_5y_6^2 \end{aligned} \tag{6-18b}$$

After cancellation of terms, Equation (6-18b) reduces to

$$\Delta = y_1y_2y_3y_4y_6 + y_1y_2y_3y_5y_6 + y_1y_2y_4y_5y_6 + y_1y_3y_4y_5y_6 \tag{6-18c}$$

The cofactor, Δ_{11}, is obtained from Δ by deleting its first row and its first column:

$$\Delta_{11} = \begin{vmatrix} y_6 & 0 & 0 & -y_6 \\ 0 & y_4 + y_5 & 0 & -y_5 \\ 0 & 0 & y_1 + y_2 + y_3 & -y_3 \\ -y_6 & -y_5 & -y_3 & y_3 + y_5 + y_6 \end{vmatrix}$$

which, when expanded and simplified, gives

$$\begin{aligned} \Delta_{11} = {} & y_1y_3y_4y_6 + y_1y_3y_5y_6 + y_1y_4y_5y_6 + y_2y_3y_4y_6 + y_2y_3y_5y_6 \\ & + y_2y_4y_5y_6 + y_3y_4y_5y_6 \end{aligned} \tag{6-19}$$

Similarly, the other cofactors of Δ are calculated and the results are:

$$\Delta_{12} = y_1y_3y_4y_6 + y_1y_3y_5y_6 \tag{6-20}$$

$$\Delta_{12'} = y_1y_3y_5y_6 \tag{6-21}$$

$$\begin{aligned}\Delta_{22} = {} & y_1y_2y_3y_4 + y_1y_2y_4y_5 + y_1y_2y_4y_6 + y_1y_3y_4y_5 \\ & + y_1y_3y_4y_6 + y_1y_2y_3y_5 + y_1y_2y_5y_6 + y_1y_3y_5y_6\end{aligned} \tag{6-22}$$

$$\Delta_{2'2'} = y_1y_2y_3y_6 + y_1y_2y_5y_6 + y_1y_3y_5y_6 \tag{6-23}$$

$$\Delta_{22'} = -y_1y_2y_5y_6 - y_1y_3y_5y_6 \tag{6-24}$$

Substituting Equations (6-18a) and (6-19) through (6-24) into (6-16), we get

$$z_{11} = \frac{\Delta_{11}}{\Delta} = \frac{\begin{aligned}y_1y_3y_4y_6 + y_1y_3y_5y_6 + y_1y_4y_5y_6 + y_2y_3y_4y_6 \\ + y_2y_3y_5y_6 + y_2y_4y_5y_6 + y_3y_4y_5y_6\end{aligned}}{y_1y_2y_3y_4y_6 + y_1y_2y_3y_5y_6 + y_1y_2y_4y_5y_6 + y_1y_3y_4y_5y_6} \tag{6-25}$$

$$\begin{aligned}z_{12} = z_{21} &= \frac{\Delta_{12} - \Delta_{12'}}{\Delta} \\ &= \frac{y_1y_3y_4y_6}{y_1y_2y_3y_4y_6 + y_1y_2y_3y_5y_6 + y_1y_2y_4y_5y_6 + y_1y_3y_4y_5y_6}\end{aligned} \tag{6-26}$$

$$\begin{aligned}z_{22} &= \frac{\Delta_{22} + \Delta_{2'2'} - 2\Delta_{22'}}{\Delta} \\ &= \frac{y_1y_2y_3y_4 + y_1y_2y_3y_5 + y_1y_2y_3y_6 + y_1y_2y_4y_5 + y_1y_2y_4y_6 + y_1y_3y_4y_5 + y_1y_3y_4y_6}{y_1y_2y_3y_4y_6 + y_1y_2y_3y_5y_6 + y_1y_2y_4y_5y_6 + y_1y_3y_4y_5y_6}\end{aligned} \tag{6-27}$$

Finally, substituting Equation (6-17) into (6-25) through (6-27) and simplifying, we obtain the following results:

$$z_{11} = \frac{\begin{aligned}(C_2G_3C_5G_6 + C_2G_4C_5G_6)s^2 + (G_1G_3C_5G_6 + G_1G_4C_5G_6 + C_2G_3G_4G_6 \\ + G_3G_4C_5G_6)s + G_1G_3G_4G_6\end{aligned}}{(G_1C_2G_3C_5G_6 + G_1C_2G_4C_5G_6)s^2 + (G_1C_2G_3G_4G_6 + G_1G_3G_4C_5G_6)s} \tag{6-28}$$

$$z_{12} = z_{21} = \frac{G_3G_4G_6}{(C_2G_3C_5G_6 + C_2G_4C_5G_6)s^2 + (C_2G_3G_4G_6 + G_3G_4C_5G_6)s} \tag{6-29}$$

$$z_{22} = \frac{\begin{aligned}(C_2G_3C_5 + C_2G_4C_5)s^2 + (C_2G_3G_4 + C_2G_3G_6 + C_2G_4G_6 \\ + G_3G_4C_5)s + G_3G_4G_6\end{aligned}}{(C_2G_3C_5G_6 + C_2G_4C_5G_6)s^2 + (C_2G_3G_4G_6 + G_3G_4C_5G_6)s} \tag{6-30}$$

(b) Short-circuit Admittance Functions

If we solve for I_1 and I_2 in Equation (6-8) using Cramer's rule, we find that

$$I_1 = \frac{\begin{vmatrix} V_i & z_{12} \\ V_o & z_{22} \end{vmatrix}}{\begin{vmatrix} z_{11} & z_{12} \\ z_{21} & z_{22} \end{vmatrix}} = \frac{z_{22}}{\Delta_z}V_i + \frac{-z_{12}}{\Delta_z}V_o \tag{6-31a}$$

$$I_2 = \frac{\begin{vmatrix} z_{11} & V_i \\ z_{21} & V_o \end{vmatrix}}{\begin{vmatrix} z_{11} & z_{12} \\ z_{21} & z_{22} \end{vmatrix}} = \frac{-z_{21}}{\Delta_z} V_i + \frac{z_{11}}{\Delta_z} V_o \tag{6-31b}$$

where

$$\Delta_z = \begin{vmatrix} z_{11} & z_{12} \\ z_{21} & z_{22} \end{vmatrix} = z_{11}z_{22} - z_{12}z_{21} \tag{6-32}$$

From Equation (6-31), it is evident that the input current I_1 at port $(1,1')$ and the output current I_2 at port $(2,2')$ may be expressed in terms of the voltages V_i and V_o. Hence, we write

$$\begin{aligned} I_1 &= y_{11}V_i + y_{12}V_o \\ I_2 &= y_{21}V_i + y_{22}V_o \end{aligned} \tag{6-33}$$

As a dual to the approach in the case of the open-circuit network functions, we now make $V_o = 0$ by short-circuiting the terminals 2 and $2'$ at port $(2,2')$. We get

$$\begin{aligned} I_1 &= y_{11}V_i \\ I_2 &= y_{21}V_i \end{aligned} \qquad (V_o = 0) \tag{6-34}$$

Solving Equation (6-34) for y_{11} and y_{21}, we find

$$y_{11} = \left. \frac{I_1}{V_i} \right|_{V_o=0} \tag{6-35}$$

$$y_{21} = \left. \frac{I_2}{V_i} \right|_{V_o=0} \tag{6-36}$$

Similarly, setting $V_i = 0$ by short-circuiting the terminals 1 and $1'$ at port $(1,1')$, we get

$$\begin{aligned} I_1 &= y_{12}V_o \\ I_2 &= y_{22}V_o \end{aligned} \qquad (V_i = 0) \tag{6-37}$$

Solving Equation (6-37) for y_{12} and y_{22}, we have

$$y_{12} = \left. \frac{I_1}{V_o} \right|_{Vi=0} \tag{6-38}$$

$$y_{22} = \left. \frac{I_2}{V_o} \right|_{Vi=0} \tag{6-39}$$

Equations (6-35), (6-36), (6-38), and (6-39) define the four network functions, y_{11}, y_{12}, y_{21}, and y_{22}, respectively, in Equation (6-33). These functions are called the *short-circuit admittance functions* of the two-port. In particular, y_{11} and y_{22} are the short-circuit driving-point admittances at ports $(1,1')$ and $(2,2')$, respectively; and y_{12} and y_{21} are the short-circuit transfer admittances of the two-port.

Comparing Equation (6-31) with (6-33), we readily obtain the following results:

$$y_{11} = \frac{z_{22}}{\Delta_z} \qquad y_{21} = \frac{-z_{21}}{\Delta_z}$$
$$y_{12} = \frac{-z_{12}}{\Delta_z} \qquad y_{22} = \frac{z_{11}}{\Delta_z} \tag{6-40}$$

where

$$\Delta_z = z_{11}z_{22} - z_{12}z_{21}$$

as defined in Equation (6-32).

Substituting Equation (6-16) into (6-32) for a reciprocal two-port, we get

$$\Delta_z = \frac{\Delta_{11}}{\Delta}\left(\frac{\Delta_{22} + \Delta_{2'2'} - 2\Delta_{22'}}{\Delta}\right) - \left(\frac{\Delta_{12} - \Delta_{12'}}{\Delta}\right)^2$$
$$= \frac{1}{\Delta^2}(\Delta_{11}\Delta_{22} + \Delta_{11}\Delta_{2'2'} - 2\Delta_{11}\Delta_{22'} - \Delta_{12}^2 + 2\Delta_{12}\Delta_{12'} - \Delta_{12'}^2)$$

Or, upon regrouping the terms, we write

$$\Delta_z = \frac{1}{\Delta^2}\{(\Delta_{11}\Delta_{22} - \Delta_{12}^2) + (\Delta_{11}\Delta_{2'2'} - \Delta_{12'}^2) - 2(\Delta_{11}\Delta_{22'} - \Delta_{12}\Delta_{12'})\} \tag{6-41}$$

Next, we recall the following identity for determinants, a consequence of the theorem of Jacobi[1]

$$\Delta_{ab}\Delta_{cd} - \Delta_{ad}\Delta_{cb} = \Delta\Delta_{abcd} \tag{6-42}$$

where a, b, c, d are positive integers less than n (n being the order of the determinant Δ), and Δ_{abcd} is the cofactor obtained from Δ by deleting rows a and c and columns b and d from Δ. Upon the application of Equation (6-42), we can rewrite (6-41) as

$$\Delta_z = \frac{1}{\Delta^2}(\Delta\Delta_{1122} + \Delta\Delta_{112'2'} - 2\Delta\Delta_{1122'}) = \frac{1}{\Delta}(\Delta_{1122} + \Delta_{112'2'} - 2\Delta_{1122'}) \tag{6-43}$$

Substituting Equations (6-16) and (6-43) into (6-40) and simplifying, we find the desired expressions for y_{11}, y_{12} $(= y_{21})$, and y_{22}:

$$y_{11} = \frac{\Delta_{22} + \Delta_{2'2'} - 2\Delta_{22'}}{\Delta_{1122} + \Delta_{112'2'} - 2\Delta_{1122'}} \tag{6-44}$$

$$y_{12} = y_{21} = \frac{\Delta_{12'} - \Delta_{12}}{\Delta_{1122} + \Delta_{112'2'} - 2\Delta_{1122'}} \tag{6-45}$$

$$y_{22} = \frac{\Delta_{11}}{\Delta_{1122} + \Delta_{112'2'} - 2\Delta_{1122'}} \tag{6-46}$$

[1]See, for example, A. C. Aitken [1], p. 98.

EXAMPLE 6-2. Refer to the same network of Example 6-1 as shown in Figure 6-3(a). The determinants to be evaluated are

$$\Delta_{1122} = \begin{vmatrix} y_4 + y_5 & 0 & -y_5 \\ 0 & y_1 + y_2 + y_3 & -y_3 \\ -y_5 & -y_3 & y_3 + y_5 + y_6 \end{vmatrix} \tag{6-47}$$

$$\Delta_{112'2'} = \begin{vmatrix} y_6 & 0 & -y_6 \\ 0 & y_1 + y_2 + y_3 & -y_3 \\ -y_6 & -y_3 & y_3 + y_5 + y_6 \end{vmatrix} \tag{6-48}$$

$$\Delta_{1122'} = -\begin{vmatrix} 0 & 0 & -y_5 \\ 0 & y_1 + y_2 + y_3 & -y_3 \\ -y_6 & -y_3 & y_3 + y_5 + y_6 \end{vmatrix} \tag{6-49}$$

After expansion and cancellation of terms, Equations (6-47) through (6-49) are simplified to (6-50) through (6-52), respectively:

$$\begin{aligned} \Delta_{1122} = {} & y_1y_3y_4 + y_1y_3y_5 + y_2y_3y_4 + y_2y_3y_5 + y_1y_4y_5 + y_2y_4y_5 + y_1y_4y_6 \\ & + y_1y_5y_6 + y_2y_4y_6 + y_2y_5y_6 + y_3y_4y_5 + y_3y_4y_6 + y_3y_5y_6 \end{aligned} \tag{6-50}$$

$$\Delta_{112'2'} = y_1y_3y_6 + y_2y_3y_6 + y_1y_5y_6 + y_2y_5y_6 + y_3y_5y_6 \tag{6-51}$$

$$\Delta_{1122'} = y_1y_5y_6 + y_2y_5y_6 + y_3y_5y_6 \tag{6-52}$$

From Equations (6-50), (6-51), and (6-52), we find that

$$\begin{aligned} \Delta_{1122} + \Delta_{112'2'} - 2\Delta_{1122'} = {} & y_1y_3y_4 + y_1y_3y_5 + y_1y_3y_6 + y_1y_4y_5 + y_1y_4y_6 \\ & + y_2y_3y_4 + y_2y_3y_5 + y_2y_3y_6 + y_2y_4y_5 \\ & + y_2y_4y_6 + y_3y_4y_5 + y_3y_4y_6 \end{aligned} \tag{6-53}$$

Thus, from Equations (6-44), (6-45) and (6-46), we get

$$\begin{aligned} y_{11} &= \frac{\Delta_{22} + \Delta_{2'2'} - 2\Delta_{22'}}{\Delta_{1122} + \Delta_{112'2'} - 2\Delta_{1122'}} \\ &= \frac{y_1y_2y_3y_4 + y_1y_2y_3y_5 + y_1y_2y_3y_6 + y_1y_2y_4y_5}{y_1y_3y_4 + y_1y_3y_5 + y_1y_3y_6 + y_1y_4y_5 + y_1y_4y_6 + y_2y_3y_4} --- \\ &\quad --- \frac{+ y_1y_2y_4y_6 + y_1y_3y_4y_5 + y_1y_3y_4y_6}{+ y_2y_3y_5 + y_2y_3y_6 + y_2y_4y_5 + y_2y_4y_6 + y_3y_4y_5 + y_3y_4y_6} \end{aligned} \tag{6-54}$$

$$\begin{aligned} y_{12} = y_{21} &= \frac{\Delta_{12'} - \Delta_{12}}{\Delta_{1122} + \Delta_{112'2'} - 2\Delta_{1122'}} \\ &= \frac{-y_1y_3y_4y_6}{y_1y_3y_4 + y_1y_3y_5 + y_1y_3y_6 + y_1y_4y_5 + y_1y_4y_6 + y_2y_3y_4 + y_2y_3y_5} --- \\ &\quad --- \overline{+ y_2y_3y_6 + y_2y_4y_5 + y_2y_4y_6 + y_3y_4y_5 + y_3y_4y_6} \end{aligned} \tag{6-55}$$

$$y_{22} = \frac{\Delta_{11}}{\Delta_{1122} + \Delta_{112'2'} - 2\Delta_{1122'}}$$

$$= \frac{y_1y_3y_4y_6 + y_1y_3y_5y_6 + y_1y_4y_5y_6 + y_2y_3y_4y_6}{y_1y_3y_4 + y_1y_3y_5 + y_1y_3y_6 + y_1y_4y_5 + y_1y_4y_6 + y_2y_3y_4 + y_2y_3y_5} \cdots$$

$$\cdots \frac{+ y_2y_3y_5y_6 + y_2y_4y_5y_6 + y_3y_4y_5y_6}{+ y_2y_3y_6 + y_2y_4y_5 + y_2y_4y_6 + y_3y_4y_5 + y_3y_4y_6} \tag{6-56}$$

Finally, substituting Equation (6-17) into (6-54) through (6-56) and simplifying, we obtain the following results:

$$y_{11} = \frac{(G_1C_2G_3C_5 + G_1C_2G_4C_5)s^2 + (G_1C_2G_3G_4 + G_1C_2G_3G_6 + G_1C_2G_4G_6 + G_1G_3G_4C_5)s + G_1G_3G_4G_6}{(C_2G_3C_5 + C_2G_4C_5)s^2 + (G_1G_3C_5 + G_1G_4C_5 + C_2G_3G_4 + C_2G_3G_6 + C_2G_4G_6 + G_3G_4C_5)s + G_1G_3G_4 + G_1G_3G_6 + G_1G_4G_6 + G_3G_4G_6} \tag{6-57}$$

$$y_{12} = y_{21} = \frac{-G_1G_3G_4G_6}{(C_2G_3C_5 + C_2G_4C_5)s^2 + (G_1G_3C_5 + G_1G_4C_5 + C_2G_3G_4 + C_2G_3G_6 + C_2G_4G_6 + G_3G_4C_5)s + (G_1G_3G_4 + G_1G_3G_6 + G_1G_4G_6 + G_3G_4G_6)} \tag{6-58}$$

$$y_{22} = \frac{(C_2G_3C_5G_6 + C_2G_4C_5G_6)s^2 + (G_1G_3C_5G_6 + G_1G_4C_5G_6 + C_2G_3G_4G_6 + G_3G_4C_5G_6)s + G_1G_3G_4G_6}{(C_2G_3C_5 + C_2G_4C_5)s^2 + (G_1G_3C_5 + G_1G_4C_5 + C_2G_3G_4 + C_2G_3G_6 + C_2G_4G_6 + G_3G_4C_5)s + (G_1G_3G_4 + G_1G_3G_6 + G_1G_4G_6 + G_3G_4G_6)} \tag{6-59}$$

6-3 OTHER NETWORK TRANSFER FUNCTIONS

(a) Transfer Impedance Function $Z_{12}(s)$

Let the passive two-port N be driven by a current generator $I_g(s)$ at port (1,1′) with a load admittance $Y_L(s)$ at port (2,2′) as shown in Figure 6-4. Then the system *as a whole* may be considered as a passive one-port driven by a

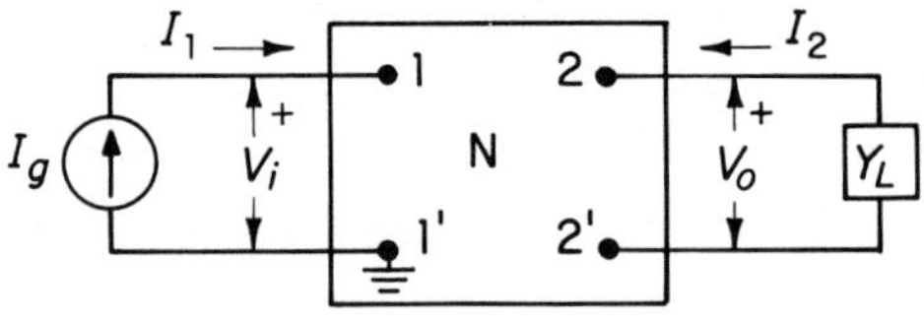

Figure 6-4 A loaded passive two-port N.

single current source I_1, and all the other currents, $I_2, I_3, \ldots, I_n$, in Equation (6-1) are zero, since they signify current sources in the system. Thus, from Equation (6-2), with $j = 2$ and $2'$, we get

$$V_2 = \frac{\Delta_{12}}{\Delta} I_1 \tag{6-60}$$

and

$$V_{2'} = \frac{\Delta_{12'}}{\Delta} I_1 \tag{6-61}$$

respectively.

But, from Figure 6-4, we see that

$$V_o = V_2 - V_{2'} \tag{6-62}$$

so that, from Equations (6-60), (6-61), and (6-62),

$$Z_{12} \triangleq \frac{V_o}{I_1} = \frac{\Delta_{12} - \Delta_{12'}}{\Delta} \tag{6-63}$$

which is the desired expression for the transfer impedance function of a loaded two-port.

It should be noted that the load $Y_L(s)$ has been included implicitly as part of Δ, although it is not shown in Equation (6-63).

(b) Voltage-ratio Transfer Function $G_{12}(s)$

The voltage-ratio transfer function $G_{12}(s)$ is defined as the ratio of the output voltage V_o to the input voltage V_i. From Equations (6-60), (6-61), and (6-62) we find that

$$V_o = \left(\frac{\Delta_{12} - \Delta_{12'}}{\Delta}\right) I_1 \tag{6-64}$$

Also, from Equation (6-2), with $j = 1$,

$$V_i = V_1 = \frac{\Delta_{11}}{\Delta} I_1 \tag{6-65}$$

Thus, dividing Equation (6-64) by (6-65) and simplifying, we get the desired result:

$$G_{12} = \frac{V_o}{V_i} = \frac{\Delta_{12} - \Delta_{12'}}{\Delta_{11}} \tag{6-66}$$

(c) Transfer Admittance Function $Y_{12}(s)$

The transfer admittance function $Y_{12}(s)$ is defined by the equation[2]

$$Y_{12} = \frac{-I_2}{V_i} \tag{6-67}$$

where I_2 is the current flowing from node 2′ to node 2 through Y_L. Thus, from Figure 6-4, we write

$$-I_2 = Y_L V_o \tag{6-68}$$

[2]The negative sign associated with I_2 in both (6-67) and (6-71) is due to the convention that I_2 is the current flowing from node 2′ to node 2 through Y_L as shown in Figure 6-4.

so that Equation (6-67) may be written as

$$Y_{12} = Y_L\left(\frac{V_o}{V_i}\right) \tag{6-69}$$

Finally, from Equations (6-69) and (6-66), we find that

$$Y_{12} = Y_L G_{12} = Y_L\left(\frac{\Delta_{12} - \Delta_{12'}}{\Delta_{11}}\right) \tag{6-70}$$

(d) Current-ratio Transfer Function $\alpha_{12}(s)$

Like $G_{12}(s)$, the current-ratio transfer function $\alpha_{12}(s)$ is defined as the ratio of the output current $-I_2$ to the input current I_1, as follows[2]:

$$\alpha_{12} \triangleq \frac{-I_2}{I_1} \tag{6-71}$$

Thus, substituting Equations (6-68) and (6-63) into (6-71), we find immediately the following result:

$$\alpha_{12} = Y_L Z_{12} = Y_L\left(\frac{\Delta_{12} - \Delta_{12'}}{\Delta}\right) \tag{6-72}$$

EXAMPLE 6-3. Consider the network of Example 6-1 as shown in Figure 6-3(a). With the load admittance Y_L at port, (2,2′),

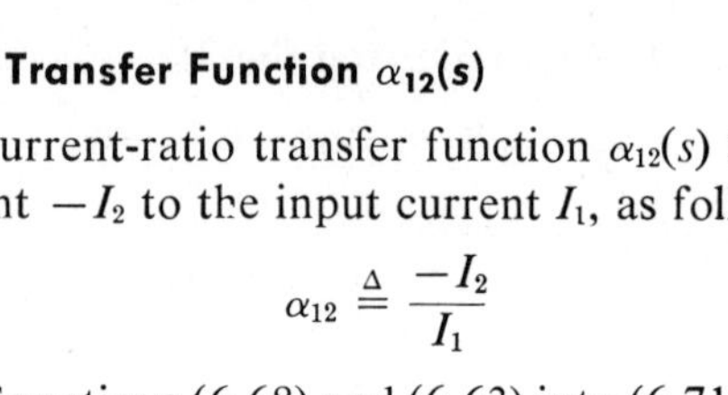

Figure 6-5 The graph of the network of Example 6-1 with a load admittance Y_L at port (2,2′).

$$\Delta = \begin{array}{c|ccccc|} & 1 & 2 & 2' & 3 & 4 \\ 1 & y_1 & 0 & 0 & -y_1 & 0 \\ 2 & 0 & y_6 + Y_L & -Y_L & 0 & -y_6 \\ 2' & 0 & -Y_L & y_4 + y_5 + Y_L & 0 & -y_5 \\ 3 & -y_1 & 0 & 0 & y_1 + y_2 + y_3 & -y_3 \\ 4 & 0 & -y_6 & -y_5 & -y_3 & y_3 + y_5 + y_6 \end{array}$$

The graph of the loaded network is shown in Figure 6-5. Thus the determinant Δ and the cofactors Δ_{11}, Δ_{12}, and $\Delta_{12'}$ are:

$$\begin{aligned} \Delta = {} & y_1y_2y_3y_4y_6 + y_1y_2y_3y_4Y_L + y_1y_2y_3y_5y_6 + y_1y_2y_3y_5Y_L \\ & + y_1y_2y_3y_6Y_L + y_1y_2y_4y_5y_6 + y_1y_2y_4y_5Y_L + y_1y_2y_4y_6Y_L \\ & + y_1y_3y_4y_5y_6 + y_1y_3y_4y_5Y_L + y_1y_3y_4y_6Y_L \end{aligned} \tag{6-73}$$

$$\begin{aligned}\Delta_{11} = {} & y_1y_3y_4y_6 + y_1y_3y_4Y_L + y_1y_3y_5y_6 + y_1y_3y_5Y_L + y_1y_3y_6Y_L \\ & + y_1y_4y_5y_6 + y_1y_4y_5Y_L + y_1y_4y_6Y_L + y_2y_3y_4y_6 \\ & + y_2y_3y_4Y_L + y_2y_3y_5y_6 + y_2y_3y_5Y_L + y_2y_3y_6Y_L \\ & + y_2y_4y_5y_6 + y_2y_4y_5Y_L + y_2y_4y_6Y_L + y_3y_4y_5y_6 \\ & + y_3y_4y_5Y_L + y_3y_4y_6Y_L \end{aligned} \tag{6-74}$$

$$\Delta_{12} = y_1y_3y_4y_6 + y_1y_3y_5y_6 + y_1y_3y_5Y_L + y_1y_3y_6Y_L \tag{6-75}$$

and

$$\Delta_{12'} = y_1y_3y_5y_6 + y_1y_3y_5Y_L + y_1y_3y_6Y_L \tag{6-76}$$

Thus, from Equations (6-75) and (6-76),

$$\Delta_{12} - \Delta_{12'} = y_1y_3y_4y_6 \tag{6-77}$$

Hence, substituting Equations (6-73) and (6-77) into (6-63) we get

$$\begin{aligned} Z_{12}(s) &= \frac{V_o}{I_1} = \frac{\Delta_{12} - \Delta_{12'}}{\Delta} \\ &= \frac{y_1y_3y_4y_6}{\left\{\begin{aligned} & y_1y_2y_3y_4y_6 + y_1y_2y_3y_4Y_L + y_1y_2y_3y_5y_6 + y_1y_2y_3y_5Y_L + y_1y_2y_3y_6Y_L \\ & + y_1y_2y_4y_5y_6 + y_1y_2y_4y_5Y_L + y_1y_2y_4y_6Y_L + y_1y_3y_4y_5y_6 \\ & + y_1y_3y_4y_5Y_L + y_1y_3y_4y_6Y_L \end{aligned}\right\}} \end{aligned} \tag{6-78}$$

To obtain $G_{12}(s)$, we substitute Equations (6-74) and (6-77) into (6-66) and find

$$\begin{aligned} G_{12}(s) &= \frac{V_o}{V_i} = \frac{\Delta_{12} - \Delta_{12'}}{\Delta_{11}} \\ &= \frac{y_1y_3y_4y_6}{\left\{\begin{aligned} & y_1y_3y_4y_6 + y_1y_3y_4Y_L + y_1y_3y_5y_6 + y_1y_3y_5Y_L + y_1y_3y_6Y_L + y_1y_4y_5y_6 \\ & + y_1y_4y_5Y_L + y_1y_4y_6Y_L + y_2y_3y_4y_6 + y_2y_3y_4Y_L + y_2y_3y_5y_6 \\ & + y_2y_3y_5Y_L + y_2y_3y_6Y_L + y_2y_4y_5y_6 + y_2y_4y_5Y_L + y_2y_4y_6Y_L \\ & + y_3y_4y_5y_6 + y_3y_4y_5Y_L + y_3y_4y_6Y_L \end{aligned}\right\}} \end{aligned} \tag{6-79}$$

Next, for $Y_{12}(s)$, we get, from Equations (6-70) and (6-79),

$$\begin{aligned} Y_{12}(s) &= \frac{-I_2}{V_i} = Y_LG_{12} \\ &= \frac{y_1y_3y_4y_6Y_L}{\left\{\begin{aligned} & y_1y_3y_4y_6 + y_1y_3y_4Y_L + y_1y_3y_5y_6 + y_1y_3y_5Y_L + y_1y_3y_6Y_L + y_1y_4y_5y_6 \\ & + y_1y_4y_5Y_L + y_1y_4y_6Y_L + y_2y_3y_4y_6 + y_2y_3y_4Y_L + y_2y_3y_5y_6 \\ & + y_2y_3y_5Y_L + y_2y_3y_6Y_L + y_2y_4y_5y_6 + y_2y_4y_5Y_L + y_2y_4y_6Y_L \\ & + y_3y_4y_5y_6 + y_3y_4y_5Y_L + y_3y_4y_6Y_L \end{aligned}\right\}} \end{aligned} \tag{6-80}$$

Similarly, for $\alpha_{12}(s)$, we find, from Equations (6-72) and (6-78),

$$\alpha_{12}(s) = \frac{-I_2}{I_1} = Y_L Z_{12}$$

$$= \frac{y_1y_3y_4y_6Y_L}{\left\{\begin{array}{l} y_1y_2y_3y_4y_6 + y_1y_2y_3y_4Y_L + y_1y_2y_3y_5y_6 + y_1y_2y_3y_5Y_L + y_1y_2y_3y_6Y_L \\ + y_1y_2y_4y_5y_6 + y_1y_2y_4y_5Y_L + y_1y_2y_4y_6Y_L + y_1y_3y_4y_5y_6 \\ + y_1y_3y_4y_5Y_L + y_1y_3y_4y_6Y_L \end{array}\right\}} \tag{6-81}$$

Finally, it is straightforward, though time-consuming, to find the expressions for Z_{12}, G_{12}, Y_{12}, and α_{12} in terms of the given element values as shown in Figure 6-3(a) with the load admittance Y_L at port (2,2′). This can be accomplished by substituting Equation (6-17) with Y_L into Equations (6-78) through (6-81), respectively, and simplifying, but we shall not attempt it here. Each of the four expressions in its final form will be a rational function in s — a ratio of two polynomials in s.

6-4 TOPOLOGICAL FORMULAS FOR Δ_{ij}

In Chapter 5, we derived the topological formula for the cofactor Δ_{jj} of the determinant Δ of the node admittance matrix Y_n in terms of the 2-tree-admittance products $T_{2_{j,1'}}(y)$, which led to the development of the topological expressions for the driving-point impedance $Z_d(s)$ and its reciprocal $Y_d(s)$. In a similar manner, we shall now give the topological formula for the asymmetrical cofactors of Δ as stated in the following theorem:

THEOREM 6-1. The (i,j) cofactor Δ_{ij} of Δ of a relaxed passive network N with n independent nodes (with node 1′ as the reference node) and without mutual inductances is given by

$$\Delta_{ij} = \sum_k T^{(k)}_{2_{ij,1'}}(y) \tag{6-82}$$

where the summation is taken over all the 2-tree admittance products of the form $T_{2_{ij,1'}}(y)$ with each containing nodes i and j in one connected part and the reference node 1′ in the other.

Proof: A complete proof of this theorem will not be given here as it is too lengthy.[3] Instead, we shall outline the major steps that are needed in the proof as follows:

We recall that in the case of Δ_{jj}, we showed that

$$\Delta_{jj} = \det A_{-j} Y_e A^T_{-j}$$

[3]For a complete proof, the interested reader may refer to Mayeda and Seshu [62].

But, in the case of Δ_{ij}, we first note that

$$\Delta_{ij} = (-1)^{i+j} M_{ij} \tag{6-83}$$

where M_{ij} is the (i,j) minor of Δ, which is the determinant obtained by deleting the ith row and the jth column from Δ. Then we show that

$$M_{ij} = \det A_{-i} Y_e A_{-j}^T \tag{6-84}$$

where, as defined in Chapter 5, the subscript in A_{-i} and in A_{-j} indicates the row being deleted from the incidence submatrix A.

Finally, we show that

$$\det A_{-i} Y_e A_{-j}^T = (-1)^{i+j} \sum_k T_{2_{ij,1'}}^{(k)}(y) \tag{6-85}$$

and, consequently, from Equations (6-83), (6-84), and (6-85), that

$$\Delta_{ij} = (-1)^{i+j} \det A_{-i} Y_e A_{-j}^T = \sum_k T_{2_{ij,1'}}^{(k)}(y) \tag{6-86}$$

which is the desired result.

It should be noted that since the 2-trees $T_{2_{i,1'}}$ of N can be obtained by finding the trees of N_{-i} by short-circuiting node i to node $1'$, it is evident that the 2-trees $T_{2_{ij,1'}}$ of N can be obtained by taking the intersection of the 2-trees $T_{2_{i,1'}}$ and the 2-trees $T_{2_{j,1'}}$; that is,

$$\sum_k T_{2_{ij,1'}}^{(k)}(y) = \left\{\sum_p T_{2_{i,1'}}^{(p)}(y)\right\} \cap \left\{\sum_q T_{2_{j,1'}}^{(q)}(y)\right\} \tag{6-87}$$

or, in shorthand notation, Equation (6-87) becomes

$$W_{ij,1'}(Y) = W_{i,1'}(Y) \cap W_{j,1'}(Y) \tag{6-88}$$

Thus Equation (6-82) may be rewritten as

$$\Delta_{ij} = W_{ij,1'}(Y) = W_{i,1'}(Y) \cap W_{j,1'}(Y) \tag{6-89}$$

which implies that Δ_{ij} may be evaluated in terms of the tree admittance products of N_{-i} and the tree admittance products of N_{-j}.

We should point out also that when $j = i$, Equation (6-82) reduces to

$$\Delta_{ii} = W_{ii,1'}(Y) = W_{i,1'}(Y) \tag{6-90}$$

which agrees with the result obtained in Chapter 5.

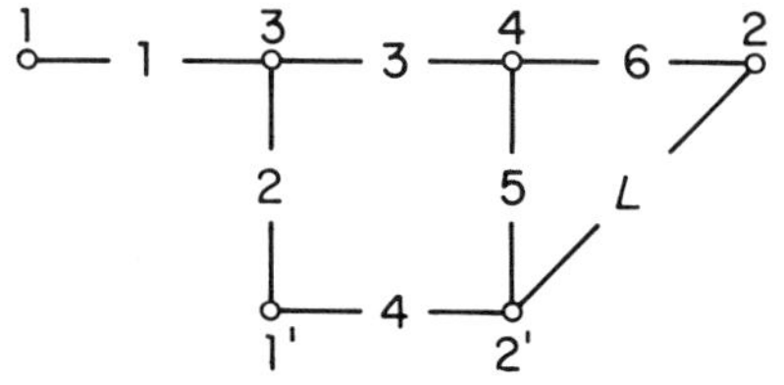

Figure 6-6 Graph G of the network of Example 6-3.

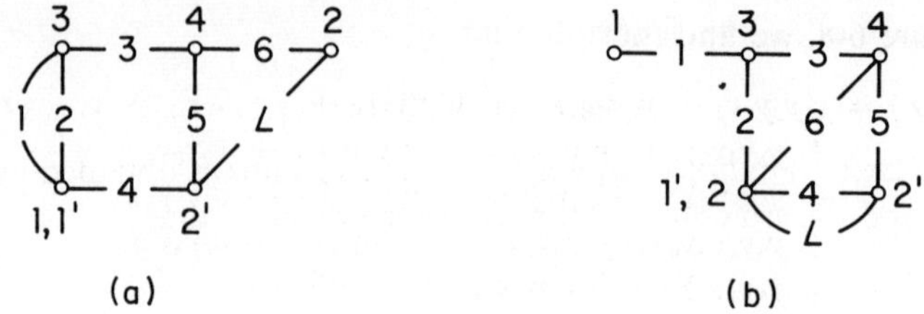

Figure 6-7 (a) Graph G_{-1} for the evaluation of Δ_{11}. (b) Graph G_{-2} for the evaluation of Δ_{22}.

EXAMPLE 6-4. Again, consider the network of Example 6-3, whose graph G is shown in Figure 6-5 and repeated in Figure 6-6 for convenience. We shall evaluate Δ_{12} topologically using Equation(6-89). The graph G_{-1}, which corresponds to Δ_{11}, is shown in Figure 6-7(a); the graph G_{-2}, which corresponds to Δ_{22}, is shown in Figure 6-7(b). The trees of G_{-1} are shown in Figure 6-8, and the trees of G_{-2} are shown in Figure 6-9.

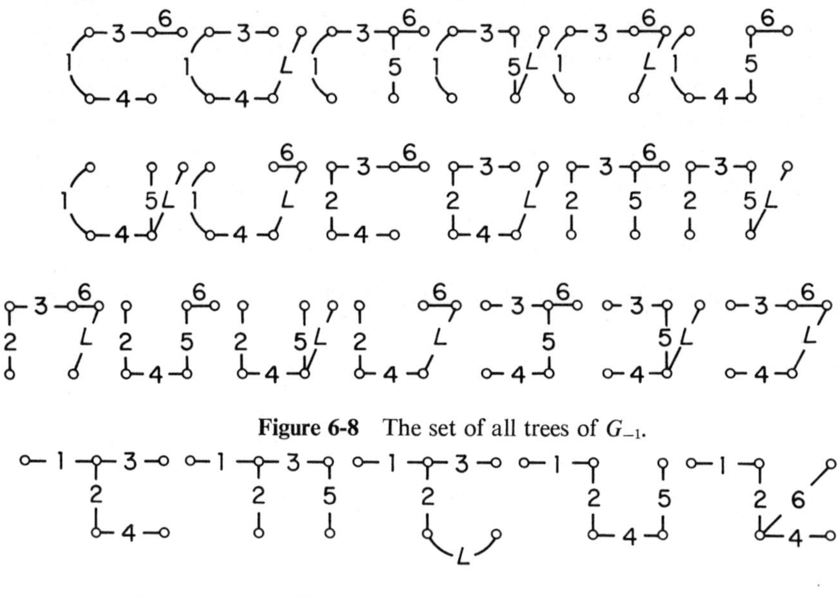

Figure 6-8 The set of all trees of G_{-1}.

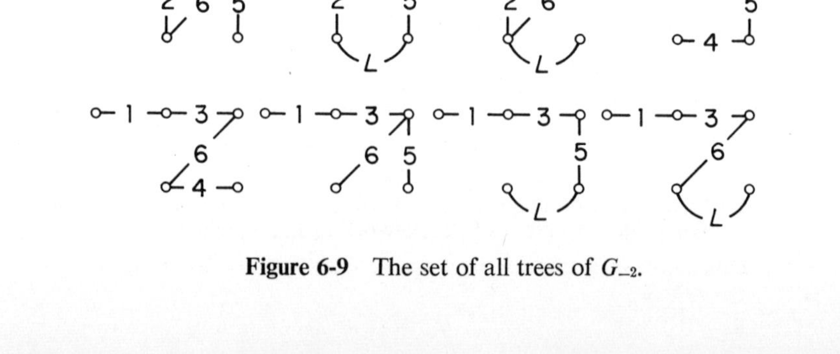

Figure 6-9 The set of all trees of G_{-2}.

From Figure 6-8, we find that

$$\begin{aligned} W_{1,1'}(Y) = {} & y_1y_3y_4y_6 + y_1y_3y_4Y_L + y_1y_3y_5y_6 + y_1y_3y_5Y_L + y_1y_3y_6Y_L \\ & + y_1y_4y_5y_6 + y_1y_4y_5Y_L + y_1y_4y_6Y_L + y_2y_3y_4y_6 \\ & + y_2y_3y_4Y_L + y_2y_3y_5y_6 + y_2y_3y_5Y_L + y_2y_3y_6Y_L \\ & + y_2y_4y_5y_6 + y_2y_4y_5Y_L + y_2y_4y_6Y_L + y_3y_4y_5y_6 \\ & + y_3y_4y_5Y_L + y_3y_4y_6Y_L \end{aligned} \tag{6-91}$$

Similarly, from Figure 6-9, we get

$$\begin{aligned} W_{2,1'}(Y) = {} & y_1y_2y_3y_4 + y_1y_2y_3y_5 + y_1y_2y_3Y_L + y_1y_2y_4y_5 + y_1y_2y_4y_6 \\ & + y_1y_2y_5y_6 + y_1y_2y_5Y_L + y_1y_2y_6Y_L + y_1y_3y_4y_5 + y_1y_3y_4y_6 \\ & + y_1y_3y_5y_6 + y_1y_3y_5Y_L + y_1y_3y_6Y_L \end{aligned} \tag{6-92}$$

Thus, from Equations (6-89), (6-91), and (6-92), we obtain

$$\Delta_{12} = W_{1,1'} \cap W_{2,1'} = y_1y_3y_4y_6 + y_1y_3y_5y_6 + y_1y_3y_5Y_L + y_1y_3y_6Y_L \tag{6-93}$$

which is indeed identical to the expression of Δ_{12} obtained by direct expansion of the cofactor as expressed in Equation (6-75).

It is evident in Example 6-4 that the use of topological formulas not only serves as a time-saving shortcut to the evaluation of various network determinants but, perhaps even more important, also gives a physical insight to the system since it relates each term in the expansion of a determinant to a part of the physical system.

6-5 TOPOLOGICAL FORMULAS FOR z_{11}, z_{12}, AND z_{22}

In a two-port network N, there are four nodes to be specified — namely, nodes 1 and 1′ at the input port (1,1′), and nodes 2 and 2′ at the output port (2,2′), as illustrated in Figure 6-4. But, for a 2-tree of the type $T_{2_{ij,1'}}$, only three nodes have been used, thus leaving the fourth one unidentified.

With very little effort, it can be shown that, in general, the following relationship holds:[4]

$$W_{ij,1'}(Y) = W_{ijk,1'}(Y) + W_{ij,k1'}(Y)$$

or, simply,

$$W_{ij,1'} = W_{ijk,1'} + W_{ij,k1'} \tag{6-94}$$

where i, j, k, and 1′ are the four terminals of N, with 1′ denoting the datum node. The symbol $W_{ijk,1'}$ denotes the sum of all the 2-tree admittance products, each containing nodes i, j, and k in one connected part and node 1′ in the other. Similarly, $W_{ij,k1'}$ is the sum of all the 2-tree admittance products, each containing nodes i and j in one connected part and nodes k and 1′ in the other.

[4]The proof of Equation (6-94) is left to the reader as a problem (see Problem 6-11).

As an illustration of Equation (6-94), we have, according to Figure 6-6,

$$W_{12,1'} = y_1y_3y_4y_6 + y_1y_3y_5y_6 + y_1y_3y_5Y_L + y_1y_3y_6Y_L$$
$$W_{122',1'} = y_1y_3y_5y_6 + y_1y_3y_5y_L + y_1y_3y_6Y_L$$

so that

$$W_{12,1'} = W_{122',1'} + W_{12,1'2'}$$

We now state and prove the next theorem.

THEOREM 6-2. With the same hypothesis and notation as stated earlier,

$$\Delta_{12} - \Delta_{12'} = W_{12,1'2'}(Y) - W_{12',1'2}(Y) \tag{6-95}$$

Proof: From Equation (6-89), we find that

$$\Delta_{12} = W_{12,1'} \quad \text{and} \quad \Delta_{12'} = W_{12',1'}$$

so that

$$\Delta_{12} - \Delta_{12'} = W_{12,1'} - W_{12',1'} \tag{6-96}$$

But, from Equation (6-94), we write

$$W_{12,1'} = W_{122',1'} + W_{12,1'2'} \tag{6-97}$$

and

$$W_{12',1'} = W_{122',1'} + W_{12',1'2} \tag{6-98}$$

Finally, substituting Equations (6-97) and (6-98) into (6-96) and canceling the like terms, we get

$$\Delta_{12} - \Delta_{12'} = W_{12,1'2'} - W_{12',1'2}$$

as stated in the theorem.

It is interesting to note that Equation (6-95) is stated by Percival[5] in the following descriptive fashion:

$$\Delta_{12} - \Delta_{12'} = W_{12,1'2'} - W_{12',1'2} =$$ 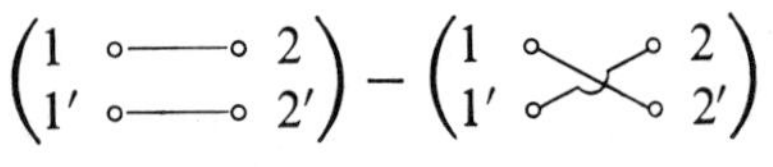

(6-99)

which illustrated the two types of 2-trees involved in the formula.

Finally, we state the topological formulas for z_{11}, z_{12}, and z_{22} in the following theorem.

THEOREM 6-3. With the same hypothesis and notation as stated earlier,

$$z_{11} = \frac{W_{1,1'}(Y)}{V(Y)} \tag{6-100a}$$

[5]Refer to W. S. Percival [67].

$$z_{12} = z_{21} = \frac{W_{12,1'2'}(Y) - W_{12',1'2}(Y)}{V(Y)} \tag{6-100b}$$

$$z_{22} = \frac{W_{2,2'}(Y)}{V(Y)} \tag{6-100c}$$

Proof: The results in Equation (6-100) are the direct consequences of substituting Equations (6-90) and (6-95) into (6-16).

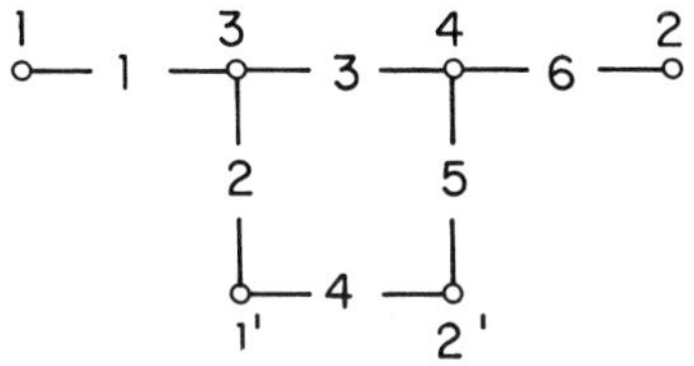

Figure 6-10 The graph G of the network used in Example 6-1.

EXAMPLE 6-5. For the purpose of comparing results, let us again consider the unloaded network used in Example 6-2. The graph G of the network is shown in Figure 6-10, and the set of all trees of G is shown in Figure 6-11. Thus, from Figure 6-11, we get

$$V(Y) = y_1y_2y_3y_4y_6 + y_1y_2y_3y_5y_6 + y_1y_2y_4y_5y_6 + y_1y_3y_4y_5y_6 \tag{6-101}$$

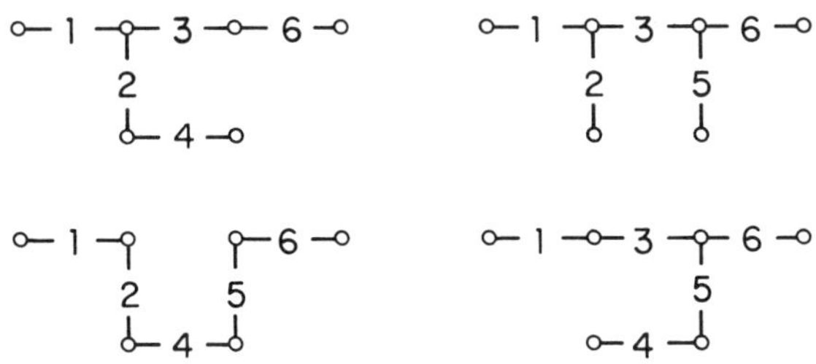

Figure 6-11 The set of all trees of G.

Next, we short-circuit vertex 1 to vertex 1′ to obtain G_{11}', which is shown in Figure 6-12. The trees of $G_{11'}$ are shown in Figure 6-13. Thus we get

$$\begin{aligned} W_{1,1'} = {} & y_1y_3y_4y_6 + y_1y_3y_5y_6 + y_1y_4y_5y_6 + y_2y_3y_4y_6 \\ & + y_2y_3y_5y_6 + y_2y_4y_5y_6 + y_3y_4y_5y_6 \end{aligned} \tag{6-102}$$

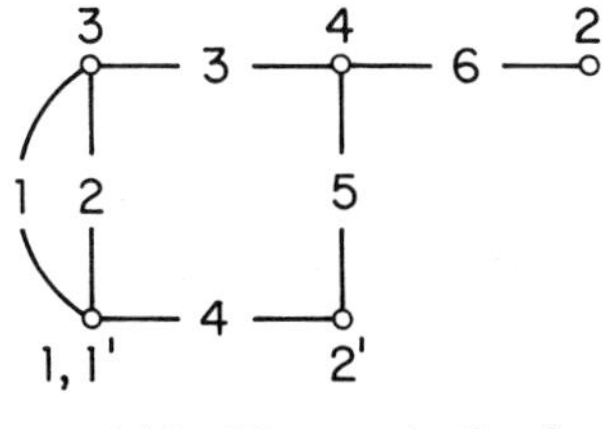

Figure 6-12 The graph G_{11}, for $W_{11'}$.

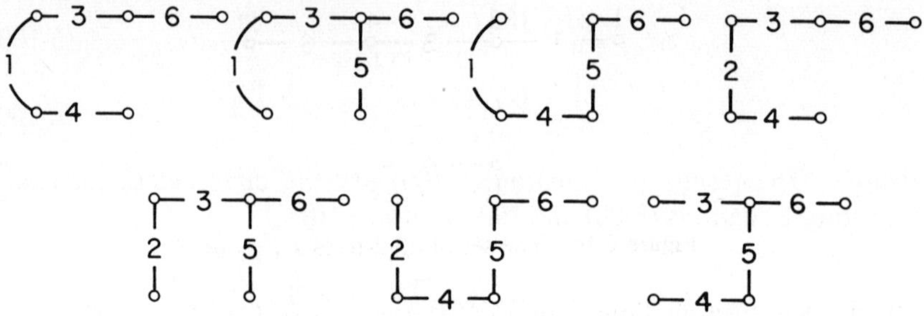

Figure 6-13 The set of all trees of $G_{11'}$.

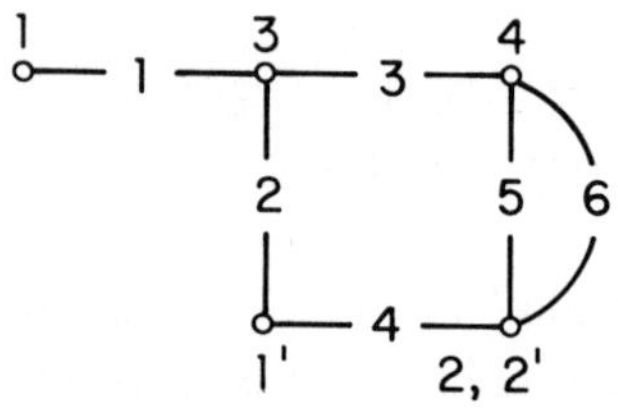

Figure 6-14 The graph $G_{22'}$ for $W_{2,2'}$.

To obtain $W_{2,2'}$, we short-circuit vertex 2 to vertex 2′ to obtain $G_{22'}$, as shown in Figure 6-14. The trees of $G_{22'}$ (which are 2-trees $T_{2_{2,2'}}$ of G) are shown in Figure 6-15. Hence we find that

$$W_{2,2'}(Y) = y_1y_2y_3y_4 + y_1y_2y_3y_5 + y_1y_2y_3y_6 + y_1y_2y_4y_5 + y_1y_2y_4y_6 + y_1y_3y_4y_5 + y_1y_3y_4y_6 \tag{6-103}$$

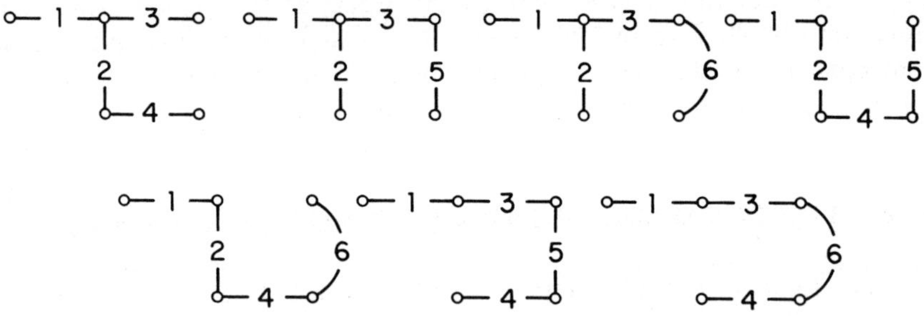

Figure 6-15 The set of all trees of $G_{22'}$.

Next, we find the set of all 2-trees $T_{2_{12,1'2'}}$ of G as shown in Figure 6-16. The set of all 2-trees $T_{2_{12,1'2}}$ for the graph G in this example happens to be an empty set. Thus,

$$W_{12,1'2'} = y_1y_3y_4y_6 \tag{6-104}$$

and

$$W_{12',1'2} = 0. \tag{6-105}$$

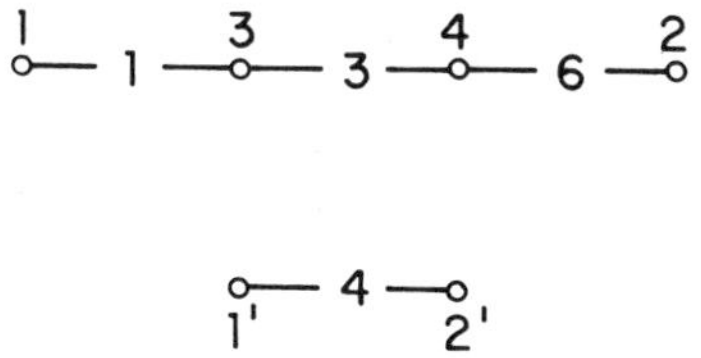

Figure 6-16 The set of all 2-trees $T_{2_{12,1'2'}}$ of G.

Finally, substituting Equations (6-101) through (6-105) into (6-100), we get

$$z_{11} = \frac{W_{1,1'}}{V}$$

$$= \frac{y_1y_3y_4y_6 + y_1y_3y_5y_6 + y_1y_4y_5y_6 + y_2y_3y_4y_6 + y_2y_3y_5y_6 + y_2y_4y_5y_6 + y_3y_4y_5y_6}{y_1y_2y_3y_4y_6 + y_1y_2y_3y_5y_6 + y_1y_2y_4y_5y_6 + y_1y_3y_4y_5y_6} \tag{6-106}$$

$$z_{12} = z_{21} = \frac{W_{12,1'2'} - W_{12',1'2}}{V}$$

$$= \frac{y_1y_3y_4y_6}{y_1y_2y_3y_4y_6 + y_1y_2y_3y_5y_6 + y_1y_2y_4y_5y_6 + y_1y_3y_4y_5y_6} \tag{6-107}$$

$$z_{22} = \frac{W_{2,2'}}{V}$$

$$= \frac{y_1y_2y_3y_4 + y_1y_2y_3y_5 + y_1y_2y_3y_6 + y_1y_2y_4y_5 + y_1y_2y_4y_6 + y_1y_3y_4y_5 + y_1y_3y_4y_6}{y_1y_2y_3y_4y_6 + y_1y_2y_3y_5y_6 + y_1y_2y_4y_5y_6 + y_1y_3y_4y_5y_6} \tag{6-108}$$

Comparing Equation (6-106) with (6-25), Equation (6-107) with (6-26), and Equation (6-108) with (6-27), we see immediately that the results obtained topologically for the open-circuit impedances agree with those obtained by direct expansion of the determinant Δ and its cofactors.

6-6 TOPOLOGICAL FORMULAS FOR y_{11}, y_{12}, AND y_{22}

From Equations (6-44), (6-45), and (6-46), we see that each of the three short-circuit admittance functions — y_{11}, y_{12}, and y_{22} — has the same common denominator:

$$\Delta_{1122} + \Delta_{112'2'} - 2\Delta_{1122'} \tag{6-109}$$

We shall relate the three terms of Equation (6-109) to the topology of the network and, consequently, derive the topological expressions for y_{11}, y_{12}, and y_{22}; but, before doing this, we introduce the definitions that follow.

DEFINITION 6-1. 3-tree. A 3-tree T_3 of a connected graph G is a subgraph of a 2-tree resulting from the removal of any one of the edges from the 2-tree.

Thus, similar to a 2-tree, a 3-tree is a subgraph of G containing three separate connected parts which, together, include all the vertices of G. In fact, a 3-tree of a graph G of v vertices has the following properties:

(a) It contains three separate connected parts.
(b) It contains all the v vertices of G.
(c) It contains $v - 3$ edges.
(d) It contains no circuits.

Since there are three separate parts in a 3-tree, certain vertices that are assigned to the three parts may be explicitly indicated in the symbol

$$T_{3_{s1,s2,s3}}$$

where s_1, s_2, and s_3 are three sets of vertices associated with the three parts of the 3-tree.

Generalizing this idea, we can state the following definition:

DEFINITION 6-2. k-tree. A *k-tree* T_k of a connected graph G is a subgraph of a $(k - 1)$-tree of G resulting from the removal of (any) one of the edges from the $(k - 1)$-tree.

Again, the symbol

$$T_{k_{s1,s2,\ldots,sk}}$$

may be used for a k-tree to explicitly indicate the k sets $(s_1, s_2, \ldots, s_k)$ of vertices associated with the k parts of the k-tree.

EXAMPLE 6-6. Consider the graph G shown in Figure 6-17(a).

Since $v = 7$, a 3-tree of G must contain four edges, and a 4-tree of G contains exactly three edges. A 3-tree of the type $T_{3_{12,36,45}}$ is shown in Figure 6-17(b), and a 4-tree of the type $T_{4_{1,2,34,5}}$ is depicted in Figure 6-17(c).

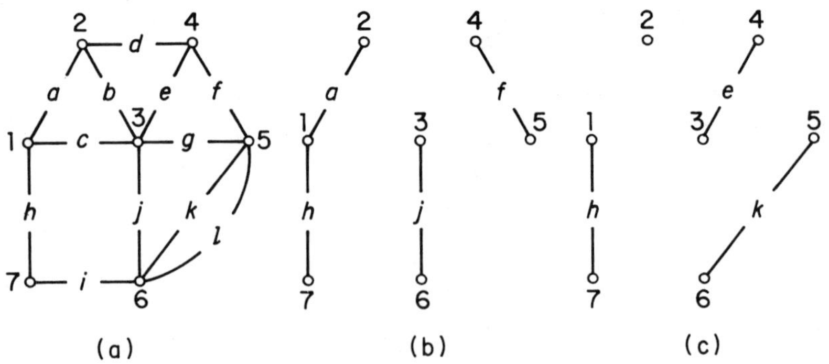

Figure 6-17 (a) A connected graph G. (b) A 3-tree $T_{3_{12,36,45}}$ of G. (c) A 4-tree $T_{4_{1,2,34,5}}$ of G.

As illustrated in Figure 6-17(c), we should point out that one or more of the k parts of a k-tree may be an isolated vertex.

We are now ready to develop a topological expression for Equation (6-109). Let us denote by $U_{a,b,c}(Y)$ the sum of all 3-tree admittance products of the form $T_{3_{a,b,c}}(y)$ with identical subscripts in both symbols to represent the same specified distribution of vertices. Then, following arguments similar to those of Theorem 6-2, we readily see that

$$\Delta_{1122} = \sum_i T^{(i)}_{3_{1,2,1'}}(y) \triangleq U_{1,2,1'}(Y) \tag{6-110}$$

$$\Delta_{112'2'} = \sum_j T^{(j)}_{3_{1,2',1'}}(y) \triangleq U_{1,2',1'}(Y) \tag{6-111}$$

$$\Delta_{1122'} = \sum_k T^{(k)}_{3_{1,22',1'}}(y) \triangleq U_{1,22',1'}(Y) \tag{6-112}$$

where 1, 1′, 2, 2′ are the four terminals of the two-port, with 1′ denoting the reference node (Figure 6-4). However, we note that, in Equations (6-110) and (6-111), only three of the four terminals have been specified. We can, therefore, further expand $U_{1,2,1'}$ and $U_{1,2',1'}$, as we did $W_{12,1'}$ and $W_{12',1'}$ in Equations (6-97) and (6-98), as follows:

$$U_{1,2,1'} = U_{12',2,1'} + U_{1,22',1'} + U_{1,2,1'2'} \tag{6-113}$$

$$U_{1,2',1'} = U_{12,2',1'} + U_{1,22',1'} + U_{1,2',1'2} \tag{6-114}$$

Substituting Equations (6-110) through (6-114) into (6-109) and simplifying, we get

$$\Delta_{1122} + \Delta_{112'2'} - 2\Delta_{1122'} = U_{12',2,1'} + U_{1,2,1'2'} + U_{12,2',1'} + U_{1,2',1'2} \tag{6-115}$$

For convenience, we shall use the shorthand notation ΣU to denote the sum on the right of Equation (6-115). Thus we define

$$\Sigma U \triangleq U_{12',2,1'} + U_{1,2,1'2'} + U_{12,2',1'} + U_{1,2',1'2} \tag{6-116}$$

Finally, we obtain the topological formulas for the short-circuit admittances, as stated in the next theorem.

THEOREM 6-4. The short-circuit admittance functions — y_{11}, y_{12}, and y_{22} — of a passive two-port network with no mutual inductances are given by

$$y_{11} = \frac{W_{2,2'}}{\Sigma U} \tag{6-117}$$

$$y_{12} = y_{21} = \frac{W_{12',1'2} - W_{12,1'2'}}{\Sigma U} \tag{6-118}$$

$$y_{22} = \frac{W_{1,1'}}{\Sigma U} \tag{6-119}$$

Proof: From Equation (6-44), we find that

$$y_{11} = \frac{\Delta_{22} + \Delta_{2'2'} - 2\Delta_{22'}}{\Delta_{1122} + \Delta_{112'2'} - 2\Delta_{1122'}} \tag{6-120}$$

But,

$$\Delta_{22} = W_{2,1'} = W_{22',1'} + W_{2,1'2'}$$
$$\Delta_{2'2'} = W_{2',1'} = W_{22',1'} + W_{2',1'2}$$

and

$$\Delta_{22'} = W_{22',1'}$$

so that

$$\Delta_{22} + \Delta_{2'2'} - 2\Delta_{22'} = W_{2,1'2'} + W_{2',1'2} = W_{2,2'} \tag{6-121}$$

Thus, from Equations (6-116), (6-120), and (6-121), we immediately obtain

$$y_{11} = \frac{W_{2,2'}}{\Sigma U}$$

which is identical to Equation (6-117).

The proofs of Equations (6-118) and (6-119) follow exactly the same pattern, and thus are left to the reader as problems.

EXAMPLE 6-7. Again, consider the network of Example 6-1. The graph G is redrawn in Figure 6-18.

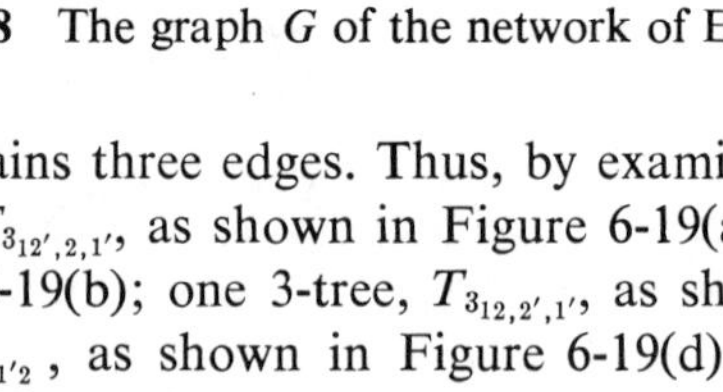

Figure 6-18 The graph G of the network of Example 6-1.

A 3-tree of G contains three edges. Thus, by examining G in Figure 6-18, we find: one 3-tree, $T_{3_{12',2,1'}}$, as shown in Figure 6-19(a); nine 3-trees, $T_{3_{1,2,1'2'}}$, as shown in Figure 6-19(b); one 3-tree, $T_{3_{12,2',1'}}$, as shown in Figure 6-19(c); and one 3-tree, $T_{3_{1,2',1'2}}$, as shown in Figure 6-19(d). Thus, from Equation (6-116) and Figure 6-19, we get

$$\begin{aligned}\Sigma U &\triangleq U_{12',2,1'} + U_{1,2,1'2'} + U_{12,2',1'} + U_{1,2',1'2} \\ &= y_1y_3y_4 + y_1y_3y_5 + y_1y_3y_6 + y_1y_4y_5 + y_1y_4y_6 + y_2y_3y_4 + y_2y_3y_5 + y_2y_3y_6 \\ &\quad + y_2y_4y_5 + y_2y_4y_6 + y_3y_4y_5 + y_3y_4y_6\end{aligned} \tag{6-122}$$

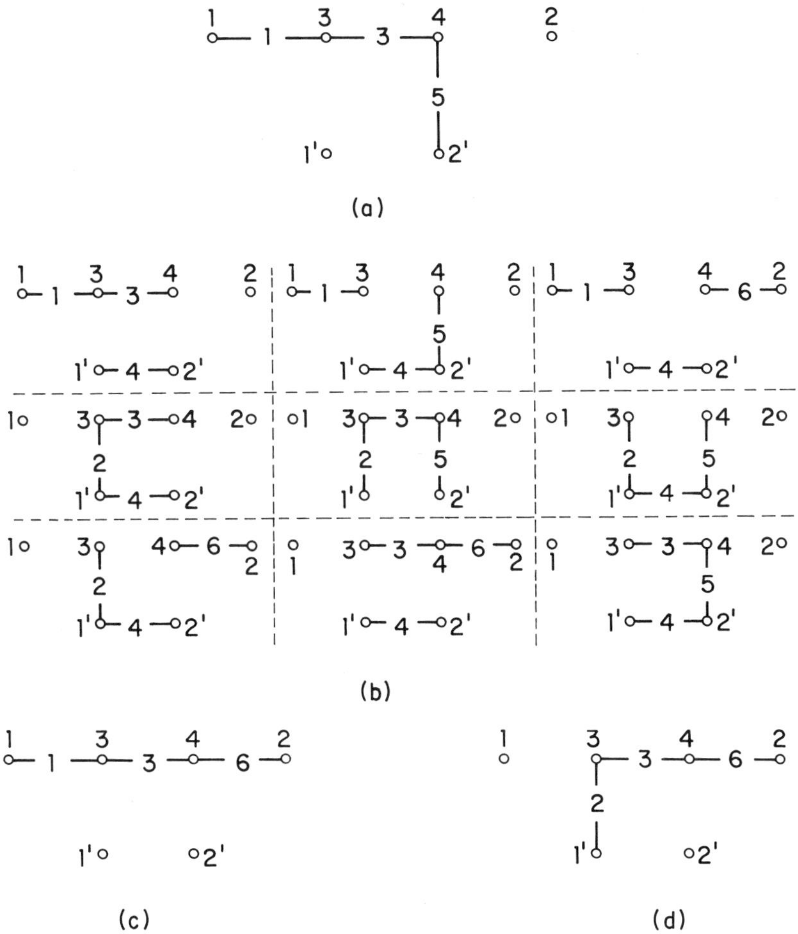

Figure 6-19 The 3-tree of G. (a) $T_{312',2,1'}$. (b) $T_{31,2,1'2'}$. (c) $T_{312,2',1'}$. (d) $T_{31,2',1'2}$.

Also, from Example 6-5, we have already obtained the expressions for $W_{1,1'}$, $W_{2,2'}$, $W_{12,1'2'}$, and $W_{12',1'2}$ [as expressed in Equations (6-102) through (6-105) respectively]. Therefore, from Equations (6-117), (6-122), and (6-103), we get

$$y_{11} = \frac{W_{2,2'}}{\Sigma U}$$

$$= \frac{y_1y_2y_3y_4 + y_1y_2y_3y_5 + y_1y_2y_3y_6 + y_1y_2y_4y_5 + y_1y_2y_4y_6 + }{y_1y_3y_4 + y_1y_3y_5 + y_1y_3y_6 + y_1y_4y_5 + y_1y_4y_6 + y_2y_3y_4 + y_2y_3y_5 +}$$

$$\frac{+ y_1y_3y_4y_5 + y_1y_3y_4y_6}{+ y_2y_3y_6 + y_2y_4y_5 + y_2y_4y_6 + y_3y_4y_5 + y_3y_4y_6} \quad (6\text{-}123)$$

Similarly, from Equations (6-118), (6-122), (6-104), and (6-105), we find that

$$y_{12} = y_{21} = \frac{W_{12',1'2} - W_{12,1'2'}}{\Sigma U}$$

$$= \frac{-y_1y_3y_4y_6}{y_1y_3y_4 + y_1y_3y_5 + y_1y_3y_6 + y_1y_4y_5 + y_1y_4y_6 + y_2y_3y_4 + y_2y_3y_5 + y_2y_3y_6 + y_2y_4y_5 + y_2y_4y_6 + y_3y_4y_6} \tag{6-124}$$

Likewise, from Equations (6-119), (6-122), and (6-102), we write

$$y_{22} = \frac{W_{1,1'}}{\Sigma U}$$

$$= \frac{y_1y_3y_4y_6 + y_1y_3y_5y_6 + y_1y_4y_5y_6 + y_2y_3y_4y_6 + y_2y_3y_5y_6 + y_2y_4y_5y_6 + y_3y_4y_5y_6}{y_1y_3y_4 + y_1y_3y_5 + y_1y_3y_6 + y_1y_4y_5 + y_1y_4y_6 + y_2y_3y_4 + y_2y_3y_5 + y_2y_3y_6 + y_2y_4y_5 + y_2y_4y_6 + y_3y_4y_5 + y_3y_4y_6} \tag{6-125}$$

Finally, comparing Equation (6-123) with (6-54), (6-124) with (6-55), and (6-125) with (6-56), we see that the topological results are the same as those obtained by direct determinant expansion.

6-7 TOPOLOGICAL FORMULAS FOR Z_{12}, Y_{12}, G_{12}, AND α_{12}

It is an easy matter to obtain topological expressions for the transfer functions Z_{12}, Y_{12}, G_{12}, and α_{12} after our discussions in the last few sections. Therefore, we simply state the results in the following theorem and leave the proofs to the reader as exercises.

THEOREM 6-5. For a passive two-port N without mutual inductances, the topological formulas for the transfer functions Z_{12}, Y_{12}, G_{12}, and α_{12} are given by

$$Z_{12}(s) = \frac{W_{12,1'2'} - W_{12',1'2}}{V} \tag{6-126}$$

$$Y_{12}(s) = Y_L\left(\frac{W_{12,1'2'} - W_{12',1'2}}{W_{1,1'}}\right) \tag{6-127}$$

$$G_{12}(s) = \frac{W_{12,1'2'} - W_{12',1'2}}{W_{1,1'}} \tag{6-128}$$

$$\alpha_{12}(s) = Y_L\left(\frac{W_{12,1'2'} - W_{12',1'2}}{V}\right) \tag{6-129}$$

where Y_L is the load admittance at port (2,2′) of N (Figure 6-4).

EXAMPLE 6-8. Let us consider once again the same loaded network N as that used in Examples 6-3 and 6-4. The graph G of N is shown in Figure 6-6;

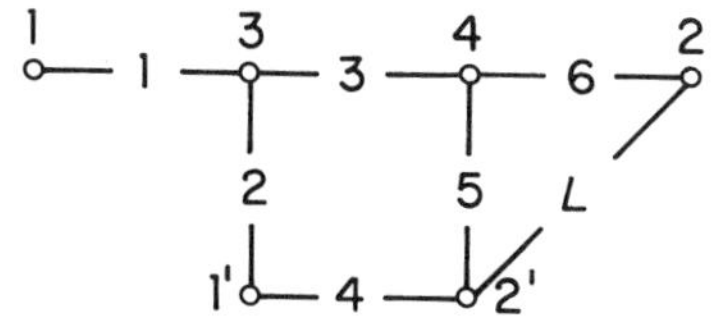

Figure 6-20 The graph G of the network N of Example 6-3.

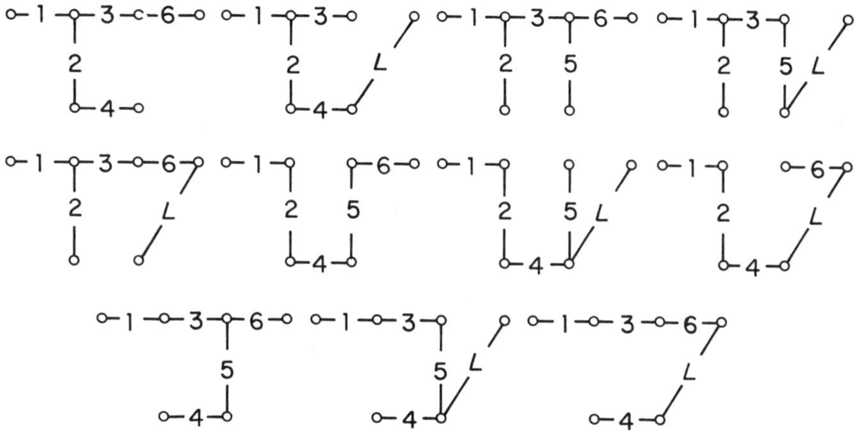

Figure 6-21 The set of all trees of G.

for convenience, is redrawn in Figure 6-20. There are eleven trees of G, as shown in Figure 6-21. Hence, we write

$$\begin{aligned} V(Y) = {} & y_1y_2y_3y_4y_6 + y_1y_2y_3y_4Y_L + y_1y_2y_3y_5y_6 + y_1y_2y_3y_5Y_L \\ & + y_1y_2y_3y_6Y_L + y_1y_2y_4y_5y_6 + y_1y_2y_4y_5Y_L + y_1y_2y_4y_6Y_L \\ & + y_1y_3y_4y_5y_6 + y_1y_3y_4y_5Y_L + y_1y_3y_4y_6Y_L \end{aligned} \tag{6-130}$$

The set of all 2-trees, $T_{2_{12,1'2'}}$, consists of only one element, which is shown in Figure 6-22, whereas the set of all 2-trees, $T_{2_{12',1'2}}$, is an empty set. Thus we find

$$W_{12,1'2'} = y_1y_3y_4y_6$$

and

$$W_{12',1'2} = 0$$

so that

$$W_{12,1'2'} - W_{12',1'2} = y_1y_3y_4y_6 \tag{6-131}$$

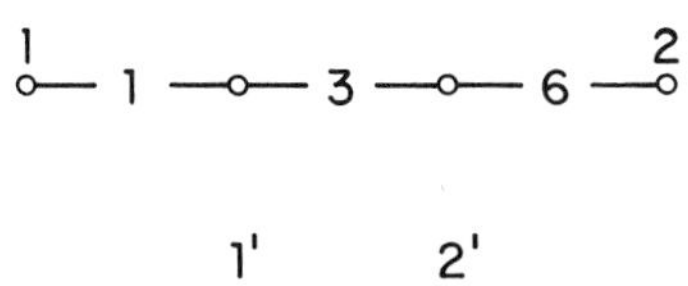

Figure 6-22 The only 2-tree $T_{2_{12,1'2'}}$ of G.

Also, from Equation (6-91) in Example 6-4, we find that

$$\begin{aligned} W_{1,1'} = {} & y_1y_3y_4y_6 + y_1y_3y_4Y_L + y_1y_3y_5y_6 + y_1y_3y_5Y_L + y_1y_3y_6Y_L \\ & + y_1y_4y_5y_6 + y_1y_4y_5Y_L + y_1y_4y_6Y_L + y_2y_3y_4y_6 \\ & + y_2y_3y_4Y_L + y_2y_3y_5y_6 + y_2y_3y_5Y_L + y_2y_3y_6Y_L \\ & + y_2y_4y_5y_6 + y_2y_4y_5Y_L + y_2y_4y_6Y_L + y_3y_4y_5y_6 \\ & + y_3y_4y_5Y_L + y_3y_4y_6Y_L \end{aligned} \tag{6-132}$$

Hence, substituting Equations (6-130) and (6-131) into (6-126), we get

$$Z_{12} = \frac{y_1y_3y_4y_6}{\begin{Bmatrix} y_1y_2y_3y_4y_6 + y_1y_2y_3y_4Y_L + y_1y_2y_3y_5y_6 + y_1y_2y_3y_5Y_L + y_1y_2y_3y_6Y_L \\ + y_1y_2y_4y_5y_6 + y_1y_2y_4y_5Y_L + y_1y_2y_3y_6Y_L + y_1y_3y_4y_5y_6 \\ + y_1y_3y_4y_5Y_L + y_1y_3y_4y_6Y_L \end{Bmatrix}} \tag{6-133}$$

Similarly, substituting Equations (6-131) and (6-132) into (6-127), we find

$$Y_{12} = \frac{y_1y_3y_4y_6Y_L}{\begin{Bmatrix} y_1y_3y_4y_6 + y_1y_3y_4Y_L + y_1y_3y_5y_6 + y_1y_3y_5Y_L + y_1y_3y_6Y_L + y_1y_4y_5y_6 \\ + y_1y_4y_5Y_L + y_1y_4y_6Y_L + y_2y_3y_4y_6 + y_2y_3y_4Y_L + y_2y_3y_5y_6 \\ + y_2y_3y_5Y_L + y_2y_3y_6Y_L + y_2y_4y_5y_6 + y_2y_4y_5Y_L + y_2y_4y_6Y_L \\ + y_3y_4y_5y_6 + y_3y_4y_5Y_L + y_3y_4y_6Y_L \end{Bmatrix}} \tag{6-134}$$

To find G_{12}, we note from Equations (6-127) and (6-128) that

$$Y_{12} = Y_LG_{12}$$

so that

$$\begin{aligned} G_{12} &= \frac{Y_{12}}{Y_L} \\ &= \frac{y_1y_3y_4y_6}{\begin{Bmatrix} y_1y_3y_4y_6 + y_1y_3y_4Y_L + y_1y_3y_5y_6 + y_1y_3y_5Y_L + y_1y_3y_6Y_L + y_1y_4y_5y_6 \\ + y_1y_4y_5Y_L + y_1y_4y_6Y_L + y_2y_3y_4y_6 + y_2y_3y_4Y_L + y_2y_3y_5y_6 + \\ + y_2y_3y_5Y_L + y_2y_3y_6Y_L + y_2y_4y_5y_6 + y_2y_4y_5Y_L + y_2y_4y_6Y_L \\ + y_2y_4y_6Y_L + y_3y_4y_5y_6 + y_3y_4y_5Y_L + y_3y_4y_6Y_L \end{Bmatrix}} \end{aligned} \tag{6-135}$$

Likewise, by comparing Equation (6-129) with (6-126), we find immediately from (6-33) that

$$\begin{aligned} \alpha_{12} &= Y_LZ_{12} \\ &= \frac{y_1y_3y_4y_6Y_L}{\begin{Bmatrix} y_1y_2y_3y_4y_6 + y_1y_2y_3y_4Y_L + y_1y_2y_3y_5y_6 + y_1y_2y_3y_5Y_L + y_1y_2y_3y_6Y_L \\ + y_1y_2y_4y_5y_6 + y_1y_2y_4y_5Y_L + y_1y_2y_4y_6Y_L + y_1y_3y_4y_5y_6 \\ + y_1y_3y_4y_5Y_L + y_1y_3y_4y_6Y_L \end{Bmatrix}} \end{aligned} \tag{6-136}$$

Finally, comparing Equations (6-133) through (6-136) with the corresponding expressions of (6-78) through (6-81) in Example 6-3, we see that the results obtained topologically in Example 6-8 agree with those obtained in Example 6-3 by direct expansion of the determinant Δ and its pertinent cofactors.

6-8 COMPUTER-AIDED TOPOLOGICAL ANALYSIS[6]

In the preceding sections, we have developed topological formulas for computing various network functions of an *RLC* two-port. We shall now show that the topological expressions developed in this chapter and in the last are well suited for computer programming.[6] More precisely, we shall show that if there exists a computer program by means of which one can compute all the trees of a connected graph with great efficiency, then any network function (driving-point or transfer of any kind) of an *RLC* two-port can be computed by the application of this program. In other words, all the 2-trees, 3-trees, etc., that are involved in the topological expressions developed in this chapter and in the last can be determined by the use of a single computer program.

To this end, let us recall that in Chapter 5 we established the fact that all the 2-trees $W_{i,j}$ of an *RLC* network N can be found by short-circuiting nodes i and j and computing the trees (using the developed computer program) of the resultant network N_1. Thus the driving-point admittance $Y_d(s)$ of N, which is the ratio of the sum of all tree admittance products of N to the sum of all tree admittance products of N_1, as well as the driving-point impedance $Z_d(s)$, which is the reciprocal of $Y_d(s)$, can be computed by a single program.

Next, let us consider the open-circuit impedances of a general *RLC* two-port. The open-circuit impedance matrix, Z_{oc}, is given by

$$Z_{oc} = \begin{bmatrix} z_{11} & z_{12} \\ z_{21} & z_{22} \end{bmatrix} = \frac{1}{V}\begin{bmatrix} W_{1,1'} & W_{12,1'2'} - W_{12',1'2} \\ W_{12,1'2'} - W_{12',1'2} & W_{2,2'} \end{bmatrix} \tag{6-137}$$

which is a matrix representation of Equations (6-100).

The terms $W_{1,1'}$ and $W_{2,2'}$ are the special cases of $W_{i,j}$ and, hence, can be computed readily by the same program. Consequently, if we can show that the terms $W_{12,1'2'} - W_{12',1'2}$ can also be calculated by the same "tree-finding" program, we have succeeded in showing that all the elements of Z_{oc} in Equation (6-137) can indeed be computed by that program.

To determine $W_{12,1'2'} - W_{12',1'2}$ by computer using a single program, consider the following 2-tree expansions:

$$W_{1,2} = W_{11',2} + W_{1,1'2} = W_{11'2',2} + W_{11',22'} + W_{12',1'2} + W_{1,1'22'} \tag{6-138}$$

$$W_{1,2'} = W_{11',2'} + W_{1,1'2'} = W_{11'2,2'} + W_{11',22'} + W_{12,1'2'} + W_{1,1'22'} \tag{6-139}$$

$$W_{2,1'} = W_{21,1'} + W_{2,11'} = W_{212',1'} + W_{21,1'2'} + W_{22',11'} + W_{2,11'2'} \tag{6-140}$$

$$W_{2',1'} = W_{2'1,1'} + W_{2',11'} = W_{2'12,1'} + W_{2'1,1'2} + W_{2'2,11'} + W_{2,11'2'} \tag{6-141}$$

[6]The material contained in this section is based on a paper of Chan and Chan [17].

From Equation (6-138), we write

$$W_{12',1'2} = W_{1,2} - W_{11'2',2} - W_{11',22'} - W_{1,1'22'} \tag{6-142}$$

From Equation (6-139), we find that

$$W_{12,1'2'} = W_{1,2'} - W_{11'2,2'} - W_{11',22'} - W_{1,1'22'} \tag{6-143}$$

From Equation (6-140) we obtain

$$W_{21,1'2'} = W_{2,1'} - W_{212',1'} - W_{22',11'} - W_{2,11'2'} \tag{6-144}$$

From Equation (6-141), we get

$$W_{2'1,1'2} = W_{2',1'} - W_{2'12,1'} - W_{2'2,1'1} - W_{2',11'2} \tag{6-145}$$

Performing the operation: (6-143) + (6-144) – (6-142) – (6-145), we have

$$2(W_{12,1'2'} - W_{12',1'2}) = W_{1,2'} + W_{2,1'} - W_{1,2} - W_{2',1'}$$

from which we find the desired expression:

$$W_{12,1'2} - W_{12',1'2} = \frac{1}{2}(W_{1,2'} + W_{2,1'} - W_{1,2} - W_{1',2'}) \tag{6-146}$$

It is evident from Equation (6-146) that the term $W_{12,1'2'} - W_{12',1'2}$ can be computed by the same "tree-finding" program developed for the computation of V since each term in the parentheses on the right-hand side of (6-146) is a special case of $W_{i,j}$ with appropriate values for i and j.

For an *RLC* grounded two-port (that is, a 3-terminal network, which is merely a special case of a general *RLC* two-port with nodes 1′ and 2′ identified by a common node 0), Equation (6-146) reduces to

$$W_{12,0} = \frac{1}{2}(W_{1,0} + W_{2,0} - W_{1,2}) \tag{6-147}$$

which again can be handled by the same program for the obvious reason.

To find the short-circuit admittances of a general *RLC* two-port, one simply observes the fact that the short-circuit admittance matrix, Y_{sc}, which is defined as

$$Y_{sc} = \begin{bmatrix} y_{11} & y_{12} \\ y_{21} & y_{22} \end{bmatrix} \tag{6-148}$$

is the inverse of the open-circuit impedance matrix Z_{oc}. Thus, once Z_{oc} is obtained by the use of a computer, the matrix $Y_{sc} = Z_{oc}^{-1}$ can be readily determined by any of the existing computer techniques.

It is worth noting that computer programs for listing all the trees of a given network as well as other programs designed for topological analysis are available.[7] One of the superior ones is the program developed by D. A. Calahan [11, 12]. The tree-finding program in Calahan's network analysis program is essentially based on the tree-listing algorithm discussed in Section 5-5 at the end of Chapter 5.

[7]See F. F. Kuo [49].

Also, it should be noted that even with the modification of the topological formulas for digital computation as discussed in this section, there are still shortcomings of the topological techniques.[7] One of the major drawbacks is the enormous number of trees and k-trees that one must compute even in a moderately complicated graph. Because of this, any computer programs for network analysis and design that are based on the topological approach are slower than most other programs using the matrix approach.[8] However, we must also point out that a unique feature that the topological method contains and the others do not is *accuracy* in computer calculation due to the fact that there are no cancellations (no subtractions present) in the expansion of determinants using the topological formulas, as stated earlier in this chapter. This accuracy requirement may, in fact, be a critical consideration in some network problems in choosing a computer program.

Finally, we might remark that a wide adoption of topological techniques in network analysis and design depends mainly upon the success in developing highly efficient computer programs to be used with the topological techniques to permit the solution of various network problems in a shorter period of time and at a lower cost than with the existing conventional methods.

6-9 SUMMARY AND REMARKS

We now summarize our discussions by listing the formulas — topological or otherwise — for the various network functions of a passive two-port without mutual magnetic coupling, as derived in this chapter as well as in Chapter 5.

(I) *Passive one-port* (refer to Figure 6-23)

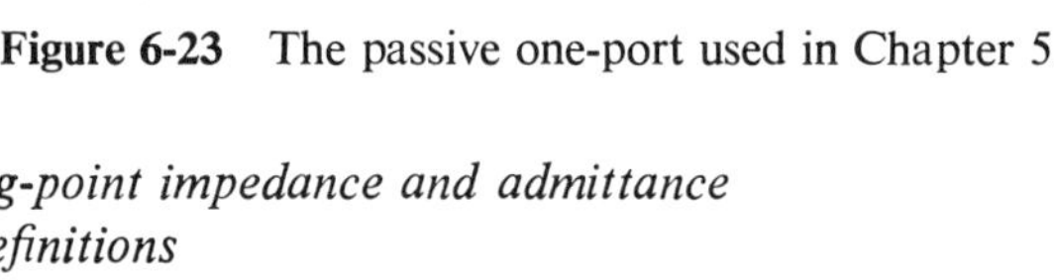

Figure 6-23 The passive one-port used in Chapter 5.

(1) *Driving-point impedance and admittance*

(a) *Definitions*

$$Z_d(s) = \frac{V_1}{I_1} \tag{5-7}$$

$$Y_d(s) = \frac{I_1}{V_1} \tag{5-8}$$

[8]Refer to W. G. Magnuson, Jr. [53].

(b) *Formulas*

$$Z_d(s) = \frac{\Delta_{11}}{\Delta} = \frac{W_{1,1'}}{V} \qquad (5\text{-}7), (5\text{-}32)$$

$$Y_d(s) = \frac{\Delta}{\Delta_{11}} = \frac{V}{W_{1,1'}} \qquad (5\text{-}8), (5\text{-}31)$$

(II) *Passive two-port* (refer to Figure 6-24)

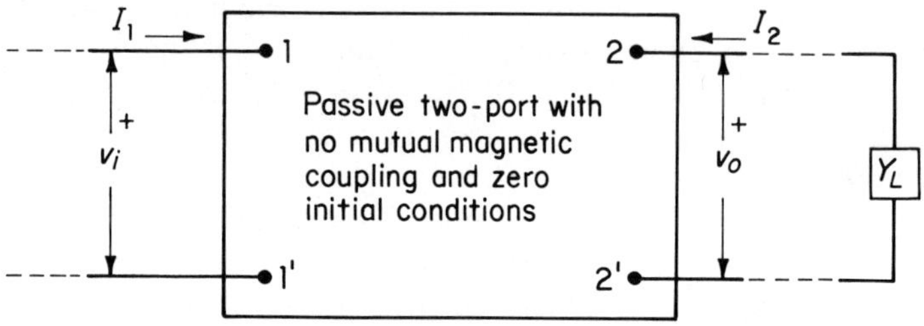

Figure 6-24 The passive two-port used in this chapter.

(1) *Open-circuit impedances*

(a) *Definitions*

$$z_{11} = \left.\frac{V_i}{I_1}\right|_{I_2=0} \qquad z_{21} = \left.\frac{V_o}{I_1}\right|_{I_2=0} \qquad (6\text{-}10), (6\text{-}11)$$

$$z_{12} = \left.\frac{V_i}{I_2}\right|_{I_1=0} \qquad z_{22} = \left.\frac{V_o}{I_2}\right|_{I_1=0} \qquad (6\text{-}13), (6\text{-}14)$$

(b) *Formulas*

$$z_{11} = \frac{\Delta_{11}}{\Delta} = \frac{W_{1,1'}}{V} \qquad (6\text{-}16a), (6\text{-}100a)$$

$$z_{12} = z_{21} = \frac{\Delta_{12} - \Delta_{12'}}{\Delta} = \frac{W_{12,1'2'} - W_{12',1'2}}{V} \qquad (6\text{-}16b), (6\text{-}100b)$$

$$z_{22} = \frac{\Delta_{22} + \Delta_{2'2'} - 2\Delta_{22'}}{\Delta} = \frac{W_{2,2'}}{V} \qquad (6\text{-}16c), (6\text{-}100c)$$

(2) *Short-circuit admittances*

(a) *Definitions*

$$y_{11} = \left.\frac{I_1}{V_i}\right|_{Vo=0} \qquad y_{21} = \left.\frac{I_2}{V_i}\right|_{Vo=0} \qquad (6\text{-}35), (6\text{-}36)$$

$$y_{12} = \left.\frac{I_1}{V_o}\right|_{Vi=0} \qquad y_{22} = \left.\frac{I_2}{V_o}\right|_{Vi=0} \qquad (6\text{-}38), (6\text{-}39)$$

(b) *Formulas*

$$y_{11} = \frac{\Delta_{22} + \Delta_{2'2'} + 2\Delta_{2,2'}}{\Delta_{1122} + \Delta_{112'2'} - 2\Delta_{1122'}} = \frac{W_{2,2'}}{\Sigma U} \qquad (6\text{-}44), (6\text{-}117)$$

$$y_{12} = y_{21} = \frac{\Delta_{12'} - \Delta_{12}}{\Delta_{1122} + \Delta_{112'2'} - 2\Delta_{1122'}}$$

$$= \frac{W_{12',1'2} - W_{12,1'2'}}{\Sigma U} \qquad (6\text{-}45), (6\text{-}118)$$

$$y_{22} = \frac{\Delta}{\Delta_{1122} + \Delta_{112'2'} - 2\Delta_{1122'}} = \frac{W_{1,1'}}{\Sigma U} \qquad (6\text{-}46), (6\text{-}119)$$

where

$$\Sigma U = U_{12',2,1'} + U_{1,2,1'2'} + U_{12,2',1'} + U_{1,2',1'2} \qquad (6\text{-}116)$$

(3) *Other network transfer functions*

(a) *Definitions*

$$Z_{12} = \frac{V_o}{I_1} \qquad Y_{12} = \frac{-I_2}{V_i} \qquad (6\text{-}63), (6\text{-}67)$$

$$G_{12} = \frac{V_o}{V_i} \qquad \alpha_{12} = \frac{-I_2}{I_1} \qquad (6\text{-}66), (6\text{-}71)$$

(b) *Formulas*

$$Z_{12} = \frac{\Delta_{12} - \Delta_{12'}}{\Delta} = \frac{W_{12,1'2'} - W_{12',1'2}}{V} \qquad (6\text{-}63), (6\text{-}126)$$

$$Y_{12} = Y_L G_{12} = Y_L\left(\frac{\Delta_{12} - \Delta_{12'}}{\Delta_{11}}\right) = Y_L\left(\frac{W_{12,1'2'} - W_{12',1'2}}{W_{1,1'}}\right) \qquad (6\text{-}70), (6\text{-}127)$$

$$G_{12} = \frac{\Delta_{12} - \Delta_{12'}}{\Delta_{11}} = \frac{W_{12,1'2'} - W_{12',1'2}}{W_{1,1'}} \qquad (6\text{-}66), (6\text{-}128)$$

$$\alpha_{12} = Y_L Z_{12} = Y_L\left(\frac{\Delta_{12} - \Delta_{12'}}{\Delta}\right) = Y_L\left(\frac{W_{12,1'2'} - W_{12',1'2}}{V}\right) \qquad (6\text{-}72), (6\text{-}129)$$

(4) *Topological formulas for computer programming*

(a) *Definitions*

$$Z_{oc} = \begin{bmatrix} z_{11} & z_{12} \\ z_{21} & z_{22} \end{bmatrix}$$ (open-circuit impedance matrix of a general passive two-port) (6-137)

$$Y_{sc} = \begin{bmatrix} y_{11} & y_{12} \\ y_{21} & y_{22} \end{bmatrix}$$ (short-circuit admittance matrix of a general passive two-port) (6-138)

(b) *Formulas*

$$Z_d(s) = \frac{W_{1,1'}}{V} \qquad \text{(one-port)} \tag{5-32}$$

$$Y_d(s) = \frac{V}{W_{1,1'}} \qquad \text{(one-port)} \tag{5-31}$$

$$Z_{oc} = \frac{1}{V}\begin{bmatrix} W_{1,1'} & W_{12,1'2'} - W_{12',1'2} \\ W_{12,1'2'} - W_{12',1'2} & W_{2,2'} \end{bmatrix}$$

$$= \frac{1}{V}\begin{bmatrix} W_{1,1'} & \frac{1}{2}(W_{1,2'} + W_{2,1'} - W_{1,2} - W_{1',2'}) \\ \frac{1}{2}(W_{1,2'} + W_{2,1'} - W_{1,2} - W_{1',2'}) & W_{2,2'} \end{bmatrix}$$

(for passive two-port without mutual inductances) (6-137), (6-146)

$$Z_{oc} = \frac{1}{V}\begin{bmatrix} W_{1,1'} & \frac{1}{2}(W_{1,0} + W_{2,0} - W_{1,2}) \\ \frac{1}{2}(W_{1,0} + W_{2,0} - W_{1,2}) & W_{2,2'} \end{bmatrix}$$

(for passive grounded two-port without mutual inductances with node 0 as the reference node) (6-147)

$$Y_{sc} = Z_{oc}^{-1}$$

Finally, we should note that the topological formulas derived in this chapter are based on the so-called Maxwell formulas[9] using the node equations. These formulas are obtained by first expressing the various network functions in terms of the determinant Δ of the node admittance matrix and its cofactors, and, in turn, relating Δ and the Δ_{ij}'s to the tree admittance products, 2-tree admittance products, etc. Thus, it is only natural for us to wonder whether it is possible to obtain a set of "dual" formulas using the loop equations instead of the node equations.

The answer to this is indeed affirmative. If, for example, we begin with the loop system of equations for a passive one-port with m loops and a voltage driver V_1:

$$\begin{bmatrix} z_{11} & z_{12} & \cdots & z_{1m} \\ z_{21} & z_{22} & \cdots & z_{2m} \\ \cdot & & & \\ \cdot & & & \\ \cdot & & & \\ z_{m1} & z_{m2} & \cdots & z_{mm} \end{bmatrix} \begin{bmatrix} I_1 \\ I_2 \\ \cdot \\ \cdot \\ \cdot \\ I_m \end{bmatrix} = \begin{bmatrix} V_1 \\ 0 \\ \cdot \\ \cdot \\ \cdot \\ 0 \end{bmatrix}$$

we can solve for I_1 using Cramer's rule:

$$I_1 = \frac{\Delta'_{11}}{\Delta'} V_1$$

[9]See J. C. Maxwell [58] for his original work.

where

$$\Delta' = \det Z_n = \det \begin{bmatrix} z_{11} & z_{12} & \cdots & z_{1m} \\ z_{21} & z_{22} & \cdots & z_{2m} \\ \cdot & & & \\ \cdot & & & \\ \cdot & & & \\ z_{m1} & z_{m2} & \cdots & z_{mm} \end{bmatrix}$$

is the determinant of the loop impedance matrix Z_n and Δ'_{11} is its (1,1) cofactor. Thus, we find

$$Z_d(s) = \frac{\Delta'}{\Delta'_{11}}$$

which is the counterpart of the expression for Z_d that we obtained in Chapter 5; namely,

$$Z_d(s) = \frac{\Delta_{11}}{\Delta}$$

Furthermore, following a parallel development by using the chord-set-impedance products, instead of the tree-admittance products, it can be shown[10] that

$$\Delta' = C[V(Z)]$$

where $C[V(Z)]$ is obtained from $V(Y)$ by replacing each tree admittance product, $T(y)$, in $V(y)$ by the product of the *impedances* of the *chord set* with respect to the same tree T.

Since a chord set is the complement of a tree, the polynomial $C[V(Z)]$ is referred to as the complementary sum of impedance produts of $V(Y)$. In a similar manner, dual expressions for the various cofactors of Δ' can be derived, and, consequently, a parallel set of topological formulas, based on Kirchhoff's rules[11] instead of on Maxwell's, can be developed. However, for present purposes, we shall be content with the brief acknowledgment of the existence of such a set of formulas and leave the details to the reader.[12]

PROBLEMS

6-1 For the passive, reciprocal, three-terminal network shown in Figure P6-1, derive the expressions for the open-circuit impedance functions z_{11}, z_{12}, and z_{22} in terms of the determinant of the node admittance matrix and its cofactors.

[10]For the derivation of this formula as well as some others which are listed as problems at the end of this chapter, see Mayeda and Seshu [62].

[11]See G. Kirchhoff [46] for his original work.

[12]Some such formulas are listed as problems at the end of this chapter. The proofs of these formulas can be found in Mayeda and Seshu [62].

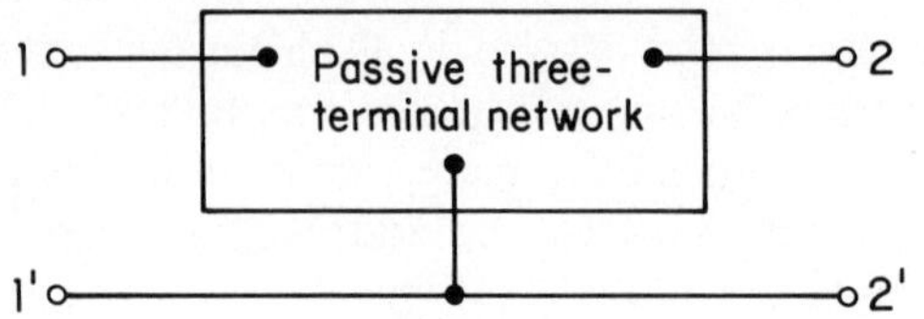

Figure P6-1

6-2 Derive the expressions for the short-circuit admittance functions y_{11}, y_{12}, and y_{22} in terms of the determinant of the node admittance matrix and its cofactors of the network shown in Figure P6-1.

6-3 Using the expressions in terms of the determinant of the node admittance matrix and its cofactors, determine the open-circuit impedance functions z_{11}, z_{12}, and z_{22} of the passive two-port shown in Figure P6-3.

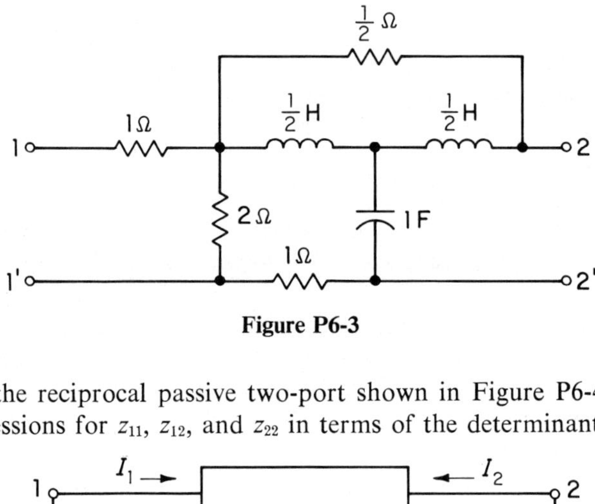

Figure P6-3

6-4 For the reciprocal passive two-port shown in Figure P6-4, derive the expressions for z_{11}, z_{12}, and z_{22} in terms of the determinant of the loop

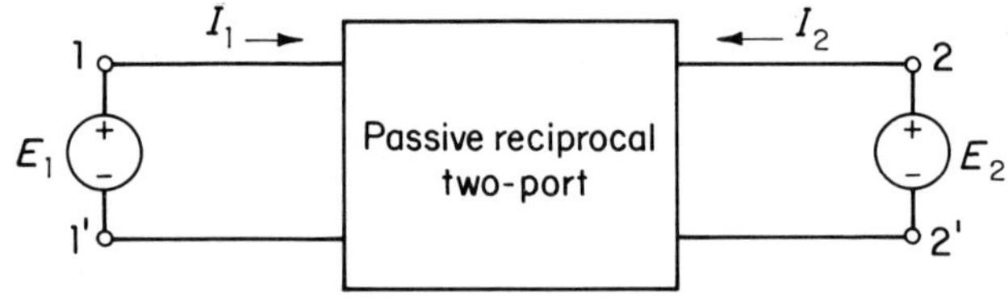

Figure P6-4

impedance matrix and its cofactors. The general expression for the current in loop k is

$$I_k = \frac{\Delta_{1k}}{\Delta'}E_1 + \frac{\Delta'_{2k}}{\Delta'}E_2 + \cdots + \frac{\Delta'_{mk}}{\Delta'}E_m$$

where Δ' is the determinant of the loop impedance matrix and Δ'_{jk} is the (j,k)th cofactor of Δ'.

6-5 Using the expressions derived in Problem 6-4, determine z_{11}, z_{12}, and z_{22} of the passive two-port shown in Figure P6-3.

6-6 Using the expressions in terms of the determinant of the node admittance matrix, determine y_{11}, y_{12}, and y_{22} of the passive two-port shown in Figure P6-6.

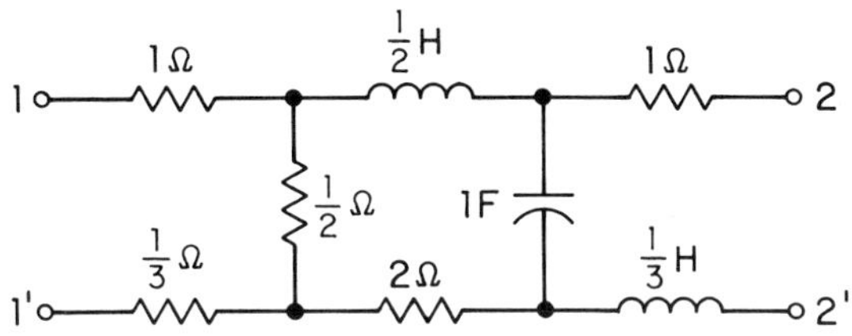

Figure P6-6

6-7 For the loaded passive three-terminal two-port shown in Figure P6-7, derive the expressions for the voltage-ratio transfer function $G_{12}(s)$ and the current-ratio transfer function $\alpha_{12}(s)$ in terms of the determinant of the node admittance matrix and its cofactors.

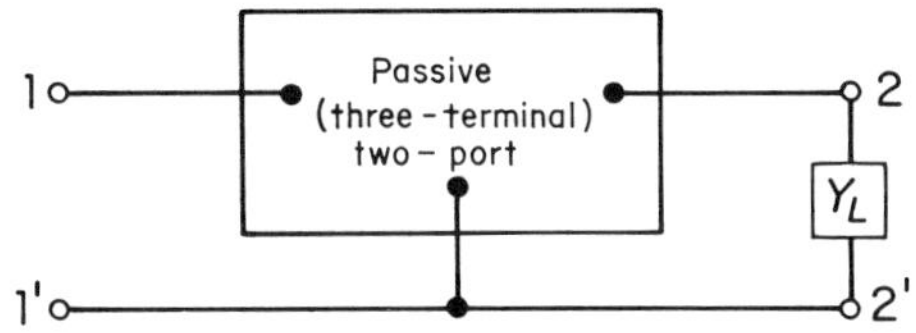

Figure P6-7

6-8 Referring to Problem 6-7, derive the expressions for the transfer impedance function $Z_{12}(s)$ and the transfer admittance function $Y_{12}(s)$.

6-9 Using the expressions derived in Problems 6-7 and 6-8, find $G_{12}(s)$, $\alpha_{12}(s)$, $Z_{12}(s)$, and $Y_{12}(s)$ of the network shown in Figure P6-9.

6-10 For the network shown in Figure P6-10, evaluate $\Delta_{12'}$ with both (a)

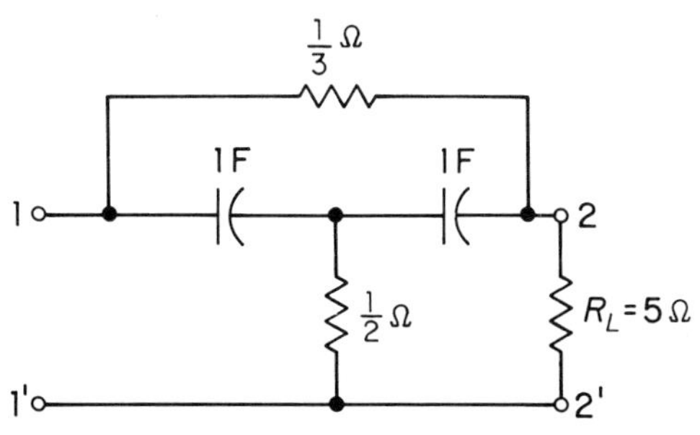

Figure P6-9

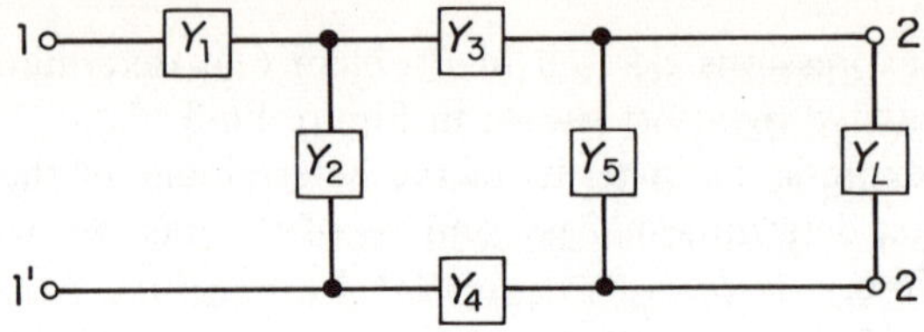

Figure P6-10

the straightforward expansion technique and (b) the topological formula.

6-11 Referring to Section 6-5, prove Equation (6-94).

6-12 Determine the open-circuit impedance functions z_{11}, z_{12}, and z_{22} of the passive two-port shown in Figure P6-12 by using the topological formulas.

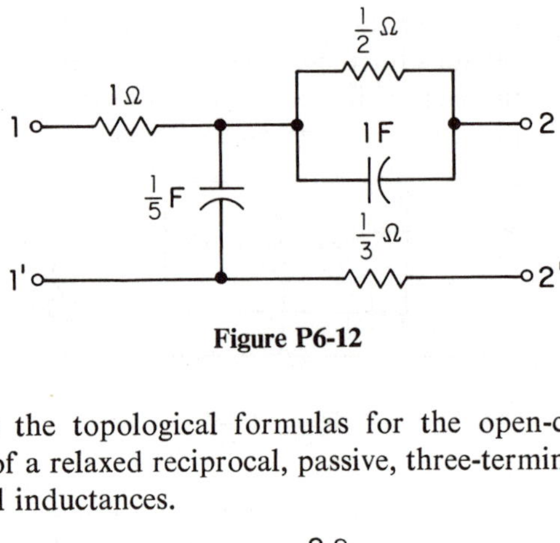

Figure P6-12

6-13 (a) Derive the topological formulas for the open-circuit impedance functions of a relaxed reciprocal, passive, three-terminal two-port without mutual inductances.

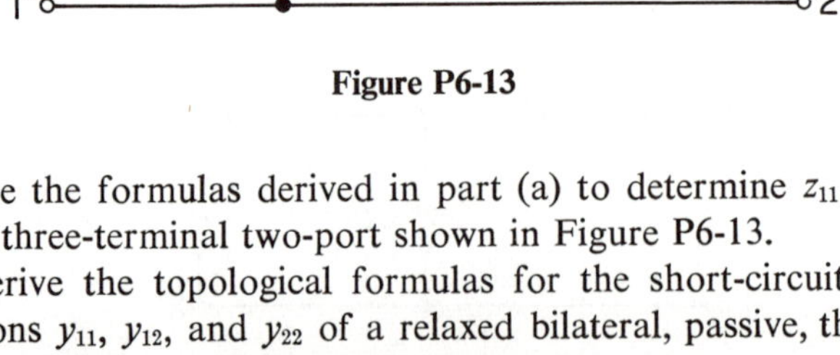

Figure P6-13

(b) Use the formulas derived in part (a) to determine z_{11}, z_{12}, and z_{22} of the three-terminal two-port shown in Figure P6-13.

6-14 (a) Derive the topological formulas for the short-circuit admittance functions y_{11}, y_{12}, and y_{22} of a relaxed bilateral, passive, three-terminal two-port without mutual inductances.

(b) Use the formulas derived in part (a) to find y_{11}, y_{12}, and y_{22} of the network shown in Figure P6-13.

6-15 Determine y_{11}, y_{12}, and y_{22} of the passive two-port shown in Figure P6-12 by using the topological formulas.

6-16 For each of the graphs in Figure P6-16, find all the 2-trees with nodes 1 and 2 in one part, and node n in another part of the 2-tree.

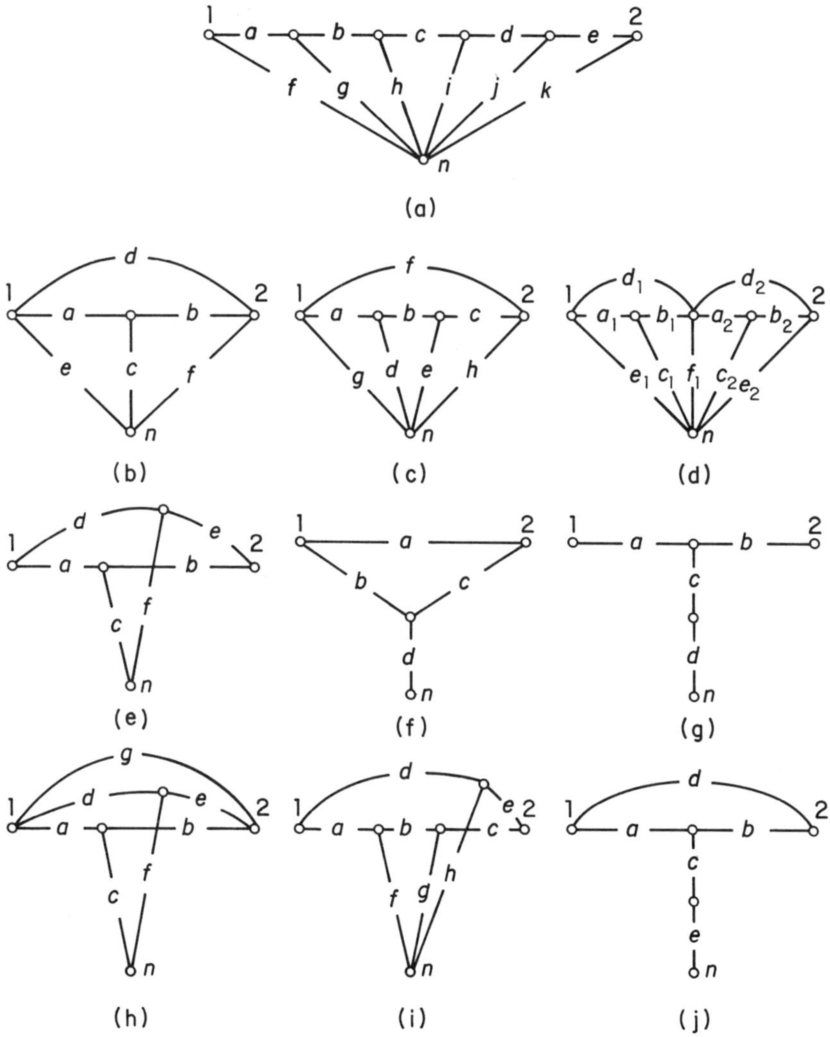

Figure P6-16

6-17 For the graph shown in Figure P6-17, obtain

(a) The set of all 3-trees $\{T_{3_{1,2,3}}\}$.

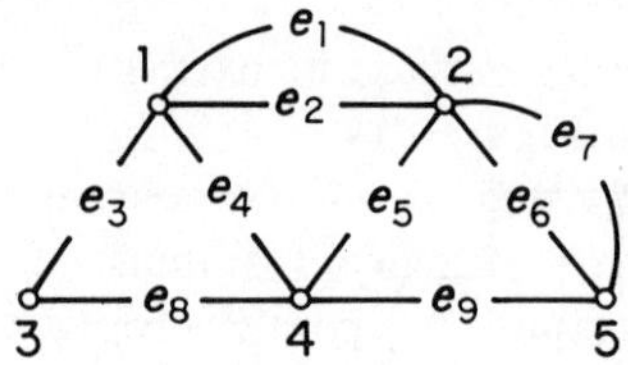

Figure P6-17

(b) The set of all 3-trees $\{T_{3_{12,3,4}}\}$.
(c) The set of all 4-trees $\{T_{4_{1,2,3,4}}\}$.
(d) The set of all 4-trees $\{T_{4_{1,23,4,5}}\}$.

6-18 Prove: (a) $y_{12} = y_{21} = \dfrac{W_{12',1'2} - W_{12,1'2'}}{\Sigma U}$

(b) $y_{22} = \dfrac{W_{1,1'}}{\Sigma U}$

where $\Sigma U \triangleq U_{12',2,1'} + U_{1,2,1'2'} + U_{12,2'1'} + U_{1,2',1'2}$

6-19 Prove that the following topological formulas for the transfer impedance function $Z_{12}(s)$, the transfer admittance function $Y_{12}(s)$, the voltage-ratio transfer function $G_{12}(s)$, and the current-ratio transfer function $\alpha_{12}(s)$ of a passive two-port without mutual inductances (Figure P6-19) are correct:

(a) $Z_{12}(s) = \dfrac{W_{12,1'2'} - W_{12',1'2}}{V}$

(b) $Y_{12}(s) = Y_L \dfrac{W_{12,1'2'} - W_{12',1'2}}{W_{1,1'}}$

(c) $G_{12}(s) = \dfrac{W_{12,1'2'} - W_{12',1'2}}{W_{1,1'}}$

(d) $\alpha_{12}(s) = Y_L\left(\dfrac{W_{12,1'2'} - W_{12',1'2}}{V}\right)$

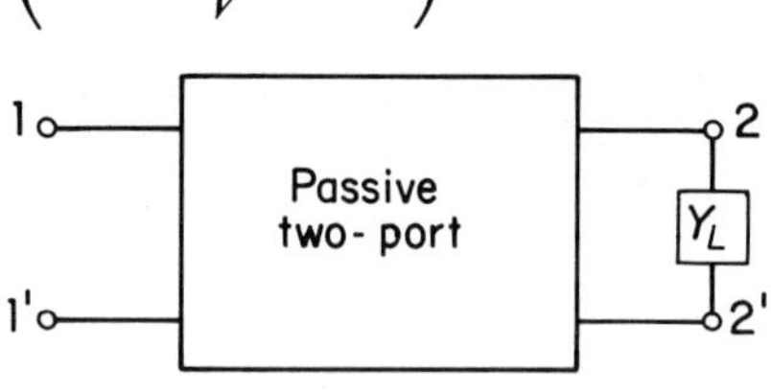

Figure P6-19

6-20 Derive the topological formula for $Z_{12}(s)$, $Y_{12}(s)$, $G_{12}(s)$, and $\alpha_{12}(s)$ of a passive three-terminal two-port, as shown in Figure P6-7.

6-21 For the network shown in Figure P6-21, determine the voltage-ratio transfer function $G_{12}(s)$.

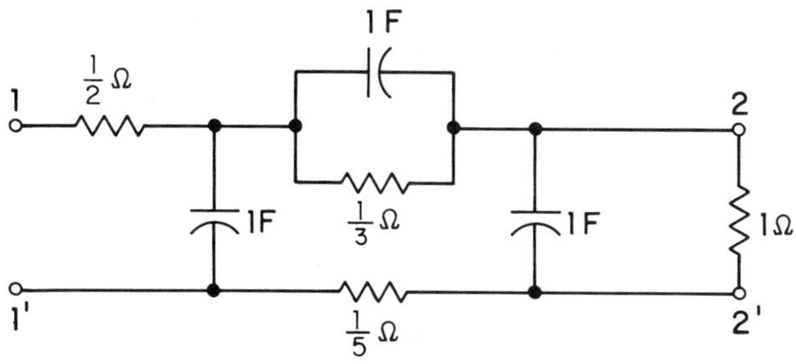

Figure P6-21

6-22 In the lattice network shown in Figure P6-22, $Z_aZ_b = R^2$. Using topological formulas, (a) find Z_{in} and (b) find V_2/I_1.

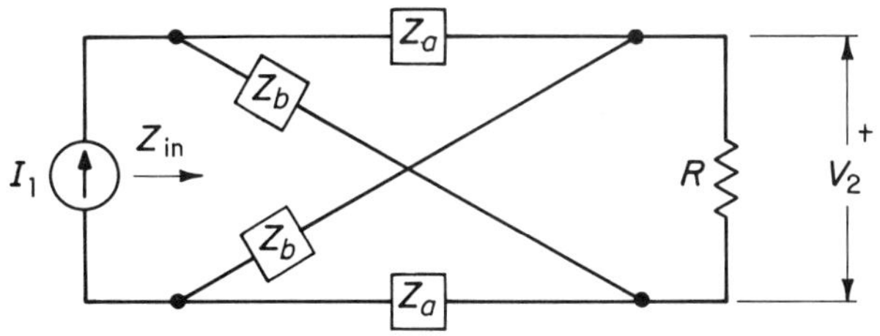

Figure P6-22

6-23 Using topological formulas, find the transfer functions in the form output/input for each of the common-ground networks shown in Figure P6-23.

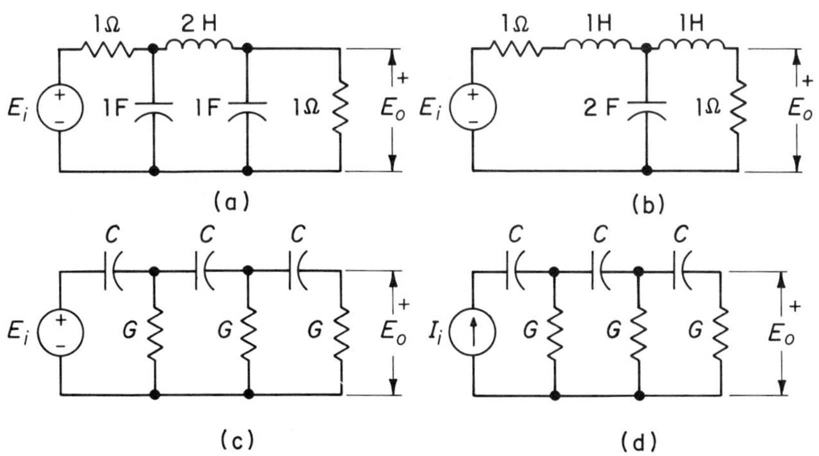

Figure P6-23

6-24 The following algorithm is proposed:
For a network of n nodes, compute (or measure) the admittance Y_1 at a chosen node pair and then coalesce (solder together) the two nodes. Compute (or measure) Y_2 at another node pair and then coalesce that node pair. Continue to Y_{n-1}. The node-basis system determinant is equal to the product; that is,

$$\Delta = Y_1 Y_2 Y_3 \cdots Y_{n-1}$$

Prove that the algorithm is valid.

6-25 For an *RLC* network N, show in detail that

$$\Delta' = (\det Z_e)\,\Delta$$

where Δ', Z_e, and Δ are the loop impedance determinant, branch impedance matrix, and node admittance determinant, respectively.

6-26 Prove that the determinant Δ' of the loop impedance matrix Z_m is equal to the expression $C[v(z)]$, which is obtained from $V(Y)$ by replacing each tree-admittance product $T(y)$ in $V(Y)$ by the product of the impedances of the chord set with respect to the same tree T. That is, prove the following:

$$\Delta' = C[V(Z)]$$

6-27 Refer to Problem 6-26. Derive a similar expression for Δ'_{11}, the (1,1) cofactor of Δ'.

6-28 Repeat Problem 6-27 for Δ'_{ij}, the (i,j) cofactor of Δ'.

6-29 Using the results obtained in Problems 6-26, 6-27, and 6-28, derive the topological expressions for the open-circuit impedances z_{11}, z_{12}, and z_{22} of an *RLC* three-terminal network.

6-30 Extending the results obtained in Problems 6-26, 6-27, and 6-28, derive the topological expressions for the short-circuit admittance y_{11}, y_{12}, and y_{22} of an *RLC* three-terminal network.

6-31 Refer to the network of Figure P6-12.
(a) Repeat Problem 6-12 using Equation (6-146) and others.
(b) Calculate the short-circuit admittances and verify that Y_{sc}, as defined by Equation (6-148), is the inverse of Z_{oc} [Equation (6-137)].

FLOW-GRAPH TECHNIQUES OF LINEAR SYSTEM ANALYSIS

CHAPTER 7

7-1 INTRODUCTION

The idea of solving a system of linear algebraic equations by the use of graphs was first introduced by Mason[1] in 1953. The main concept is to associate the linear system under study with a graph, called *signal-flow graph*, as defined by Mason and then to solve the system topologically using the properties of this special type of graph.

Although signal-flow graphs have proved to be very useful for the analysis of engineering problems because they display intuitively the causal relationships between the variables of a system under study, the expansion of Mason's gain formula is not simple and its derivation is very complicated. For these reasons, Coates [24] introduced a modified version of Mason's graph, called the *flow graph*, and gave a rigorous derivation of a simpler gain formula. In 1960, Desoer [26] presented a different derivation of the Coates formula, considerably simpler than Coates' original derivation since only a few of the determinant properties were employed in the development. Recently, Chan and Bapna [16] made use of the concepts of a tree and a 2-tree and modified the Coates gain formula still further. Immediately following the Chan-Bapna development, Chan and Mai [19] introduced a new graph (quite different from the Mason and Coates graphs), and gave a simple derivation of a new gain formula for the solution of the system variables.

[1]Mason has two papers on this subject; see S. J. Mason [55, 56].

In this chapter, we shall, in general, refer to all the aforementioned graphs as flow graphs and distinguish one from the other by identifying each with its author. The four different flow graphs which will be discussed are the Mason graphs, the Coates graph, the modified Coates graph, and the Chan-Mai graph, each of which yields a gain formula for the system under study. We shall devote one section to the discussion of each of these flow-graph techniques. Then we shall examine, very briefly, the graph reduction technique which is based on the Mason graph. Finally, in the summary, we shall give a comparison of the gain formulas derived from the four flow graphs and point out the interrelationships between them.

7-2 THE MASON GRAPH

The discussion presented in this section is based on two papers by Samuel J. Mason [55, 56]. It refers to the Mason graph (or signal-flow graph as Mason defined it) and its application to the solution of a system of linear equations which may arise — for example, from an electrical network or a linear feedback system. Our main concern is to present his method of solving a linear system under study so that we can compare this method with the following ones and appreciate the advantages and disadvantages of each technique over the others. As mentioned in the introduction, the derivation of Mason's formula is very complicated and, for this reason, we shall not attempt to include it here.

DEFINITION 7-1. Signal-flow Graph G_m (Mason Graph) and Related Terms. A *signal-flow graph* G_m of a system of linear equations is a network with junction points called *nodes*, which are connected by directed line segments called *branches*. Signals travel along branches only in the direction described by the arrows of the branches. A signal x_k, traveling along a branch between x_k and x_j, is multiplied by the *gain* of the branch, g_{kj}, so that a signal of $g_{kj}x_k$ is delivered at node x_j, as indicated in Figure 7-1. An *input node* (*source*) is a

$$x_k \circ\!\!-\!\!-\!\!-\; g_{kj} \;-\!\!-\!\!\!\longrightarrow\!\circ\, x_j \qquad \Longleftrightarrow \qquad x_j = g_{kj}\, x_k$$

Figure 7-1 Representation of a simple equation by a flow graph.

node that contains only outgoing branches; an *output node* (*sink*) is a node that has only incoming branches. A *path* is a continuous unidirectional succession of branches, all of which are traveling in the same direction; a *forward path* is a path from the input node to the output node along which all nodes are encountered exactly once; and a *feedback path* (*loop*) is a closed path that originates from and terminates at the same node, and along which all other nodes are encountered exactly once. (The trivial case is a *self-loop*, which contains exactly one node and one branch.) A *path gain* is the product of all the branch gains of the path; and, similarly, a *loop gain* is the product of all the branch gains of the branches in a loop.

The procedure for obtaining the Mason graph from a system of linear algebraic equations may be described in the following steps:

(1) Arrange all the equations of the system in such a way that the jth dependent (output) variable x_j in the jth equation is expressed explicitly in terms of the other variables. Thus, if the system under study is given by

$$\begin{aligned} a_{11}x_1 + a_{12}x_2 + \cdots + a_{1n}x_n &= b_1 \\ a_{21}x_1 + a_{22}x_2 + \cdots + a_{2n}x_n &= b_2 \\ &\vdots \\ a_{n1}x_1 + a_{n2}x_2 + \cdots + a_{nn}x_n &= b_n \end{aligned} \tag{7-1}$$

where $b_1, b_2, \ldots, b_n$ are inputs (sources) and $x_1, x_2, \ldots, x_n$ are outputs, the equations may be rewritten as

$$\begin{aligned} x_1 &= \frac{1}{a_{11}}b_1 - \frac{a_{12}}{a_{11}}x_2 - \frac{a_{13}}{a_{11}}x_3 - \cdots - \frac{a_{1n}}{a_{11}}x_n \\ x_2 &= \frac{1}{a_{22}}b_2 - \frac{a_{21}}{a_{22}}x_1 - \frac{a_{23}}{a_{22}}x_3 - \cdots - \frac{a_{2n}}{a_{22}}x_n \\ &\vdots \\ x_n &= \frac{1}{a_{nn}}b_n - \frac{a_{n1}}{a_{nn}}x_1 - \frac{a_{n2}}{a_{nn}}x_2 - \cdots - \frac{a_{n-1,n-1}}{a_{nn}}x_{n-1} \end{aligned} \tag{7-2}$$

(2) The number of input nodes in the flow graph is equal to the number of nonzero b's; that is, each of the source nodes corresponds to a nonzero b_j.

(3) To each of the output nodes is associated one of the dependent variables, $x_1, x_2, \ldots, x_n$.

(4) The value of the variable represented by a node is equal to the sum of all the incoming signals (Figure 7-1).

(5) The value of the variable represented by any node is transmitted onto all branches leaving the node.

It is a simple matter to write the equations from the flow graph since every node, except the source nodes, of the graph represents an equation, and the equation associated with node k, for example, is obtained by equating to x_k the sum of all incoming branch gains multiplied by the values of the variables from which these branches originate.

Mason's general gain formula is now stated in the following theorem, but the derivation of the formula is omitted because of its complexity.

THEOREM 7-1. Let T be the overall graph gain and T_k be the gain of the kth forward path from the source to the sink. Then

$$T = \frac{1}{\Delta'} \sum_k T_k \Delta_k \tag{7-3}$$

where

$$\Delta' = 1 - \sum_m p_{m1} + \sum_m p_{m2} - \sum_m p_{m3} + \cdots + (-1)^j \sum_m p_{mj}$$

with

p_{m1} = loop gain (the product of all the branch gains around a loop) or the gain of a *first-order* loop

p_{m2} = product of the *loop* gains of the mth set of two nontouching loops (these are called *second-order* loops)

p_{m3} = product of the loop gains of the mth set of three nontouching loops (*third-order* loops)

p_{mj} = product of the loop gains of the mth set of j nontouching loops (*loops of order j*)

Δ_k = the value of Δ for that subgraph of the graph obtained by removing the kth forward path along with all those branches touching the path

T_k = product of the branch gains in the kth forward path.

EXAMPLE 7-1. Let us determine x_2 of the following system using Mason's general gain formula [Equation (7-3)].

$$\begin{aligned} a_{11}x_1 + a_{12}x_2 + a_{13}x_3 &= b_1 \\ a_{22}x_2 + a_{23}x_3 &= 0 \\ a_{31}x_1 \qquad\qquad + a_{33}x_3 &= b_3 \end{aligned} \tag{7-4}$$

First, we rewrite the three equations of (7-4) as

$$\begin{aligned} x_1 &= \frac{1}{a_{11}}b_1 - \frac{a_{12}}{a_{11}}x_2 - \frac{a_{13}}{a_{11}}x_3 \\ x_2 &= -\frac{a_{23}}{a_{22}}x_3 \\ x_3 &= \frac{1}{a_{33}}b_3 - \frac{a_{31}}{a_{33}}x_1 \end{aligned} \tag{7-5}$$

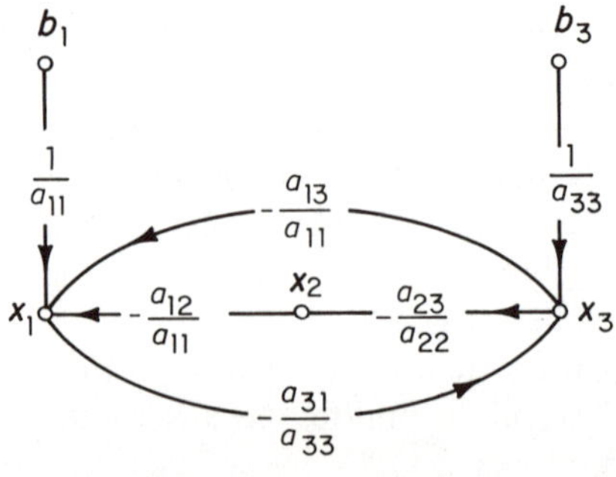

Figure 7-2 The Mason graph G_m of the system of Equation (7-5).

Then we draw the signal-flow graph (Mason graph) G_m of the system given in (7-5), as shown in Figure 7-2. By Theorem 7-1, we obtain the following quantities of Equation (7-3):

$$\begin{aligned}\Sigma p_{m1} &= \left(-\frac{a_{13}}{a_{11}}\right)\left(-\frac{a_{31}}{a_{33}}\right) + \left(-\frac{a_{23}}{a_{22}}\right)\left(-\frac{a_{12}}{a_{11}}\right)\left(-\frac{a_{31}}{a_{33}}\right) \\ &= \frac{a_{13}a_{31}}{a_{11}a_{33}} - \frac{a_{12}a_{23}a_{31}}{a_{11}a_{22}a_{33}} \\ &= \frac{a_{13}a_{22}a_{31} - a_{12}a_{23}a_{31}}{a_{11}a_{22}a_{33}} \\ \Delta' &= 1 - \frac{a_{13}a_{22}a_{31} - a_{12}a_{23}a_{31}}{a_{11}a_{22}a_{33}} \\ &= \frac{a_{11}a_{22}a_{33} - a_{13}a_{22}a_{31} + a_{12}a_{23}a_{31}}{a_{11}a_{22}a_{33}}\end{aligned}$$

For the forward path from b_1 to x_2,

$$T_1 = \left(\frac{1}{a_{11}}\right)\left(-\frac{a_{31}}{a_{33}}\right)\left(-\frac{a_{23}}{a_{22}}\right) = \frac{a_{23}a_{31}}{a_{11}a_{22}a_{33}}$$

and, since there are no nontouching loops with respect to G_1,

$$\Delta_1 = 1 - 0 = 1$$

For the forward path from b_3 to x_2,

$$T_3 = \left(\frac{1}{a_{33}}\right)\left(-\frac{a_{23}}{a_{22}}\right) = -\frac{a_{23}}{a_{22}a_{33}}$$

and, for the same reason as in the case of Δ_1, we have

$$\Delta_3 = 1 - 0 = 1$$

Thus, for the input b_1 alone,

$$(x_2)_{b_1} = \frac{1}{\Delta'}T_1\Delta_1 b_1$$

and, for the input b_3 alone,

$$(x_2)_{b_3} = \frac{1}{\Delta'}T_3\Delta_3 b_3$$

Finally, by the principle of superposition, the final expression for x_2 is

$$\begin{aligned}x_2 &= (x_2)_{b_1} + (x_2)_{b_3} \\ &= \frac{1}{\Delta'}T_1\Delta_1 b_1 + \frac{1}{\Delta'}T_3\Delta_3 b_3 \\ &= \frac{1}{\Delta'}(T_1\Delta_1 b_1 + T_3\Delta_3 b_3) \\ &= \frac{a_{11}a_{22}a_{33}}{a_{11}a_{22}a_{33} - a_{13}a_{22}a_{31} + a_{12}a_{23}a_{31}}\left(\frac{a_{23}a_{31}}{a_{11}a_{22}a_{33}}\cdot 1\cdot b_1 - \frac{a_{23}}{a_{22}a_{33}}\cdot 1\cdot b_3\right)\end{aligned}$$

or

$$x_2 = \frac{a_{23}a_{31}b_1 - a_{11}a_{23}b_3}{a_{11}a_{22}a_{33} - a_{13}a_{22}a_{31} + a_{12}a_{23}a_{31}} \tag{7-6}$$

We shall illustrate the Mason flow-graph technique by a more complicated system, as given in the next example.

EXAMPLE 7-2. Consider the following system of equations:

$$\begin{aligned} x_1 &= bx_2 + dx_3 + mx_0 \\ x_2 &= ax_1 + kx_3 + fx_4 \\ x_3 &= cx_1 + hx_4 + nx_0 \\ x_4 &= ex_2 + gx_3 \end{aligned} \tag{7-7}$$

the Mason graph of which is shown in Figure 7-3.

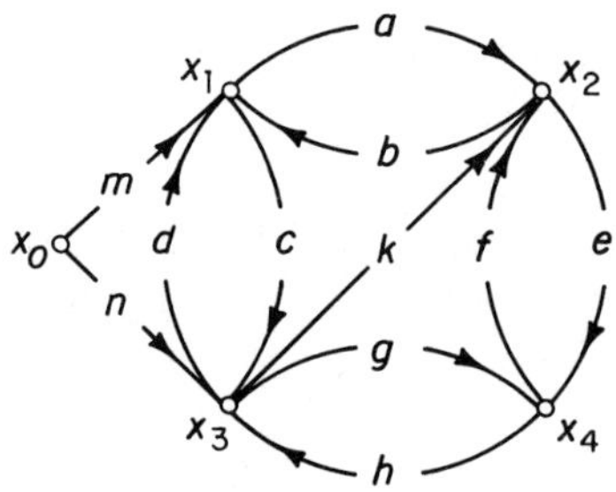

Figure 7-3 The Mason graph of the system of Equation (7-7).

We shall calculate the gain x_1/x_0 by Mason's gain formula [Equation (7-3)]. By inspection, we see that there are eight first-order loops (*ab*, *cd*, *ef*, *gh*, *bck*, *ehk*, *aehd*, and *bcgf*), two second-order loops (*abgh* and *cdef*), and no loops of order 3 or higher. Thus, by Theorem 7-1, we have

$$\begin{aligned} \Sigma P_{m1} &= ab + cd + ef + gh + bck + ehk + aehd + bcgf \\ \Sigma P_{m2} &= abgh + cdef \end{aligned}$$

so that

$$\begin{aligned} \Delta' &= 1 - \Sigma P_{m1} + \Sigma P_{m2} \\ &= 1 - (ab + cd + ef + gh + bck + ehk + aehd + bcgf) + abgh + cdef \end{aligned} \tag{7-8}$$

Also, there are four forward paths from x_0 to x_1. They are listed, along with their corresponding Δ_k, as follows:

$$\begin{array}{lll} \text{Path 1:} & G_1 = m & \Delta_1 = 1 - (ef + gh + ehk) \\ \text{Path 2:} & G_2 = nd & \Delta_2 = 1 - ef \\ \text{Path 3:} & G_3 = nkb & \Delta_3 = 1 - 0 = 1 \\ \text{Path 4:} & G_4 = ngfb & \Delta_4 = 1 - 0 = 1 \end{array} \tag{7-9}$$

Finally, substituting Equations (7-8) and (7-9) into Equation (7-3) of Theorem 7-1, we obtain the desired gain expression:

$$\frac{x_1}{x_0} = \frac{1}{\Delta'} \sum_{k=1}^{4} G_k \Delta_k$$

$$= \frac{m - m(ef + gh + ehk) + nd - ndef + nkb + ngfb}{1 - (ab + cd + ef + gh + bck + ehk + aehd + bcgf) + (abgh + cdef)} \tag{7-10}$$

For a quick check of the correctness of the result, we let

$$a = b = c = d = e = f = g = h = k = n = 1 \quad \text{and} \quad m = 2 \tag{7-11}$$

and substituting these values in Equation (7-10), we have

$$\frac{x_1}{x_0} = \frac{2 - 2(3) + 1 - 1 + 1 + 1}{1 - (8) + (2)} = \frac{-2}{-5} = \frac{2}{5} \tag{7-12}$$

Next, substituting Equation (7-11) into (7-7) and rearranging the terms, we write

$$\begin{aligned} x_1 - x_2 - x_3 \qquad &= 2x_0 \\ x_1 - x_2 + x_3 + x_4 &= 0 \\ x_1 \qquad - x_3 + x_4 &= -x_0 \\ x_2 + x_3 - x_4 &= 0 \end{aligned} \tag{7-13}$$

Applying Cramer's rule, we have

$$x_1 = \frac{\begin{vmatrix} 2 & -1 & -1 & 0 \\ 0 & -1 & 1 & 1 \\ -1 & 0 & -1 & 1 \\ 0 & 1 & 1 & -1 \end{vmatrix} x_0}{\begin{vmatrix} 1 & -1 & -1 & 0 \\ 1 & -1 & 1 & 1 \\ 1 & 0 & -1 & 1 \\ 0 & 1 & 1 & -1 \end{vmatrix}} = \frac{2x_0}{5} \quad \text{or} \quad \frac{x_1}{x_0} = \frac{2}{5}$$

which yields the same result as Equation (7-12) by Mason's flow-graph technique.

7-3 THE COATES GRAPH

Although Mason's signal-flow graphs constitute a very useful technique for the analysis of linear systems, the derivation of his gain formula is omitted in most texts for a very obvious reason — the inherent complexity of the derivation. Consequently, not many teachers are able to include this derivation in their lectures and the interested reader naturally feels dissatisfied. In 1959 C. L. Coates [24] modified Mason's signal-flow graph and referred to his new one as "flow graph." As a result of such modification, he has not only succeeded in showing that Mason's general gain formula is not the simplest expansion, but has also presented a rigorous derivation of a new gain formula that is considerably simpler than Mason's, besides being easier to expand.

However, the derivation of Coates' gain formula is still not very simple, and thus is unsuitable for class presentation. Realizing this, C. A. Desoer [26] presented an independent derivation of Coates' formula, which will be included in this section after the introduction of the few following definitions.

DEFINITION 7-2. Flow Graph (Coates Graph) G_c. A *flow graph*, G_c, is a graph of weighted oriented branches, each of which has a positive direction and a weight (called the *branch gain*) interconnected at points called nodes, defining uniquely a system of linear algebraic equations. The flow graph corresponding to a system of n linear algebraic equations in n unknowns,

$$\begin{aligned} a_{11}x_1 + a_{12}x_2 + \cdots + a_{1n}x_n - b_1 &= 0 \\ a_{21}x_1 + a_{22}x_2 + \cdots + a_{2n}x_n - b_2 &= 0 \\ &\vdots \\ a_{n1}x_1 + a_{n2}x_2 + \cdots + a_{nn}x_n - b_n &= 0 \end{aligned} \tag{7-14}$$

has n nodes $x_1, x_2, \ldots, x_n$ (representing the n variables $x_1, x_2, \ldots, x_n$, respectively) plus a source node 1. If, in the system of Equation (7-14), $a_{kj} \neq 0$, there is a branch directed from node x_j to node x_k with branch gain a_{kj}; if $b_k \neq 0$, there is a branch directed from the source node 1 to node x_k with branch gain $-b_k$.

EXAMPLE 7-3. For the case $n = 2$, assuming that all the a_{ij}'s and the b_k's are nonzero ($i,j,k = 1,2$), the system of Equation (7-14) becomes

$$\begin{aligned} a_{11}x_1 + a_{12}x_2 - b_1 &= 0 \\ a_{21}x_1 + a_{22}x_2 - b_2 &= 0 \end{aligned} \tag{7-15}$$

and the flow graph representing Equation (7-15) is shown in Figure 7-4.

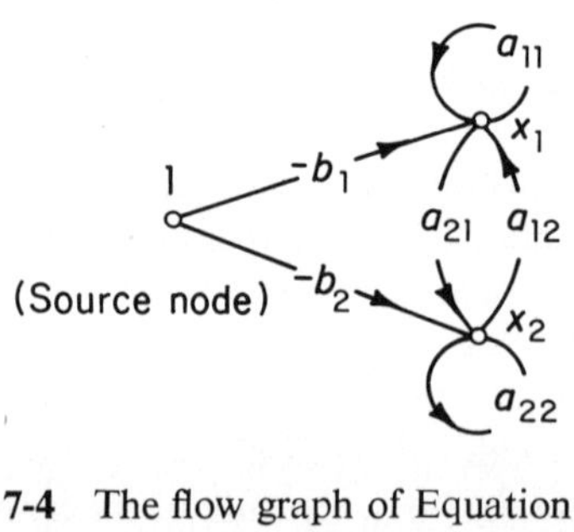

Figure 7-4 The flow graph of Equation (7-15).

From Example 7-3, it is seen that if the flow graph is known, the system of equations represented by the graph can be readily obtained by noting the following two simple observations:

(a) Each node of the graph, except for the source node, represents an equation in the system.

(b) Since the incoming branches of a node, say x_k, are identified by their branch gain, $a_{k1}, a_{k2}, \ldots, a_{kn}$ and $-b_k$, the kth equation of the system is obtained by equating to zero the sum of the products of the branch gains of all the incoming branches of x_k times the respective nodes from which these branches originate; that is,

$$a_{k1}x_1 + a_{k2}x_2 + \cdots + a_{kn}x_n + (-b_k)\,1 = 0$$

Next, let us recall Cramer's rule for solving a system of n linear algebraic equations in n unknowns [Equation (7-14)] written in matrix form, $AX = B$:

$$\begin{bmatrix} a_{11} & a_{12} & \cdots & a_{1n} \\ a_{21} & a_{22} & \cdots & a_{2n} \\ \cdot & & & \\ \cdot & & & \\ \cdot & & & \\ a_{n1} & a_{n2} & \cdots & a_{nn} \end{bmatrix} \begin{bmatrix} x_1 \\ x_2 \\ \cdot \\ \cdot \\ \cdot \\ x_n \end{bmatrix} = \begin{bmatrix} b_1 \\ b_2 \\ \cdot \\ \cdot \\ \cdot \\ b_n \end{bmatrix} \tag{7-16}$$

If the determinant Δ of the coefficient matrix A is different from zero, Cramer's rule states that

$$x_j = \frac{\sum_{k=1}^{n} \Delta_{kj} b_k}{\Delta} \qquad (j = 1, 2, \ldots, n) \tag{7-17}$$

where Δ_{kj} is the cofactor of the element a_{kj} of Δ.

Relating the quantities in Equation (7-17) topologically to the subgraphs of the flow graphs (some of which are yet to be defined), Coates was able to derive his gain formula based on the modification he made on the Mason graph. But, before discussing Coates' gain formula and its derivation, it is necessary to introduce the following definitions:

DEFINITION 7-3. Subgraph G_{co}. The *subgraph* G_{co} of the flow graph G_c is the subgraph of G_c by deleting the source node 1 and all its (outgoing) branches.

DEFINITION 7-4. Connection and Connection Gain. A *connection* of the flow graph G_{co} is a subgraph of G_{co} such that

(a) Each node of G_{co} is included.

(b) Each node has only one branch terminating at it and one branch originating from it.

The *connection gain* of a connection is the product of the branch gains of all the branches of that connection. It is denoted by $C(G_{co})$.

DEFINITION 7-5. Directed Loop and Forward Path. A *directed loop* is a circuit of directed branches such that the directions of all its branches have the same sense. Similarly, a *forward path* is a path of directed branches in which the direction of all the branches have the same sense from the initial node to the final node.

DEFINITION 7-6. One-connection and One-connection Gain. A *one-connection* from node p to node q of graph G_c is a subgraph of G_c containing all the nodes of G_c such that

(a) No branch of the subgraph terminates at p and only one branch originates from p.

(b) No branch of the subgraph originates from q and only one branch terminates at q.

(c) All other nodes of the subgraph have exactly one incoming and one outgoing branch.

The *one-connection gain* of a one-connection is similarly defined as is the connection gain of a connection.

Thus, a one-connection is a forward path together with, possibly, some directed loops.

EXAMPLE 7-4. Consider the flow graph G_c of Figure 7-4. The corresponding subgraph G_{co} is shown in Figure 7-5(a); the two connections of G_{co} are given in Figure 7-5(b); and the two one-connections of G_c are depicted in Figure 7-5(c). The connection gains of the two connections of Figure 7-5(b) are $a_{12}a_{21}$ and $a_{11}a_{22}$. Similarly, the one-connection gains of the two one-connections of Figure 7-5(c) are $-b_1a_{22}$ and $-b_2a_{11}$.

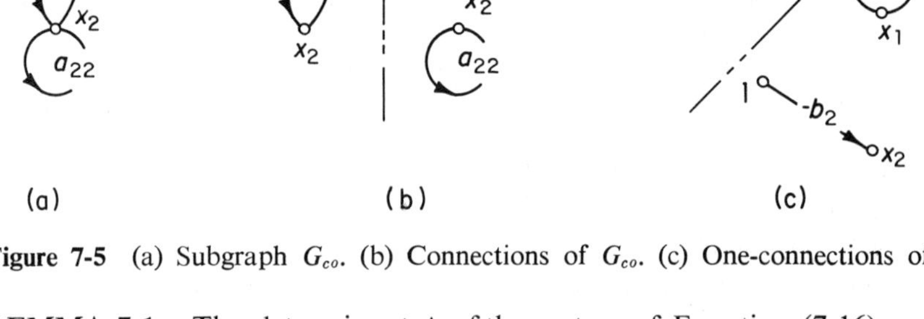

Figure 7-5 (a) Subgraph G_{co}. (b) Connections of G_{co}. (c) One-connections of G_c.

LEMMA 7-1. The determinant Δ of the system of Equation (7-16) can be evaluated topologically from graph G_{co} of its flow graph G_c by the formula

$$\Delta = (-1)^n \sum_p (-1)^{L_p} C(G_{co})_p \tag{7-18}$$

where L_ρ is the number of directed loops in the ρth connection; $C(G_{co})_\rho$ is the connection gain of the ρth connection of G_{co}; and the summation (on ρ) is taken over all connections of G_{co}.

Proof: From the theory of determinants, it is known that

$$\Delta = \sum_{p} (\text{sgn } P)\, a_{1i_1}a_{2i_2}\ldots a_{ni_n} \tag{7-19}$$

where the summation is taken over all the $n!$ permutations of the form $P = (i_1, i_2, i_3, \ldots, i_n)$ of the integers $(1, 2, \ldots, n)$ and (sgn P) of a permutation is $+1$ or -1, depending on whether the permutation P is even or odd.[2]

Now, consider a particular product π in the sum given in Equation (7-19). Let π' be the set of all branches whose gains appear in the product π. Since π is a product of factors a_{ki_k} with k running from 1 to n, there is exactly one branch of π' that terminates at each of the n nodes of G_{co}. Since $i_1, i_2, \ldots, i_n$ is a permutation of $(1, 2, \ldots, n)$, there is exactly one branch of π' originating from each node of G_{co}.

Conversely, given an arbitrary connection of G_{co}, since, by definition, it contains one branch terminating at each of the nodes of G_{co} and one branch leaving each of the same nodes, then its connection gain can be written as

$$a_{1i_1}a_{2i_2}\ldots a_{ni_n}$$

where $i_1, i_2, \ldots, i_n$ is a permutation of the integers $1, 2, \ldots, n$. Hence, it will appear as one of the products in Equation (7-19).

Therefore, by simply listing all the connections of G_{co} and finding all the connection gains — that is, $C(G_{co})$ — one obtains all the terms of the sum in Equation (7-19). As for the sign of each term (sgn P), consider again a particular product π of the sum given in Equation (7-19) and its associated connection π' of the graph G_o. Clearly, π' is a collection of directed loops. For the sake of simplicity, let us assume that π' consists of three nontouching directed loops having n_1, n_2, n_3 nodes, respectively. Then, $n_1 + n_2 + n_3 = n$ since, by definition, all nodes of G_{co} are included in a connection. Now we relabel the nodes of π' so that, when we traverse along the first loop of π' in the positive direction, the nodes $1, 2, \ldots, n_1$ will be traversed in that order, and similarly for the other two loops. The branch-gain product of the first loop is then

$$a_{1n_1}a_{21}a_{32}\ldots a_{n_1,n_1-1}$$

Note that the factors of this product are ordered so that their row subscripts occur in their natural order as required by Equation (7-19). Hence the sign assigned to π is that assigned to the permutation defined by the column subscripts:

1st loop: $n_1, 1, 2, 3, \ldots, n_1 - 1$
2nd loop: $n_1 + n_2, n_1 + 1, n_1 + 2, \ldots, n_1 + n_2 - 1$
3rd loop: $n_1 + n_2 + n_3, n_1 + n_2 + 1, n_1 + n_2 + 2, \ldots, n_1 + n_2 + n_3 - 1$

[2]For the definition of an even or odd permutation see, for example, F. E. Hohn [43].

To rearrange this permutation in the natural order,

$$(n_1 - 1) + (n_2 - 1) + (n_3 - 1) = n_1 + n_2 + n_3 - 3 = n - 3$$

interchanges between adjacent symbols are required. Hence the sign of the permutation (sgn P) is $(-1)^{n-3} = (-1)^n(-1)^{-3} = (-1)^n(-1)^3$. Note that there are three directed loops in the product π. It is clear, then, that if the connection π' has L loops, the sign would have been $(-1)^n(-1)^L$. Therefore,

$$\Delta = (-1)^n \sum_{\rho} (-1)^{L_\rho} C(G_{co})_\rho$$

This completes the proof.

LEMMA 7-2. The expression

$$\sum_{k=1}^{n} \Delta_{kj} b_k$$

in Equation (7-17) can be evaluated topologically from the flow graph G_c of the system of Equation (7-16) by the formula

$$\sum_{k=1}^{n} \Delta_{kj} b_k = (-1)^n \sum_{\tau} (-1)^{L_\tau} C(G_c;\ 1 \to x_j)_\tau \tag{7-20}$$

where $C(G_c;\ 1 \to x_j)$ is the one-connection gain of the τth one-connection from node 1 to node x_j of G_c, L_τ is the number of directed loops in the τth one-connection, and the summation on τ is over all one-connections of G_c.

Proof: Let us first begin our proof by considering one b_k in particular; that is, let us evaluate

$$\sum_{k=1}^{n} \Delta_{kj} b_k$$

by assuming that there is only one branch with gain $-b_k$ connecting the source node 1 to the rest of the graph. Note that $\Delta_{kj} b_k$ is equal to the determinant obtained by replacing the jth column of Δ by a column of zeros, except for the element of the kth row which is b_k; that is,

$$\Delta_{kj} b_k = \begin{vmatrix} a_{11} & a_{12} & \cdots & a_{1,j-1} & 0 & a_{1,j+1} & \cdots & a_{1n} \\ a_{21} & a_{22} & \cdots & a_{2,j-1} & 0 & a_{2,j+1} & \cdots & a_{2n} \\ \cdot & & & & \cdot & & & \cdot \\ \cdot & & & & \cdot & & & \cdot \\ \cdot & & & & \cdot & & & \cdot \\ a_{k1} & a_{k2} & \cdots & a_{k,j-1} & b_k & a_{k,j+1} & \cdots & a_{kn} \\ \cdot & & & & \cdot & & & \cdot \\ \cdot & & & & \cdot & & & \cdot \\ \cdot & & & & \cdot & & & \cdot \\ a_{n1} & a_{n2} & \cdots & a_{n,j-1} & 0 & a_{n,j+1} & \cdots & a_{nn} \end{vmatrix}$$

This determinant will not change its value if all the elements of the kth row, except b_k, are replaced by zero. The result is denoted by the determinant Δ''. Thus,

$$\Delta'' = \begin{vmatrix} a_{11} & a_{12} & \cdots & a_{1,j-1} & 0 & a_{1,j+1} & \cdots & a_{1n} \\ a_{21} & a_{22} & \cdots & a_{2,j-1} & 0 & a_{2,j+1} & \cdots & a_{2n} \\ \cdot & & & & \cdot & & & \cdot \\ \cdot & & & & \cdot & & & \cdot \\ \cdot & & & & \cdot & & & \cdot \\ a_{k-1,1} & a_{k-1,2} & \cdots & a_{k-1,j-1} & 0 & a_{k-1,j+1} & \cdots & a_{k-1,n} \\ 0 & 0 & \cdots & 0 & b_k & 0 & \cdots & 0 \\ a_{k+1,1} & a_{k+1,2} & \cdots & a_{k+1,j-1} & 0 & a_{k+1,j+1} & \cdots & a_{k+1,n} \\ \cdot & & & & \cdot & & & \cdot \\ \cdot & & & & \cdot & & & \cdot \\ \cdot & & & & \cdot & & & \cdot \\ a_{n1} & a_{n2} & \cdots & a_{n,j-1} & 0 & a_{n,j+1} & \cdots & a_{nn} \end{vmatrix}$$

Now let us investigate what effect the above operation has on the original graph G_c. Obviously, replacing the jth column of Δ by a column of zeros means the deletion of all the outgoing branches from node x_j; replacing the kth row of Δ by a row of zeros means the deletion of all the incoming branches to node x_k of graph G_c; putting b_k in the a_{kj} position means adding a branch from node x_j to node x_k with branch gain b_k. Let us call this modified graph G_c'. Then, if graph G_c is as shown in Figure 7-6, graph G_c' will be the graph depicted in Figure 7-7. Thus, on the basis of Lemma 7-1,

$$\Delta'' = (-1)^n \sum_\gamma (-1)^{L_\gamma} C(G_c')_\gamma$$

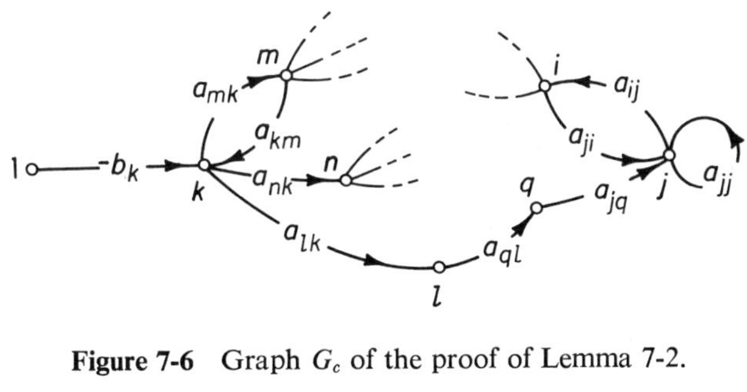

Figure 7-6 Graph G_c of the proof of Lemma 7-2.

where $C(G_c')_\gamma$ is the connection gain of the γth connection of G_c'; L_γ is the number of directed loops of the γth connection; and the summation is taken over all connections of G_c'.

In order to interpret this result in terms of graph G_c, let us examine the graph G_c' as shown in Figure 7-7. Each one of the connections of G_c' includes the branch b_k because it is the only branch leaving node x_j. In each one of

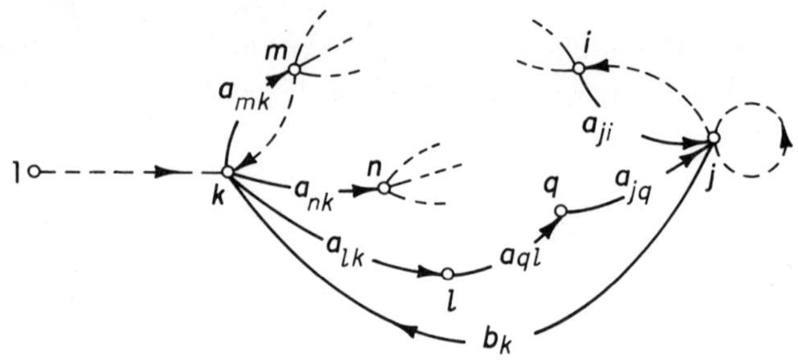

Figure 7-7 Graph G_c' of the proof of Lemma 7-2.

these connections, let us disconnect branch b_k at node j and identify the "hanging" node of b_k as node 1 (as it was originally in graph G_c). Furthermore, let us change the branch gain from b_k to $-b_k$. Then, in each case, the resulting configuration is equal to a one-connection from 1 to x_j of G_c. Since, by displacing the branch b_k, one directed loop has been "opened," the number of directed loops in the one-connection of G_c is one less than that of the original connection of G_c'. This will result in a change of sign, which will cancel the one caused by the change of sign of the branch gain b_k. Consequently,

$$\Delta_{kj} b_k = \Delta'' = (-1)^n \sum_{\tau} (-1)^{L_\tau} C(G_c; 1 \rightarrow x_j)_\tau$$

It is evident that if there is more than one b_k in the system, Equation (7-20) will result, as asserted by the lemma. The proof is now complete.

We shall state and prove Coates' gain formula in the following theorem:

THEOREM 7-2. In a system of n linear algebraic equations in n unknowns,

$$\sum_{j=1}^{n} a_{ij} x_j = b_i \qquad (i = 1, 2, \ldots, n)$$

The value of x_j is given by

$$x_j = \frac{\sum_{\tau} (-1)^{L_\tau} C(G_c; 1 \rightarrow x_j)_\tau}{\sum_{\rho} (-1)^{L_\rho} C(G_{co})_\rho} \tag{7-21}$$

where

$C(G_c; 1 \rightarrow x_j)_\tau$ is the one-connection gain of the τth one-connection from 1 to x_j of G_c.

L_τ is the number of directed loops in the τth one-connection.

$C(G_{co})_\rho$ is the connection gain of the ρth connection.

L_ρ is the number of directed loops in the ρth connection.

The summations are over all one-connections of G_c (on τ) and all connections of G_{co} (on ρ).

Proof: Combining Equations (7-17) and (7-18) of Lemma 7-1 with Equation (7-20) of Lemma 7-2, we immediately have the result as expressed in Equation (7-21). Hence the theorem.

EXAMPLE 7-5. Consider the same system [Equation (7-4)] as in Example 7-1; namely,

$$\begin{aligned} a_{11}x_1 + a_{12}x_2 + a_{13}x_3 &= b_1 \\ a_{22}x_2 + a_{23}x_3 &= 0 \\ a_{31}x_1 \qquad\qquad + a_{33}x_3 &= b_3 \end{aligned}$$

The Coates graph, G_c, is shown in Figure 7-8 and graph G_{co} in Figure 7-9. The set of all connections of G_{co} is depicted in Figure 7-10 and the set of all

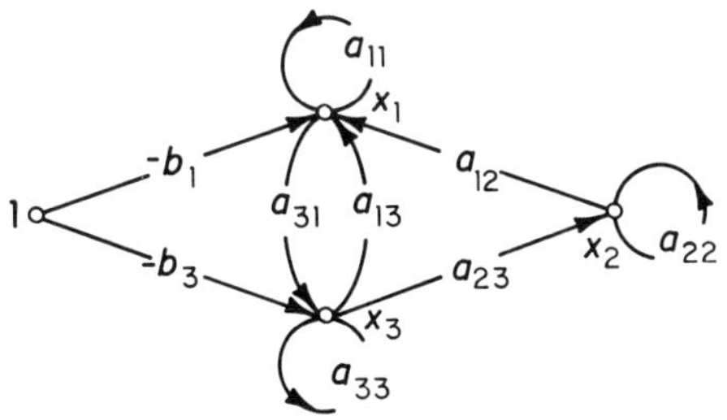

Figure 7-8 The Coates graph G_c of the system of Equation (7-4).

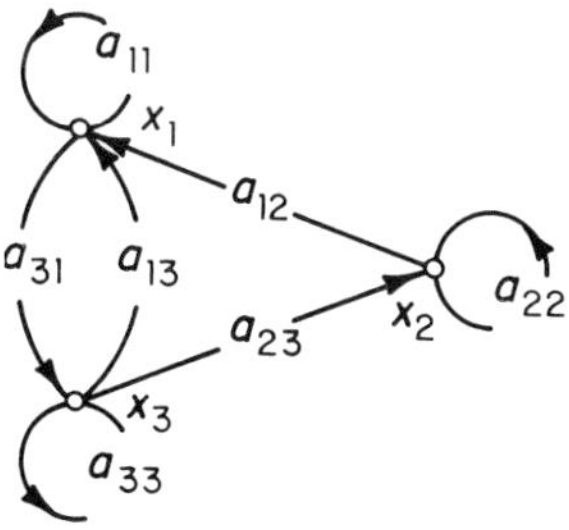

Figure 7-9 Graph G_{co} of system of Equation (7-4).

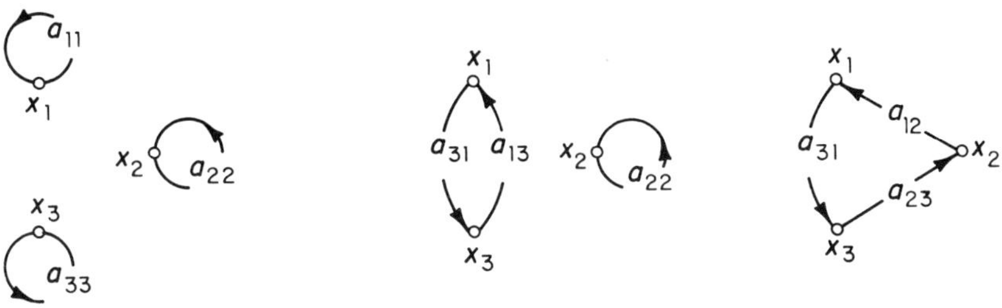

Figure 7-10 The set of all connections of G_{co}.

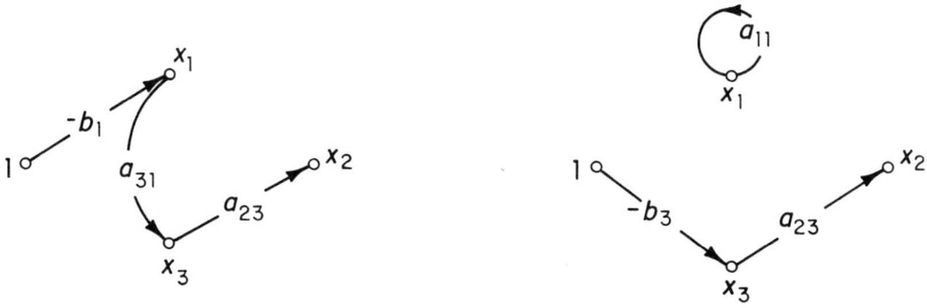

Figure 7-11 The set of all one-connections of G_c.

one-connections of G_c in Figure 7-11. Thus, by Equation (7-21), with $j = 2$, we get

$$x_2 = \frac{(-1)^0(-b_1)a_{31}a_{23} + (-1)^1(-b_3)a_{23}a_{11}}{-a_{11}a_{22}a_{33} + a_{13}a_{31}a_{22} - a_{12}a_{31}a_{23}}$$

$$= \frac{a_{23}a_{31}b_1 - a_{11}a_{23}b_3}{a_{11}a_{22}a_{33} - a_{13}a_{22}a_{31} + a_{12}a_{23}a_{31}} \tag{7-22}$$

which agrees with the result [Equation (7-6)] found by Mason's method in Example 7-1.

7-4 THE MODIFIED COATES GRAPH BY CHAN AND BAPNA

The material discussed in this section is based solely on a paper of Chan and Bapna [16]. The most significant contribution of this paper is the considerable simplification of Desoer's derivation of the Coates gain formula [26]. To this end, the concepts of a loop set and a 2-loop set, analogous to the concepts of a tree and a 2-tree, are introduced in deriving the modified gain formula.

As will be seen later in the development of this section, due to the similarity of the definitions of the loop set and the 2-loop set, the application, as well as the derivation, of the modified formula is significantly simpler than the Coates formula with regard to the evaluation of the terms both in the determinant Δ and in the cofactor.

We shall first introduce a few definitions.

DEFINITION 7-7. Modified Coates Graph G_l. A *modified Coates graph* G_l is an oriented graph representing a system of equations

$$\sum_{j=1}^{n} a_{ij}x_j = b_i \qquad (i = 1, 2, \ldots, n)$$

such that the variable x_j is represented as a *vertex* (also called x_{ij}), and the coefficient a_{ij} of the variable x_j as an *edge* with a gain a_{ij}. Furthermore, edge a_{ij} in G_l is connected between vertices x_i and x_j, with an arrow directed from x_j to x_i. Finally, a *source vertex* 1 is included in G_l such that, for each constant b_k in the system, there is an edge with a gain b_k in G_l which originates from 1 and terminates at x_k.

DEFINITION 7-8. Graph G_{lo}. *Graph* G_{lo} is the subgraph of G_l when the source vertex 1 is deleted along with its incident edges. Thus the modified flow graph G_l is essentially the same as G_c (the Coates graph) except for the b_k's that are entered in the graph with the negative signs omitted; that is, G_{lo} is exactly the same as G_{co} in Coates' method.

DEFINITION 7-9. Loop Set L and Its Product P. A *loop set* L is a subgraph of G_{lo} containing all the vertices of G_{lo} with each vertex having one incoming and one outgoing edge. The product P of the gains of all the edges in L is called a *loop-set product.*

DEFINITION 7-10. Graph G_{lk}. *Graph* G_{lk} is a subgraph of G_l such that all the outgoing edges from a *reference vertex* x_k are removed and the source vertex 1 is short-circuited to x_k.

DEFINITION 7-11. 2-loop Set L_2 and Its Product P_2. A *2-loop set* L_2 is a subgraph of G_{lk} containing all the vertices of G_{lk}, with each vertex having one incoming and one outgoing edge. The *product* P_2 of the gains of all the edges in L_2 is called a *2-loop-set product.* Obviously, due to the fact that the graph G_{lo} in Definition 7-8 is exactly identical to that of Definition 7-3, the loop set L of Definition 7-9 is the same as the connection in Definition 7-4.

Now, consider a linear system of n independent equations in n unknowns,

$$\sum_{j=1}^{n} a_{ij} = b_i \qquad (i = 1, 2, \ldots, n)$$

the matrix form of which is given by

$$[a_{ij}]x] = b]$$

where $[a_{ij}]$ is the coefficient matrix, $x]$ is the unknown column matrix, and $b]$ is the constant column matrix. The solution of the variable x_j using Cramer's rule is given by

$$x_j = \frac{1}{\Delta} \sum_{i=1}^{n} \Delta_{ij} b_i$$

where $\Delta = \det\,[a_{ij}]$ is the coefficient determinant, and Δ_{ij} is the cofactor of the element a_{ij} of Δ.

LEMMA 7-3. Each loop-set product p_j corresponds to a term in the expansion of Δ.

Proof: Consider any product term g_j of the determinant Δ which is expanded by rows,

$$g_j = a_{1j_1} a_{2j_2} \ldots a_{nj_n} \qquad (a_{ij_r}, r = 1, 2, \ldots, n)$$

where the first subscript i of a_{ik} indicates the row in which the element a_{ik} is located, and the second subscript k identifies the column that a_{ik} is in. As the expansion of the determinant is carried out by the ascending order of rows, the first subscript i of the elements of the product term g_j is fixed, whereas each of the second subscripts $j_1, j_2, \ldots, j_n$ takes a value from 1 to n, where no two subscripts of j_r are equal. Therefore, g_j represents a possible permutation of the second subscripts j_r (with $r = 1, 2, \ldots, n$) of Δ. The total number of terms in Δ is equal to the number of all possible permutations of the columns of Δ.

Now, in g_j, there are n elements, each of which is in a distinct row and in a distinct column. The elements in row i of Δ represent the incoming edges at vertex x_j in the corresponding graph G_{lo}; the elements in column j of Δ correspond to the outgoing edges of vertex x_j in G_{lo}. Hence, g_j corresponds to a graph that has one incoming and one outgoing edge from every vertex. Moreover, g_j contains elements from each row and each column; that is, the corresponding graph of g_j includes all the vertices of graph G_{lo}. Therefore, g_j corresponds to a loop-set product P_j of graph G_{lo} and the proof is complete.

LEMMA 7-4. The sign of the loop-set product P_j of a loop set L_j of graph G_{lo} is given by $(-1)^{N+N_L}$ where N is the number of edges and N_L is the number of loops in L_j.

Proof: Recall a property of a determinant: If a product term has all of its elements on the principal diagonal of the determinant, then the sign of that product term, sign P, is $+1$.

In a loop set L, all the self-loops correspond to the diagonal elements of Δ, but the elements other than the self-loops correspond to off-diagonal elements in the determinant. Since each loop set has one element in each column, we can interchange any two elements (with respect to either columns or rows) at a time so as to bring all nonzero elements to fall on the principal diagonal. Then sign P can be determined by knowing the total number of interchanges required to place all the off-diagonal nonzero elements on the principal diagonal.

A loop with two edges, a_{ij} and a_{ji}, needs one interchange to put both elements on the principal diagonal, as shown in Figure 7-12. For a loop with

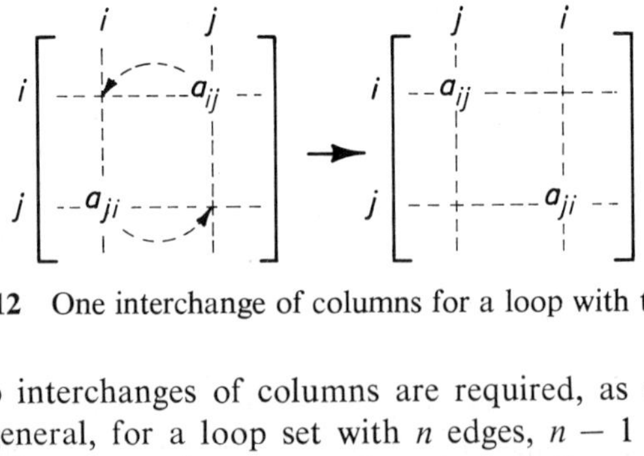

Figure 7-12 One interchange of columns for a loop with two edges.

three edges, two interchanges of columns are required, as shown in Figure 7-13. Thus, in general, for a loop set with n edges, $n - 1$ interchanges are required to put all nonzero elements on the principal diagonal.

Let N be the number of vertices in graph G_{lo} and N_L be the number of loops in a loop set L_j. In a loop set, the number of vertices and the number

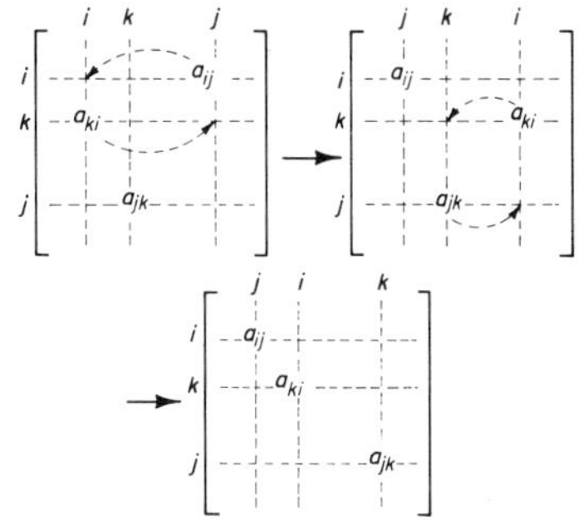

Figure 7-13 Two interchanges of columns for a loop with three edges.

of edges are equal (by definition). Since each loop set must contain all of the vertices of G_{lo}, we write

$$N = \Sigma l_1 + 2\Sigma l_2 + 3\Sigma l_3 + \cdots + n\Sigma l_n \tag{7-23}$$

where l_i is the loop with i vertices in any loop set L_j and $n \leq N$ is the largest number of vertices contained in any of all the loops in L_j. Also, for the same loop set L_j,

$$N_L = \Sigma l_1 + \Sigma l_2 + \Sigma l_3 + \cdots + \Sigma l_n \tag{7-24}$$

But the number of interchanges I required to bring all the elements in the product P_j on the principal diagonal is

$$I = \Sigma l_2 + 2\Sigma l_3 + 3\Sigma l_4 + \cdots + (n-1)\Sigma l_n \tag{7-25}$$

From Equations (7-23) and (7-24), we find that

$$N - N_L = \Sigma l_2 + 2\Sigma l_3 + \cdots + (n-1)\Sigma l_n \tag{7-26}$$

which is the same as Equation (7-25). Consequently, since sign $P_j = I$, we have

$$\text{sign } P_j = (-1)^{N-N_L} = (-1)^{N+N_L}$$

as asserted in the lemma.

LEMMA 7-5. Each product term in the expansion of

$$\sum_{i=1}^{n} \Delta_{ij} b_i$$

corresponds to a 2-loop-set product P_2.

Proof: From Cramer's rule, the foregoing summation is obtained by replacing column j (associated with x_j in the determinant of the coefficient matrix) by the column $\{b_1, b_2, \ldots, b_n\}$, which is the column matrix $b]$ in the matrix equation of the given system, and evaluating the resultant determinant.

Note that the ith row of $[a_{ij}]$ represents the incoming edges at vertex x_j in the corresponding graph, and the jth column of $[a_{ij}]$ represents the outgoing edges from vertex x_j in G_l. Now, the column corresponding to vertex x_j has been replaced by the column $\{b_1, b_2, \ldots, b_n\}$, and hence we remove the outgoing edges from vertex x_j in G_l and short-circuit x_j to the source vertex 1, where the edges $b_1, b_2, \ldots, b_n$ are originating.

In this process, the incoming edges at vertex x_j remain the same (except for the self-loop) and the outgoing ones are those due to (source) vertex 1. It should be remembered that vertex 1 does not have any incoming edges. The resulting subgraph G_{lj} is similar to subgraph G_{lo} as far as the number of vertices, edges, etc., is concerned. Hence, the rest of the proof is similar to the proof of Lemma 7-3.

LEMMA 7-6. The sign of the 2-loop-set product P_{2j} of the jth 2-loop set L_{2j} of graph G_{lk} is given by $(-1)^{N+N_{L_2}}$ where N is the number of edges and N_{L_2} is the number of loops in L_2.

The proof of Lemma 7-6 is identical to that of Lemma 7-4.

We now state the gain formula in the following theorem:

THEOREM 7-3. In a system of n linear equations in n unknowns,

$$\sum_{j=1}^{n} a_{ij}x_j = b_i \qquad (i = 1, 2, \ldots, n)$$

the variable x_j is given by

$$x_j = \frac{\sum_{\text{all } P_2} (-1)^{N_{L_2}} P_2}{\sum_{\text{all } P} (-1)^{N_L} P} \tag{7-27}$$

where N_{L_2} is the number of loops in a 2-loop set L_2 and N_L is the number of loops in a loop set L.

Proof: Each of the two subgraphs, G_{lo} and G_{lj}, has N vertices. Therefore, loop set L and 2-loop set L_2 also have N vertices. The number of vertices is equal to the number of edges in L or L_2. Thus, in light of Lemmas 7-3 through 7-6,

$$x_j = \frac{\sum_{i=1}^{n} \Delta_{ij} b_i}{\Delta} = \frac{\sum_{\text{all } P_2} (-1)^{N+N_{L_2}} P_2}{\sum_{\text{all } P} (-1)^{N+N_L} P} = \frac{(-1)^N \sum_{\text{all } P_2} (-1)^{N_{L_2}} P_2}{(-1)^N \sum_{\text{all } P} (-1)^{N_L} P}$$

or

$$x_j = \frac{\sum_{\text{all } P_2} (-1)^{N_{L_2}} P_2}{\sum_{\text{all } P} (-1)^{N_L} P}$$

where N_{L_2} is the number of loops in a 2-loop set L_2, and N_L is the number of loops in a loop set L.

The next theorem will enable us to determine the number of loop sets and 2-loop sets in a given flow graph.

THEOREM 7-4. The number of loop sets in graph G_{lo} is equal to Δ_1, where

$$\Delta_1 = \sum_{i=1}^{n} a_{1j_1}a_{2j_2}\ldots a_{nj_n}$$

is obtained from Δ by replacing each nonzero element a_{ij} by 1 and defining the sign of each term as $+1$ in the expansion of the resultant "determinant."

Proof: We note that if in Δ all nonzero elements are replaced by 1 and the resulting determinant is expanded, each of the nonzero terms in the expansion of Δ will be equal to either $+1$ or -1. We also note that each nonzero term in the expansion of Δ is the product of N nonzero elements of Δ and is correspondent to a loop-set product P. Hence each loop set must correspond to a $+1$ in the expansion of Δ_1 as defined in the theorem.

EXAMPLE 7-6. For the benefit of comparing results, let us consider again the linear system of Equation (7-4). The modified Coates graph G_l is shown in Figure 7-14(a), and graph G_{lo} is depicted in Figure 7-14(b). The set of all loops is shown in Figure 7-15. The graph G_{l2}, with respect to x_2, is depicted in Figure 7-16. Thus, the set of all 2-loop sets of G_{l2} is as shown in Figure 7-17. Finally, by Equation (7-27) of Theorem 7-3, we get

$$x_2 = \frac{(-1)^2 a_{11}a_{23}b_3 + (-1)^1 a_{31}a_{23}b_1}{(-1)^3 a_{11}a_{22}a_{33} + (-1)^2 a_{13}a_{31}a_{22} + (-1)^1 a_{12}a_{31}a_{23}} \tag{7-28}$$

or

$$x_2 = \frac{a_{23}a_{31}b_1 - a_{11}a_{23}b_3}{a_{11}a_{22}a_{33} - a_{13}a_{22}a_{31} + a_{12}a_{23}a_{31}}$$

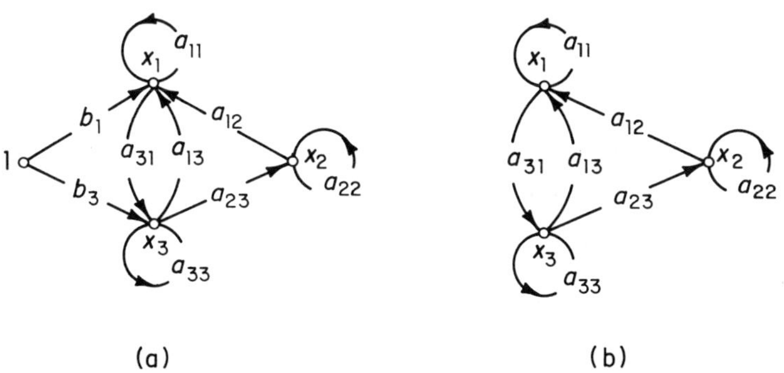

Figure 7-14 (a) The modified Coates graph G_l. (b) Graph G_{lo} of the system of Equation (7-4).

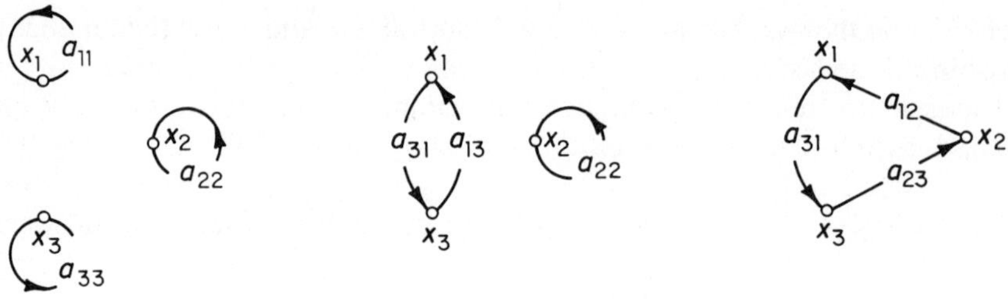

Figure 7-15 The set of all loop sets of G_{lo}.

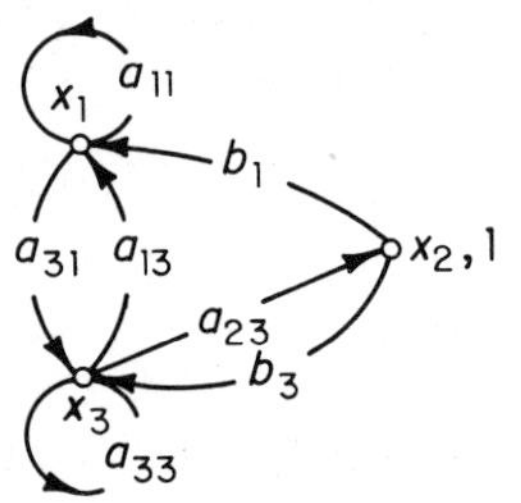

Figure 7-16 Graph G_{l2} with respect to x_2.

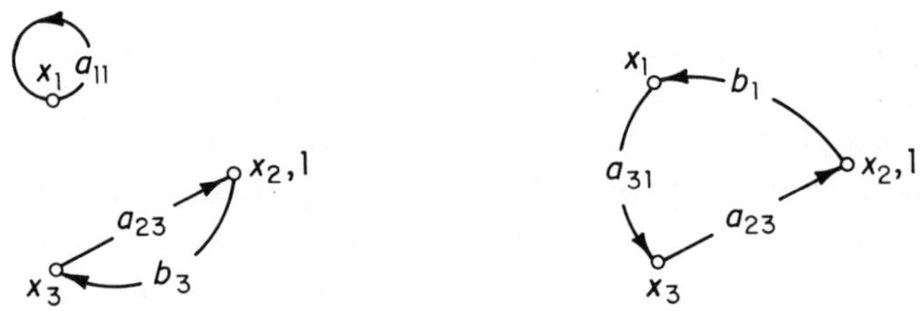

Figure 7-17 The set of all 2-loop sets of G_{l2}.

These equations are identical to Equations (7-6) and (7-22), which were obtained by the Mason and the Coates gain formulas, respectively.

7-5 THE CHAN-MAI GRAPH

As in the preceding section, our entire discussion will be based on a single paper.[3] The new flow graph which we are going to present may be considered a "dual" of the modified Coates graph due to the parallelism among the characteristics of the two graphs. Because of the symmetry which prevails in all the subgraphs involved, the derivation of the gain formula from the newly defined flow graph is the simplest among the four methods presented in this chapter. The number of definitions necessary for the development of the gain formula is also minimal due to the simplicity of this flow graph.

[3]Refer to Chan and Mai [19].

DEFINITION 7-12. Chan-Mai Graph G_s. A *Chan-Mai graph* G_s is an oriented graph representing a system of simultaneous linear algebraic equations

$$\sum_{j=1}^{n} a_{ij}x_j = b_i \qquad (i = 1, 2, \ldots, n)$$

in such a way that each constant b_i is represented by a *constant node* b_i and each variable x_j by a *variable node* x_j; and that if $a_{ij} \neq 0$, there is an oriented edge with a gain a_{ij} directed from node x_j to node b_i. Furthermore, the nodes of G_s are arranged in such a way that all the constant nodes, $b_1, b_2, \ldots, b_n$, are in the top row (with vertex b_j explicitly marked $b_j = 0$ when b_j has a zero value), and all the variable nodes $x_1, x_2, \ldots, x_n$ are in the bottom row, so that nodes with the same subscript are connected by vertical branches.

DEFINITION 7-13. Graph G_{sj}. A *graph* G_{sj} is a graph obtained from G_s by deleting all the outgoing edges of the node x_j and then adding an oriented edge with a gain b_i if $b_i \neq 0$, directed from node x_j to node b_i, where $i = 1, 2, \ldots, n$.

DEFINITION 7-14. Separation S. A *separation* S of a given graph (G_s or G_{sj}) is a subgraph that satisfies the following three requirements:

(a) It contains all the nodes of the graph (G_s or G_{sj}).
(b) Each variable node x_j has exactly one outgoing branch.
(c) Each constant node b_i has exactly one incoming branch (Figure 7-18).

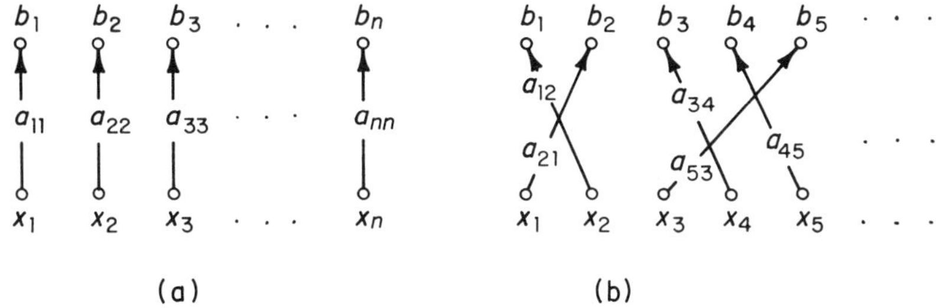

Figure 7-18 (a) Vertical branches of a separation. (b) Nonvertical branches of a separation.

DEFINITION 7-15. Separation Product S_p. A *separation product* S_p is the product of the gains of all the branches in S. A separation product of G_s is denoted by $S_p(G_s)$ and that of G_{sj} by $S_p(G_{sj})$.

DEFINITION 7-16. Symmetric Pair. A *symmetric pair* is a pair of branches in a separation S, one branch of which is from node x_i to node b_j and the other branch from node x_j to b_i for any $i \neq j$ [Figure 7-19(a)].

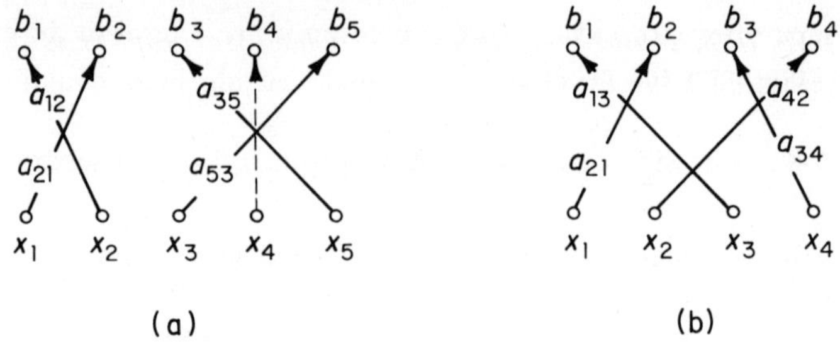

Figure 7-19 (a) Symmetric pairs in a separation. (b) Asymmetric branches in a separation.

DEFINITION 7-17. Asymmetric Branch. An *asymmetric branch* is a branch in a separation S that does not belong to any symmetric pair within S. [Figure 7-19(b)].

LEMMA 7-7. With the same notation as in Lemma 7-1, each term g_j in the expansion of Δ corresponds to a separation product S_{p_j} of G_s and vice versa.

Proof: Note that any product-term g_j of Δ, when expanded by rows, has the form

$$g_j = a_{1j_1}a_{2j_2} \ldots a_{nj_n} \qquad (a_{ij_r},\ i = 1, 2, \ldots, n;\ j_1, j_2, \ldots, j_n \text{ is a permutation of the set } r = 1, 2, \ldots, n)$$

where the first subscript i of a_{ij} identifies the row, and the second subscript j indicates the column in which the element a_{ij} is located. As the expansion of the determinant is carried out in the ascending order of rows, the first subscript i of the elements of the product term g_j is fixed; each of the second subscripts $j_1, j_2, \ldots, j_n$ takes a value from 1 to n, where no two subscripts of j_r are equal. Therefore, g_j represents a possible permutation of the second subscripts j_r, (with $r = 1, 2, \ldots, n$) of g_j. The total number of terms in Δ is equal to the number of all possible permutations of the columns of $[a_{ij}]$.

In g_j there are n elements, each of which is in a distinct row and in a distinct column. The elements in row i of $[a_{ij}]$ represent the incoming branches at node b_i, whereas the elements in column j correspond to the outgoing branches at node x_j of the flow graph G_s. Hence, g_j corresponds to a subgraph of G_s in which every constant node b_i has exactly one incoming branch and every variable node x_j has exactly one outgoing branch. Furthermore, g_j contains elements from each row and each column of $[a_{ij}]$. Therefore, the corresponding graph of g_j includes all the constant and variable nodes of G_s and, hence, satisfies all three requirements in the definition of a separation. Hence, g_j corresponds to a separation product S_{p_j} of the flow graph G_s. Finally, by application of the same argument in the reverse order, it is seen that each separation product S_{p_j} of G_s corresponds to a term in the expansion of Δ. This completes the proof.

LEMMA 7-8. The sign of the separation product S_{pj} of a separation S_j of G_s is given by

$$\begin{aligned} \text{sign } S_{p_j} &= (-1)^{n_s+n_a-1} && \text{for } n_a \neq 0 \\ &= (-1)^{n_s} && \text{for } n_a = 0 \end{aligned}$$

where n_s is the number of symmetric pairs and n_a is the number of asymmetric branches in S_j.

Proof: Observe that in a separation S_j all the vertical edges correspond to the diagonal elements of Δ, but all the symmetric pairs and asymmetric branches correspond to the off-diagonal elements. If there are no asymmetric branches in S_j, $n_a = 0$, the sign of S_{p_j} is then determined by the number of symmetric pairs. Since each pair, a_{ij} and a_{ji}, needs one interchange of variable nodes (or constant nodes) to put them both in the vertical position, it therefore requires n_s interchanges of variable nodes to put all the edges in n_s symmetric pairs in the vertical position; that is, the sign of S_{p_j} is $(-1)^{n_s}$ (see Figure 7-20).

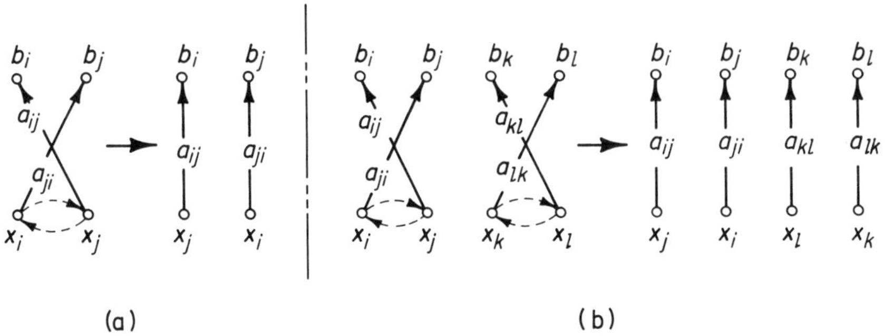

Figure 7-20 (a) One interchange of variables nodes. (b) Two interchanges of variable nodes.

If there are asymmetric branches in S_j, the number of such branches must be greater than or equal to 3 (that is, $n_a \geqq 3$), which is almost self-evident. For $n_3 = 3$, it requires two interchanges of variable nodes to place all of them in the vertical position (Figure 7-21); for $n_a = 4$, three interchanges (Figure 7-22); for $n_a = 5$, four interchanges; and so on. Thus, in general, for an S_j with n_a asymmetric branches, $n_a - 1$ interchanges of variable nodes (or constant nodes) are necessary to put all of them in the vertical position. This implies that the sign of S_{p_j} is $(-1)^{n_a-1}$. Together, with the possible presence of n_s symmetric pairs in the same S_j, the sign of S_{p_j} is therefore given by $(-1)^{n_s+n_a-1}$. Hence the lemma.

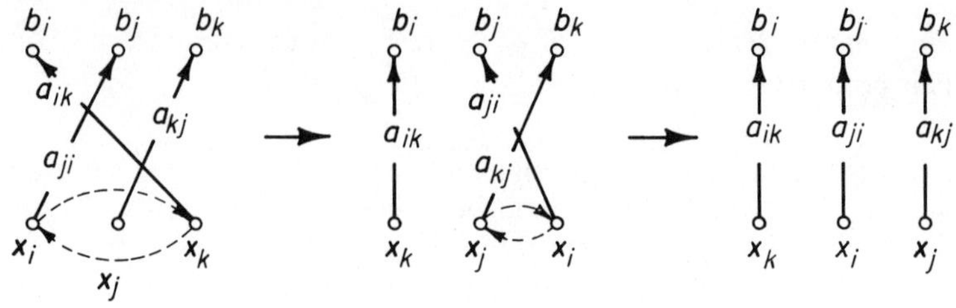

Figure 7-21 Two interchanges of variable nodes for $n_a = 3$.

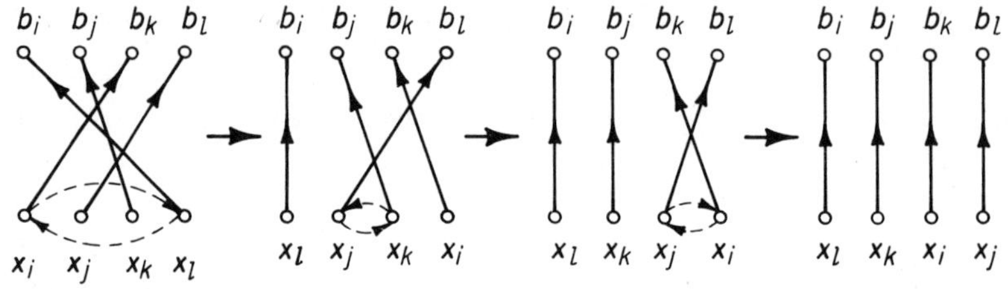

Figure 7-22 Three interchanges of variables nodes for $n_a = 4$.

LEMMA 7-9. The set of all the product-terms of Δ_j is equal to the set of all the separation products of the corresponding graph G_{sj}. The sign of $S_{p_k}(G_{sj})$ is equal to $(-1)^{n_s+n_a-1}$ for $n_a \neq 0$, and is equal to $(-1)^{n_s}$ for $n_a = 0$, where n_a is the number of asymmetric branches and n_s is the number of symmetric pairs of S_{p_k}.

Proof: Recall that Δ_j is the determinant obtained from Δ by replacing the jth column of Δ with elements of the constant column matrix. But the elements in the jth column of Δ correspond to all outgoing branches from the variable node x_j. Thus, replacing the jth column of Δ with the constant column in forming Δ_j corresponds to deleting all the outgoing branches of node x_j in G_s and adding the oriented branches with gains b_i from node x_j to node b_i for all $b_i \neq 0$ ($i = 1, 2, \ldots, n$) in obtaining G_{sj}. This completes the proof.

Finally, combining Lemmas 7-7, 7-8, and 7-9, we have the gain formula as stated in the following theorem:

THEOREM 7-5. In a linear system of n simultaneous independent algebraic equations in n unknowns,

$$\sum_{j=1}^{n} a_{ij}x_j = b_i \qquad (i = 1, 2, \ldots, n)$$

the value of the variable x_j is given by

$$x_j = \frac{\sum_{\tau} (\text{sign } S_{p_\tau}) S_{p_\tau}(G_{sj})}{\sum_{\rho} (\text{sign } S_{p_\rho}) S_{p_\rho}(G_s)} \tag{7-29}$$

where

$$\begin{aligned} \text{sign } S_{p_k} &= (-1)^{n_{sk}+n_{ak}-1} \qquad \text{for } n_{a_k} \neq 0 \quad (k = \tau \text{ or } \rho) \\ &= (-1)^{n_{sk}} \qquad \text{for } n_{a_k} = 0 \end{aligned}$$

Here, n_{a_τ} is the number of asymmetric branches in the τth separation of the graph G_{sj}; n_{s_τ}, the number of symmetric pairs in the τth separation of G_{sj}; $S_{p_\tau}(G_{sj})$, the separation product of the τth S of G_{sj}; n_{a_ρ}, the number of asymmetric branches in the ρth separation of the flow graph G_s; n_{s_ρ}, the number of symmetric pairs in the ρth S of G_s; and $S_{p_\rho}(G_s)$, the separation product of the ρth S of G_s. The summations of the separation products, $S_{p\rho}(G_s)$ and $S_{p\tau}(G_{sj})$, are taken over all separations of G_s and G_{sj}, respectively.

EXAMPLE 7-7. Once again, consider the system of Equation (7-4). The Chan-Mai graph G_s is depicted in Figure 7-23. The set of all separations of G_s is shown in Figure 7-24.

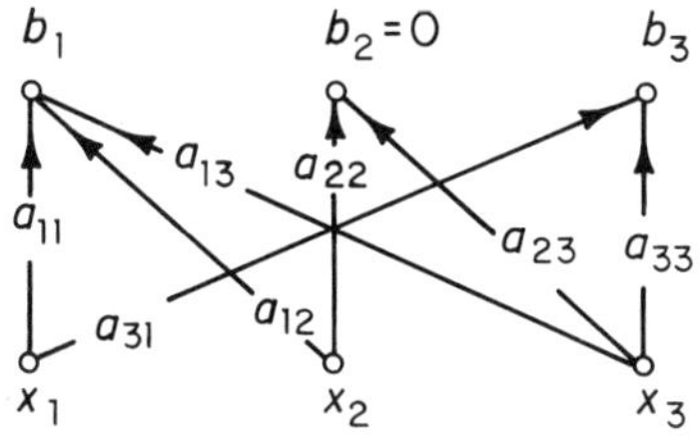

Figure 7-23 The Chan-Mai graph G_s of the system of Equation (7-4).

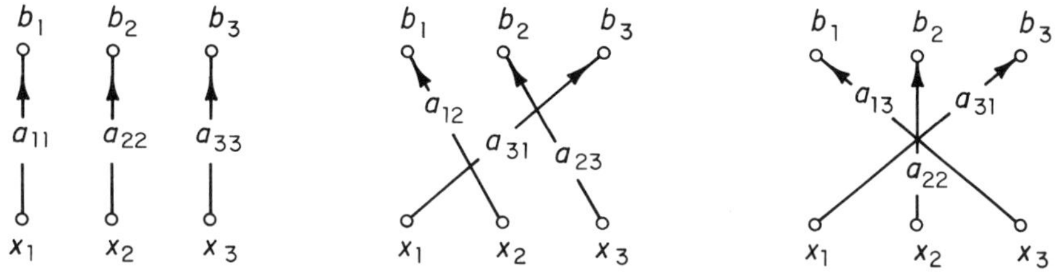

Figure 7-24 The set of all the separations of G_s.

We shall compute the value of x_2 by the Chan-Mai gain formula [Equation (7-29)]. First, we delete in G_s all the outgoing edges from node x_2 and add one edge with a gain b_1 from node x_2 to node b_1, and another edge with a gain b_3 from x_2 to node b_3. The resulting graph G_{s2} and the set of all separations of G_{s2} are shown in Figures 7-25 and 7-26, respectively.

By Theorem 7-5, we find that

$$\begin{aligned} x_2 = \frac{\Delta_2}{\Delta} &= \frac{(-1)^{1+1-1}a_{11}a_{23}b_3 + (-1)^{0+3-1}a_{23}a_{31}b_1}{(-1)^{0+3-1}a_{11}a_{22}a_{33} + (-1)^{0+3-1}a_{12}a_{23}a_{31} + (-1)^{1+1-1}a_{13}a_{22}a_{31}} \\ &= \frac{a_{23}a_{31}b_1 - a_{11}a_{23}b_3}{a_{11}a_{22}a_{33} + a_{12}a_{23}a_{31} - a_{13}a_{22}a_{31}} \end{aligned} \tag{7-30}$$

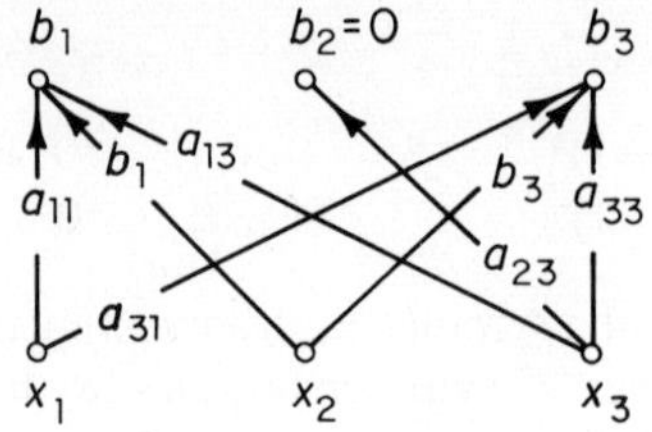

Figure 7-25 The graph G_{s2}.

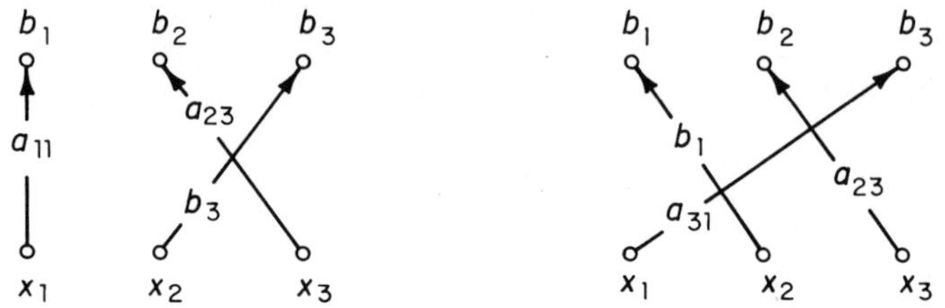

Figure 7-26 The set of all the separations of G_{s2}.

which again is the same as Equations (7-6), (7-22) and (7-28), obtained by the preceding three gain formulas (Examples 7-1, 7-5, and 7-6).

7-6 GRAPH TRANSFORMATIONS AND GRAPH REDUCTION

In this section we shall first establish the interrelationships between flow graphs by means of graph transformations and then discuss very briefly some of the elementary rules that can be employed in reducing a considerably complicated flow graph to a simple one, thus obtaining the gain of the system under study.

The graph-reduction techniques to be presented here apply to a Mason graph. If the given flow graph is not a Mason graph, it can be easily transformed into a Mason equivalent since all the four different flow graphs discussed in the preceding sections are derived from the same system of equations (illustrated in Examples 7-1, 7-6, and 7-7); consequently, there obviously should exist an intimate relationship between any two of them. To illustrate this important point, let us, for example, show how a Mason graph can be transformed into a modified Coates equivalent, and vice versa, leaving other graph transformations as exercises (Problems 7-29 through 7-31).

THEOREM 7-6. A Mason graph G_m can be transformed into a modified Coates equivalent G_l by the following procedure:

(a) Subtract 1 from the gain of each existing self-loop.

(b) Add a self-loop with gain -1 to each node devoid of a self-loop.

(c) Multiply by $-b_k$ the gain of the branch at the kth source node b_k ($k = 1, 2, \ldots, s$, where s is the number of source nodes) and then combine all the (s) source nodes into one source node (now denoted by 1).

THEOREM 7-7. A modified Coates graph G_l can be transformed into a Mason equivalent G_m by the following procedure:

(a) Add 1 to the gain of each existing self-loop.

(b) Add a self-loop with gain 1 to each node devoid of a self-loop except the source node 1.

(c) Break the source node 1 into s source nodes (s being the degree of the source node); identify the new s source nodes by $b_1, b_2, \ldots, b_s$ with the gain of the corresponding s branches multiplied by

$$-\frac{1}{b_1}, -\frac{1}{b_2}, \ldots, -\frac{1}{b_s}$$

respectively, so that the new gains of these branches are all equal to $+1$, keeping the branch orientations unchanged.

Theorems 7-6 and 7-7 are self-evident if we carefully compare the definitions of the Mason and the modified Coates graphs.

EXAMPLE 7-8. Let us consider the Mason graph G_m as shown in Figure 7-2 and obtain the modified Coates equivalent G_l of this graph using the procedure stated in Theorem 7-6. The resultant modified Coates equivalent G_l is shown in Figure 7-27.

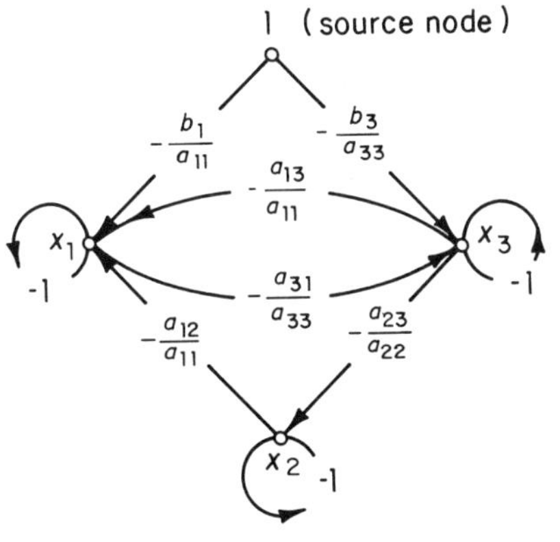

Figure 7-27 The modified Coates equivalent G_l of the Mason graph G_m in Example 7-1.

To verify that this graph is truly the Coates equivalent, we shall write the system of equations directly from this resultant graph and then prove that the system of equations so obtained is indeed the same as that given in Example 7-1 — that is, the system of Equation (7-4). Thus, recalling how a modified

Coates graph is formulated from a given system of equations (Definition 7-7), we write from the graph shown in Figure 7-27,

$$\text{For node } x_1: \quad -x_1 - \frac{a_{12}}{a_{11}} x_2 - \frac{a_{13}}{a_{11}} x_3 = \frac{-b_1}{a_{11}} \tag{7-31a}$$

$$\text{For node } x_2: \quad -x_2 - \frac{a_{23}}{a_{22}} x_3 = 0 \tag{7-31b}$$

$$\text{For node } x_3: \quad -\frac{a_{31}}{a_{33}} x_1 \qquad -x_3 = \frac{-b_3}{a_{33}} \tag{7-31c}$$

Next, multiplying Equation (7-31a) by $-a_{11}$, (7-31b) by $-a_{22}$, (7-31c) by $-a_{33}$, and rearranging the terms, we get

$$a_{11}x_1 + a_{12}x_2 + a_{13}x_3 = b_1 \tag{7-32a}$$

$$a_{22}x_2 + a_{23}x_3 = 0 \tag{7-32b}$$

$$a_{31}x_1 \qquad + a_{33}x_3 = b_3 \tag{7-32c}$$

which is identical to Equation (7-4), as expected.

EXAMPLE 7-9. Consider the modified Coates graph G_l shown in Figure 7-14(a). We now follow the procedure stated in Theorem 7-7 and obtain the Mason equivalent of this graph. The resultant graph G_m is depicted in Figure 7-28. Now we write the system of equations directly from this resultant graph:

$$x_1 = (a_{11} + 1)x_1 + a_{12}x_2 + a_{13}x_3 - b_1 \tag{7-33a}$$

$$x_2 = (a_{22} + 1)x_2 + a_{23}x_3 \tag{7-33b}$$

$$x_3 = a_{31}x_1 + (a_{33} + 1)x_3 - b_3 \tag{7-33c}$$

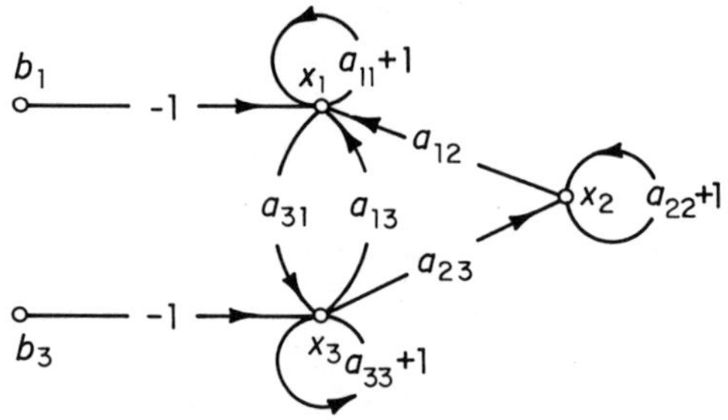

Figure 7-28 The Mason equivalent G_m of the modified Coates graph G_l in Example 7-6.

Finally, simplifying the three parts of Equation (7-33) and rearranging the terms, we obtain the following system:

$$a_{11}x_1 + a_{12}x_2 + a_{13}x_3 = b_1 \tag{7-34a}$$

$$a_{22}x_2 + a_{23}x_3 = 0 \tag{7-34b}$$

$$a_{31}x_1 \qquad + a_{33}x_3 = b_3 \tag{7-34c}$$

which is seen to be identical to the system of Equation (7-4) given in Example 7-6. With the idea of the transformation from one flow graph to another (between any two of the four types discussed), we shall next introduce, only very briefly, the concept of the reduction of a flow graph. The graph-reduction techniques to be presented here are based on the Mason graph and, hence, are *valid for the Mason graph only*.

Four basic operations can be performed to reduce a Mason graph to a simpler equivalent graph: (a) the combination of cascading branches into a single branch; (b) the combination of parallel branches into a single branch; (c) the elimination of a node; and, (d) the elimination of a self-loop. These four operations are summarized in Table 7-1.

TABLE 7-1

Four Basic Operations of Graph Reduction

	Description of operation	Original graph	Equivalent graph after operation
(a)	Combining cascading branches	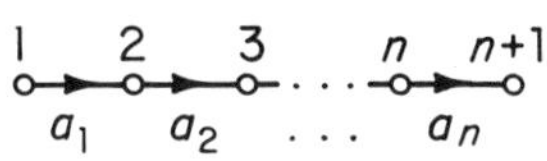	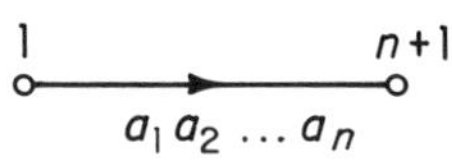
(b)	Combining parallel branches	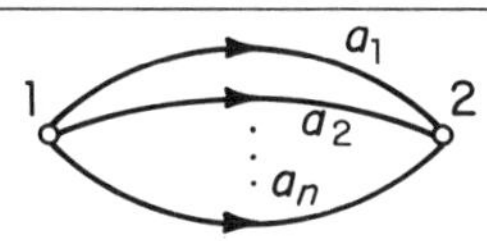	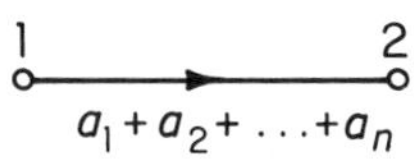
(c)	Eliminating a node	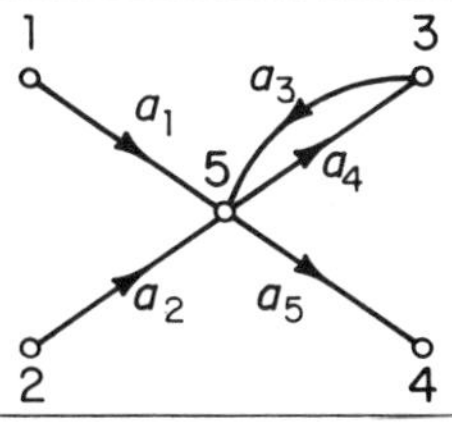	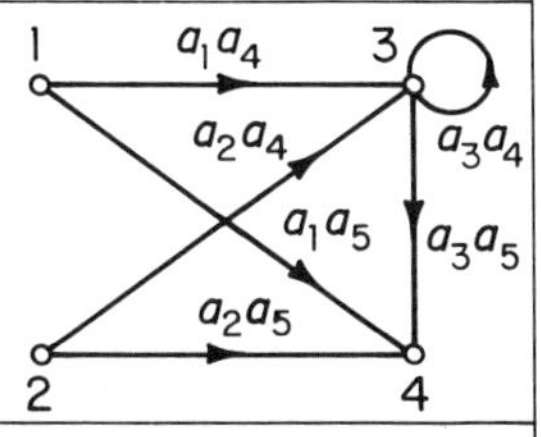
(d)	Eliminating a self-loop	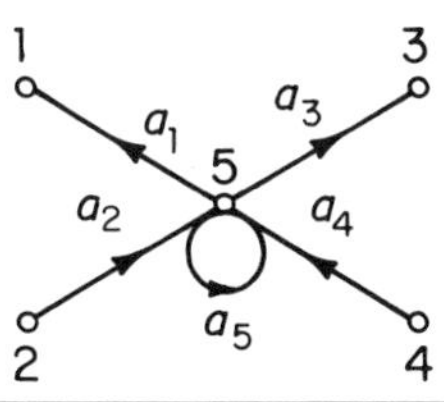	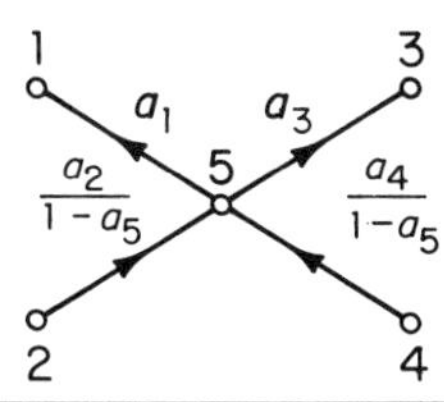

The four operations described in Table 7-1 can be proved most readily by simply writing the corresponding equations. This will be left as a problem (see Problem 7-6).

It should also be pointed out that these four operations may be combined to produce new rules for the reduction of flow graphs. The following examples will serve to illustrate the ideas introduced.

EXAMPLE 7-10. Consider the graph shown in Figure 7-28, which is the Mason equivalent of the modified Coates graph in Example 7-6. We shall apply the four operations described in Table 7-1 to reduce this Mason equivalent to the simplest form and then obtain the gain of the system for x_2. The

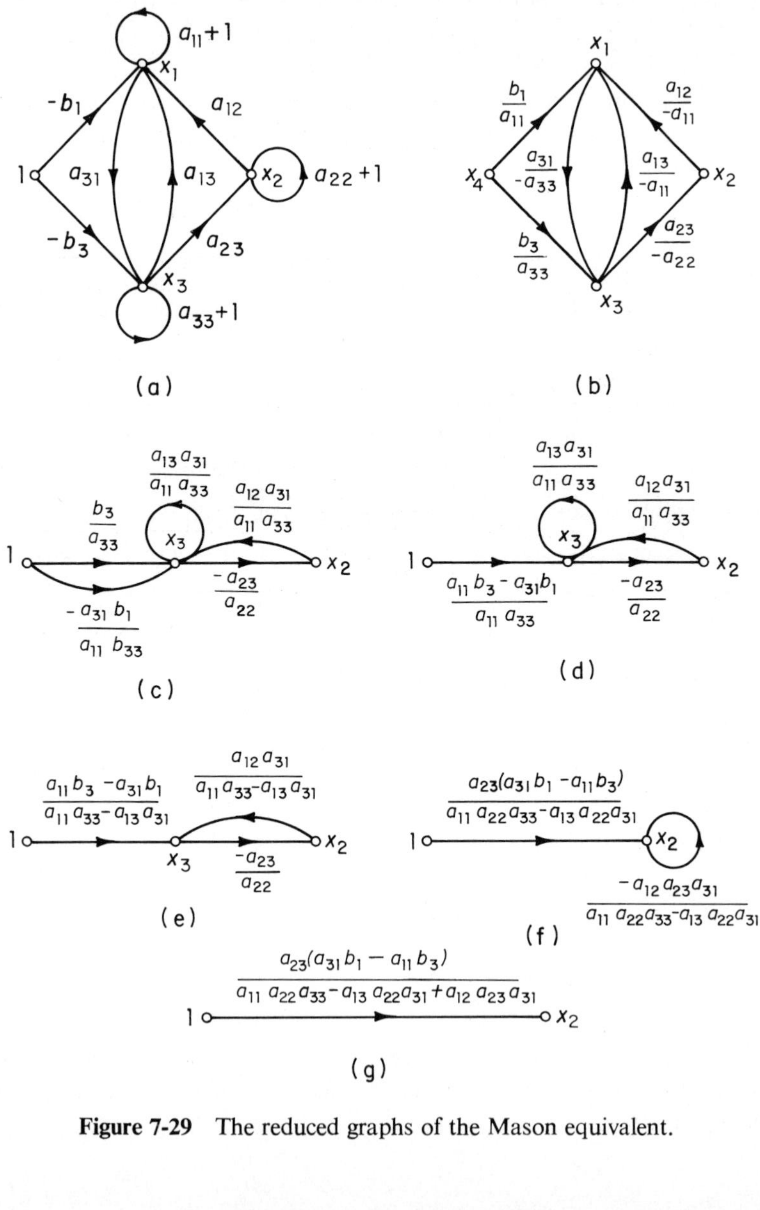

Figure 7-29 The reduced graphs of the Mason equivalent.

reduced graphs are shown in Figure 7-29 and the steps are listed as follows:

(a) Combination of the two source nodes b_1 and b_3 into one (denoted by 1), on the basis of the principle of superposition, with the gains of the two branches now equal to $-b_1$ and $-b_3$.
(b) Elimination of the self-loops.
(c) Elimination of node x_1.
(d) Combination of the two parallel branches between nodes 1 and x_3.
(e) Elimination of the self-loop at node x_3.
(f) Elimination of node x_3.
(g) Elimination of the self-loop at node x_2.

We note, in this example, that step (a) is performed so that both source nodes, b_1 and b_3, can be treated simultaneously. This step is obviously valid from the principle of superposition. The graph shown in Figure 7-29(g) is in the simplest form since, from this graph, we write immediately

$$x_2 = \frac{a_{23}(a_{31}b_1 - a_{11}b_3)}{a_{11}a_{22}a_{33} - a_{13}a_{22}a_{31} + a_{12}a_{23}a_{31}} \tag{7-35}$$

which is the desired result, and is, of course, identical to Equations (7-6), (7-22), (7-28), and (7-30), as expected.

Finally, we suggest that the reader go through the operations step by step so that he will appreciate more fully the applications of the rules involved.

7-7 FORMULATION OF THE FLOW GRAPH DIRECTLY FROM A LINEAR NETWORK

In this section we shall show how a flow graph can be obtained directly from a given linear network. Since any of the four flow graphs discussed in the preceding sections can be transformed into one of the other three (Problems 7-29 through 7-32), we shall formulate the algorithm for obtaining the Chan-Mai graph from a given linear network, leaving the formulation of the algorithm for obtaining the other three flow graphs as exercises (Problems 7-33 and 7-34).

Based on the theory developed in Section 7-5, an algorithm for obtaining the flow graph (either G_s or G_{sj}) from a linear network can be readily formulated. The procedures are stated in terms of loop analysis and node analysis as follows:

(A) Loop Analysis

(a) Graph G_s. The flow graph G_s is obtained from a given linear network N in the following steps:

(1) Label the loop currents in N and assign the same direction (either

clockwise or counterclockwise) to each of the loop currents, $I_1, I_2, \ldots, I_n$, where n is the total number of independent loops in N.

(2) Place the constant nodes $b_1, b_2, \ldots, b_n$ on the top row and the variable nodes $I_1, I_2, \ldots, I_n$ on the bottom row so that node I_j is vertically under node b_j $(j = 1, 2, \ldots, n)$.

(3) Place a directed branch from node I_j to node b_j $(i = 1, 2, \ldots, n)$ with the branch gain equal to the sum of all the impedances around loop j.

(4) For each branch between loops i and j $(i \neq j)$ in N, there is a symmetric pair between nodes i and j in G, each branch of the pair having the branch gain equal to the negative of the sum of all the impedances common to loops i and j in N.

The resulting graph is the desired graph G_s in terms of loop analysis.

(b) Graph G_{sj}. The graph G_{sj} is obtained from N directly by the modification of graph G_s, as stated in the following steps:

(1) Delete from G_s all the outgoing branches at node I_j.

(2) If there are voltage sources in loop i $(i = 1, 2, \ldots, n)$ in N, there is a directed branch in G_{sj} from node I_j to node b_i with the branch gain equal to the algebraic sum of all the voltage sources around i in N [a positive (or negative) sign is associated with a voltage source if it is a voltage rise (or drop) in the direction of the loop current I_i].

The resulting graph is the desired graph G_{sj} in terms of loop analysis.

The justification of the steps in the foregoing procedure is obvious from the fact that there exists a one-to-one correspondence between the positions of the elements in the determinant Δ of the loop impedance matrix and the edges of the flow graph G_s. Similar observations are noted in the case of G_{sj}, which corresponds to the determinant Δ_j resulting from Δ by replacing the elements in the jth column by the elements in the voltage-source vector. The following example serves to illustrate the ideas involved in the procedure.

EXAMPLE 7-11. Consider the bridged-T network N with a resistive load R_5 and a voltage source $E(s)$ as shown in Figure 7-30. The graph G_s corresponding to N is obtained by following the four steps stated in (a) of the algorithm for the loop analysis as shown in Figure 7-31.

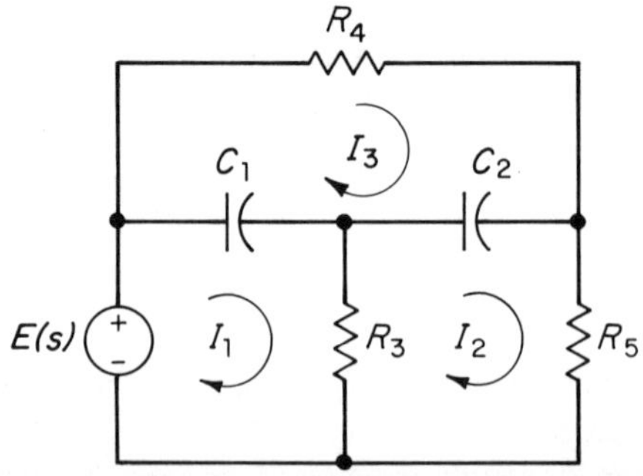

Figure 7-30 The network N of Example 7-11.

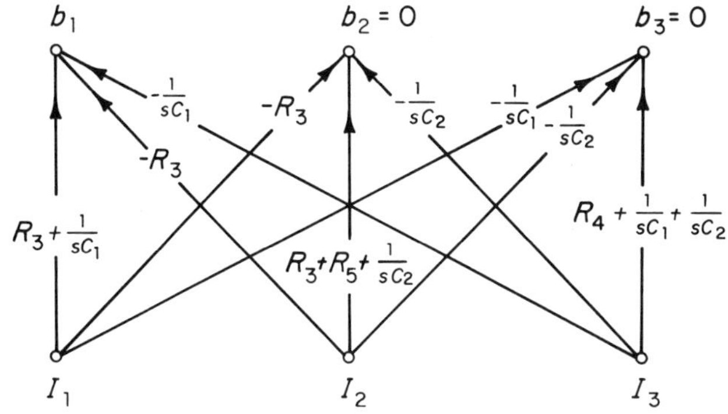

Figure 7-31 Graph G_s corresponding to N of Figure 7-30.

Suppose that it is required to determine the loop current I_2 in N. Following the two steps in (b), graph G_{s2} is obtained as shown in Figure 7-32. Note that, in this example, E is the only voltage source that is in loop 1 ($i = 1$). Thus, to obtain G_{s2} ($i = 2$), all the outgoing edges at node I_2 of G_s are deleted [step (1) of (b)] and the directed edge from node I_2 to node b_1 with edge gain E is placed between the two nodes [step (2) of (b)], as illustrated in Figure 7-32.

The sets of all separations of G_s and of G_{s2} are shown in Figures 7-33 and 7-34, respectively. Finally, by Theorem 7-5,

$$I_2 = \frac{\Delta_2}{\Delta} = \frac{(-1)^{1+1-1}(-R_3)\left(R_4 + \frac{1}{sC_1} + \frac{1}{sC_2}\right)E + (-1)^{0+3-1}\left(-\frac{1}{sC_1}\right)\left(-\frac{1}{sC_2}\right)E}{\begin{aligned}\Big\{&(-1)^{0+3-1}\left(R_3 + \frac{1}{sC_1}\right)\left(R_3 + R_5 + \frac{1}{sC_2}\right)\left(R_4 + \frac{1}{sC_1} + \frac{1}{sC_2}\right) \\ &+ (-1)^{1+1-1}\left(R_3 + \frac{1}{sC_1}\right)\left(-\frac{1}{sC_2}\right)\left(-\frac{1}{sC_2}\right) \\ &+ (-1)^{0+3-1}(-R_3)\left(-\frac{1}{sC_1}\right)\left(-\frac{1}{sC_2}\right) \\ &+ (-1)^{1+1-1}(-R_3)\left(R_4 + \frac{1}{sC_1} + \frac{1}{sC_2}\right)(-R_3) \\ &+ (-1)^{0+3-1}\left(-\frac{1}{sC_1}\right)(-R_3)\left(-\frac{1}{sC_2}\right) \\ &+ (-1)^{1+1-1}\left(-\frac{1}{sC_1}\right)\left(R_3 + R_5 + \frac{1}{sC_2}\right)\left(-\frac{1}{sC_1}\right)\Big\}\end{aligned}}$$

which, after simplification, becomes

$$I_2 = \frac{\{C_1C_2R_3R_4s^2 + (C_1 + C_2)R_3s + 1\}E}{{}_1C_2CR_3R_4R_5s^2 + \{(C_1 + C_2)R_3(R_4 + R_5) + C_2R_4R_5\}s + (R_4 + R_5)}$$

(B) Node Analysis

(a) Graph G_s. The flow graph G_s is obtained from a given linear network N in the following steps:

(1) Choose a reference node in N and label the other nodes $V_1, V_2, \ldots, V_n$, where $n + 1$ is the total number of nodes in N.

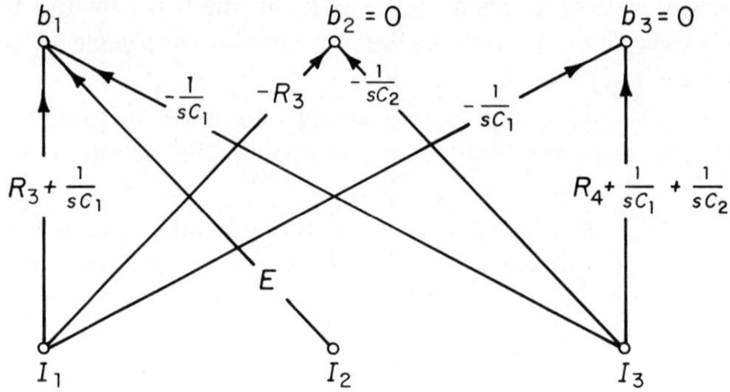

Figure 7-32 Graph G_{s2} for determining I_2 in N of Figure 7-27.

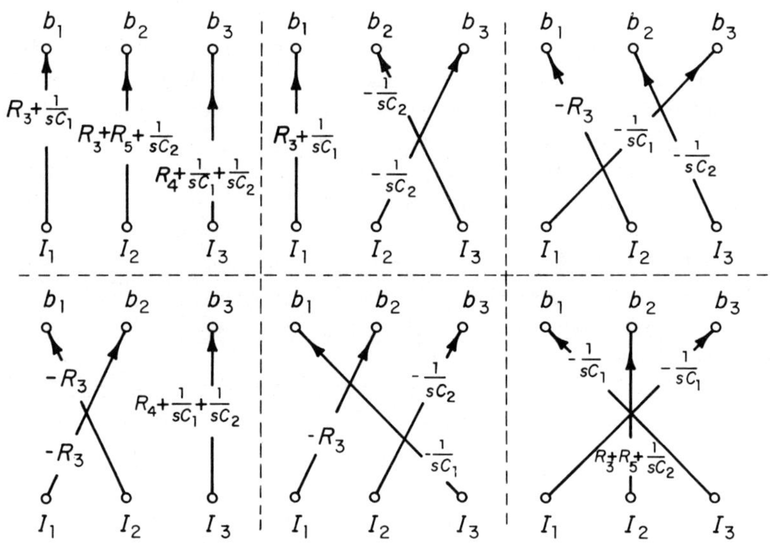

Figure 7-33 The set of all separations of G_s of Figure 7-28.

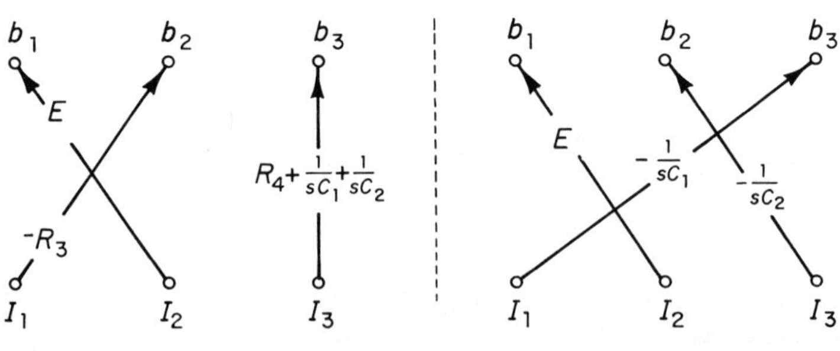

Figure 7-34 The set of all separations of G_{s2} of Figure 7-32.

(2) Place the constant nodes $b_1, b_2, \ldots, b_n$ on the top row and the variable nodes $V_1, V_2, \ldots, V_n$ on the bottom row so that node V_j is vertically under node b_j ($j = 1, 2, \ldots, n$).

(3) Place a directed branch from node V_j to node b_j ($j = 1, 2, \ldots, n$) with the branch gain equal to the sum of all the admittances connected to node V_j.

(4) For each branch connected between nodes V_i and V_j ($i \neq j$) in N, there is a symmetric pair between nodes V_i and V_j in G_s, each branch of the pair having the branch gain equal to the negative of the sum of all the admittances connected between nodes V_i and V_j in N.

The resulting graph is the desired graph G_s in terms of node analysis.

(b) Graph G_{sj}. The graph G_{sj} is obtained from N directly by the modification of G_s as stated in the following steps:

(1) Delete from G_s all the outgoing edges at node V_j.

(2) If there are current sources connected to node i ($i = 1, 2, \ldots, n$) in N, there is a directed branch in G_{sj} from node V_j to node b_i with the branch gain equal to the sum of all the current sources connected to node V_j in N. [A positive (negative) sign is associated with a current source if it is toward (away from) node V_j.]

The resulting graph is the desired graph G_{sj} in terms of node analysis.

It is unnecessary to remark on the justification of the steps stated for the node analysis as the parallelism that is evident there between procedures (A) and (B) can be readily compared to the parallelism between the developments of the loop analysis and the node analysis of a linear network.

The following example will illustrate the application of procedure (B).

EXAMPLE 7-12. Consider the *RC* ladder network N with the load G_5 and the current source I shown in Figure 7-35. Suppose we are interested in the

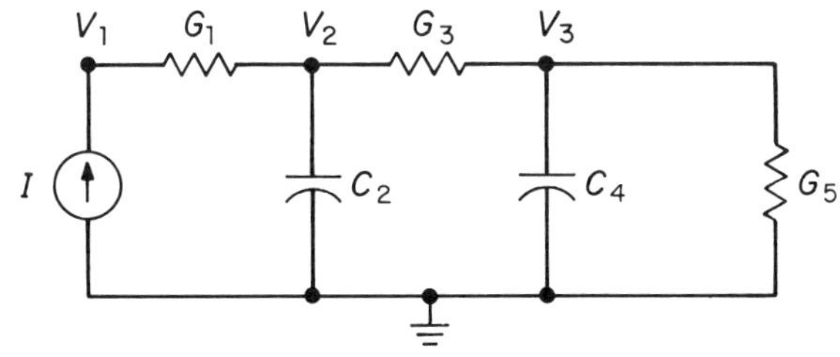

Figure 7-35 The network N of Example 7-12.

output voltage V_3 across the load G_5. The two graphs necessary for the node analysis are G_s and G_{s3}, which are obtained directly from N following the steps in procedure (B). These two graphs, G_s and G_{s3}, are shown in Figures 7-36 and 7-37, respectively.

The remaining task of this analysis problem is merely to obtain the sets of separations of G_s and of G_{s3} and, finally, to evaluate V_3 by the application of

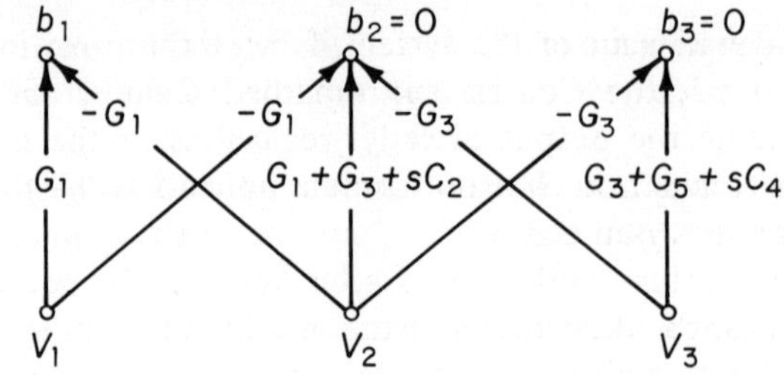

Figure 7-36 Graph G_s corresponding to N of Figure 7-35.

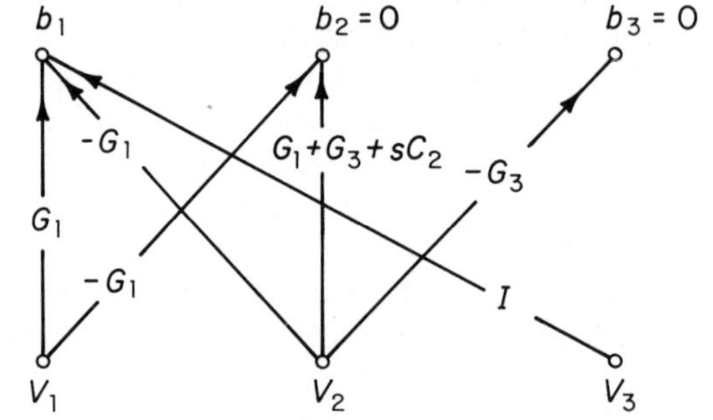

Figure 7-37 Graph G_{s3} for determining V_3 in N of Figure 7-35.

Theorem 7-5. The details are not given here since the steps will be straightforward and parallel to those illustrated in Example 7-11 [once the graphs G_s and G_{sj} are obtained from the network].

7-8 SUMMARY

In this chapter, we have presented four flow-graph gain formulas and introduced the graph-reduction technique. The gain formulas are direct methods for the determination of the value of a variable, whereas the reduction technique, as the name implies, is a step-by-step approach to the answer. Our emphasis has been clearly on the gain formula since the latter is included in most control systems texts.

The Mason and the Coates formulas are the classical ones in the theory of flow graphs and are known to most readers who have interest in flow-graph techniques; the modified Coates and the Chan-Mai formulas, on the other hand, are the most recent developments in the area. Inasmuch as the Coates graph and the modified Coates graph are essentially the same except for the derivations of the gain formulas,[4] we have chosen to emphasize the latter.

Since the Mason formula computes the ratio of a specified output over *one* particular input, the principle of superposition must be employed in order

[4]The interested reader is referred to Chan and Chan [18].

to obtain the overall gain of the system if more than one input is involved. On the other hand, the Coates, the modified Coates, and the Chan-Mai formulas determine the output directly, regardless of the number of inputs present. In this respect, the Mason formula appears to be the least powerful of the four methods discussed.

As mentioned earlier, and also as seen later in the developments in the preceding sections, the derivations of the modified Coates and the Chan-Mai formulas are simple and easy to follow, whereas the two classical ones — especially the Mason formula — are difficult to derive. Other formulas are also possible and the interested reader is encouraged to make an attempt to see whether still better formulas may be derived.

However, since the development of the Mason graph and its gain formula is based on the cause-effect relationships (Figure 7-1) between the inputs and outputs of the system under study, the method provides an excellent physical insight of the system; and, more important, it allows one to apply the graph-reduction technique to the graph as no other flow-graph methods can do. It is because of this significant advantage of Mason's that the method still enjoys its popularity, especially in control systems, regardless of the complexity involved in the derivation of its gain formula.

As in the case of the k-tree topological approach, computer programs are also available for the calculation of network functions, using flow-graph gain formulas;[5] but, because there exist cancellations in the terms of the gain expressions, the flow-graph techniques are not "minimum effort" ones as are the k-tree formulas.[6]

In conclusion, one more example will be included to give a close comparison of the application of the four gain formulas so that the reader can better appreciate the similarities and differences, as well as the advantages and disadvantages, of these methods.

EXAMPLE 7-13. Given the block diagram of a feedback control system with tachometer feedback, as shown in Figure 7-38, determine the closed-loop transfer function.

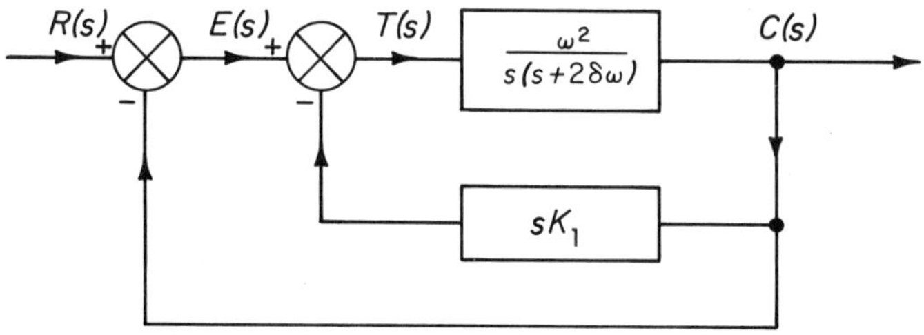

Figure 7-38 Block diagram with tachometer feedback.

[5]See W. W. Happ [40, 41]; Carpenter and Happ [13]; and Dunn and Chan [28].
[6]Refer to, for example, Equation (42) of Seshu and Balabanian [75].

The equations describing this system are:

$$E(s) = R(s) - C(s)$$
$$T(s) = E(s) - sk_1C(s)$$
$$C(s) = \frac{\omega^2}{s(s + 2\delta\omega)}T(s)$$

(1) Mason's Method

The Mason graph is shown in Figure 7-39. We have

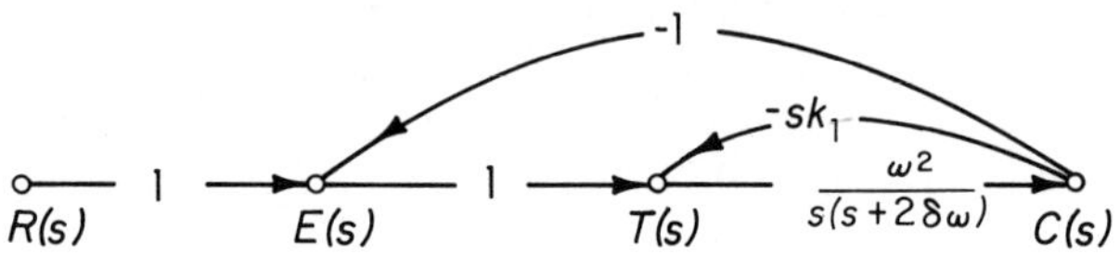

Figure 7-39 The Mason graph G_m.

$$\Sigma P_{m_1} = (-1)\cdot 1\cdot\frac{\omega^2}{s(s + 2\delta\omega)} + (-sk_1)\frac{\omega^2}{s(s + 2\delta\omega)}$$
$$= \frac{-\omega^2 - sk_1\omega^2}{s(s + 2\delta\omega)} = \frac{-(1 + sk_1)\,\omega^2}{s(s + 2\delta\omega)}$$
$$\Sigma P_{m_2} = \Sigma P_{m_3} = \cdots = 0$$
$$\Delta' = 1 - \Sigma P_{m_1} = 1 + \frac{\omega^2(1 + sk_1)}{s(s + 2\delta\omega)} = \frac{s^2 + (2\delta\omega + k_1\omega^2)\,s + \omega^2}{s(s + 2\delta\omega)}$$
$$T_1 = 1\cdot 1\cdot\frac{\omega^2}{s(s + 2\delta\omega)} = \frac{\omega^2}{s(s + 2\delta\omega)}$$
$$\Delta_1 = 1$$

and

$$T_1\Delta_1 = \frac{\omega^2}{s(s + 2\delta\omega)}$$

Hence,

$$T = \frac{C(s)}{R(s)} = \frac{1}{\Delta'}\sum_k T_k\Delta_k = \frac{s(s + 2\delta\omega)}{s^2 + (2\delta\omega + k_1\omega^2)s + \omega^2}\cdot\frac{\omega^2}{s(s + 2\delta\omega)}$$
$$= \frac{\omega^2}{s^2 + (2\delta\omega + k_1\omega^2)s + \omega^2}$$

(2) Coates' Method

Rearranging the system of equations we get

$$\begin{aligned} E(s) + C(s) \qquad\qquad &= R(s) \\ E(s) - sk_1C(s) \qquad - T(s) &= 0 \\ C(s) - \frac{\omega^2}{s(s + 2\delta\omega)}\,T(s) &= 0 \end{aligned}$$

The Coates graph G_c is shown in Figure 7-40. The graph G_{co} is illustrated in Figure 7-41. The set of all connections of G_{co} is depicted in Figure 7-42. The set of all the one-connections of G_c is shown in Figure 7-43.

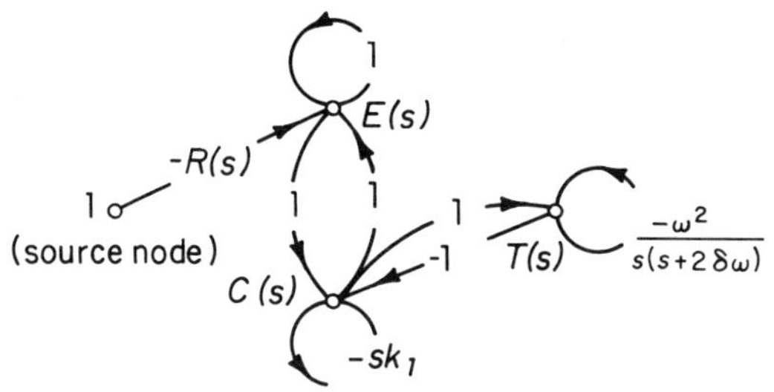

Figure 7-40 The Coates graph G_c.

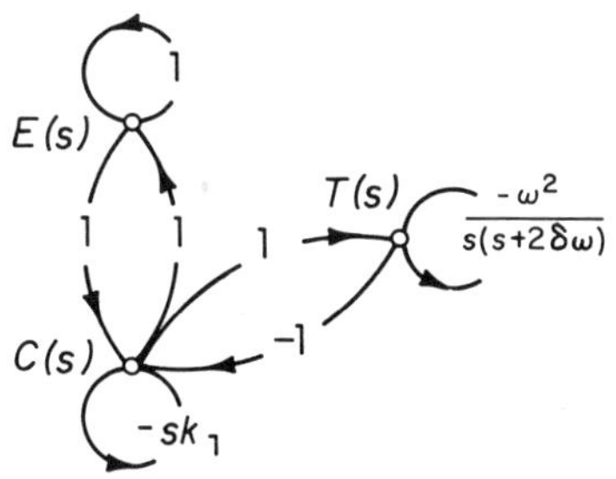

Figure 7-41 Graph G_{co} of Coates' method.

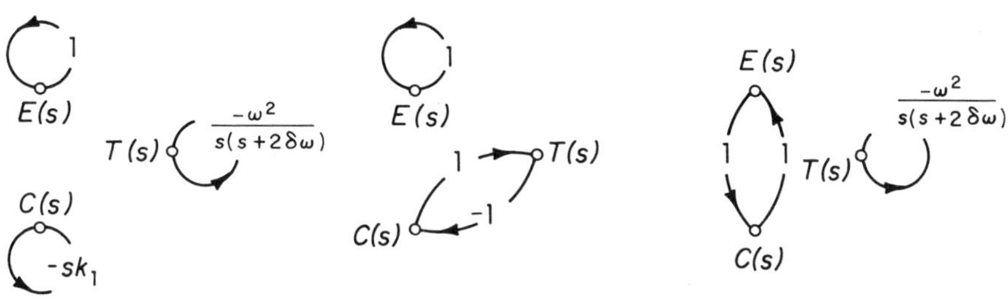

Figure 7-42 The set of all connections of G_{co}.

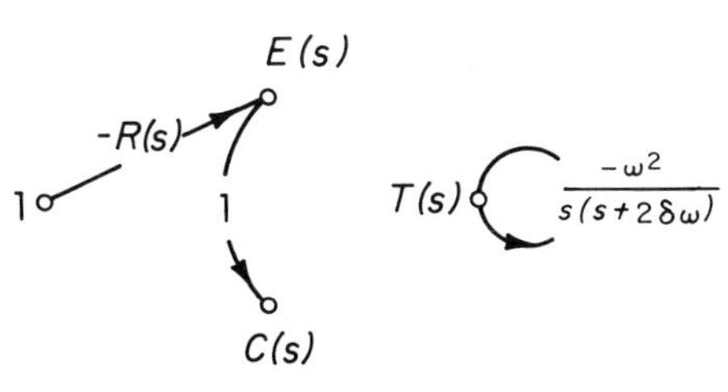

Figure 7-43 The set of all the connections of G_c.

Thus,

$$C(s) = \frac{(-1)[-R(s)]\cdot 1\cdot \dfrac{-\omega^2}{s(s+2\delta\omega)}}{(-1)1\cdot(-sk_1)\cdot\dfrac{-\omega^2}{s(s+2\delta\omega)} + 1\cdot 1\cdot(-1) + 1\cdot 1\cdot \dfrac{-\omega^2}{s(s+2\delta\omega)}}$$

$$= \frac{\omega^2 R(s)}{sk_1\omega^2 + s(s+2\delta\omega) + \omega^2} = \frac{\omega^2 R(s)}{s^2 + (2\delta\omega + k_1\omega^2)s + \omega^2}$$

or

$$\frac{C(s)}{R(s)} = \frac{\omega^2}{s^2 + (2\delta\omega + k_1\omega^2)s + \omega^2}$$

(3) The Modified Coates Method

The modified Coates graph G_l of the system is shown in Figure 7-44, and the graph G_{lo} is shown in Figure 7-45. The graph G_{lc} with respect to node $C(s)$ is depicted in Figure 7-46 and the set of all loop sets of G_{lo} in Figure 7-47.

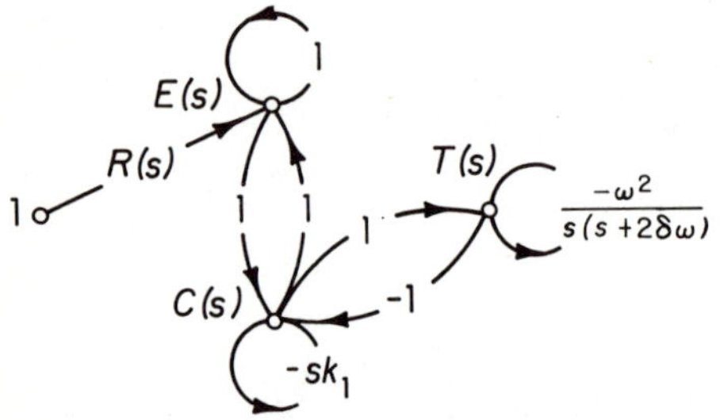

Figure 7-44 The modified Coates graph G_l.

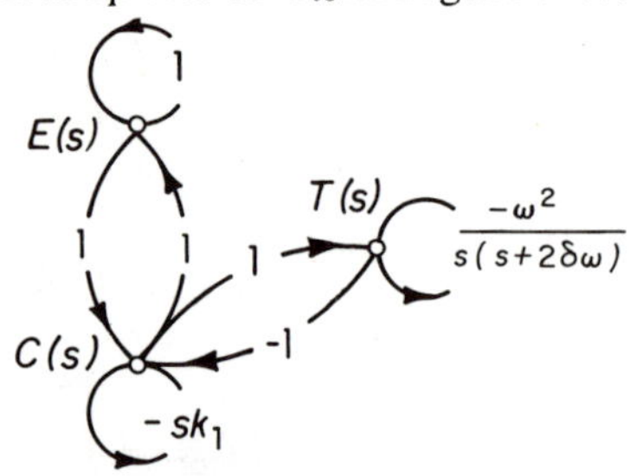

Figure 7-45 Graph G_{lo} of the modified Coates method.

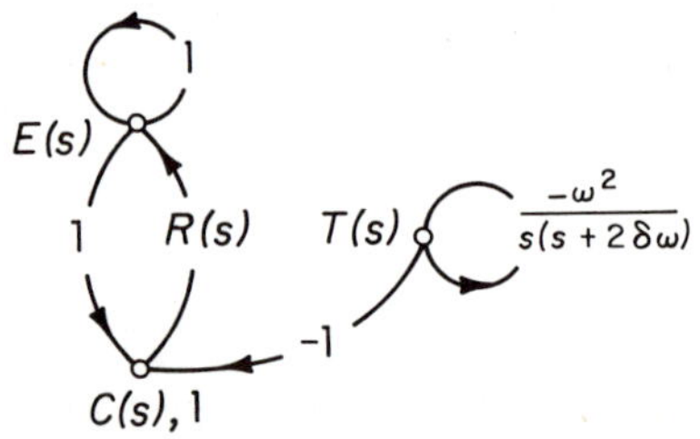

Figure 7-46 Graph G_{lc} with respect to node $C(s)$.

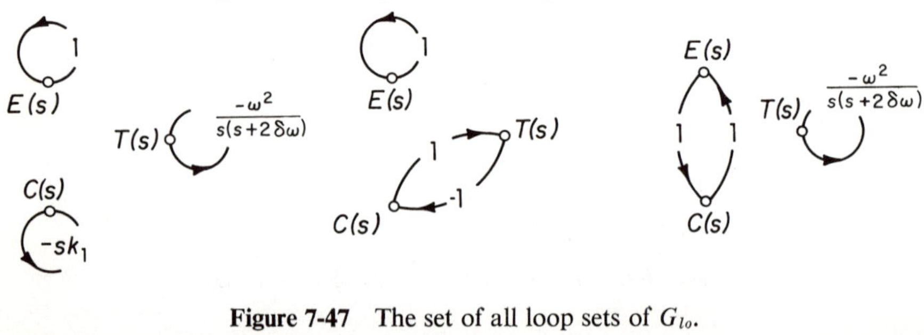

Figure 7-47 The set of all loop sets of G_{lo}.

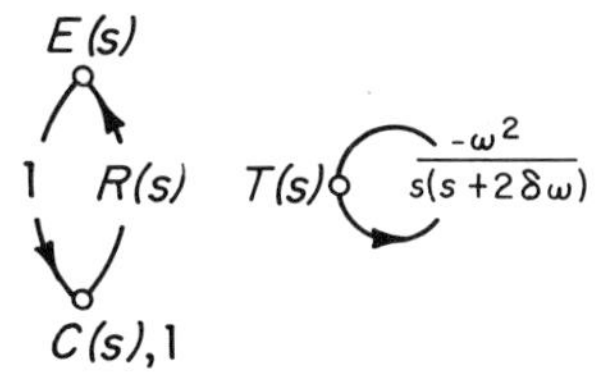

Figure 7-48 The only 2-loop set of G_{lc}.

Next, the set of all the 2-loop sets of G_{lc} is illustrated in Figure 7-48. Therefore, we write

$$C(s) = \frac{(-1)^2 \cdot 1 \cdot R(s) \cdot \dfrac{-\omega^2}{s(s+2\delta\omega)}}{(-1)1 \cdot (-sk_1) \cdot \dfrac{-\omega^2}{s(s+2\delta\omega)} + 1 \cdot 1 \cdot (-1) + 1 \cdot 1 \cdot \dfrac{-\omega^2}{s(s+2\delta\omega)}}$$

$$= \frac{\omega^2 R(s)}{sk_1\omega^2 + s(s+2\delta\omega) + \omega^2} = \frac{\omega^2 R(s)}{s^2 + (2\delta\omega + k_1\omega^2)s + \omega^2}$$

or

$$\frac{C(s)}{R(s)} = \frac{\omega^2}{s^2 + (2\delta\omega + k_1\omega^2)s + \omega^2}$$

(4) The Chan-Mai Method

The Chan-Mai graph G_s of the system is shown in Figure 7-49 and the set of all separations of G_s in Figure 7-50. To solve for $C(s)$, the graph G_{s2} required

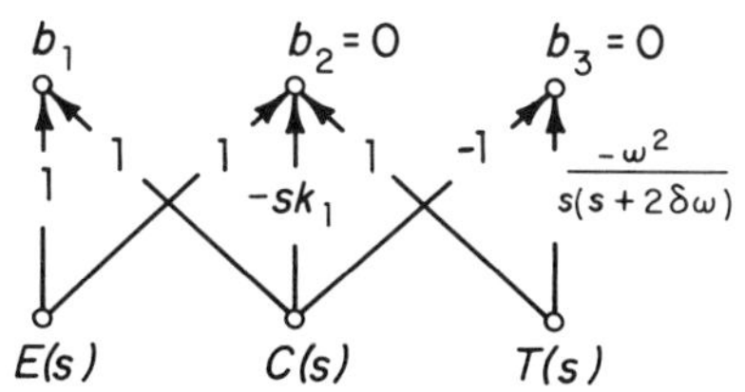

Figure 7-49 The Chan-Mai graph G_s.

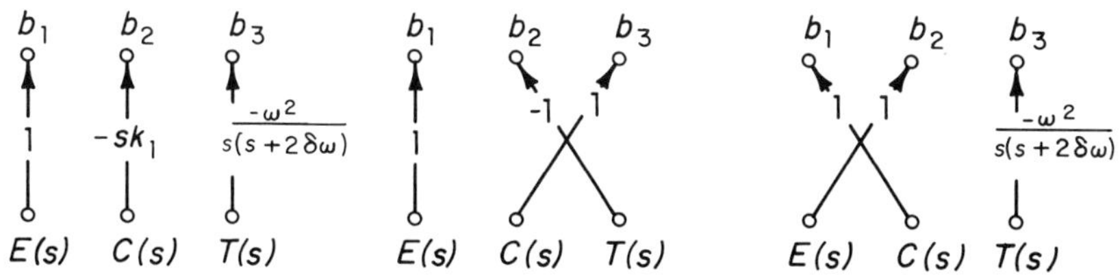

Figure 7-50 The set of all separations of G_s.

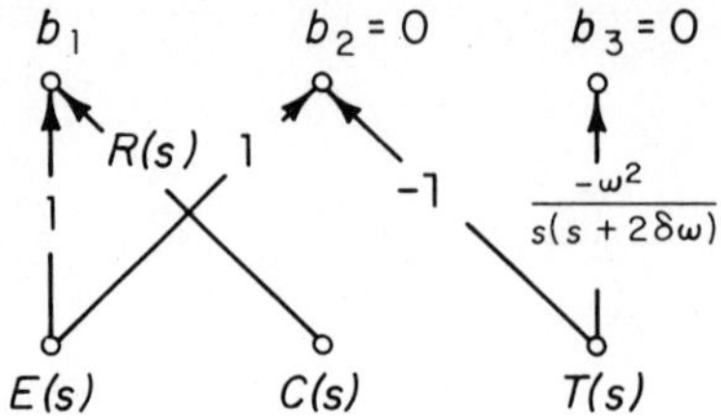

Figure 7-51 Graph G_{s2}.

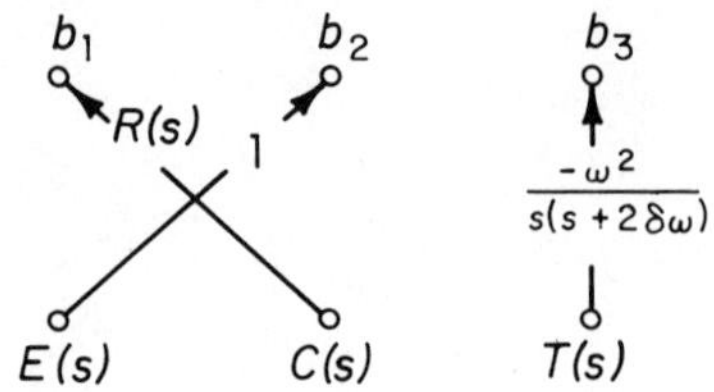

Figure 7-52 The only separation of G_{s2}.

is illustrated in Figure 7-51 and the set of all the separations of G_{s2} in Figure 7-52. Then, by Theorem 7-5,

$$C(s) = \frac{\Delta_2}{\Delta} \frac{(-1)^{1+1-1}R(s)(1)\dfrac{-\omega^2}{s(s+2\delta\omega)}}{\left\{(-1)^{0+3-1}(1)(-sk_1)\dfrac{-\omega^2}{s(s+2\delta\omega)} + (-1)^{1+1-1}(1)(1)(-1) + (-1)^{1+1-1}(1)(1)\dfrac{-\omega^2}{s(s+2\delta\omega)}\right\}}$$

$$= \frac{\omega^2 R(s)}{sk_1\omega^2 + s(s+2\delta\omega) + \omega^2}$$

Finally, the closed-loop transfer function is

$$\frac{C(s)}{R(s)} = \frac{\omega^2}{s^2 + (2\delta\omega + k_1\omega^2)s + \omega^2}$$

as expected.

PROBLEMS

7-1 Draw the Mason graph of each of the following systems of equations:
(a) The a_{ij}'s are nonzero constants and the x_j's are unknowns:

$$\begin{aligned} a_{11}x_1 \qquad\qquad\qquad\qquad\qquad &+ a_{14}x_4 = b_1 \\ a_{21}x_1 + a_{22}x_2 + a_{23}x_3 + \qquad &\qquad\;\; = 0 \\ a_{31}x_1 \qquad\qquad + a_{33}x_3 \qquad &\qquad\;\; = b_3 \\ a_{42}x_2 \qquad\qquad\qquad &+ a_{44}x_4 = 0 \end{aligned}$$

(b) The variables are w, x, y, and z:

$$\begin{aligned} w - 5x - 3y \qquad\;\; &= 2 \\ 2w + x - 6y + 4z &= 5 \\ 2x - y + 3z &= 0 \\ x + 5y - z &= 3 \end{aligned}$$

7-2 From the Mason graph shown in Figure P7-2, find the system of equations.

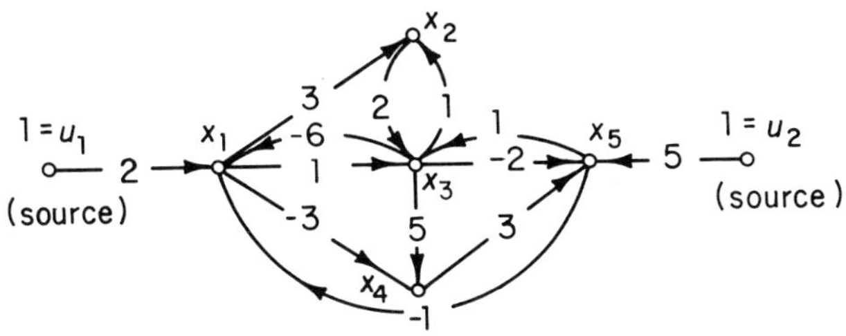

Figure P7-2

7-3 Using Mason's general gain formula, solve the following system of equations for the value of x_3, assuming that the a_{ij}'s are nonzero:

$$\begin{aligned} a_{11}x_1 + a_{12}x_2 + a_{13}x_3 &= b_1 \\ a_{21}x_1 + a_{22}x_2 + a_{23}x_3 &= b_2 \\ a_{32}x_2 + a_{33}x_3 &= 0 \end{aligned}$$

7-4 Solve the same system of equations as given in Problem 7-3 for the value of x_3 by using (a) the Coates formula and (b) the modified Coates formula.

7-5 Solve the same system of equations as given in Problem 7-3 for the value of x_3 by using the Chan-Mai formula.

7-6 Prove the validity of the four basic operations of graph reduction described in Table 7-1.

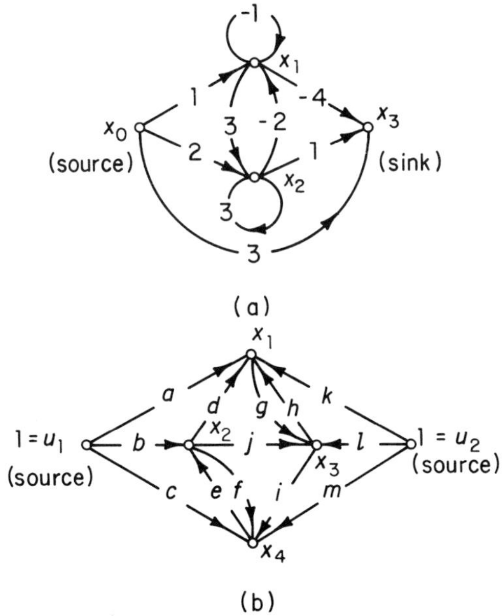

Figure P7-7

7-7 For each of the two Mason graphs shown in Figure P7-7, obtain the simultaneous equations associated with the (signal-flow) graph.

7-8 For the following system of two equations, obtain the (a) Mason graph, (b) Coates graph, (c) modified Coates graph, and (d) Chan-Mai graph associated with the system. Treat x_1 and x_2 as independent variables.

$$\begin{aligned} 6x_1 + 10x_2 + 2x_3 + x_4 &= 0 \\ 4x_1 + x_2 + x_3 + x_4 &= 0 \end{aligned}$$

7-9 Repeat Problem 7-8 but with x_1 and x_3 treated as independent variables.

7-10 Find x_6/x_1 by absorbing nodes x_2, x_3, x_4, and x_5 successively for the Mason graph shown in Figure P7-10.

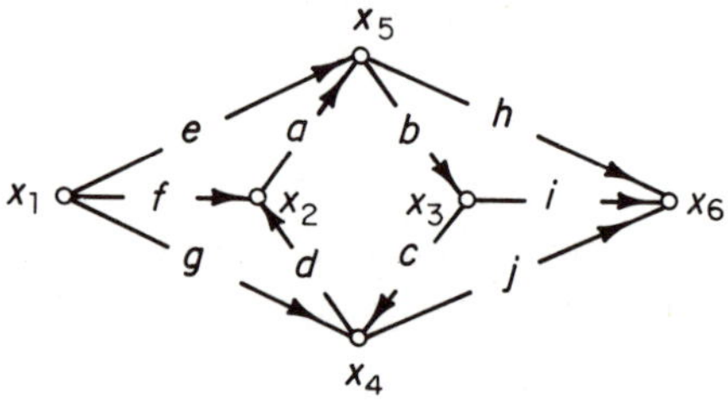

Figure P7-10

7-11 Solve the following system of equations for I_3 by using Mason's graph-reduction technique:

$$\begin{aligned} \left(1 + \frac{2}{s}\right)I_1 - \frac{2}{s}I_2 \qquad\qquad &= E \\ \frac{2}{s}I_1 + \left(2 + \frac{5}{s}\right)I_2 - \frac{3}{s}I_3 &= 0 \\ -\frac{3}{s}I_2 + \left(s + 5 + \frac{3}{s}\right)I_3 &= 0 \end{aligned}$$

7-12 For each of the networks shown in Figure P7-12, draw a suitable Mason graph and find the required network functions (a) V_o/V_i for the network in Figure P7-12(a); and (b) V_d/V_1 and V_e/V_1 for the network in Figure P7-12(b).

7-13 Refer to Figure P7-13. It is desired to find E_1 and I_2 in part (a), and I_2 in part (b). Draw suitable Mason graphs, but do not solve for the various expressions. Choose trees as follows: Z_a, Z_b, and E_2 for part (a); Z_b, Z_c, and E_2 for part (b).

7-14 Repeat Problem 7-12 using (a) the Coates graph and (b) the modified Coates graph.

7-15 Repeat Problem 7-12 using the Chan-Mai graph.

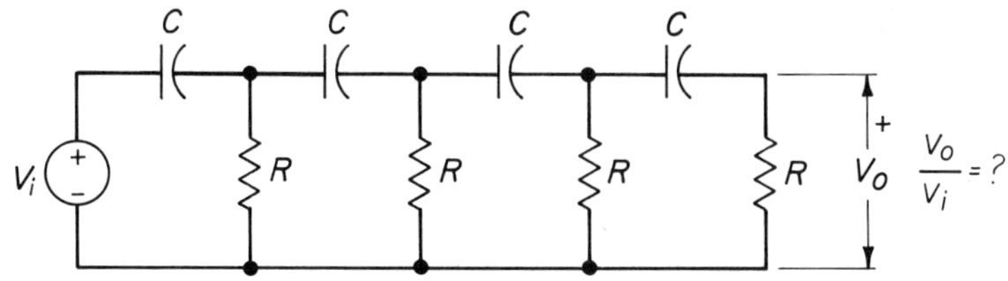

(a)

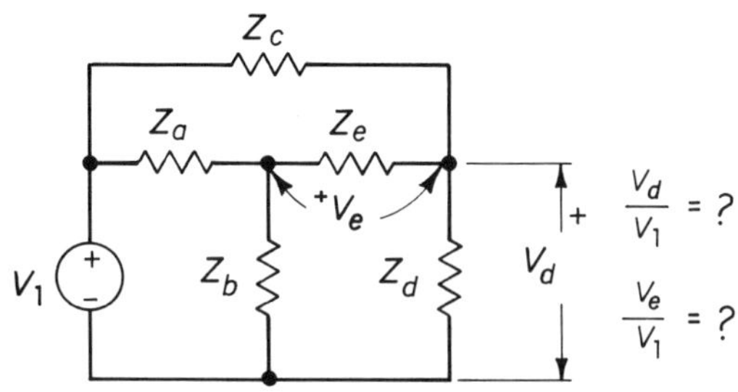

(b)

Figure P7-12

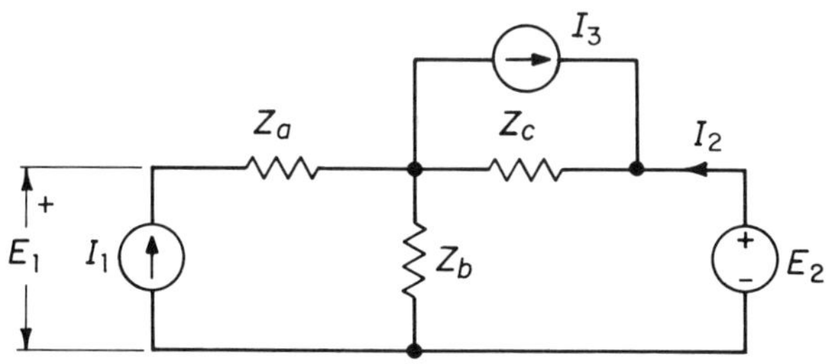

(a)

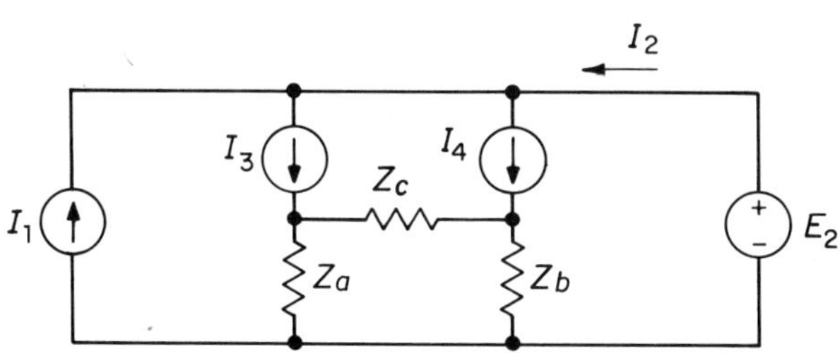

(b)

Figure P7-13

7-16 Find the output voltage of the bridged-T network depicted in Figure P7-16 by using (a) Mason's formula, (b) the Coates formula, (c) the modified Coates formula, and (d) the Chan-Mai formula.

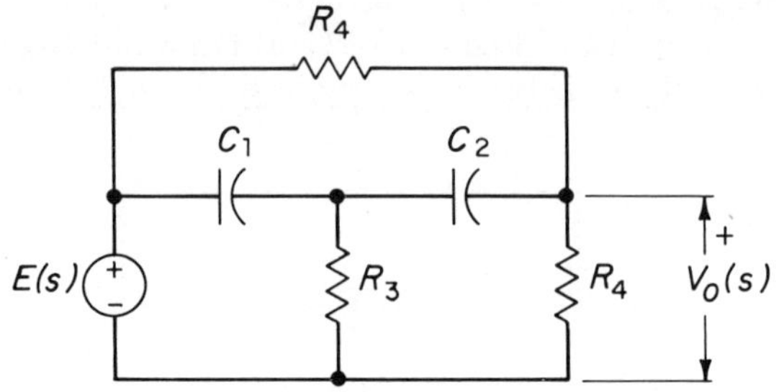

Figure P7-16

7-17 The equivalent circuit of some common collector transistor amplifiers at midband frequencies is shown in Figure P7-17. Determine its voltage gain, v_o/v_i, by using (a) the Mason formula, (b) the Coates formula, (c) the modified Coates formula, and (d) the Chan-Mai formula.

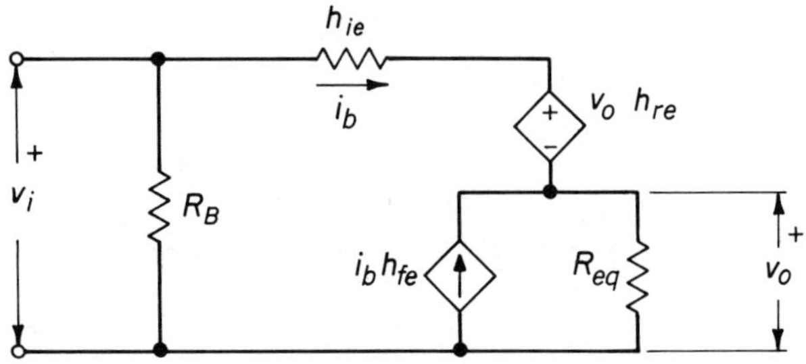

Figure P7-17

7-18 Repeat Problem 7-17 using any conventional technique. Compare your answers.

7-19 Some feedback amplifiers have a midband frequency equivalent circuit as shown in Figure P7-19. Determine its voltage gain topologically by

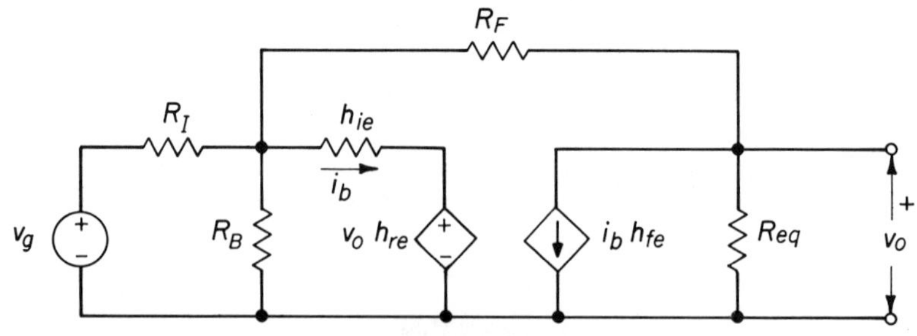

Figure P7-19

using (a) the Mason formula, (b) the Coates formula, (c) the modified Coates formula, and (d) the Chan-Mai formula.

7-20 Repeat Problem 7-19 using any conventional technique and compare your answers.

7-21 Refer to the network of Figure P7-21. (a) Draw any suitable flow graph and find Z_{in}. (b) As K becomes very large, what is Z_{in} approximately?

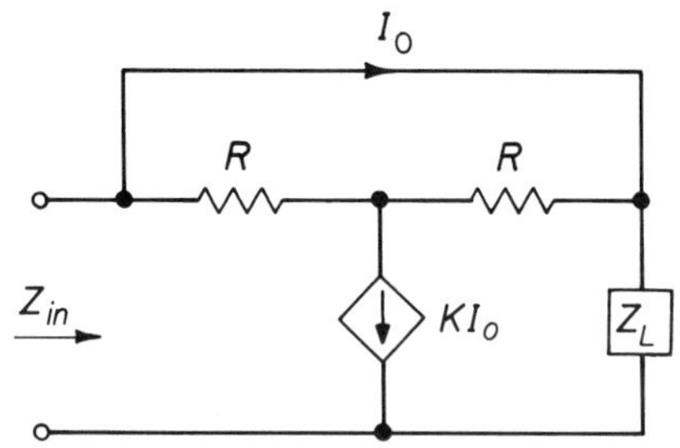

Figure P7-21

7-22 For the network shown in Figure P7-22, find E_o/E_i by any flow-graph technique. If both α_1 and α_2 are equal to 1, simplify the expression for E_o/E_i. Can you think of any practical applications for this network?

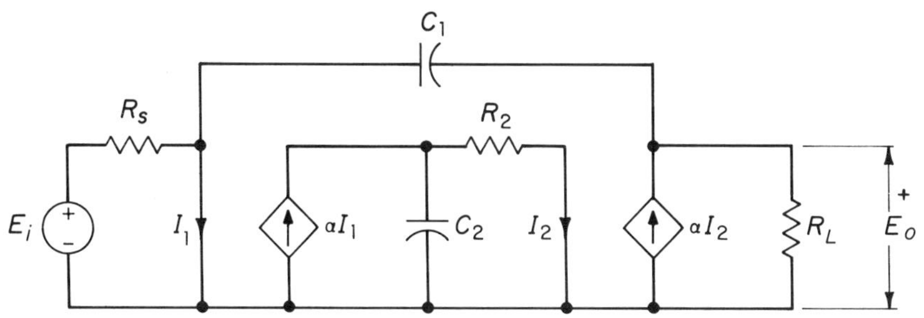

Figure P7-22

7-23 In Figure P7-23, two active networks are connected in cascade. Using any flow graph method, find an expression for E_o/E_s in terms of Z_s, Z_L, and the h parameters of the networks.

Remarks: The definition of the h parameters can be found in most electronics texts.

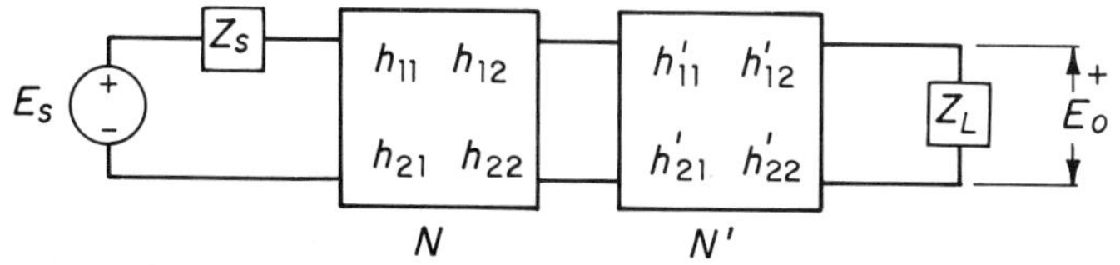

Figure P7-23

7-24 For each of the (equivalent) networks shown in Figure P7-24, draw a suitable flow graph (of your choice) and find the required network function. (a) E_o/E_s. (b) Z_{in}.

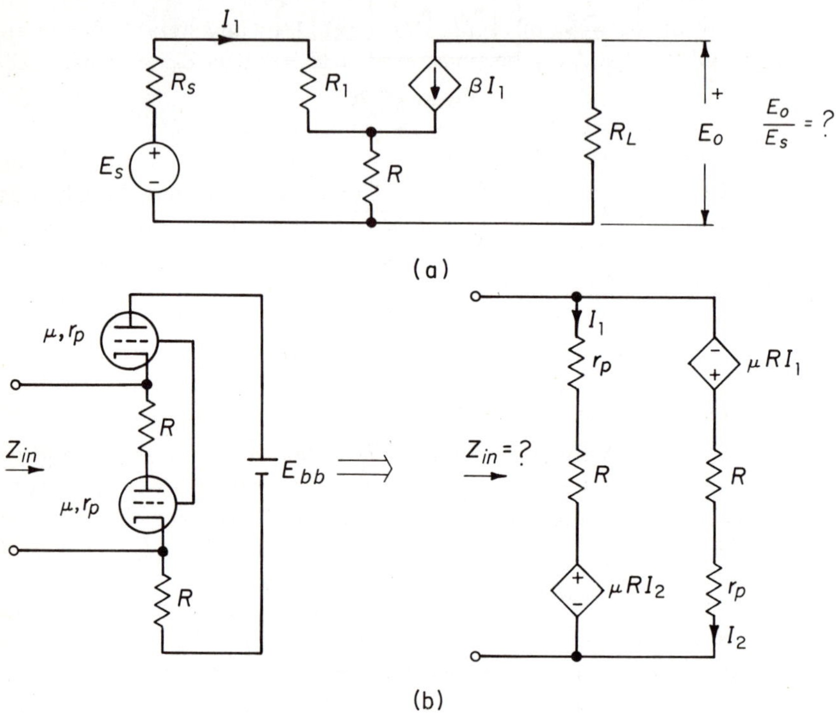

Figure P7-24

7-25 Given the T-equivalent network of a transistor (Figure P7-25). Find the h parameters for common-emitter, common-collector, and common-base configurations using one flow graph (of any type) for each configuration.

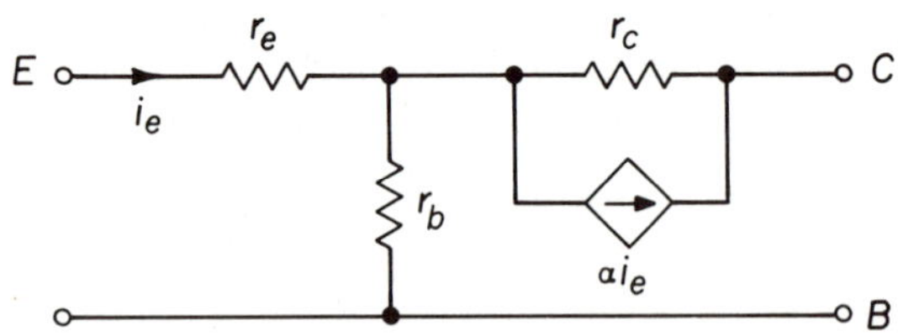

Figure P7-25

7-26 Using flow-graph techniques, find (a) Z_{in} and (b) E_o/E_s of the network shown in Figure P7-26.

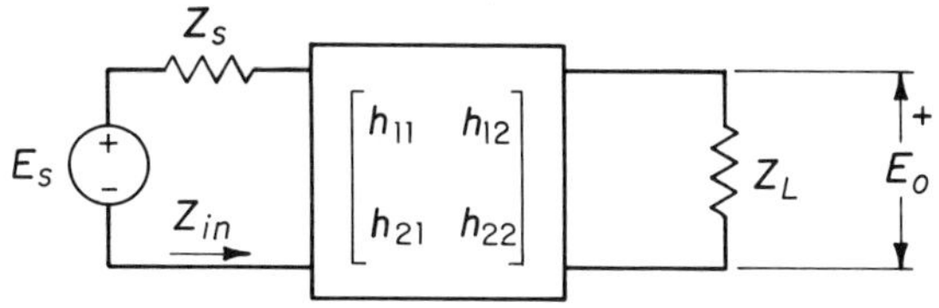

Figure P7-26

7-27 Compare the four flow-graph gain formulas — the Mason formula, the Coates formula, the modified Coates formula, and the Chan-Mai formula — by listing the advantages and disadvantages of each formula over the others. Make your comments as complete as possible.

7-28 Compare the flow-graph techniques as a whole with the conventional techniques for solving a linear system. Make your comments as complete as possible.

7-29 Suggest a systematic procedure for transforming a Coates graph into (a) a modified Coates graph and (b) a Chan-Mai graph.

7-30 Suggest a systematic procedure for transforming a Mason graph into (a) a modified Coates graph and (b) a Chan-Mai graph.

7-31 Suggest a systematic procedure for transforming a modified Coates graph into (a) a Mason graph, (b) a Coates graph, and (c) a Chan-Mai graph.

7-32 Suggest a systematic procedure for transforming a Chan-Mai graph into (a) a Mason graph, (b) a Coates graph, and (c) a modified Coates graph.

7-33 Give an algorithm for obtaining directly from a given linear network (a) the Mason graph, (b) the Coates graph, and (c) the modified Coates graph using loop analysis.

7-34 Repeat Problem 7-32 using node analysis.

7-35 Obtain from the network of Figure P7-12(b) directly using loop analysis (a) the Mason graph, (b) the Coates graph, (c) the modified Coates graph, and (d) the Chan-Mai graph.

7-36 Repeat Problem 7-34 using node analysis.

7-37 Repeat Problem 7-34 for each of the networks shown in Figure P7-13.

7-38 Repeat Problem 7-35 for each of the networks shown in Figure P7-13.

CHAPTER 8

TOPOLOGICAL ANALYSIS OF ACTIVE NETWORKS

8-1 INTRODUCTION

We have presented the topological analysis of passive one-ports and two-ports without mutual magnetic coupling, and have derived the topological formulas for these networks. We have also discussed various flow graphs for the analysis of linear systems and have developed a topological gain formula based on each of the flow graphs introduced.

In Chapter 6 we pointed out that, even though the topological formulas (which are expressed in terms of the node admittance products as developed in Chapters 5 and 6) are based on Maxwell's rules[1] using the node equations, another set of topological formulas may be developed, based on Kirchhoff's rules,[2] using the loop equations. The relationships between these two sets of formulas are easy to see since they are dual sets of formulas, as pointed out in the Summary of Chapter 6, and discussed in great detail in a paper by Mayeda [62]. Also, the relationships between any two of the flow graphs discussed in Chapter 7 are quite evident because it is not difficult to recognize the one-to-one correspondence among the terms in the two gain formulas for any two different flow graphs. However, the relationships between either the Maxwell or the Kirchhoff formulas and the formulas derived from the flow graphs are, by no means, obvious.

[1]See J. C. Maxwell [57].
[2]See G. Kirchhoff [46].

The objectives of this chapter are twofold. The first is to point out that the Maxwell formulas (or the Kirchhoff formulas) and the Coates formulas (or any other flow-graph formulas) can be obtained from one another; and the second objective, which is the main concern of this chapter, is to derive some of the topological formulas for the network functions of active networks allowing, also, mutual inductances.

Although the extension of the Maxwell topological formulas to active networks with mutual inductances has been investigated by a number of researchers,[3] our entire development is based mainly on the work of Mayeda [62] and Chen [20]. The reason is that the former is considered to be one of the classical and well-known works on topological analysis, whereas the latter is one of the latest developments in the area and is, perhaps, the best among the existing topological methods for the analysis of active networks.

8-2 THE WEIGHTED DIRECTED GRAPH OF CHEN

One of the most significant advantages of the topological formulas developed in Chapters 5 and 6 for the passive one-ports and two-ports over the conventional methods is that there is no cancellation of terms in the topological formulas, and thus errors that might occur in the cancellation process will be avoided. However, all the flow-graph formulas discussed in Chapter 7 have the same drawback as the conventional analysis methods because of the possibility of cancellation of terms. As one of the valuable contributions, Chen modifies the flow graphs and defines a weighted directed graph upon which a new flow-graph technique is developed so as to permit the evaluation of general network determinants by means of the "directed trees" of the new graph, avoiding the problem of cancellation[4] as well as that of sign determination for each term. The following are the definitions that will be needed for the development of the theory.

DEFINITION 8-1. Weighted Directed Graph $G(V,E,f)$ and Related Terms. A *weighted directed graph* (or simply *digraph*) is a 3-tuple, $G(V,E,f)$, consisting of a set V of elements called nodes, together with a set E of ordered pairs, (i,j), i and $j \in V$, called edges of G; the node i is called the *initial node* and node j the *terminal node*. For any $i \in V$, the symbols $\rho(i)$ and $\rho^*(i)$ denote the numbers of the sets of edges of G having i as the initial and terminal nodes, respectively. Finally, f is a mapping function with domain in E and range in the

[3]See W. S. Percival [68], S. J. Mason [55], C. L. Coates [23], A. Talbot [77], A. Nathan [65], and D. P. Brown [7].

[4]There will be no cancellation provided that all of the elements in a general network (loop or node) determinant are distinct; see W. K. Chen [20].

complex field such that $f(i,j)$ represents the weight associated with the edge (i,j) of G; and the mapping of an isolated node is defined to be 1.

For convenience, the idea of the mapping function will be extended from a single edge (i,j) in E to a subgraph R of G such that

$$f(R) = \Pi f((t,k))$$

where the product is taken over all (t,k) in R. This is expressed in the next definition.

DEFINITION 8-2. Mapping of a Weighted Directed Subgraph. The *mapping*, $f(R)$, of a weighted directed subgraph R is the product $\Pi f((t,k))$ taken over all (t,k) in R; that is, $\Pi f((t,k))$ is the product of the weights (gains) of all the directed edges of R.

DEFINITION 8-3. Sectional Graph. The *sectional graph* $G[A]$ of $G(V,E,f)$ is the subgraph whose node set is A (A being a subset of V) and whose edges are all those edges in G that connect two nodes in A.

DEFINITION 8-4. Directed Circuit and Its Length. A *directed circuit* is a circuit of directed edges such that the orientations of all its edges are in the same direction (either all in the clockwise or all in the counterclockwise direction). The circuit is said to be of *length m* if the number of edges in the circuit is m. A directed circuit of length 1 is called a (directed) *self-loop*.

DEFINITION 8-5. Labeled Digraph and Associated Undirected Graph. A *labeled digraph* $G(V,E,f)$ is a weighted directed graph in which every node is labeled by one of the integers from 1 through $\mu(V)$, where $\mu(V)$ is the total number of nodes of G. The *associated undirected graph* G_u of G is the graph whose nodes are the same as those in G and whose edges are those obtained from G by first omitting the edge orientations and then replacing each pair of parallel edges of the same weight by a single edge. If there are parallel edges with different weights in G, the undirected graph G_u of G is not defined.

The following example illustrates the definitions introduced thus far.

EXAMPLE 8-1. Consider the digraph of Figure 8-1. We should now point out that a digraph is used to represent the topology of an active network in

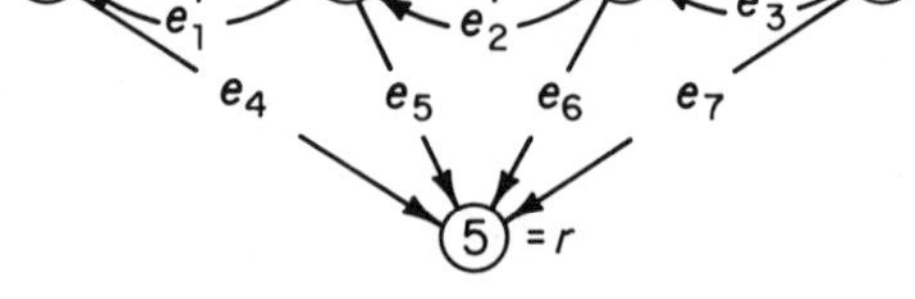

Figure 8-1 A digraph G.

which each passive element, say element j, is denoted by a pair of directed edges, e_j and $e_{j'}$, such that both edges have the same admittance $e_j = e_{j'} = y_j$. Thus, if the network of interest is a passive one, the corresponding digraph G may be "reduced" to an undirected graph G_u as described in Definition 8-5. On the other hand, if the network is an active network, the corresponding digraph cannot be "reduced" to an undirected graph G_u since G_u is defined only when the network is passive.

By inspection of Figure 8-1, the edge (1,2) of G is e_1, and the edge (2,1) is $e_{1'}$. For e_1, node 1 is the initial node, and node 2 the terminal node. For node 1, $\rho(1) = 2$ and $\rho^*(1) = 1$; but for node 5, $\rho(5) = 0$ and $\rho^*(5) = 4$. As for the mapping of the edges, we have $f((2,3)) = e_2$, $f((4,5)) = e_7$, and so forth.

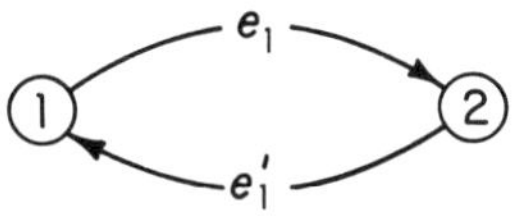

Figure 8-2 A directed circuit.

A directed circuit of G is shown in Figure 8-2. Let this directed circuit be denoted by C. Then C is a subgraph of G and is of length 2 since there are two edges, e_1 and $e_{1'}$, contained in C. Also, by Definition 8-2,

$$\begin{aligned} f(C) &= \Pi f((i,j)) \qquad (i,j = 1,2) \\ &= f((1,1))\cdot f((1,2))\cdot f((2,1))\cdot f((2,2)) \\ &= 1\cdot e_1\cdot e_{1'}\cdot 1 \\ &= e_1\cdot e_{1'} \end{aligned}$$

Next, let A be the subset $\{1,2,3\}$ of the set $V = \{1,2,3,4,5\}$. Then, the sectional graph $G[A]$ of G is the subgraph of Figure 8-3. Finally, the associated undirected graph G_u of G is shown in Figure 8-4. Note that G is a labeled digraph since each of its five nodes are labeled by the integers 1 through 5 [$\mu(V) = 5$ for G].

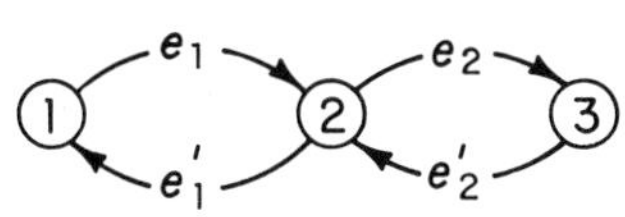

Figure 8-3 The sectional graph [GA].

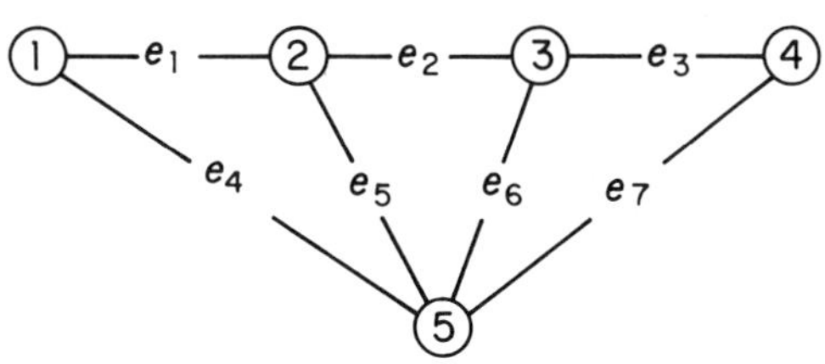

Figure 8-4 The associated undirected graph G_u.

8-3 NETWORK TOPOLOGY VERSUS FLOW-GRAPH TOPOLOGY[5]

In this section, we shall first show how a digraph can be obtained from the node admittance matrix of a given network, and then we shall establish the relationships between the Maxwell topological formulas and the flow-graph formulas. Although such relationships are not obvious at first sight, the existence of such relationships is quite logical since both sets of formulas deal with simultaneous linear algebraic equations. A major difference between the application of the Maxwell (or Kirchhoff) formulas and the flow-graph techniques is that the former may be used only when the network is given (so that the k-trees may be obtained from the network graph), whereas the latter can always be applied when the system of equations are known. The relationships we shall develop here will enable one to transfer some of the simplification techniques from one set of formulas to the other.

Consider the node admittance matrix Y_n of order $n - 1$ of a given n-node network N where Y_n is of the form

$$Y_n = \begin{bmatrix} \sum_{x=1}^{n-1} y_{1x} & -y_{12} & -y_{13} & \cdots & -y_{1(n-1)} \\ -y_{21} & \sum_{x=1}^{n-1} y_{2x} & -y_{23} & \cdots & -y_{2(n-1)} \\ \cdot & & \cdot & \cdots & \\ \cdot & & \cdot & \cdots & \\ \cdot & & \cdot & \cdots & \\ -y_{(n-1)1} & -y_{(n-1)2} & -y_{(n-1)3} & \cdots & \sum_{x=1}^{n-1} y_{(n-1)x} \end{bmatrix} \tag{8-1}$$

It is interesting to note that y_{ii}, which is not the (i,i) element of Y_n, but, rather, the ith element in the sum

$$\sum_{x=1}^{n-1} y_{ix}$$

which is the (i,i) element of Y_n, is equal to the sum of all the entries of the ith row of Y_n for any $i = 1, 2, \ldots, n - 1$.

The procedure for associating a digraph[6] $G(V,E,f)$ with Y_n is outlined in the following steps:

(1) There are exactly n nodes in G (that is, $\mu(V) = n$) and each node is labeled by one of the integers 1 through $n - 1$ (with the nth node labeled by r as the *reference node*) such that node k of G corresponds to the kth row of Y_n.

[5]The discussion presented in this section is based mainly on the paper of Chen and Quisenberry [22].

[6]Throughout the entire chapter, the digraphs we shall be dealing with will be the *labeled digraphs* only.

(2) If $y_{ij} \neq 0$ $(i \neq j)$, there is a directed edge $(i,j) \in E$ such that $f((i,j)) = y_{ij}$.
(3) If $y_{ii} \neq 0$, there is a directed edge $(i,r) \in E$ such that $f((i,r)) = y_{ii}$.

The foregoing procedure not only serves as a method for obtaining G from Y_n, but also enables one to write Y_n from G when G is the specified quantity. Since a one-to-one correspondence exists between the terms in the expansion of any two of the flow-graph gain formulas developed in the preceding chapter, it will suffice to pick any one of these formulas as a representative one and show how one can arrive at a network topological formula (either a Maxwell or a Kirchhoff formula) from it, and, conversely, to establish the relationships between the two sets of formulas. Thus, there will be no loss of generality if we only establish such relationships between the Coates gain formula and a Maxwell formula. To this end, we introduce the following definitions:

DEFINITION 8-6. Directed Tree T_d^r. A subgraph T_d^r of $G(V,E,f)$ is said to be a *directed tree* of G if and only if

(a) The associated undirected graph of T_d^r is a tree of the associated undirected graph G_u of G

(b) For all $i \in V_r$, $\rho(i) = 1$, where r is the reference node of G and V_r is the subset of V with node r removed, and $\rho(r) = 0$.

Note that in the symbol T_d^r, T_d shows that it is a directed tree, and the superscript r indicates that node r is the reference node of the tree.

EXAMPLE 8-2. Let us consider the digraph G of Figure 8-1. Let node 5 be the reference node. Then, four of the directed trees of G are shown in Figure 8-5.

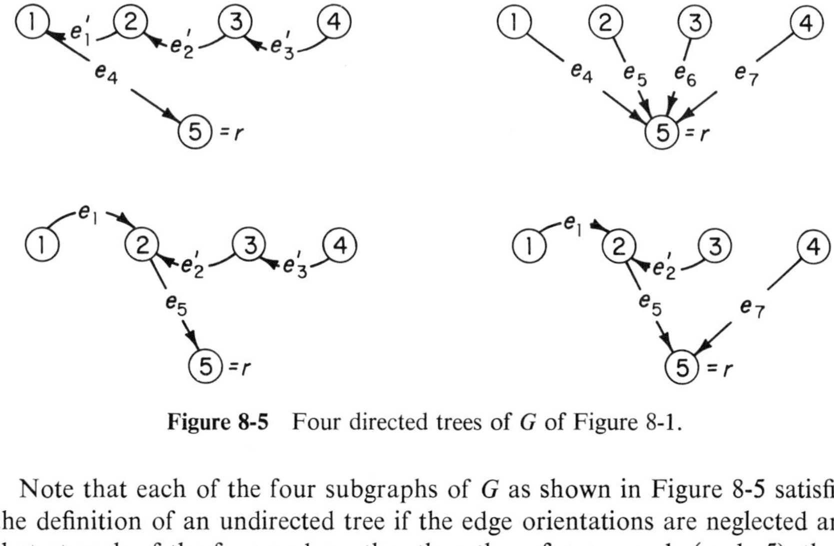

Figure 8-5 Four directed trees of G of Figure 8-1.

Note that each of the four subgraphs of G as shown in Figure 8-5 satisfies the definition of an undirected tree if the edge orientations are neglected and that at each of the four nodes, other than the reference node (node 5), there is exactly one outgoing edge. Moreover, there is no outgoing edge at the reference node so that $\rho(5) = 0$. Thus, all of the four subgraphs satisfy both conditions stated in Definition 8-6, and hence are directed trees of G.

We shall next outline the procedure for obtaining the Coates graph from Y_n of a given network. Using the new notation introduced earlier, let the Coates graph be denoted by $G_c(V_c,E_c,f_c)$. Then G_c can be obtained as follows:

(1) There are $n - 1$ nodes in G_c [that is, $\mu(V_c) = n - 1$] and each node is labeled by one of the integers 1 through $n - 1$ such that node k of G_c corresponds to the kth row of Y_n.

(2) If $y_{ji} \neq 0\,(i \neq j)$, there is a directed edge (i,j) at node j such that $f_c((i,j)) = -y_{ji}$ for i and $j = 1, 2, \ldots, n - 1\ (i \neq j)$.

(3) If $y_{ii} \neq 0$, there is a self-loop $(i,i)_i$ at node i such that $f_c((i,i)_i) = y_{ii}$ for $i = 1, 2, \ldots, n - 1$.

EXAMPLE 8-3. Consider the simple bridged-T network N shown in Figure 8-6. The node admittance matrix Y_n is given by

$$Y_n = \begin{array}{c} \begin{array}{ccc} \quad 1 \quad & \quad 2 \quad & \quad 3 \quad \end{array} \\ \left[\begin{array}{ccc} (y_1 + y_4) & -y_1 & -y_4 \\ -y_1 & (y_1 + y_2 + y_3) & -y_3 \\ -y_4 & -y_3 & (y_3 + y_4 + y_5) \end{array}\right]\begin{array}{c} 1 \\ 2 \\ 3 \end{array} \end{array} \tag{8-2}$$

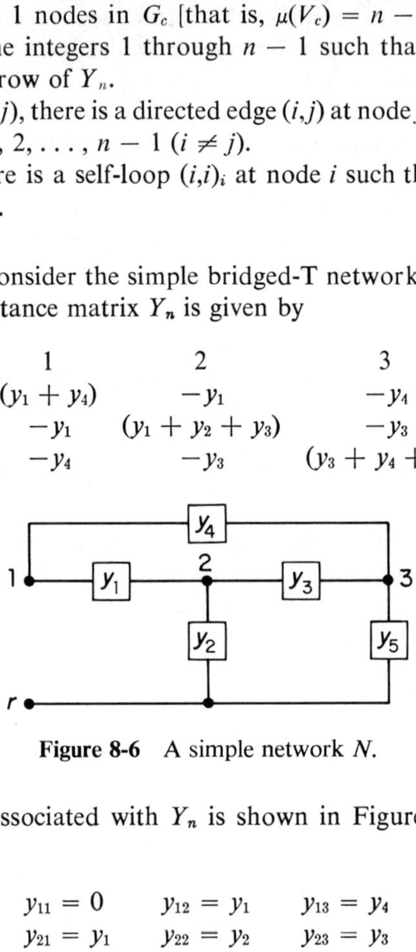

Figure 8-6 A simple network N.

The Coates graph associated with Y_n is shown in Figure 8-7. Note that, in this example,

$$\begin{array}{lll} y_{11} = 0 & y_{12} = y_1 & y_{13} = y_4 \\ y_{21} = y_1 & y_{22} = y_2 & y_{23} = y_3 \\ y_{31} = y_4 & y_{32} = y_3 & y_{33} = y_5 \end{array}$$

so that Y_n in Equation (8-2) may be expressed in the form

$$Y_n = \begin{bmatrix} (y_{11} + y_{12} + y_{13}) & -y_{12} & -y_{13} \\ -y_{21} & (y_{21} + y_{22} + y_{23}) & -y_{23} \\ -y_{31} & -y_{32} & (y_{31} + y_{32} + y_{33}) \end{bmatrix} \tag{8-3}$$

as defined in Equation (8-1).

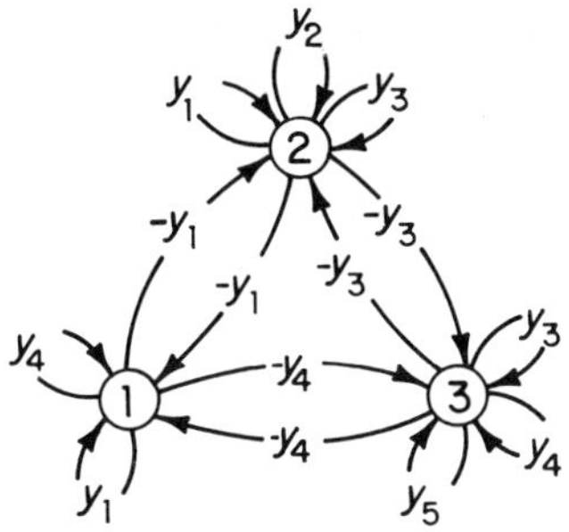

Figure 8-7 The Coates graph $G_c(V_c,E_c,f_c)$ associated with Y_n in Equation (8-2).

Let us now recall a Coates formula for evaluating a system determinant as discussed in Chapter 7. From Equation (7-18), we write

$$\Delta = \sum_{\rho} (-1)^{n+L_\rho} C(G_0)_\rho \tag{8-4}$$

which, in terms of the notation introduced in this chapter, may be written as[7]

$$\det Y_n = \sum_{i} (-1)^{q_i} f_c(h^{(i)}) \tag{8-5}$$

where $h^{(i)}$ is a connection[8] in G_c, q_i is the number of *even components*[9] in $h^{(i)}$, and the summation is over all $h^{(i)}$ in G_c.

Next, we shall define the following:

DEFINITION 8-7. Cyclic Connection. A connection $h^{(i)}$ of G_c is said to be *cyclic* if it contains only self-loops of the form

$$(i_1,i_1)_{k_1}(i_2,i_2)_{k_2}\cdots(i_p,i_p)_{k_p}$$

such that the subscripts of the mapping $f(h)$ of h (connection gain of h) form a cyclic order, $y_{ab}y_{bc}\cdots y_{qp}y_{pa}$, where p is the number of edges in h and $abc\cdots qp$ is a permutation of $i_1, i_2, \ldots, i_p$.

EXAMPLE 8-4. Let us consider the general form of Y_n [Equation (8-3)] of a 3-node network N. If all the y_{ij} are different from zero ($i,j = 1, 2, 3$), the Coates graph G_c corresponding to Y_n will be the one shown in Figure 8-8. Two cyclic connections of G_c are depicted in Figure 8-9 with the corresponding mappings (connection gains) written directly below the connections.

Let us consider the general case where G_c is the Coates graph associated with a network with any (finite) number of nodes. Let B_1 be the set of all

[7]The transformation from Equation (8-4) to (8-5) is quite simple and is left as an exercise.

[8]Refer to Section 7-3 (Definition 7-5) for the definition of a connection.

[9]An even component is a component which contains an even number of edges. A component in $h^{(i)}$ is a "connected piece" (or, more formally, a maximally connected subgraph (Definition 4-7) of $h^{(i)}$.

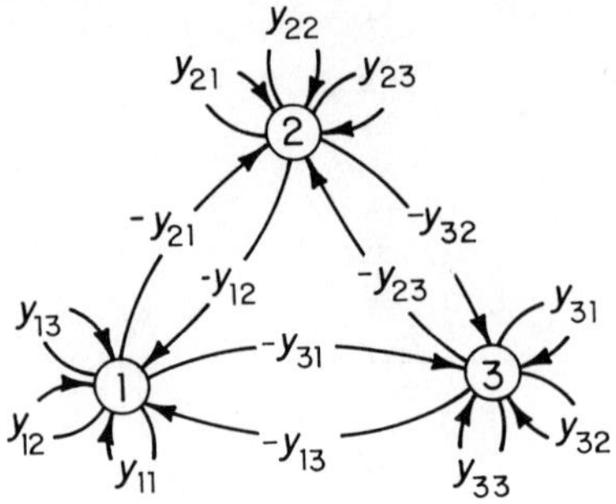

Figure 8-8 The Coates graph G_c associated with Y_n in Equation (8-3).

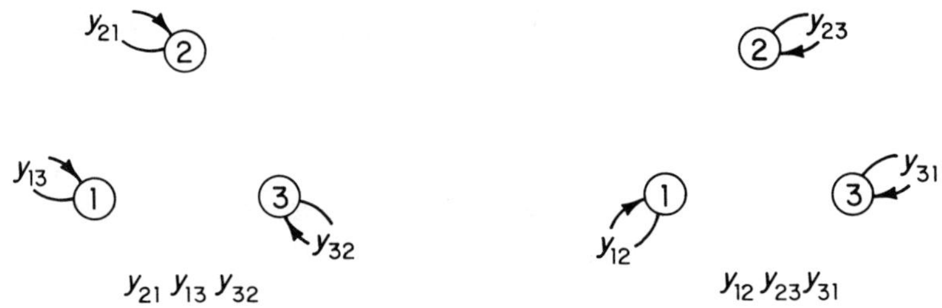

Figure 8-9 Two cyclic connections of G_c.

connections in G_c, and let B_2 be the subset of B_1 such that each $h_2^{(k)}$ in B_2 satisfies one of the following three conditions:

(a) It is a *cyclic connection.*

(b) It *contains at least one directed circuit of length* $\geqq 2$.

(c) It has the gain (product of the edge gains) equal to one of the gains of the connections in (b).

Also, let $T_d^{r(k)}$ be an element of the set $\{T_d^{r(k)}\}$ of all directed trees in G. Then we state the following theorem without proof:[10]

THEOREM 8-1. With the same notation as defined previously,

$$\sum_k (-1)^{q_k} f(h_2^{(k)}) = 0 \tag{8-6}$$

where $h_2^{(k)}$ is an element in B_2, and the summation is over all $h_2^{(k)}$ in B_2. Moreover, there exists a one-to-one correspondence between the elements $h_3^{(m)}$ in B_3 and the elements $T_d^{r(j)}$ in $\{T_d^{r(j)}\}$ such that if $h_3^{(p)}$ and $T_d^{r(p)}$ are corresponding elements, then

$$f_c(h_3^{(p)}) = f(T_d^{r(p)}) \tag{8-7}$$

where $B_3 = B_1 - B_2$, and $h_3^{(p)}$ is a connection in B_3.

The next theorem is a direct consequence of Theorem 8-1.

[10]The proof of both Theorems 8-1 and 8-2 can be found in Chen and Quisenberry [22].

THEOREM 8-2.[10] With the same notation as in Theorem 8-1,

$$\Delta = \det Y_n = \sum_p f_c(h_3^{(p)}) = \sum_p f(T_d^{r(p)}) \tag{8-8}$$

where $\det Y_n$ is the determinant of the node admittance matrix Y_n of the given network N; $f(T_d^{r(p)})$ is a directed-tree admittance product of the directed tree $T_d^{r(p)}$ in the digraph G of N; $h_3^{(p)}$ is a connection in B_3 of the Coates graph G_c obtained from Y_n. The first summation is over all $h_3^{(p)}$ in B_3 and the second is over all $T_d^{r(p)}$ in G.

Thus, $\det Y_n$ of a given N may be evaluated either by forming the sum of all the directed-tree admittance products or by the Coates formula, from which the sum of all the connection gains is obtained. The former is more direct since there will be no cancellations involved in the process, as the following example will illustrate.

EXAMPLE 8-5. In this example, we shall evaluate $\det Y_n$ of the Y_n in Equation (8-3) by various methods. We shall first evaluate it by direct expansion; next by the Coates formula; then by the directed-tree method; and, finally, by Equation (5-18), as stated in Theorem 5-2.

(a) By Direct Determinant Expansion

$$\det Y_n = \det \begin{bmatrix} (y_{11} + y_{12} + y_{13}) & -y_{12} & -y_{13} \\ -y_{21} & (y_{21} + y_{22} + y_{23}) & -y_{23} \\ -y_{31} & -y_{32} & (y_{31} + y_{32} + y_{33}) \end{bmatrix}$$

or

$$\begin{aligned} \det Y_n = {} & (y_{11} + y_{12} + y_{13})(y_{21} + y_{22} + y_{23})(y_{31} + y_{32} + y_{33}) \\ & - y_{13}y_{32}y_{21} - y_{12}y_{23}y_{31} - y_{31}y_{13}(y_{21} + y_{22} + y_{23}) \\ & - y_{23}y_{32}(y_{11} + y_{12} + y_{13}) - y_{21}y_{12}(y_{31} + y_{32} + y_{33}) \end{aligned} \tag{8-9}$$

which has 38 terms after being fully expanded. But, after cancellation of terms,

$$\begin{aligned} \det Y_n = {} & y_{11}y_{21}y_{31} + y_{11}y_{21}y_{32} + y_{11}y_{21}y_{33} + y_{11}y_{22}y_{31} + y_{11}y_{22}y_{32} \\ & + y_{11}y_{22}y_{33} + y_{11}y_{23}y_{31} + y_{11}y_{23}y_{33} + y_{12}y_{22}y_{31} \\ & + y_{12}y_{22}y_{32} + y_{12}y_{22}y_{33} + y_{12}y_{23}y_{33} + y_{13}y_{21}y_{33} \\ & + y_{13}y_{22}y_{32} + y_{13}y_{22}y_{33} + y_{13}y_{23}y_{33} \end{aligned} \tag{8-10}$$

(b) By the Coates Formula

The Coates graph G_c is shown in Figure 8-8. Since there are three self-loops at each node, there are $3 \times 3 \times 3 = 27$ connections formed by the nine self-loops. In addition, there are three directed circuits (loops) of length

2, and each of the three self-loops at each node forms a connection with the directed circuit of length 2 located opposite the self-loop. Thus, there are nine such connections (each of which consists of a circuit of length 2 and a self-loop). Finally, there are two directed circuits of length 3 forming two connections, thus giving a total number of $27 + 9 + 2 = 38$ connections. This is exactly the same number of terms in the expansion of Equation (8-9). It is a simple matter to verify that these two sets of 38 terms can be equated term by term. Hence, after cancellation, only 16 connections will remain; and, consequently, by Equation (7-18) in Theorem 7-2, det Y_n is exactly the same as in Equation (8-10).

(c) By the Directed-tree Method of Chen

First we obtain the digraph G from Y_n as shown in Figure 8-10. The set

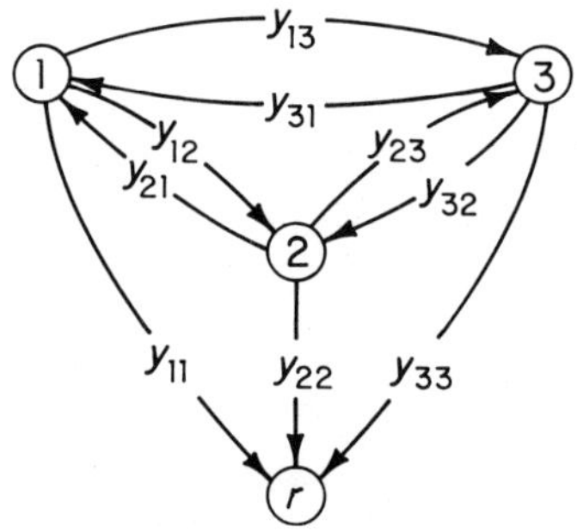

Figure 8-10 The digraph G for Y_n.

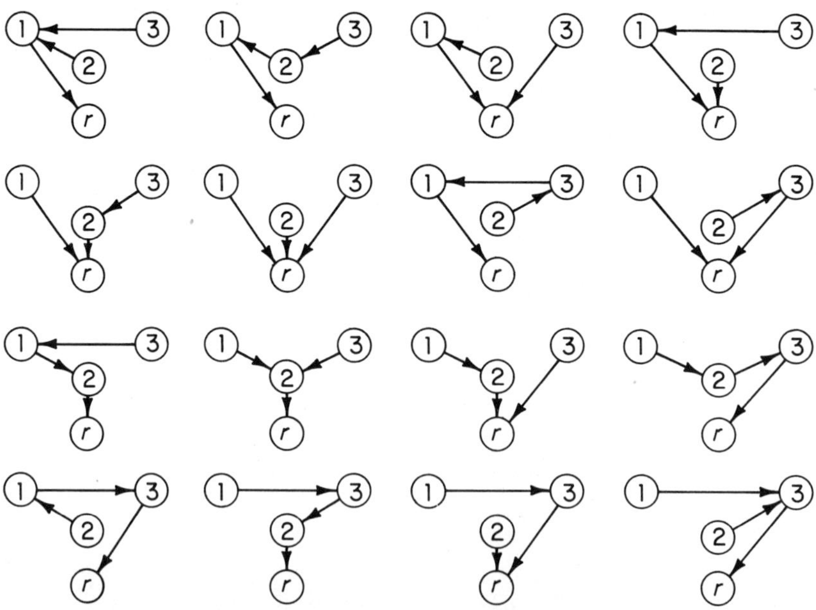

Figure 8-11 The set of all directed trees of G.

of all directed trees of G is shown in Figure 8-11. Thus, summing up the directed-tree admittance products, we get the same expression for Y_n as given by Equation (8-10).[11]

(d) By the Maxwell Formula [Equation (5-18)]

The network N, which given a Y_n as presented by Equation (8-3), is drawn in Figure 8-12 — provided, of course, that $y_{ij} = y_{ji}$ for all $i \neq j\ (i,j = 1, 2, 3)$. The nonoriented graph G_n of N is shown in Figure 8-13. The total number t of trees of G_n is given by

$$t = C(6,3) - K_3 = \frac{6\cdot 5\cdot 4}{3\cdot 2} - 4 = 20 - 4 = 16$$

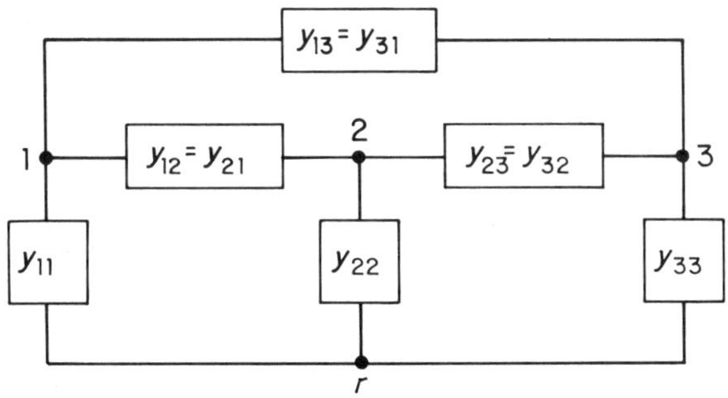

Figure 8-12 The passive network N that yields the Y_n as given by Equation (8-3).

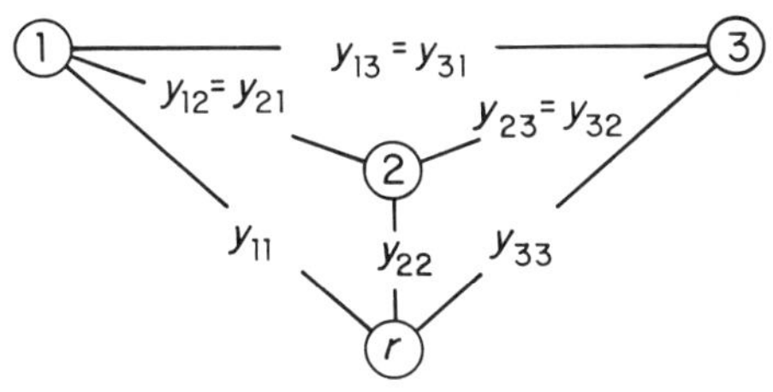

Figure 8-13 Graph G_n of N.

where $C(6,3)$ is the combination of six objects taken three at a time, and K_3 is the number of circuits of length 3. The set of all trees of G_n is depicted in Figure 8-14. Thus, by Equation (5-18),

$$\begin{aligned}\det Y_n &= \sum_i T^{(i)}(y) \\ &= y_{11}y_{12}y_{13} + y_{11}y_{12}y_{23} + y_{11}y_{12}y_{33} + y_{11}y_{22}y_{13} + y_{11}y_{22}y_{23} \\ &\quad + y_{11}y_{22}y_{33} + y_{11}y_{23}y_{13} + y_{11}y_{23}y_{33} + y_{12}y_{22}y_{13} + y_{12}y_{22}y_{23} \\ &\quad + y_{12}y_{22}y_{33} + y_{12}y_{23}y_{33} + y_{13}y_{12}y_{33} + y_{13}y_{22}y_{23} + y_{13}y_{22}y_{33} \\ &\quad + y_{13}y_{23}y_{33}\end{aligned}$$

keeping in mind that $y_{ij} = y_{ji}$ for all $i,j = 1, 2, 3$ has been assumed.

[11]Note that when the matrix of a given network determinant is symmetric, there exists a one-to-one correspondence between the directed trees of the digraph and the trees of the corresponding undirected graph.

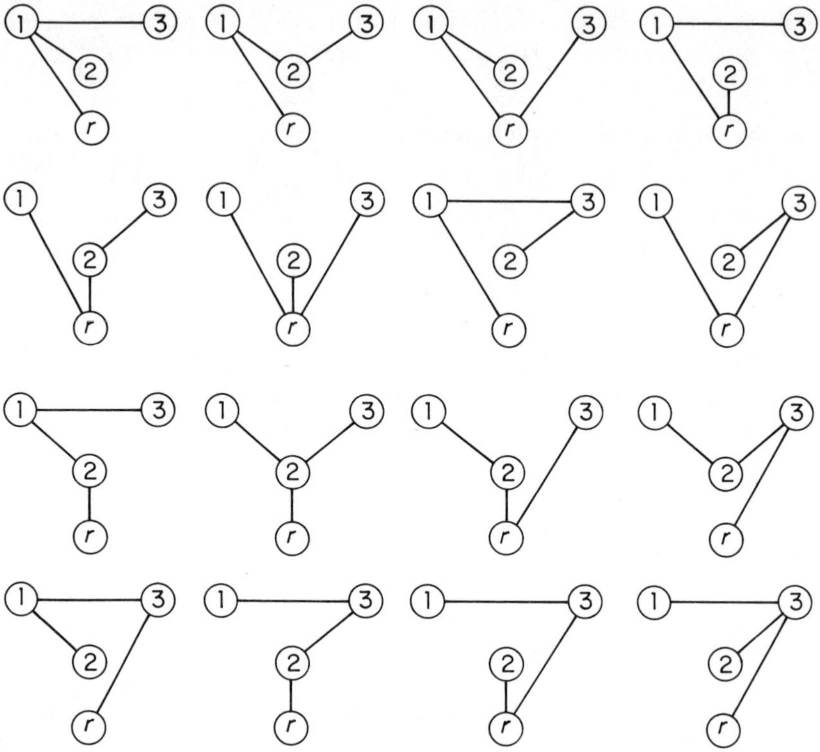

Figure 8-14 The set of all trees of G_n.

Let us now identify the directed trees obtained in (c) by the method of Chen with the uncanceled connections obtained by the Coates formula. These uncanceled connections of G_c are shown in Figure 8-15.

Finally, we shall examine the connections of G_c that cancel. The 22 canceled connections of G_c are shown in Figure 8-16 where they are paired off systematically with (ki) canceled by (kj), $(k = 1, 2, \ldots, 11; i,j = a, b,$ or $c)$. Note that the letter "a" indicates that the connection $h_2^{(k)}$ belongs to type (a) of the set B_2 of connections defined in Theorem 8-1, and, similarly, for the letters "b" and "c" in Figure 8-16.

From Example 8-5, we observe also a very important fact: The nonoriented graph G_n of the network N of Figure 8-12 is precisely the associated undirected graph G_u of the digraph G (as defined in Definition 8-5). This implies the following corollary:

COROLLARY 8-1. For a passive network without mutual inductances, Equation (8-8) in Theorem 8-2 becomes

$$\det Y_n = \sum_p f_c(h_3^{(p)}) = \sum_i T^{(i)}(y) \tag{8-11}$$

where the notation is the same as in Equation (8-8) in Theorem 8-2.

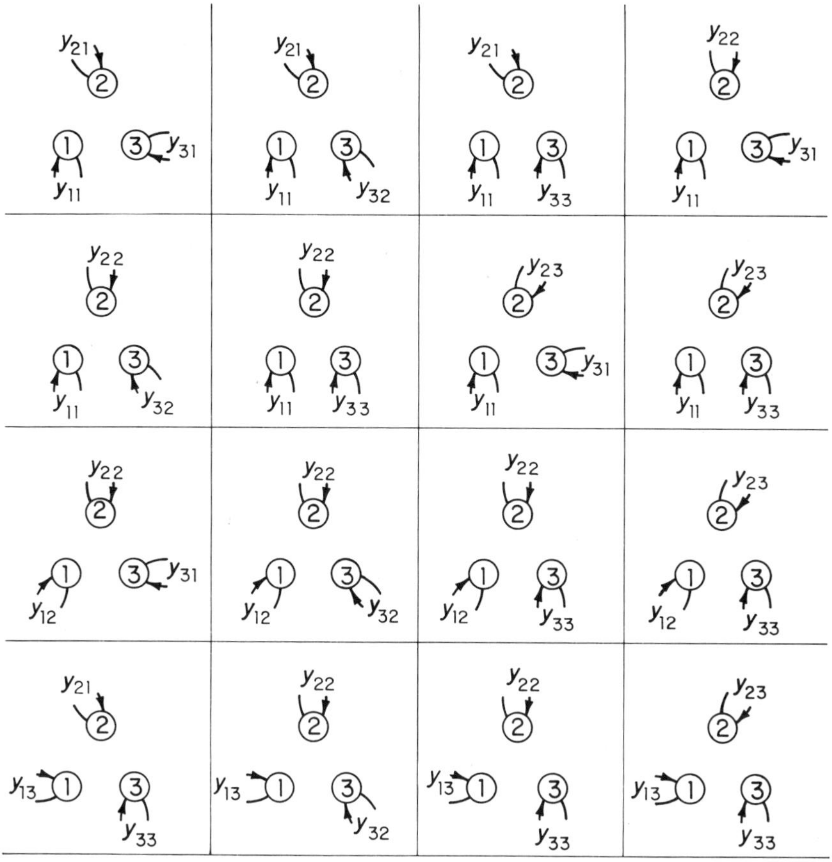

Figure 8-15 The uncanceled connections of G_c.

Evidently, Equation (8-11) is the same as Equation (5-18) in Theorem 5-2. Thus, Corollary 8-1 is merely a special case of Theorem 8-2. We should also point out the important fact that Equation (8-11) relates the well-known Maxwell formula [Equation (5-18)] to the Coates formula [Equation (7-18)] such that the terms (trees) in the Maxwell formula are precisely those uncanceled terms (connections) in the Coates formula.

We should also point out that the method of Chen as expressed in Theorem 8-2, using digraphs and directed trees, is the more general formula of which the Maxwell formula [Equation (5-18)] is a special case that can be used when the network N under study is a passive one without mutual inductances. If N is an active network and contains mutual inductances, Theorem 8-2 may be considered as a logical extension of the Maxwell formula since both formulas avoid cancellations[12] and/or the directed trees in Equation (8-8) also

[12]Here again, there will be no cancellation only if all the elements of the general network determinant are distinct.

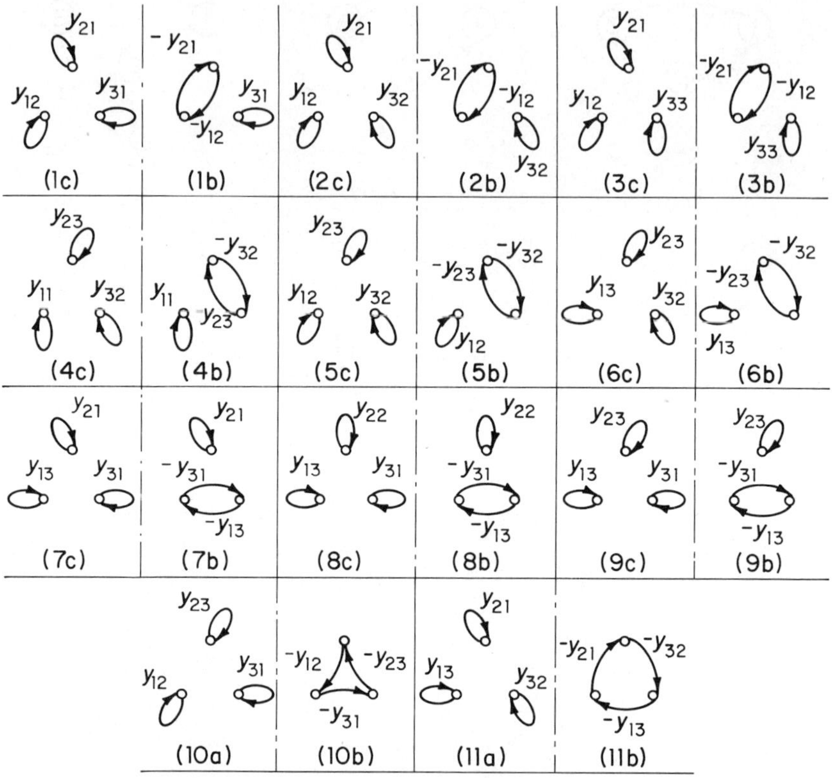

Figure 8-16 The 22 canceled connections of G_c.

correspond to the uncanceled connections in the associated Coates graph.

In order to show the relationship between the uncanceled one-connections in the Coates formula and the subgraphs of the corresponding digraph, we shall need the following definition.

DEFINITION 8-8. Directed 2-tree. A *directed 2-tree*, T_{2d}^{ij}, of a digraph $G(V,E,f)$ is a subgraph of G such that (a) the corresponding undirected graph of T_{2d}^{ij} is a 2-tree of the associated undirected graph G_u of G, and (b) each of the two connected components of T_{2d}^{ij} is a directed tree for some sectional graph (see Definition 8-3) of G. The two nodes i and j, in T_{2d}^{ij} serve as the reference nodes for the two directed subgraphs that, together, constitute T_{2d}^{ij}.

EXAMPLE 8-6. Consider the digraph G of Figure 8-10. Take $i = 1$ and $j = r$. Some of the 2-trees T_{2d}^{ir} are shown in Figure 8-17. Similarly, three of the 2-trees T_{2d}^{r} are depicted in Figure 8-18.

DEFINITION 8-9. Cyclic One-connection. Let $h_{ij} = p_{ij} \cup h^*$ be a one-connection from nodes i to j in G_c where p_{ij} is the unique directed path from i to

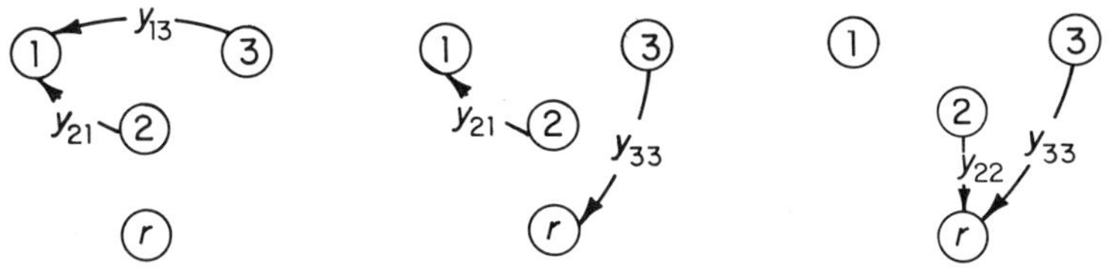

Figure 8-17 Three of the 2-trees T_{2d}^{ir} of G of Figure 8-10.

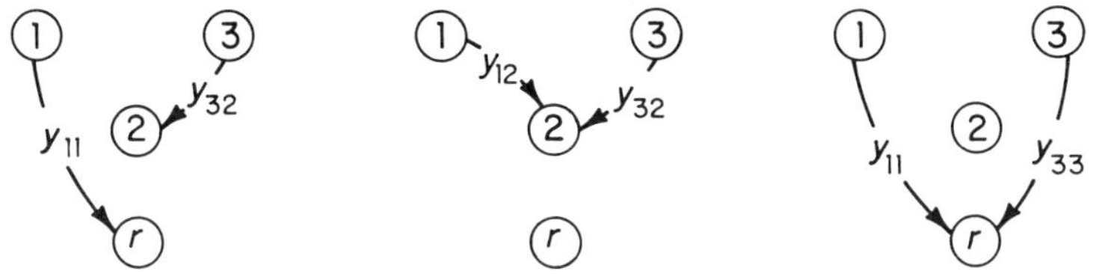

Figure 8-18 Three of the 2-trees T_{2d}^{2r} of G of Figure 8-10.

j in h_{ij}, and h^* is the complement of p_{ij} with respect to h_{ij} (such that $p_{ij} \cup h^* = h_{ij}$ and $p_{ij} \cap h^* = \phi$). Then h_{ij} is a *cyclic one-connection* and h^* is a cyclic connection of $G_c^{p_{ij}}$ where $G_c^{p_{ij}}$ is obtained from G_c by the removal of all the edges incoming to and outgoing from p_{ij}, and of all the isolated nodes. Let C_1 be the set $\{h_{ij}^{(k)}\}$ or all one-connections of the form h_{ij} in G_c, and let C_2 be the subset of C_1 such that each h_{ij} in C_2 satisfies one of the following three conditions:

(a) It is *a cyclic one-connection.*
(b) It *contains at least one directed circuit of length* ≥ 2.
(c) Its gain is equal to that of one of the connections in (b).

Also, let $T_{2d}^{ij(k)}$ be an element of the set $\{T_{2d}^{ij(k)}\}$ of all directed 2-trees with reference nodes i and j in G. Then we state the following theorem for the canceled one-connections corresponding to Theorem 8-1 for the canceled connections.

THEOREM 8-3. With the preceding notation,

$$\sum_k (-1)^{q_{2k}-1} f_c(h_{ij2}^{(k)}) = 0 \tag{8-12}$$

where $h_{ij2}^{(k)}$ is an element in C_2 (which is a one-connection of the form h_{ij}); q_{2k} is the number of even components in $h_{ij2}^{(k)}$, and the summation is on k over all $h_{ij2}^{(k)}$ in C_2. Furthermore, there exists a one-to-one correspondence between the elements $h_{ij3}^{(k)}$ in C_3 and the elements $T_{2d}^{ij(k)}$ in $\{T_{2d}^{ij(k)}\}$ such that if $h_{ij3}^{(k)}$ corresponds to $T_{2d}^{ij(k)}$, then

$$(-1)^{q_{3k}-1} f_c(h_{ij3}^{(k)}) = f(T_{2d}^{ij(k)}) \tag{8-13}$$

where $C_3 = C_1 - C_2$, and q_{3k} is the number of even components in $h_{ij3}^{(k)}$.

A proof of Theorem 8-3 can be found in a paper of Chen.[13] Again, a direct consequence of this theorem is the following. (Also, see Theorem 8-2 for the parallelism.)

THEOREM 8-4.[13] The (i,j)-cofactor Δ_{ij} of the node determinant $\Delta \equiv \det Y_n$ of an active network N with mutual inductances is given by

$$\Delta_{ij} = \sum_k (-1)^{q_{3k}-1} f_c(h_{ij3}^{(k)}) = \sum_k f(T_{2d}^{ij(k)}) \tag{8-14}$$

where the first summation is over all $h_{ij3}^{(k)}$ in C_3, and the second over all $T_{2d}^{ij(k)}$ in G, and all other notations are the same, as defined in Theorem 8-3.

As a special case of Theorem 8-4, we may state the next corollary for a passive network without mutual inductances.

COROLLARY 8-2. For a passive network N without mutual inductances, the (i,j)-cofactor Δ_{ij} of the node determinant $\Delta \equiv \det y_n$ is given by

$$\Delta_{ij} = \sum_k (-1)^{q_{3k}-1} f_c(h_{ij3}^{(k)}) = \sum_k f(T_{2ij,r}^{(k)}) \tag{8-15}$$

where $f(T_{2ij,r}^{(k)})$ is a 2-tree admittance product of the 2-tree $T_{2ij,r}^{(k)}$ of N with node i in one component and node j in the other, and node r as the datum node (that is, the node with respect to which Y_n is written). All other notations are the same as in Theorem 8-4.

Note that Equation (8-15) is identical to Equation (6-82) in Theorem 6-1, which, of course, is a Maxwell formula for the evaluation of a general (i,j)-cofactor Δ_{ij} of the node determinant $\Delta \equiv \det Y_n$ of a passive network with no mutual inductances.

In this section we have discussed the topological formulas of Chen for the evaluation of the node determinant Δ and the cofactors Δ_{ij} of an active network containing mutual inductances. We have also shown how these formulas may be reduced to the corresponding Maxwell formulas when the network under study is a passive one with no mutual inductances. Furthermore, we have established the intimate relationships (one-to-one correspondence) between (a) the Coates graph G_c and the corresponding digraph G, (b) the uncanceled connections $h_3^{(k)}$ of G_c and the directed trees $T_d^{(r)k}$ of G, and (c) the uncanceled one-connections $h_{ij3}^{(k)}$ of G_c and the directed 2-trees $T_{2d}^{ij(k)}$ of G. In the next section we shall discuss some of the applications of these topological formulas to active networks allowing, also, the presence of mutual inductances.

[13]See W. K. Chen [20] for the proofs of both Theorems 8-3 and 8-4.

8-4 DIGRAPH ANALYSIS OF ACTIVE NETWORKS

Based on the theory of digraphs developed in the preceding section, we shall discuss in this section some of the applications of the digraphs to the analysis of active networks. We shall first formulate a simple procedure for the construction of the associated digraph of a network consisting of resistors, capacitors, inductors, transformers, transistors, and tubes as elements. Examples will then be given to illustrate some of the important ideas involved.

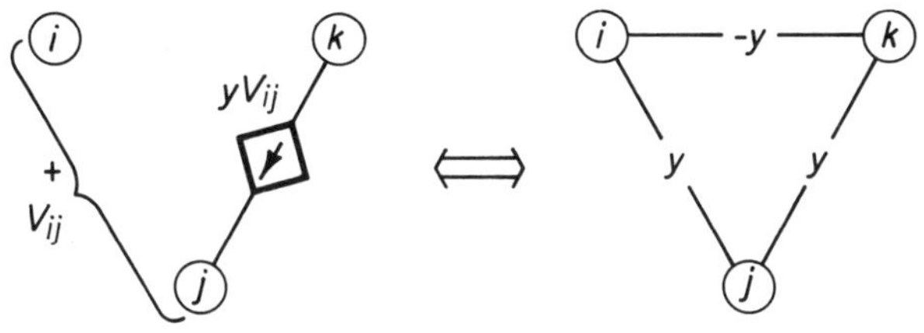

(a) (b)

Figure 8-19 A dependent source and its corresponding graph.

Consider a dependent current source, the equivalent circuit of which is shown in Figure 8-19(a). This obviously is the equivalent circuit of a vacuum tube triode with node i identified by the grid, node j by the cathode, and node k by the plate. We see that the dependent current source is connected between nodes j and k with the positive current away from node k. Thus the quantity $yV_{ij} = y(V_i - V_j)$ will enter the node-analysis matrix equation $Y_nV_n = I_n$, as

$$\begin{array}{c} \\ \cdot \\ \cdot \\ \cdot \\ j \\ \cdot \\ \cdot \\ \cdot \\ k \\ \cdot \\ \cdot \\ \cdot \end{array}
\underbrace{\overset{\begin{array}{ccccc}\cdots & i & \cdots & j & \cdots\end{array}}{\begin{bmatrix} & & & & \\ & & & & \\ & & & & \\ & -y & & y & \\ & & & & \\ & & & & \\ & & & & \\ & y & & -y & \\ & & & & \\ & & & & \\ & & & & \end{bmatrix}}}_{Y_n}
\underbrace{\begin{bmatrix} \cdot \\ \cdot \\ \cdot \\ V_i \\ \cdot \\ \cdot \\ \cdot \\ V_j \\ \cdot \\ \cdot \\ \cdot \end{bmatrix}}_{V_n} = I_n$$

where only those elements that are relevant to the discussion are shown in Y_n. Finally, recalling Steps (2) and (3) of the procedure for associating a digraph with Y_n (Section 8-3), and observing that the (j,i) element of Y_n is $-y$, the (k,i) element of Y_n is y, and the (k,j) element of Y_n is $-y$, it is evident that in the digraph there is an edge from j to i with a value $-(-y)$, an edge from k to i with a value $-(y)$, and an edge from k to j with a value $-(-y)$, as depicted in Figure 8-19(b).

This very same idea can also be applied to derive the digraph associated with a transistor as well as that for other active elements.[14] The procedure for obtaining the associated digraph from a given active network is outlined in the following steps:

(1) Obtain an equivalent network from the given network by replacing all the active elements by their corresponding equivalent circuits.

(2) Dependent sources are replaced by their corresponding graphs as illustrated in Figure 8-19. This is accomplished by first removing the dependent sources and then superimposing their corresponding graphs on the equivalent circuit obtained in Step (1), as shown in Figure 8-19.

(3) Replace each passive element by a pair of oppositely directed edges with equal weights — the weight of each edge is the admittance of the passive element.

(4) Combine the parallel edges (that is, edges oriented in the same direction between the same pair of nodes), if any, such that the weight of the resultant edge is the sum of the weights of the parallel edges.

(5) Choose a reference node, denoted by r, and remove all the outgoing edges from r (since these edges will not be contained in any of the directed trees and directed 2-trees).

Note that we have not considered mutual inductances in the foregoing

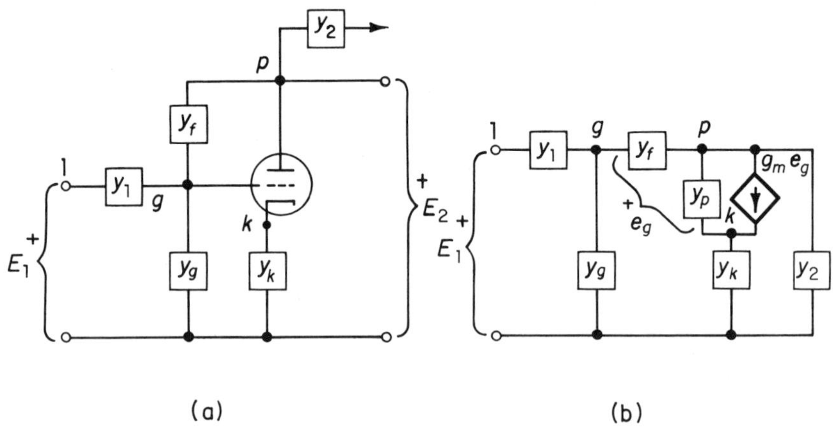

Figure 8-20 (a) A voltage feedback amplifier. (b) Its equivalent network.

[14]Some of the applications of this idea are listed as problems at the end of this chapter.

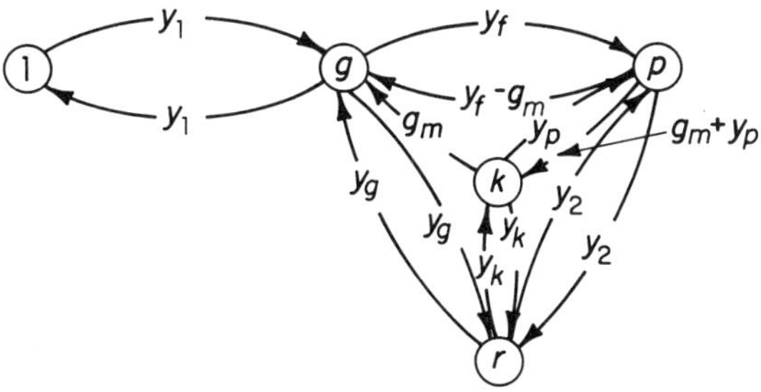

Figure 8-21 The associated digraph of the equivalent network.

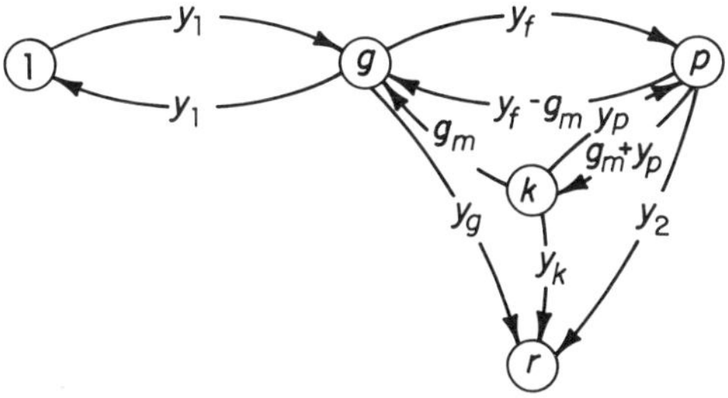

Figure 8-22 The simplified digraph G.

procedure, because if the given network consists of mutual inductances it will be much simpler to draw the associated digraph from the node admittance matrix instead of trying to obtain the graph directly from the network. An example will be given later in this section to illustrate how this procedure may be accomplished. But, first, let us study the following example:

EXAMPLE 8-7. Consider the general voltage-feedback amplifier shown in Figure 8-20(a); its equivalent network is shown in Figure 8-20(b). The associated digraph of the equivalent network is depicted in Figure 8-21. Now, with the reference node identified by r and the outgoing edges removed from r, we obtain the simplified digraph, as shown in Figure 8-22. With the simplified digraph G on hand, it is straightforward to evaluate the node determinant Δ and its cofactors. This is left to the reader as an exercise.

EXAMPLE 8-8. Consider the network N of Figure 8-23 with mutual inductances. The node admittance matrix Y_n is related to the current-source vector (column matrix) I_n and the node-voltage vector V_n by the node-analysis matrix equation $I_n = Y_n V_n$:

$$\begin{bmatrix} I_1 \\ 0 \\ 0 \\ 0 \end{bmatrix} = \begin{bmatrix} G_1 + G_7 & -G_1 & 0 & 0 \\ -G_1 & G_1 + G_2 + G_3 & -G_3 & 0 \\ 0 & -G_3 & G_3 + y_5 & y \\ 0 & 0 & 0 & G_6 + y_4 \end{bmatrix} \begin{bmatrix} V_1 \\ V_2 \\ V_3 \\ V_4 \end{bmatrix}$$

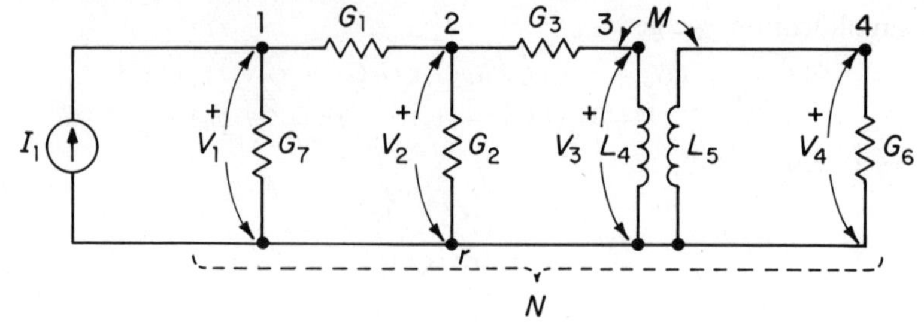

Figure 8-23 A passive network N with mutual inductances.

where

$$y_4 = \frac{L_4}{s\Delta} \qquad y_5 = \frac{L_5}{s\Delta} \qquad y = \frac{-M}{s\Delta} \qquad \Delta = L_4L_5 - M^2(\neq 0)$$

since

$$\begin{bmatrix} I_3 \\ I_4 \end{bmatrix} = \begin{bmatrix} y_5 & y \\ y & y_4 \end{bmatrix} \begin{bmatrix} V_3 \\ V_4 \end{bmatrix}$$

The associated digraph G is shown in Figure 8-24,[15] and the corresponding

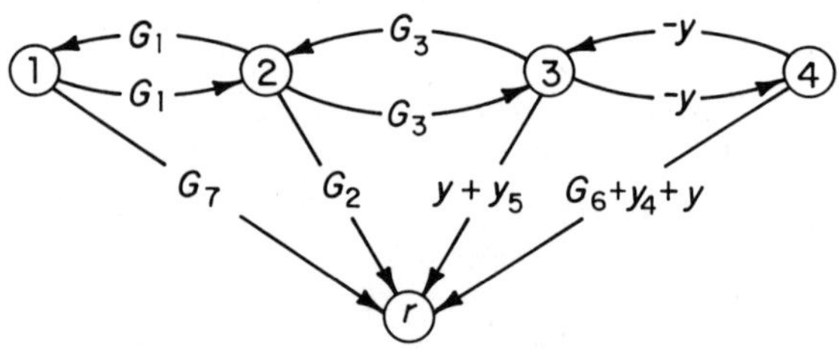

Figure 8-24 The associated digraph G of N.

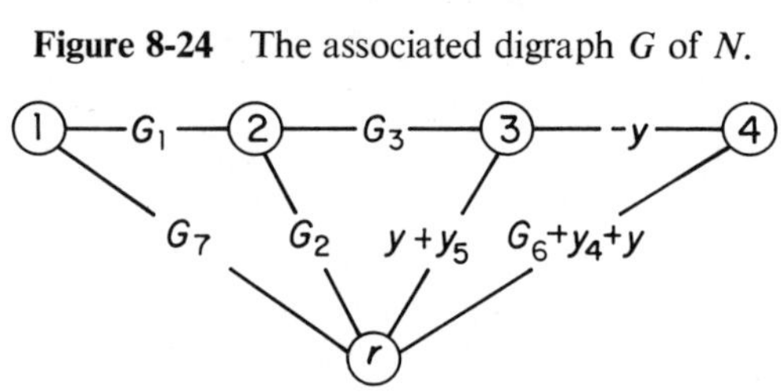

Figure 8-25 The corresponding undirected graph G_u of G.

undirected graph G_u in Figure 8-25. By Corollary 8-1, the sum of all tree-admittance products of G_u is det Y_n:

$$\begin{aligned}
\det Y_n = (G_6 + y_4 + y)\{&G_7G_1G_3 + G_2G_1G_3 + (y_5 + y)G_1G_3 + G_7G_2G_3 \\
&+ G_7(y_5 + y)G_1 + G_7G_2(y_5 + y) + G_7(y_5 + y) + G_2(y_5 + y)G_1 \\
&- y(G_1G_7 + G_3G_7 + G_2G_7 + G_1G_3 + G_1G_2)\} \\
&- y(y + y_5)\{(G_2 + G_3)G_7 + (G_2 + G_3)G_1 + G_1G_7\} \\
&- yG_3G_2(G_1 + G_7) - yG_3G_1G_7
\end{aligned}$$

[15]The derivation of the digraph of an ideal transformer is left as a problem (see Problem 8-6).

After simplification, we get

$$\det Y_n = (G_6 + y_4)\{(G_1 + G_7)(G_2G_3 + G_3y_5 + G_2y_5) + G_1G_7(G_3 + y_5)\} - y^2(G_2 + G_3) - y^2G_1G_7$$

Next, by Corollary 8-2,

$$\Delta_{14} = \sum_k T_{214,r}^{(k)}(y) = G_1G_3(-y)$$

and

$$\Delta_{13} = \sum_j T_{213,r}^{(j)}(y) = G_1G_3(-y) + G_1G_3(G_6 + y_4 + y) = G_1G_3(G_6 + y_4)$$

The sets[16] of 2-trees $\{T_{214,r}^{(k)}\}$ and $\{T_{213,r}^{(j)}\}$ corresponding to Δ_{14} and Δ_{13} are shown in Figures 8-26 and 8-27, respectively.

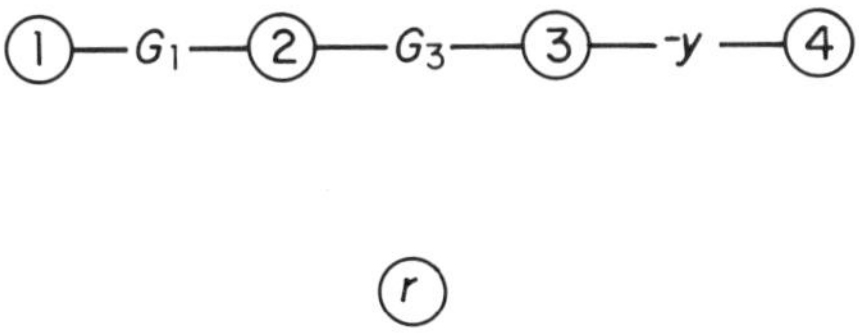

Figure 8-26 The set of all 2-trees $\{T_{214,r}^{(k)}\}$ of G_u.

Figure 8-27 The set of all 2-trees $\{T_{213,r}^{(j)}\}$ of G_u.

From Examples 8-7 and 8-8, we note that the digraph analysis of Chen still produces terms that cancel. However, in the case of active elements such as tubes and transistors, this method does avoid redundancies.

8-5 VOLTAGE GRAPH AND CURRENT GRAPH OF MAYEDA

In the preceding section we presented the digraph analysis of active networks developed by Chen. We shall devote the remainder of this chaper to a concise discussion of the two-graph method of Mayeda for the development

[16]Note the difference between the symbol $T_{2ij,r}^{(k)}$ which denotes the kth 2-tree (with vertices i and j in one part and vertex r, the reference vertex, in the other part) of the graph of a *passive* network and the symbol $T_{2d}^{ij(k)}$ which represents the kth directed 2-tree (with vertex i being the reference vertex in one part and vertex j, the reference vertex, in the other part) of the digraph of an *active* network.

of topological formulas for active networks. As in Chen's analysis, the assumptions of reciprocity and no magnetic coupling, which were made in Chapters 5 and 6, are now removed so that nonreciprocal elements as well as mutual inductances will be included.

DEFINITION 8-10. Transadmittance $y_{ij,mn}$. A *transadmittance* $y_{ij,mn}$ is a network element that satisfies the equation

$$y_{ij,mn}v_{mn}(s) = i_{ij}(s) \tag{8-16}$$

where $v_{mn}(s)$ is the voltage transform across vertices m and n with positive polarity mark at m, and $i_{ij}(s)$ is the current transform through the edge connected between vertices i and j from i to j in the network.

We should point out that Equation (8-16) is merely a generalization of the well-known voltage-current relationship

$$y_{ij}v_{ij} = i_{ij} \tag{8-17}$$

for the passive element y_{ij}, since, if $i = m$ and $j = n$, Equation (8-16) becomes

$$y_{ij,ij}v_{ij} = i_{ij} \tag{8-18}$$

in which case the second set of subscripts is the same as the first in $y_{ij,ij}$ and, hence, may be omitted. Thus we write

$$y_{ij} \equiv y_{ij,ij} \tag{8-19}$$

and Equation (8-18) reduces to (8-17).

EXAMPLE 8-9. The topological representation of the transadmittance $y_{ij,mn}$ is illustrated in Figure 8-28. The general representation of $y_{ij,mn}$ is shown in (a); the representation of the transconductance $g_{PK,GK}$ is depicted in (b); and the representation of a passive element $y_{ij,ij}$ is illustrated in (c), which, by Equation (8-19), is identical to y_{ij} as shown in (d).

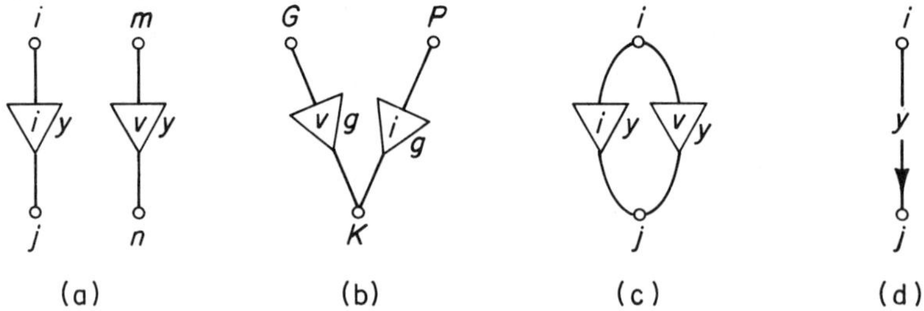

Figure 8-28 Topological representation of (a) $y_{ij,mn}$, (b) $g_{PK,GK}$, (c) $y_{ij,ij}$, and (d) y_{ij}.

Note that in Figure 8-28 an edge with a v inside a hollow arrowhead is an element which is used to represent the location of the edge voltage between the two vertices. Likewise, an edge with an i inside a hollow arrowhead is an element used to represent the location of the edge current through the edge from one vertex to the other. The arrowhead in an edge indicates the edge orientation.

We now summarize our discussion on the topological representations of various transadmittances and, hence, various electrical networks in Table 8-1.

TABLE 8-1

Topological Representations of Various Networks

Type of Elements	*Mathematical Expressions*	*Network Representation*	*Topological Representation*
Resistance, Capacitance, Self-inductance	$y_{ij}v_{ij} = i_{ij}$	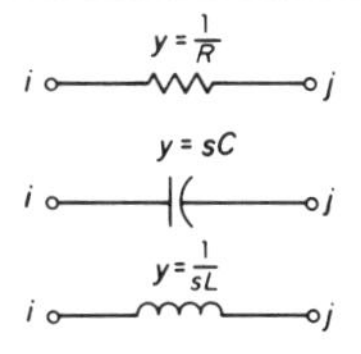	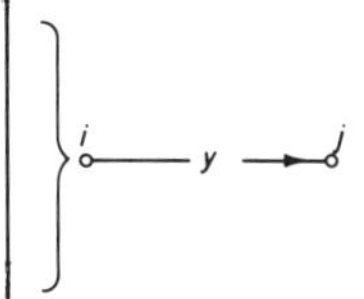
Inverse Mutual Inductance	$i_{ij} = \left(\frac{L_{mn}}{s\Delta}\right)v_{ij} + \left(-\frac{M_{ij,mn}}{s\Delta}\right)v_{mn}$ $= \Gamma_{mn}v_{ij} + \Gamma_{ij,mn}v_{mn}$ $i_{mn} = \left(-\frac{M_{mn,ij}}{s\Delta}\right)v_{ij} + \left(\frac{L_{ij}}{s\Delta}\right)v_{mn}$ $= \Gamma_{mn,ij}v_{ij} + \Gamma_{ij}v_{mn}$ where $\Delta = L_{ij}L_{mn} - M_{ij,mn}^2$	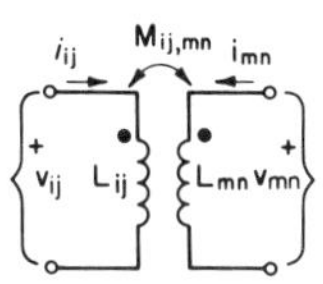	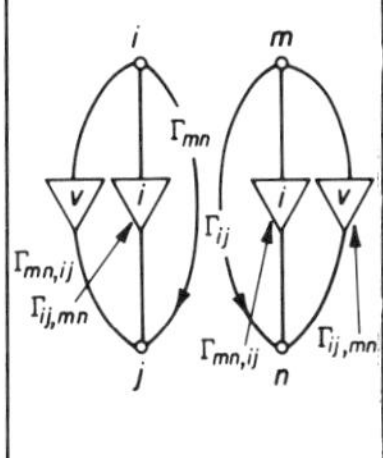
Vacuum Tube	$g_{mPK,GK}v_{GK} = i_{PK}$ (current source) $\frac{1}{r_{PPK}}v_{PK} = i_{PK}$ (in r_p)	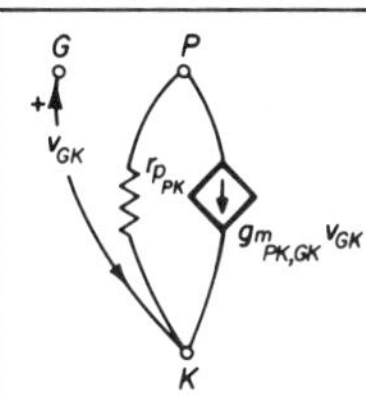	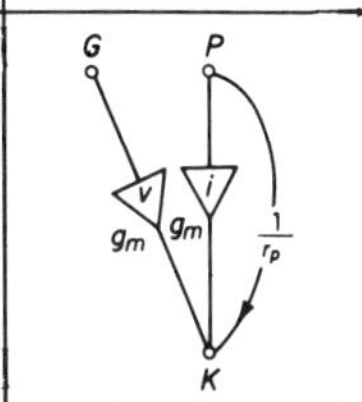
Transistor	$i_{EJ} = \frac{1}{R_{EJ}}v_{EJ} = g_{EJ}v_{EJ}$ $i_{RJC} = \frac{1}{R_{JC}}v_{JC} = g_{JC}v_{JC}$ $i_{RJB} = \frac{1}{R_{JB}}v_{JB} = g_{JB}v_{JB}$	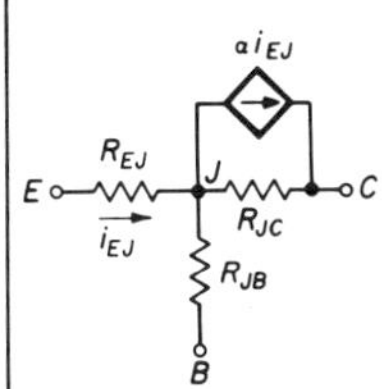	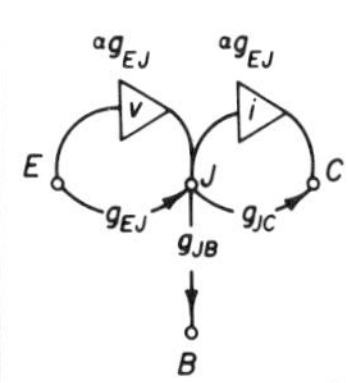

Since, in general, the locations of the edge voltage v_{mn} is different from that of the edge current i_{ij}, it is logical to represent the edge voltages and edge currents by two separate graphs such that one graph specifies the locations of all the edge voltages and the other the locations of all the edge currents in the given network. Thus we define the following terms:

DEFINITION 8-11. Voltage Graph, Current Graph, and Resultant Graph. An electrical network consisting of elements of the form $y_{ij,mn}$ that satisfies Equation (8-16) is represented by two graphs called *voltage graph* G_v and *current graph* G_i such that G_v contains all the edges with the v inside the arrowheads and the passive edges, and G_i contains all the edges with the i inside the arrowheads and the passive edges. The graph $\underline{G}$, obtained by superimposing G_v and G_i, is called the *resultant graph* of the given network.

The procedure for obtaining the resultant graph and the voltage and current graphs of a given network N is outlined in the following steps:

(1) Represent the nodes of N by the same number of vertices and enter the "passive" edges between vertices according to the way the passive elements of N are connected between the nodes.

(2) Represent the active elements by their corresponding topological representations as shown in Table 8-1. This gives the resultant graph $\underline{G}$ of N.

(3) To obtain G_v, simply remove from $\underline{G}$ all the edges with the i's inside the arrowheads, and the remainder is G_v.

(4) Similarly, to obtain G_i we remove from $\underline{G}$ all the v edges, and the remainder gives G_i.

EXAMPLE 8-10. Consider the network N shown in Figure 8-29, which contains a transistor and magnetic coupling as well as some passive elements. As the first step, we obtain the partial graph as shown in Figure 8-30. Next,

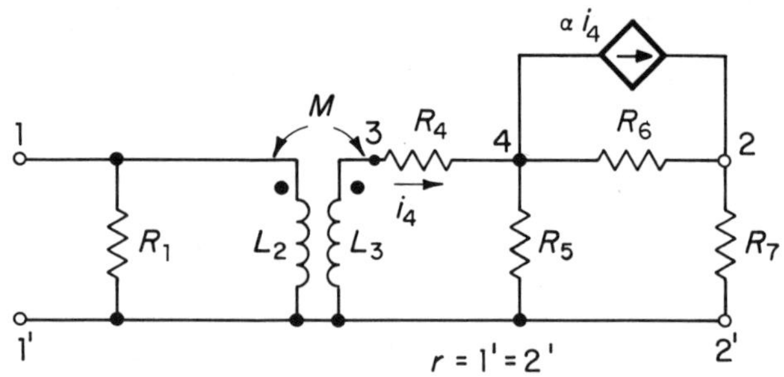

Figure 8-29 An active network N with magnetic coupling.

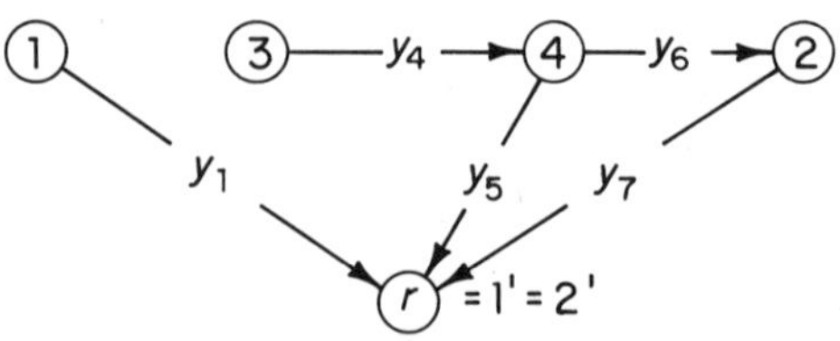

Figure 8-30 The partial graph containing "passive" edges only.

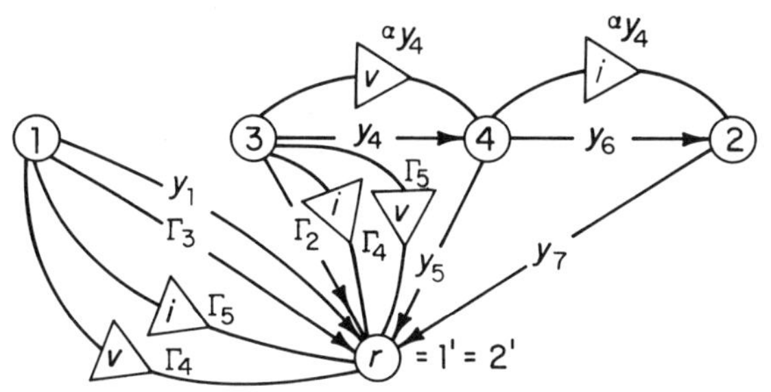

$$\begin{cases} y_i = \frac{1}{R_i}, \; i = 1,4,5,6,7 \\ \Gamma_2 = \frac{L_2}{s\Delta} \\ \Gamma_3 = \frac{L_3}{s\Delta} \\ \Gamma_4 = \Gamma_5 = \frac{-M}{s\Delta} \\ \Delta = L_2 L_3 - M^2 \end{cases}$$

Figure 8-31 The resultant graph $\underline{G}$ of N.

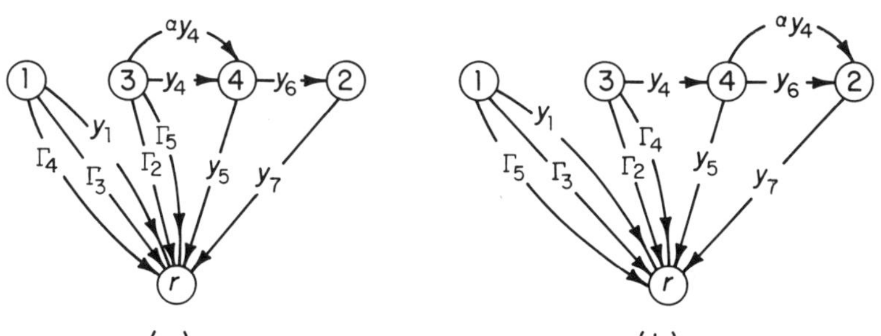

Figure 8-32 (a) The voltage graph G_v. (b) The current graph G_i of $\underline{G}$.

we enter the topological representations of the transformer and the transistor to form the resultant graph $\underline{G}$ as illustrated in Figure 8-31. Finally, omitting the i edges, we find G_v, as shown in Figure 8-32(a); and omitting the v edges, we obtain G_i, as illustrated in Figure 8-32(b).

Note that in the voltage graph and current graph we have omitted the symbols v and i for the v edges and the i edges, respectively, since, in G_v, all the edges are "voltage" edges because they all belong to the voltage graph, and similar comments may be made for the current graph G_i. The orientation of a passive edge is entirely arbitrary, but the orientation of a pair of edges for the transadmittance $y_{ij,mn}$ is fixed by the order in which it appears; that is, for $y_{ij,mn}$, if we assign the orientation of the i edge to be directed from i to

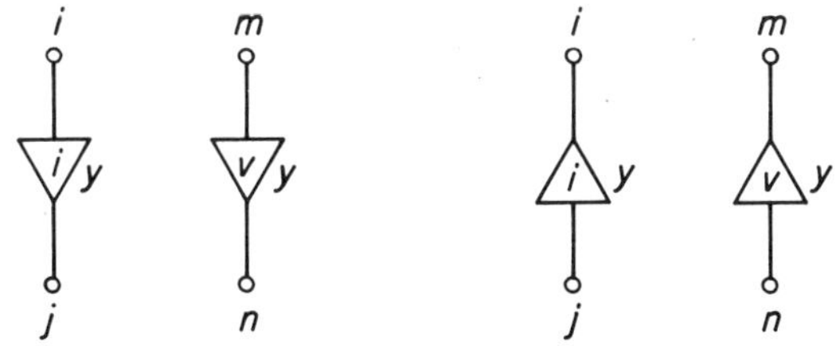

Figure 8-33 Illustrations for the orientation of the pair of edges representing $y_{ij,mn}$.

j, then the orientation of the v edge is fixed and must be directed from m to n. If, on the other hand, we assign the orientation of the v edge to be directed from n to m, then the orientation of the i edge must be from j to i, accordingly. In other words, when the orientation of one of the edges of the pair is assigned, the orientation of the other is automatically fixed. This is illustrated in Figure 8-33.

8-6 SIGN PERMUTATION AND PRINCIPAL TREE

As seen in Chapters 5 and 6, much of the simplicity of the topological formulas derived for passive reciprocal networks without mutual inductance is due to the fact that the edge admittance matrix $Y_e(s)$ is a diagonal matrix. However, in the case of active networks with magnetic coupling, Y_e will no longer be a diagonal matrix. This point is illustrated in the following example:

EXAMPLE 8-11. Consider the network N shown in Figure 8-34(a). The equivalent circuit and the (ordinary) graph G are shown in Figure 8-34(b)

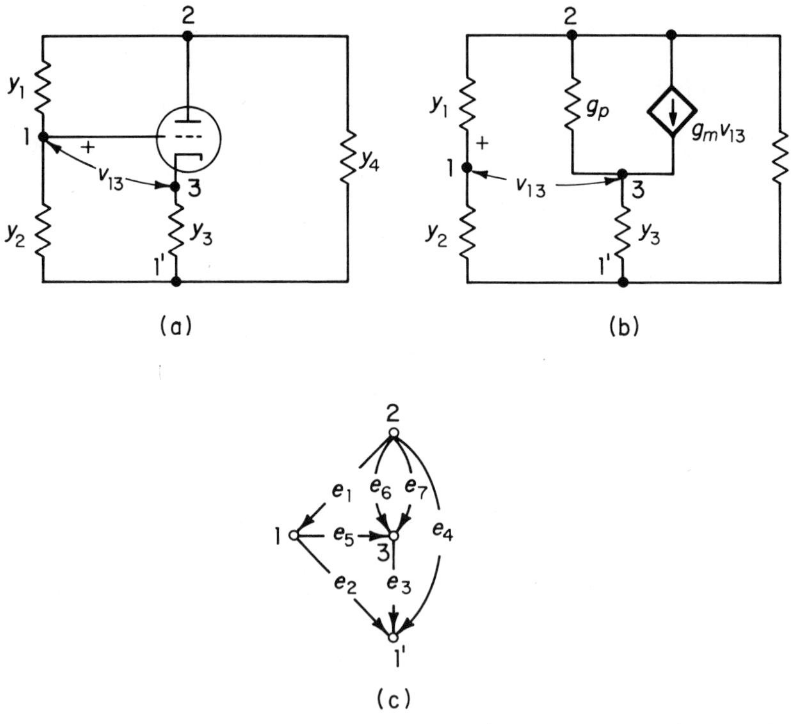

Figure 8-34 (a) An active network N. (b) The equivalent circuit. (c) The graph G of N.

and (c), respectively. The current-voltage matrix equation for N is given by $I_e(s) = Y_c(s)V_e(s)$:

$$\begin{bmatrix} i_1 \\ i_2 \\ i_3 \\ i_4 \\ i_5 \\ i_6 \\ i_7 \end{bmatrix} = \begin{bmatrix} y_1 & 0 & 0 & 0 & 0 & 0 & 0 \\ 0 & y_2 & 0 & 0 & 0 & 0 & 0 \\ 0 & 0 & y_3 & 0 & 0 & 0 & 0 \\ 0 & 0 & 0 & y_4 & 0 & 0 & 0 \\ 0 & 0 & 0 & 0 & 0 & 0 & 0 \\ 0 & 0 & 0 & 0 & 0 & g_p & 0 \\ 0 & 0 & 0 & 0 & g_m & 0 & 0 \end{bmatrix} \begin{bmatrix} v_1 \\ v_2 \\ v_3 \\ v_4 \\ v_5 \\ v_6 \\ v_7 \end{bmatrix} \tag{8-20}$$

where the currents i_j ($j = 1, 2, \cdots, 7$) and the voltages v_j ($j = 1, 2, \cdots, 7$) are the edge currents and edge voltages, respectively, with edge orientations as specified in G [Figure 8-34(c)].

In order to simplify the derivations of the topological formulas for active networks with mutual inductances, the voltage graph and the current graph of Mayeda's method discussed here are introduced so that Y_e of the Mayeda graph $\underline{G}$ will again be a diagonal matrix. For the development Mayeda's theory, a few more definitions are needed.

DEFINITION 8-12. Voltage Incidence Submatrix A_v and Current Incidence Submatrix A_i. The matrices A_v and A_i, called the *voltage incidence submatrix* and the *current incidence submatrix*, respectively, are the incidence submatrices of the voltage graph G_v and the current graph G_i, respectively, with the *same reference vertex*.

EXAMPLE 8-12. Let us consider the same network N as in Example 8-11. The Mayeda graph $\underline{G}$, the voltage graph G_v, and the current graph G_i of N

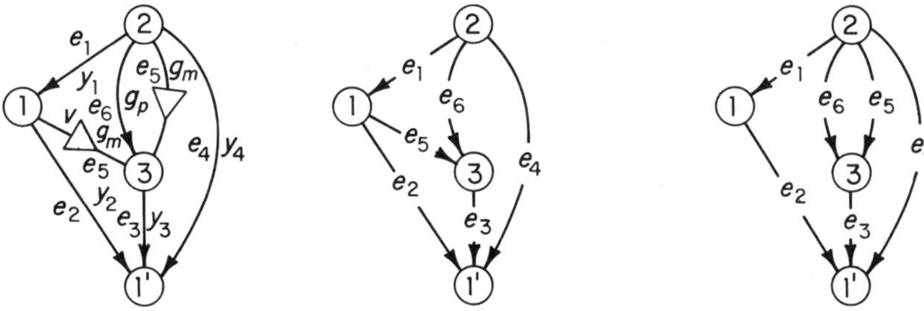

Figure 8-35 (a) Resultant graph $\underline{G}$. (b) Voltage graph G_v. (c) Current graph G_i of N.

are depicted in Figure 8-35(a), (b), and (c), respectively. The incidence submatrices A_v of G_v and A_i of G_i with the same reference vertex 1′ are given by

$$A_v = \begin{array}{c} \\ 1 \\ 2 \\ 3 \end{array} \overset{\begin{array}{cccccc} e_1 & e_2 & e_3 & e_4 & e_5 & e_6 \end{array}}{\begin{bmatrix} -1 & 1 & 0 & 0 & 1 & 0 \\ 1 & 0 & 0 & 1 & 0 & 1 \\ 0 & 0 & 1 & 0 & -1 & -1 \end{bmatrix}} \tag{8-21}$$

and

$$A_i = \begin{matrix} \\ 1 \\ 2 \\ 3 \end{matrix}\begin{matrix} e_1 & e_2 & e_3 & e_4 & e_5 & e_6 \\ \left[\begin{matrix} -1 \\ 1 \\ 0 \end{matrix}\right. & \begin{matrix} 1 \\ 0 \\ 0 \end{matrix} & \begin{matrix} 0 \\ 0 \\ 1 \end{matrix} & \begin{matrix} 0 \\ 1 \\ 0 \end{matrix} & \begin{matrix} 0 \\ 1 \\ -1 \end{matrix} & \left.\begin{matrix} 0 \\ 1 \\ -1 \end{matrix}\right] \end{matrix} \tag{8-22}$$

Before we go on with the development of Mayeda's theory, the following remarks are in order. We note, from Example 8-12, that in Mayeda's modified graph $\underline{G}$, the quantity g_m is represented by two edges — the v edge and the i edge — both denoted by the same symbol e_5 (or g_m). Next, we note that in both graphs, G_v and G_i, there is an edge identified by e_5. The reason for this notation is that, in Mayeda's development, the two edges denoted by the same symbol, e_5 (or g_m), are considered to be the *same edge, occupying different positions.* This is, in fact, one of the most important concepts in Mayeda's method of developing the topological formulas for active networks!

Let us now come back to the discussion of Mayeda's theory. Consider a general linear network N, and let A_v and A_i be the incidence submatrices of G_v and G_i of $\underline{G}$ of N, respectively, with vertex r as the reference vertex, such that column e_j of A_v and A_i represents edge e_j of $\underline{G}$ for all j. Then, by Kirchhoff's current law, we write

$$A_i I_e = J_n \tag{8-23}$$

where I_e is the column matrix of edge currents and J_n the column matrix of source currents applied to N and each of these currents is directed *from* vertex r. [The formulation of Equation (8-23) will be illustrated in an example to be given later.] Also, the node transformation matrix equation [the Laplace transform of Equation (4-36)] is given by

$$V_e = A_v^T V_n \tag{8-24}$$

where V_e is the column matrix of edge voltages and V_n the column matrix of node voltages with respect to vertex r. Since each element in N satisfies the voltage-current relationship given in Equation (8-16), we write

$$Y_e V_e = I_e \tag{8-25}$$

where Y_e, defined by

$$Y_e = \begin{bmatrix} y_1 & 0 & \cdots & 0 \\ 0 & y_2 & \cdots & 0 \\ \cdot & & & \\ \cdot & & & \\ \cdot & & & \\ 0 & 0 & \cdots & y_e \end{bmatrix} \tag{8-26}$$

is the edge admittance matrix of $\underline{G}$ and is thus a diagonal matrix of order e

(e being the number of edges of $\underline{G}$). Combining Equations (8-23), (8-24), and (8-25), we get

$$Y_n V_n = J_n \tag{8-27}$$

where

$$Y_n \triangleq A_i Y_e A_v^T \tag{8-28}$$

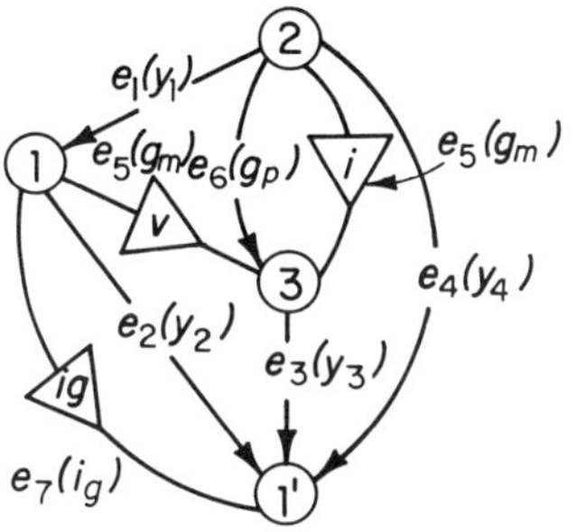

Figure 8-36 The new graph $\underline{\underline{G}}$ obtained from G by adding e_7 across vertices 1 and 1′.

EXAMPLE 8-13. Let the network N be the active network shown in Figure 8-34(a). Furthermore, let N be driven by a current source i_g connected in parallel with y_2 between nodes 1 and 1′. The edge e_7, representing i_g, is then added in parallel with edge e_2 between vertices 1 and 1′ of the resultant graph $\underline{G}$ of Figure 8-35(a). The new graph $\underline{\underline{G}}$ is shown in Figure 8-36. The edge admittance matrix Y_e of $\underline{G}$ (not $\underline{\underline{G}}$) is

$$Y_e = \begin{array}{c} \begin{matrix} e_1 & e_2 & e_3 & e_4 & e_5 & e_6 \end{matrix} \\ \begin{bmatrix} y_1 & 0 & 0 & 0 & 0 & 0 \\ 0 & y_2 & 0 & 0 & 0 & 0 \\ 0 & 0 & y_3 & 0 & 0 & 0 \\ 0 & 0 & 0 & y_4 & 0 & 0 \\ 0 & 0 & 0 & 0 & g_m & 0 \\ 0 & 0 & 0 & 0 & 0 & g_p \end{bmatrix} \end{array} \tag{8-29}$$

Note once again that Y_e is of order 6 (not 7) since g_m in $\underline{G}$ is a *single edge* (not two) occupying two different positions! The incidence submatrices of G_i and G_v are given by Equations (8-22) and (8-21), respectively. Hence, by Equations (8-22) and (8-23), we get $A_i I_e = J_n$ as follows:

$$\begin{bmatrix} -1 & 1 & 0 & 0 & 0 & 0 \\ 1 & 0 & 0 & 1 & 1 & 1 \\ 0 & 0 & 1 & 0 & -1 & -1 \end{bmatrix} \begin{bmatrix} i_1 \\ i_2 \\ i_3 \\ i_4 \\ i_5 \\ i_6 \end{bmatrix} = \begin{bmatrix} -i_g \\ 0 \\ 0 \end{bmatrix} \tag{8-30}$$

Note that A_i is of order 3×6 and I_e is of order 6×1 so that J_n is a 3×1 column matrix. [In general, J_n is a $(v-1) \times 1$ matrix with each row identifying a vertex of $\underline{G}$.] Note also that the (1,1) element of J_n in Equation (8-28) is $-i_g$ because i_g is the "source" current applied to $\underline{G}$ connecting at vertex 1. The minus sign signifies that i_g is directed *toward* the reference vertex r $(=1')$. [Current i_g would be positive if it were directed from vertex r as defined in Equation (8-23)].

Next, by Equations (8-21), (8-22), (8-28) and (8-29), we write $Y_n = A_i Y_e A_v^T$ as follows:

$$Y_n = \begin{bmatrix} -1 & 1 & 0 & 0 & 0 & 0 \\ 1 & 0 & 0 & 1 & 1 & 1 \\ 0 & 0 & 1 & 0 & -1 & -1 \end{bmatrix} \begin{bmatrix} y_1 & 0 & 0 & 0 & 0 & 0 \\ 0 & y_2 & 0 & 0 & 0 & 0 \\ 0 & 0 & y_3 & 0 & 0 & 0 \\ 0 & 0 & 0 & y_4 & 0 & 0 \\ 0 & 0 & 0 & 0 & g_m & 0 \\ 0 & 0 & 0 & 0 & 0 & g_p \end{bmatrix} \begin{bmatrix} -1 & 1 & 0 \\ 1 & 0 & 0 \\ 0 & 0 & 1 \\ 0 & 1 & 0 \\ 1 & 0 & -1 \\ 0 & 1 & -1 \end{bmatrix}$$

or

$$Y_n = \left[\begin{array}{c|c|c} y_1 + y_2 & -y_1 & 0 \\ \hline -y_1 + g_m & y_1 + y_4 + g_p & -g_m - g_p \\ \hline -g_m & -g_p & y_3 + g_m + g_p \end{array}\right] \tag{8-31}$$

Thus Equation (8-27) becomes

$$\left[\begin{array}{c|c|c} y_1 + y_2 & -y_1 & 0 \\ \hline -y_1 + g_m & y_1 + y_4 + g_p & -g_m - g_p \\ \hline -g_m & -g_p & y_3 + g_m + g_p \end{array}\right] \begin{bmatrix} v_{1r} \\ v_{2r} \\ v_{3r} \end{bmatrix} = \begin{bmatrix} -i_g \\ 0 \\ 0 \end{bmatrix} \tag{8-32}$$

From Equation (8-31) we note that the node admittance matrix Y_n of an active network is no longer a symmetric matrix.

We shall develop a topological expression of the determinant Δ of Y_n, similar to Equation (5-18) as stated in Theorem 5-2 for a passive network with no mutual inductances. But, first, let us state the following definition.

DEFINITION 8-13. Common Tree T_c (Complete Tree) and Common-tree Admittance Product $T_c(y)$. A *common tree* (or *complete tree*) of G is a subgraph of $\underline{G}$ consisting of $v-1$ edges, such that these edges constitute a tree in both G_v and G_i of $\underline{G}$. The product of the admittances of all the edges of a common tree is called the *common-tree admittance product* of that tree.

EXAMPLE 8-14. Consider the graphs $\underline{G}$, G_i, and G_v of the network N in Example 8-12, as shown in Figure 8-35. Two of the common trees of $\underline{G}$ are $T_c^{(1)} = \{e_2,e_3,e_4\}$ and $T_c^{(2)} = \{e_1,e_2,e_5\}$, as shown in Figure 8-37 (a) and (b), respectively. The common-tree admittance products of the two common trees $T_c^{(1)}$ and $T_c^{(2)}$ are $y_2y_3y_4$ and $y_1y_2g_m$, respectively.

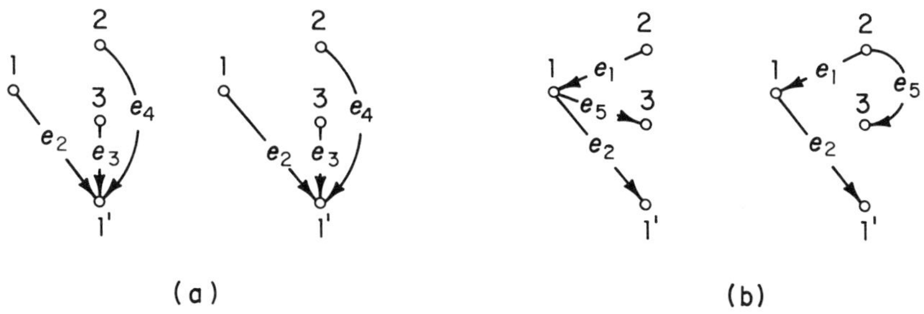

Figure 8-37 Two common trees of $\underline{G}$: (a) $T_c^{(1)} = \{e_2,e_1,e_1\}$. (b) $T_c^{(2)} = \{e_1,e_2,e_1\}$.

Observe that since common tree $T_c^{(1)}$ contains passive edges only, it appears in the same form as a tree of both G_v and G_i. Common tree $T_c^{(2)}$, on the other hand, contains an active element (e_5); hence, it is a tree of G_v in one form and a tree of G_i in a different form (due to the different positions edge e_5 occupies), as illustrated in Figure 8-37(b).

We are now in a position to state and prove the next theorem.

THEOREM 8-5. Let Y_n be the node admittance matrix of a general linear network N, which may contain nonreciprocal elements and mutual inductances. Then the determinant Δ of Y_n is given by

$$\Delta = \det Y_n = \sum_i \epsilon_i \cdot T_c^{(i)}(y) \tag{8-33}$$

where $T_c^{(i)}(y)$ is the common-tree admittance product of the common tree $T_c^{(i)}$; ϵ_i is the sign (either $+1$ or -1) of $T_c^{(i)}(y)$ determined by the product of the major of A_i and the corresponding major of A_v, and the summation is over all common trees $T_c^{(i)}$ of the resultant graph G of N.

Proof: The proof of this theorem is quite similar to that of Theorem 5-2. By the Binet-Cauchy theorem (5-1), the determinant Δ of Y_n can be expressed as

$$\Delta = |(A_iY_e)A_v^T| = \Sigma \text{ corresponding majors of } (A_iY_e) \text{ and } A_v^T$$

Now, let $e_{j_1}, e_{j_2}, \ldots, e_{j_{v-1}}$ be the set of $v - 1$ columns of A_i that forms a major of A_i. Then, following a similar argument given in the proof of Theorem 5-2, we see that

$$\text{major of } A_iY_e = y_{j_1}y_{j_2}\ldots y_{j_{v-1}} \times (\text{major of } A_i) \tag{8-34}$$

Thus we write

$$\Delta = \sum y_{j_1}y_{j_2}\ldots y_{j_{v-1}} \times (\text{major of } A_i) \times (\text{corresponding major of } A_v^T) \tag{8-35}$$

But, for a major of A_i to be nonzero, the set of edges corresponding to the columns of that major must constitute a tree of G_i (Theorem 2-12). Similarly, the $v - 1$ edges corresponding to the columns of a nonzero major of A_v form a tree of G_v. Finally, we note that the value of a nonzero major of either A_i or A_v is ± 1. Hence the theorem.

We next define the following terms:

DEFINITION 8-14. Principal Vertex. The *principal vertex* of an edge e_i of a tree T with reference to vertex r is the one endpoint of e_i such that the unique path between that endpoint and vertex r contains e_i.

EXAMPLE 8-15. Let T be tree of a graph G with reference vertex r as illustrated in Figure 8-38. For edge e_1, vertex 1 is the principal vertex; for e_2, the principal vertex is 2; for e_3, it is 3; and for e_4, it is 4.

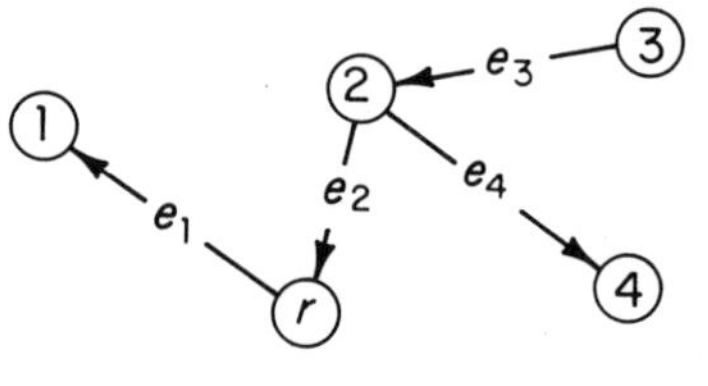

Figure 8-38 A tree T of a graph G with reference vertex r.

Since the path between any pair of vertices in a given tree is always unique, there is exactly one principal vertex for any edge in the tree.

DEFINITION 8-15. Sign Permutation and Its Value. The *sign permutation* for a common tree T_c is a $2 \times (v - 1)$ "matrix" with rows corresponding to G_i and G_v and columns corresponding to vertices such that

$$\text{The } (1,j) \text{ entry} = \begin{cases} y_k & \text{if the principal vertex of } y_k \text{ is } j \text{ in } T_{c_i} \\ & \text{and the orientation of } y_k \text{ is away from } j \\ \bar{y}_k & \text{if the principal vertex of } y_k \text{ is } j \text{ in } T_{c_i} \\ & \text{and the orientation of } y_k \text{ is toward } j \end{cases}$$

where T_{c_i} is the tree in G_i corresponding to T_c.

The $(2,j)$ entry is defined in a similar manner except that T_{c_i} is replaced by T_{c_v} in the foregoing definition, T_{c_v} being the tree in G_v corresponding to the same tree T_c. Finally, the *value* of a sign permutation is defined as $(-1)^{p+q}$, where p is the number of interchanges of columns required in one row of the sign permutation to make the rows identical (neglecting the superscript minus signs) and q is the total number of superscript minus signs in the sign permutation.

EXAMPLE 8-16. The sign permutation of the two common trees in Example 8-14 are:

For $T_c^{(1)}$:

$$\begin{matrix} & 1 & 2 & 3 \\ G_i & y_2 & y_4 & y_3 \\ G_r & y_2 & y_4 & y_3 \end{matrix} = (-1)^{0+0} = +1$$

and

For $T_c^{(2)}$:

$$\begin{matrix} & 1 & 2 & 3 \\ G_i & y_2 & y_1 & \bar{g}_m \\ G_v & y_2 & y_1 & \bar{g}_m \end{matrix} = (-1)^{0+2} = +1$$

DEFINITION 8-16. Principal Tree. A *principal tree* is a tree in which all the edges are connected to the reference vertex r.

Thus, a principal tree is a star with its "center" at the reference vertex r. The next theorem shows the relationship between the determinant of the incidence submatrix A_{T_1} of a tree T_1 and that of the incidence submatrix A_{T_p} of the principal tree T_p.

THEOREM 8-6. With the notation as just defined, any tree T_1 can be reduced to a principal tree and

$$\det A_{T_1} = \det A_{T_p} \tag{8-36}$$

Proof: Since in any tree T_1 there exists at least one vertex j ($\neq r$) of degree 1, let that edge incident with j be denoted by $e_k(j,j')$ with endpoints j and j', and let the corresponding column in A_{T_1} be denoted by k. If j' is not r, then the (j,k) entry of column k is ± 1 and the (j',k) entry is ∓ 1; all other entries in column k are zero. Since vertex j is of degree 1 and e_k is the only edge incident with it, all the entries of row j (corresponding to vertex j) are zero except the (j,k) entry. Hence, if row j is added to row j', the resultant row j' is the same as the original row j', with the exception of the (j',k) entry, which has been reduced to zero. Let the resultant matrix be A_{T_2}. Then

$$\det A_{T_1} = \det A_{T_2}$$

since adding a row to another does not change the value of the determinant. Note that A_{T_2} is the incidence submatrix of a new graph T_2 that is obtained by breaking edge e_k from T_1 at vertex j' and reconnecting e_k to vertex r so that T_2 is a new tree with e_k connected between vertices j and r. Note also that the orientation of e_k in the new tree T_2 remains unchanged by this operation.

Next, consider the subgraph of $T_2 - e_k$, which obviously is a tree (of some subgraph) since it is a subgraph of T_2. The incidence submatrix with the same reference vertex r is obtained by deleting from A_{T_2} row j and column k. The same process can now be repeated with another edge $e_{k'}$ with endpoints

p and p', where p is of degree 1 and $p' \neq r$. This results in a tree T_3 with incidence submatrix A_{T_3} and the same reference vertex r. Thus we write

$$\det A_{T_1} = \det A_{T_2} = \det A_{T_3}$$

as asserted. This completes the proof.

The following is an immediate consequence of Theorem 8-6 and needs no proof.

COROLLARY 8-3. Let T be a common tree of $\underline{G}$ and T_i and T_v be the corresponding trees in G_i and G_v, respectively. Also, let A_{T_i} and A_{T_v} be the incidence submatrices of T_i and T_v, respectively. Then

$$\det A_{T_i} \det A_{T_v} = \det A_{T_{pi}} \det A_{T_{pv}} \tag{8-37}$$

where $A_{T_{pi}}$ and $A_{T_{pv}}$ are the incidence matrices of the principal trees corresponding to T_i and T_v, respectively.

The following properties of the incidence submatrix A_{T_p} of a principal tree T_p are obvious:

(a) There is exactly one nonzero entry in each row and each column.

(b) If the (j,k) entry is nonzero, vertex j corresponding to row j is the principal vertex of edge e_k corresponding to column k.

We shall use these two properties of A_{T_p} to prove the following theorem:

THEOREM 8-7. With the same notation as in Corollary 8-3 and Definition 8-15,

$$\det A_{T_i} \det A_{T_v} = (-1)^{p+q} \tag{8-38}$$

Proof: It is evident that the values of $\det A_{T_{pi}}$ and $\det A_{T_{pv}}$ are either $+1$ or -1. Thus, from Equation (8-37) we get

$$\det A_{T_i} \det A_{T_v} = \pm 1 \tag{8-39}$$

It is $+1$ if $\det A_{T_{pi}} = \det A_{T_{pv}}$, and -1 if $\det A_{T_{pi}} = -\det A_{T_{pv}}$. But, from property (a) of a principal tree, if each row containing a "-1" is multiplied by -1 to make the entry positive, we get

$$\det A_{T_{pi}} = (-1)^{q_i} \det A'_{T_{pi}} \qquad \text{and} \qquad \det A_{T_{pv}} = (-1)^{q_v} \det A'_{T_{pv}}$$

so that

$$\det A_{T_{pi}} \det A_{T_{pv}} = (-1)^{q_i+q_v} \det A'_{T_{pi}} \det A'_{T_{pv}} \tag{8-40}$$

where q_i is the number of negative entries in the rows of $A_{T_{pi}}$, q_v the number of negative entries in the rows of $A_{T_{pv}}$, $A'_{T_{pi}}$ is obtained by replacing all the "-1" entries by the "$+1$" in $A_{T_{pi}}$, and $A'_{T_{pv}}$ is obtained by replacing all the

"-1" entries with "$+1$" in $A_{T_{pv}}$. Furthermore, by property (b) of a principal tree and Definition 8-15, we see that

$$q_i + q_v = q$$

so that Equation (8-40) becomes

$$\det A_{T_{pi}} \det A_{T_{pv}} = (-1)^q \det A'_{T_{pi}} \det A'_{T_{pv}}$$

Finally, if we interchange the columns of $A'_{T_{pi}}$ so as to make the two matrices, $A'_{T_{pi}}$ and $A'_{T_{pv}}$, identical, we get

$$\det A_{T_{pi}} \det A_{T_{pv}} = (-1)^q(-1)^p = (-1)^{q+p} \tag{8-41}$$

where p is the number of interchanges of columns of $A'_{T_{pi}}$ to make the resultant matrix equal to $A'_{T_{pv}}$. The theorem is now proved.

The next theorem is a consequence of Theorems 8-5 and 8-7.

THEOREM 8-8. With the same notation as in Theorems 8-5 and 8-7,

$$\Delta = \sum_i \text{(value of sign permutation of } T_c^{(i)})T_c^{(i)}(y) \equiv V \tag{8-42}$$

where V denotes the algebraic sum of all the common-tree admittance products with the sign ϵ_i of each common-tree admittance product $T_c^{(i)}(y)$ equal to the value of the sign permutation of that common tree $T_c^{(i)}$.

We shall now give one example to illustrate the application of Theorem 8-8.

EXAMPLE 8-17. Let us evaluate the determinant Δ of the admittance matrix Y_n given by Equation (8-31), using the topological formula given by Equation (8-42). There are $C(6,3) = 20$ combinations of six edges taken three at a time. Listing all these 20 combinations and eliminating those that contain circuits in either G_v or G_i, we get 12 common trees of $\underline{G}$ (Figure 8-35) of which 8 contain only passive edges: (e_1,e_2,e_3), (e_1,e_2,e_6), (e_1,e_3,e_4), (e_1,e_3,e_6), (e_1,e_4,e_6), (e_2,e_3,e_4), (e_2,e_3,e_6), and (e_2,e_4,e_6). Those that contain active edges are: (e_1,e_2,e_5), (e_1,e_3,e_5), (e_1,e_4,e_5), and (e_2,e_4,e_5). The common-tree admittance products of the common trees that contain passive edges only are always positive. (Why?) Hence, we only need to compute the values of the sign permutations of those common trees that contain active edges in order to determine the signs of the corresponding common-tree admittance products. This computation is summarized in Table 8-2.

Thus, by Theorem 8-8, we get

$$\begin{aligned}\Delta \equiv \det Y_n = y_1y_2y_3 &+ y_1y_2g_m + y_1y_2g_p + y_1y_3y_4 + y_1y_3g_m \\ &+ y_1y_3g_p + y_1y_4g_m + y_1y_4g_p + y_2y_3y_4 \\ &+ y_2y_3g_p + y_2y_4g_m + y_2y_4g_p\end{aligned} \tag{8-43}$$

To check the result, the reader is urged to expand $\det Y_n$, either by direct expansion or by any of the flow-graph techniques discussed in Chapter 7, and see whether both methods yield the same answer.

TABLE 8-2

Computation of Values of Sign Permutations of Common-tree Admittance Products

Common Tree	*Common-Tree Admittance Product*	*Computation of Value of Sign Permutation*
(e_1,e_2,e_5)	$y_1y_2g_m$	$\begin{matrix} & 1 & 2 & 3 \\ i & y_2 & y_1 & \bar{g}_m \\ v & y_2 & y_1 & \bar{g}_m \end{matrix} = (-1)^{0+2} = +1$
(e_1,e_3,e_5)	$y_1y_3g_m$	$\begin{matrix} i \\ v \end{matrix}\begin{pmatrix} \bar{y}_1 & g_m & y_3 \\ g_m & y_1 & y_3 \end{pmatrix} = (-1)^{1+1} = +1$
(e_1,e_4,e_5)	$y_1y_4g_m$	$\begin{matrix} i \\ v \end{matrix}\begin{pmatrix} \bar{y}_1 & y_4 & \bar{g}_m \\ \bar{y}_1 & y_4 & \bar{g}_m \end{pmatrix} = (-1)^{0+4} = +1$
(e_2,e_4,e_5)	$y_2y_4g_m$	$\begin{matrix} i \\ v \end{matrix}\begin{pmatrix} y_2 & y_4 & \bar{g}_m \\ y_2 & y_4 & \bar{g}_m \end{pmatrix} = (-1)^{0+2} = +1$
(e_1,e_2,e_3)	$y_1y_2y_3$	(passive), +1
(e_1,e_2,e_6)	$y_1y_2g_p$	(passive), +1
(e_1,e_3,e_4)	$y_1y_3y_4$	(passive), +1
(e_1,e_3,e_6)	$y_1y_3g_p$	(passive), +1
(e_1,e_4,e_6)	$y_1y_4g_p$	(passive), +1
(e_2,e_3,e_4)	$y_2y_3y_4$	(passive), +1
(e_2,e_3,e_6)	$y_2y_3g_p$	(passive), +1
(e_2,e_4,e_6)	$y_2y_4g_p$	(passive), +1

8-7 TWO-GRAPH ANALYSIS OF ACTIVE NETWORKS

In this section, Mayeda's "two-graph" method will be applied to general linear networks which include restricted classes of networks such as the *RLC* passive networks without magnetic coupling as well as the active networks consisting of tubes, transistors, transformers, along with passive elements.

We shall first derive the topological formulas for the open-circuit admittance functions in considerable detail. Then, the short-circuit admittance functions and other network functions are obtained in a like manner.

(a) Open-circuit Admittance Functions

Consider a general linear one-port N with terminals i and j (r being the reference node). Let a passive element y_{ij} [$= y_{ij,ij}$, from Equation (8-19)] be connected to N across the terminals i and j with orientation from i to j. Let us denote the graph of N by $\underline{G}$ and the graph of N', which is N plus y_{ij}, by $\underline{G}'$ as shown in Figure 8-39. Since there is no external source as far as N' is concerned, Equation (8-27) may be written as

$$Y_n'V_n = 0 \tag{8-44}$$

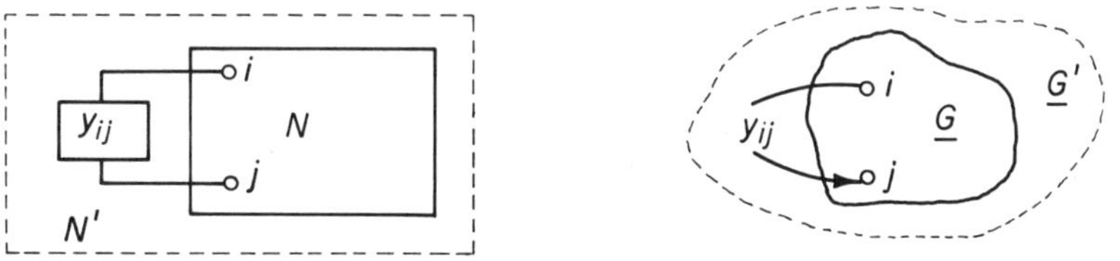

Figure 8-39 (a) A general linear one-port N with external element of y_{ij}. (b) Graph $\underline{G}$ corresponding to N and graph $\underline{G}'$ corresponding to N'.

where Y_n' is the node admittance matrix of N' and Equation (8-44) represents a homogeneous system of $v - 1$ node equations in $v - 1$ unknowns for N', and where v is the number of vertices of N' (and of N, also). This system has nontrivial solutions if and only if $\Delta_1 = \det Y_n' = 0$; that is,

$$\Delta_1 = \det Y_n' = \det (A_iY_eA_v^T) = 0 \tag{8-45}$$

Also, from Equation (8-33),

$$\Delta_1 = \sum_k \epsilon_k T_c^{(k)}(y) \tag{8-46}$$

where the summation is over all common-tree admittance products of N'. Thus, Equations (8-45) and (8-46) imply

$$\Delta_1 = \sum_{k_1} \epsilon_{k_1} T_c^{(k_1)}(y) + \sum_{k_2} \epsilon_{k_2} T_c^{(k_2)}(y) = 0 \tag{8-47}$$

where the first summation is over all common-tree admittance products, none of which contains element y_{ij}; and the second summation is over all common-tree admittance products, each of which contains element y_{ij}.

Since the first summation is the determinant Δ of the node admittance matrix Y_n of N, we get, from Equation (8-42),

$$\sum_{k_1} \epsilon_{k_1} T_c^{(k_1)}(y) = \det Y_n = V \tag{8-48}$$

where V is defined in Theorem 8-8.

Now let us examine the second summation in Equation (8-47). Since all of the common-tree admittance products contain element y_{ij}, if y_{ij} is removed from these common trees, each of the terms in the summation becomes a "common 2-tree" with node i in one part and the reference node j in the other. Thus, we define the following:

DEFINITION 8-17. Common 2-tree and Common-2-tree Admittance Product. A *common* 2-*tree* $T_{c2_{i,j}}$ is a subgraph that is a 2-tree in both the current graph G_i and the voltage graph G_v of the resultant graph $\underline{G}$, such that vertices i and j are in two different components of the 2-tree in both G_i and G_v. In general, $T_{c2_{i,j;m,n}}$ is a common 2-tree if it is a 2-tree in both G_i and G_v such that vertices i and j are in different components of the 2-tree in G_i, and vertices m and n are in different components of the 2-tree in G_v. A *common 2-tree admittance product* is the product of the admittances of all the edges in a common 2-tree. The symbol W is used to denote the sum of all 2-trees of like properties (similar to the W in the passive case).

Definition 8-17 is evidently a logical extension of the concepts of a 2-tree and a 2-tree admittance product defined for the case of passive networks without mutual inductance. This extension is based on the introduction of the concept of a transadmittance $y_{ij,mn}$ as defined by Equation (8-16).

We now state and prove the following theorem:

THEOREM 8-9. With the same notation as previously defined,

$$Y_{o_{i,j}} \triangleq -y_{ij} = \frac{V}{W_{i,j}} \tag{8-49}$$

where $W_{i,j}$ is defined by

$$\begin{aligned} W_{i,j} &\triangleq \left(\frac{1}{y_{ij}}\right) \sum_{k_2} \epsilon_{k_2} T_c^{(k_2)}(y) \\ &= \frac{1}{y_{ij}} \sum_{i_2} \epsilon_{i_2} \times \left\{ \begin{matrix} \text{common 2-tree admittance} \\ \text{product } T_{c2}^{(i_2)}(y) \text{ containing } y_{ij} \end{matrix} \right\} \end{aligned} \tag{8-50}$$

and V has been defined in Theorem 8-8.

Proof: The theorem follows directly from Equations (8-47) and (8-48) and the definition of $W_{i,j}$ as given by Equation (8-50). Substituting Equation (8-48) into (8-47) we get

$$\Delta_1 = V + \sum_{k_2} \epsilon_{k_2} T_c^{(k_2)}(y) = 0 \tag{8-51}$$

Next, by Equation (8-50), we rewrite (8-51) as

$$V + y_{ij} W_{i,j} = 0$$

Solving for $-y_{ij}$, we get

$$-y_{ij} = \frac{V}{W_{i,j}}$$

as asserted in the theorem.

Note that $-y_{ij}$ is defined as the open-circuit driving-point admittance between nodes i and j of the given network N and is denoted by $Y_{o_{i,j}}$. This can be represented topologically by the definition of transadmittance as shown in Figure 8-40.

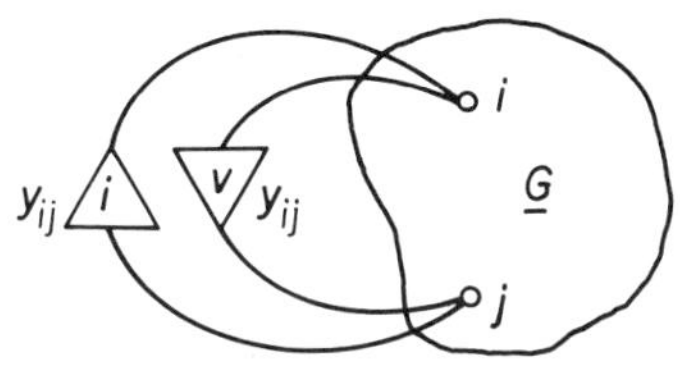

Figure 8-40 Topological representation of $Y_{o_{i,j}}$.

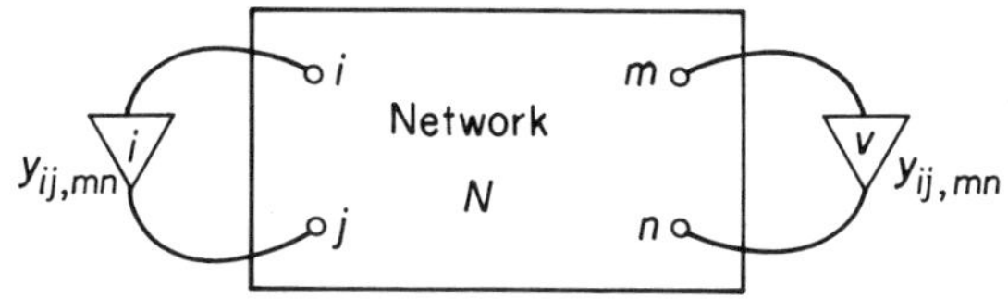

Figure 8-41 A network N with $y_{ij,mn}$.

Next, let us consider the transadmittance $y_{ij,mn}$ with respect to two node pairs (i,j) and (m,n) of a general network N, as depicted in Figure 8-41. Similar to the development for the admittance y_{ij} $(= y_{ij,ij})$, the determinant Δ_1 for Y_n' of network N' (which is N plus $y_{ij,mn}$) is given by

$$\Delta_1 = \sum_{q_1} \epsilon_{q_1} T_c^{(q_1)}(y) + \sum_{q_2} \epsilon_{q_2} T_c^{(q_2)}(y) = 0 \tag{8-52}$$

where the first summation is over all common-tree admittance products not containing $y_{ij,mn}$, and the second is over all common-2-tree admittance products containing $y_{ij,mn}$.

Now, if we define $W_{i,j;m,n}$ as

$$W_{i,j;m,n} \triangleq \frac{1}{y_{ij,mn}} \sum_{q_2} \epsilon_{q_2} T_c^{(q_2)}(y) \tag{8-53}$$

we immediately have the next theorem.

THEOREM 8-10. With the same notation as defined previously,

$$Y_{o_{i,j;m,n}} \triangleq -y_{i,j,mn} = \frac{V}{W_{i,j;m,n}} \tag{8-54}$$

where $-y_{ij,mn}$ is defined as an open-circuit transfer admittance function and is denoted by $Y_{o_{ij,mn}}$.

The proof of Theorem 8-10 is almost identical to that of Theorem 8-9 and, hence, is left as an exercise.

It should be noted that Equation (8-49) is merely a special case of Equation (8-54), for, if $m = i$ and $n = j$, Equation (8-54) becomes

$$Y_{o_{ij,ij}} = -y_{ij,ij} = \frac{V}{W_{i,j;i,j}} \tag{8-55}$$

which is identical to Equation (8-49) in light of Equation (8-47) and Definition 8-17.

EXAMPLE 8-18. Consider the network N shown in Figure 8-42. Suppose we want to compute $W_{1,0;2,0}$. The first step is to add an edge $y_{10,20}$ to G to form $\underline{G}'$, as illustrated in Figure 8-43(a). We then obtain G'_i of G'_v of $\underline{G}'$ as shown in Figure 8-43(b) and (c). There are exactly four common trees that contain $y_{10,20}$. The admittance products of these four common trees are:

$$y_1y_3y_{10,20} \qquad y_1g_my_{10,20} \qquad y_1g_py_{10,20} \qquad y_3g_my_{10,20}$$

They are shown in *i-v* pairs in Figure 8-44.

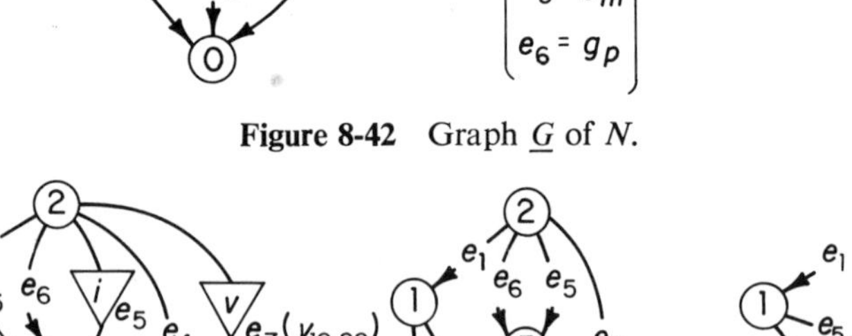

Figure 8-42 Graph $\underline{G}$ of N.

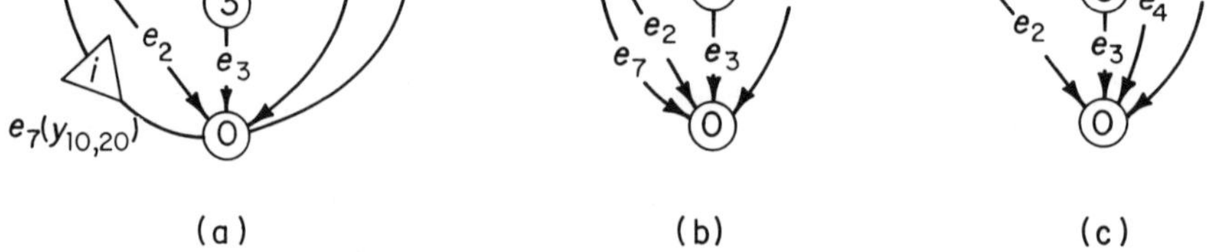

Figure 8-43 (a) Graph $\underline{G}'$. (b) Graph G'_i. (c) Graph G'_v.

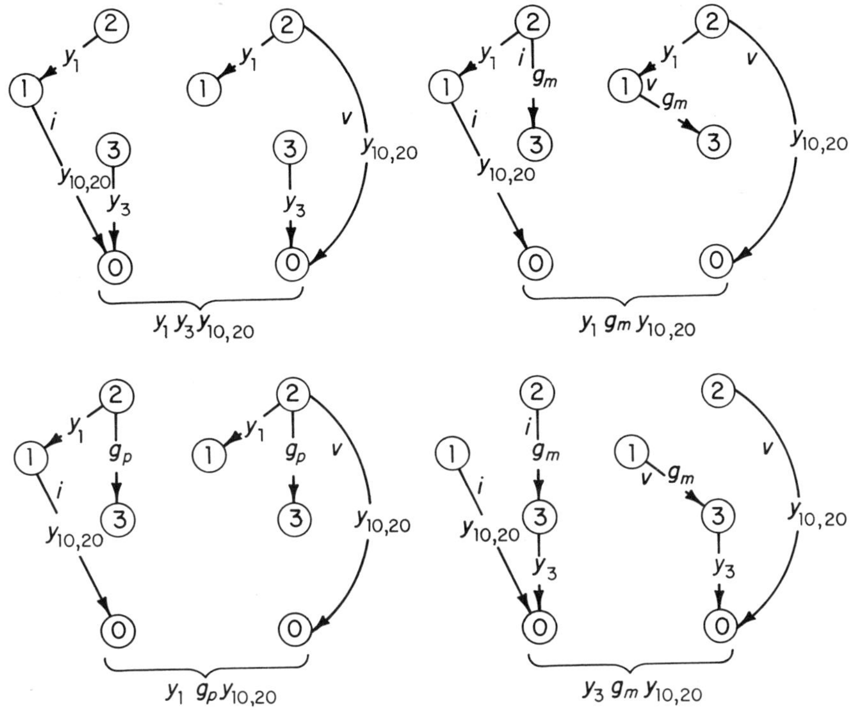

Figure 8-44 The four common trees containing $y_{10,20}$.

TABLE 8-3

Evaluation of Sign Permutation of the Four Common-tree Admittance Products

Common Tree	*Admittance Product*	*Value of Sign Permutation*
(e_1, e_3, e_7)	$y_1 y_3 y_{10,20}$	$\begin{matrix} \\ i \\ v \end{matrix}\begin{pmatrix} 1 & 2 & 3 \\ y_{10,20} & y_1 & y_3 \\ \bar{y}_1 & y_{10,20} & y_3 \end{pmatrix} = (-1)^{1+1} = +1$
(e_1, e_5, e_7)	$y_1 g_m y_{10,20}$	$\begin{matrix} i \\ v \end{matrix}\begin{pmatrix} y_{10,20} & y_1 & \bar{g}_m \\ \bar{y}_1 & y_{10,20} & \bar{g}_m \end{pmatrix} = (-1)^{1+3} = +1$
(e_1, e_6, e_7)	$y_1 g_p y_{10,20}$	$\begin{matrix} i \\ v \end{matrix}\begin{pmatrix} y_{10,20} & y_1 & \bar{g}_p \\ \bar{y}_1 & y_{10,20} & \bar{g}_p \end{pmatrix} = (-1)^{1+3} = +1$
(e_3, e_5, e_7)	$y_3 g_m y_{10,20}$	$\begin{matrix} i \\ v \end{matrix}\begin{pmatrix} y_{10,20} & g_m & y_3 \\ g_m & y_{10,20} & y_3 \end{pmatrix} = (-1)^{1+0} = -1$

The evaluation of the sign permutations of these four common trees is illustrated in Table 8-3. Thus, by Equation (8-53), we get

$$W_{1,0;2,0} = y_1 y_3 + y_1 g_m + y_1 g_p - y_3 g_m \tag{8-56}$$

Following a development similar to that given in Section 6-2, we can obtain the topological formulas for the open-circuit admittance functions for two-port networks with the help of Equations (8-49) and (8-54) as follows:

From the node-analysis matrix equation with terminals 1, 1′, 2, and 2′ and vertex 1′ as reference, as shown in Figure 8-45,

$$\begin{bmatrix} V_i \\ V_o \end{bmatrix} = \begin{bmatrix} \dfrac{\Delta_{11}}{\Delta} & \dfrac{\Delta_{21} - \Delta_{2'1}}{\Delta} \\ \dfrac{\Delta_{12} - \Delta_{12'}}{\Delta} & \dfrac{\Delta_{22} + \Delta_{2'2'} - \Delta_{22'} - \Delta_{2'2}}{\Delta} \end{bmatrix} \begin{bmatrix} I_1 \\ I_2 \end{bmatrix} \tag{8-57}$$

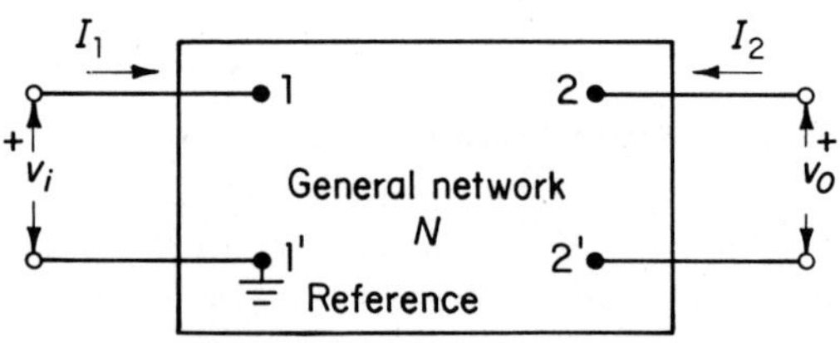

Figure 8-45 A general two-port N with terminal voltages and currents defined by Equations (8-57) through (8-60).

we write the open-circuit admittance functions as

$$\begin{aligned}
\left.\frac{I_1}{V_i}\right|_{I_2=0} &\triangleq y_{11',11'} = y_{11'} = \frac{\Delta}{\Delta_{11}} \\
\left.\frac{I_2}{V_i}\right|_{I_1=0} &\triangleq y_{22',11'} = \frac{\Delta}{\Delta_{21} - \Delta_{2'1}} \\
\left.\frac{I_1}{V_o}\right|_{I_2=0} &\triangleq y_{11',22'} = \frac{\Delta}{\Delta_{12} - \Delta_{12'}} \\
\left.\frac{I_2}{V_o}\right|_{I_1=0} &\triangleq y_{22',22'} = y_{22'} = \frac{\Delta}{\Delta_{22} + \Delta_{2'2'} - \Delta_{22'} - \Delta_{2'2}}
\end{aligned} \tag{8-58}$$

Substituting Equations (8-49) and (8-54) into (8-58), we get immediately the following topological formulas for the open-circuit admittance functions:

$$\begin{aligned}
\left.\frac{I_1}{V_i}\right|_{I_2=0} &= y_{11'} = \frac{V}{W_{1,1'}} \\
\left.\frac{I_2}{V_i}\right|_{I_1=0} &= y_{22',11'} = \frac{V}{W_{2,2';1,1'}} \\
\left.\frac{I_1}{V_o}\right|_{I_2=0} &= y_{11',22'} = \frac{V}{W_{1,1';2,2'}} \\
\left.\frac{I_2}{V_o}\right|_{I_1=0} &= y_{22'} = \frac{V}{W_{2,2'}}
\end{aligned} \tag{8-59}$$

The determinant Δ and the relevant cofactors are related to the topological quantities by

$$\begin{aligned}
\Delta &= V \\
\Delta_{11} &= W_{1,1'} \\
\Delta_{21} - \Delta_{2'1} &= W_{2,2';1,1'} \\
\Delta_{12} - \Delta_{12'} &= W_{1,1';2,2'} \\
\Delta_{22} + \Delta_{2'2'} - \Delta_{22'} - \Delta_{2'2} &= W_{2,2'}
\end{aligned}$$

Expressing the topological formulas in matrix form, Equation (8-57) may be rewritten as

$$\begin{bmatrix} V_i \\ V_o \end{bmatrix} = \begin{bmatrix} \dfrac{W_{1,1'}}{V} & \dfrac{W_{2,2';1,1'}}{V} \\ \dfrac{W_{1,1';2,2'}}{V} & \dfrac{W_{2,2'}}{V} \end{bmatrix} \begin{bmatrix} I_1 \\ I_2 \end{bmatrix}$$

(b) Short-circuit Admittance Functions

Let N_s be the network obtained from a given network N by coalescing nodes 1 and 1′ as depicted in Figure 8-46(a). The (resultant) graph $\underline{G}_s$ of N_s is shown in Figure 8-46(b). If N contains v nodes, then $\underline{G}_s$ contains $v - 1$

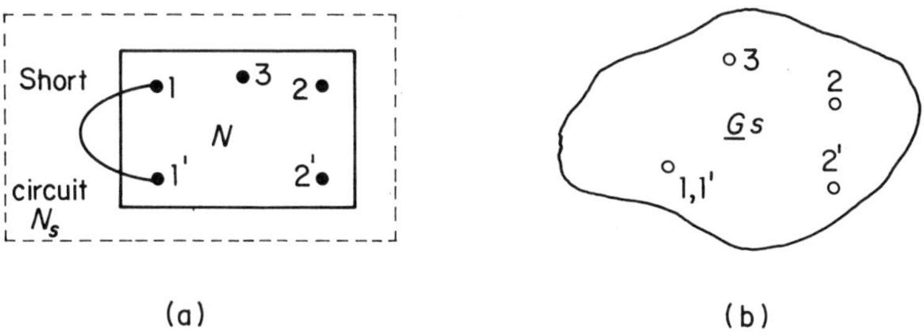

Figure 8-46 (a) A network N with nodes 1 and 1′ short-circuited. (b) Graph $\underline{G}_s$ corresponding to N_s (which is N with nodes 1 and 1′ short-circuited).

nodes, and a common tree T_c in $\underline{G}_s$ contains $v - 2$ edges. Thus the corresponding trees T_i and T_v of the current graph G_{s_i} and voltage graph G_{s_v} are both 2-trees of the type $T_{2_{1,1'}}$ in the (resultant) graph $\underline{G}$ of N. Hence, the sum V_s of all common-tree admittance products of $\underline{G}_s$ is equal to the sum $W_{1,1'}$ of all common-2-tree admittance products of G; that is,

$$V_s = W_{1,1'} \tag{8-60}$$

It is evident from Equation (8-60) that the effect of shorting nodes 1 and 1′ on the node admittance matrix Y_n $(= A_i Y_e A_v^T)$ of N is the deletion of row 1 and column 1 in Y_n, where row 1 of A_i corresponds to vertex 1 in G_i and row 1 of A_v to vertex 1 in G_v of $\underline{G}$.

From the node-analysis equations of N_s, we write

$$\begin{bmatrix} V_{31'} \\ V_{22'} \end{bmatrix} = \begin{bmatrix} \dfrac{\Delta'_{33}}{\Delta'} & \dfrac{\Delta'_{23} - \Delta'_{2'3}}{\Delta'} \\ \dfrac{\Delta'_{32} - \Delta'_{32'}}{\Delta'} & \dfrac{\Delta'_{22} + \Delta'_{2'2'} - \Delta'_{22'} - \Delta'_{2'2}}{\Delta'} \end{bmatrix} \begin{bmatrix} J_{31'} \\ J_{22'} \end{bmatrix}$$

where Δ' is the determinant of the node admittance matrix of N_s. We see that Δ' is equal to the cofactor Δ_{11} of Y_n of N with node $1'$ as the reference node. Also, since $\Delta_{11} = W_{1,1'}$, it follows that

$$\Delta' = W_{1,1'} \tag{8-61}$$

Similarly,

$$\Delta'_{22} + \Delta'_{2'2'} - \Delta'_{22'} - \Delta'_{2'2} = \Delta_{1122} + \Delta_{112'2'} - \Delta_{1122'} - \Delta_{112'2} \tag{8-62}$$

Following a similar development used for the passive networks in Chapter 6, we now define the following:[17]

$$U_{1,1';2,2'} \triangleq \Delta_{1122} + \Delta_{112'2'} - \Delta_{1122'} - \Delta_{112'2} \tag{8-63}$$

Noting the fact that

$$W_{2,2'} = \Delta_{22} + \Delta_{2'2'} - \Delta_{22'} - \Delta_{2'2} \tag{8-64}$$

we write

$$U_{1,1';2,2'} = \frac{1}{y_{11'}y_{22'}} \sum_i \epsilon_i \times \begin{Bmatrix} \text{common-tree admittance product} \\ T_c^{(i)}(y), \text{ which contains both } y_{11'} \\ \text{and } y_{22'} \text{ of } \underline{G}' \text{ of } N' \end{Bmatrix} \tag{8-65}$$

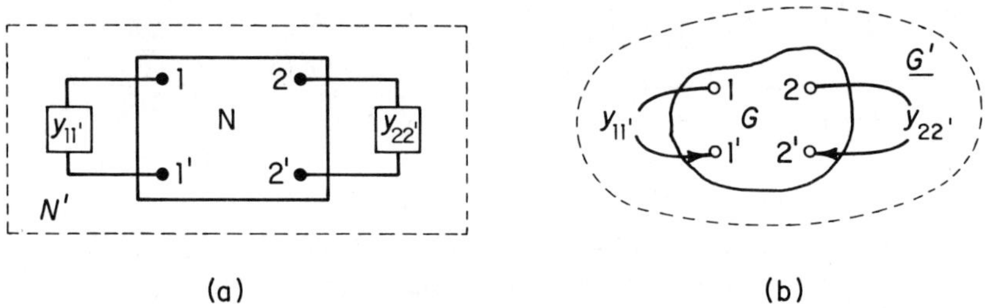

Figure 8-47 (a) Networks N and N'. (b) The corresponding graphs $\underline{G}$ and G'.

where $\underline{G}'$ and N' are related to G and N as illustrated in Figure 8-47. Finally, from the matrix equations of the short-circuit admittance equations

$$\begin{bmatrix} I_1 \\ I_2 \end{bmatrix} = \frac{1}{\Delta_{1122} + \Delta_{112'2'} - \Delta_{1122'} - \Delta_{112'2}} \times \begin{bmatrix} (\Delta_{22} + \Delta_{2'2'} - \Delta_{22'} - \Delta_{2'2}) & -(\Delta_{21} - \Delta_{2'1}) \\ -(\Delta_{12} - \Delta_{12'}) & (\Delta_{11}) \end{bmatrix} \begin{bmatrix} V_i \\ V_o \end{bmatrix} \tag{8-66}$$

[17]The symbol U in Equation (8-63) denotes the fact that it is an expression in connection with (complete) 3-trees. The first pair of subscripts identifies the pair of terminals at port $(1,1')$ of the two-port and the second pair denotes the two terminals at port $(2,2')$.

we write immediately the corresponding topological matrix equation as

$$\begin{bmatrix} I_1 \\ I_2 \end{bmatrix} = \frac{1}{U_{1,1';2,2'}} \begin{bmatrix} W_{2,2'} & -W_{2,2';1,1'} \\ -W_{1,1';2,2'} & W_{1,1'} \end{bmatrix} \begin{bmatrix} V_i \\ V_o \end{bmatrix} \tag{8-67}$$

where I_1, I_2, V_i, and V_o are defined in Figure 8-45.

EXAMPLE 8-19. Let us consider once again the network N of Figure 8-34(a) and determine $U_{1,1';2,2'}$ by the foregoing topological formulas. Our first step is to connect to $\underline{G}$ of N edge $y_{11'}$ between vertices 1 and 1′, and edge $y_{2,1'}$ between vertices 2 and 1′ to form graph $\underline{G}'$, as illustrated in Figure 8-48.

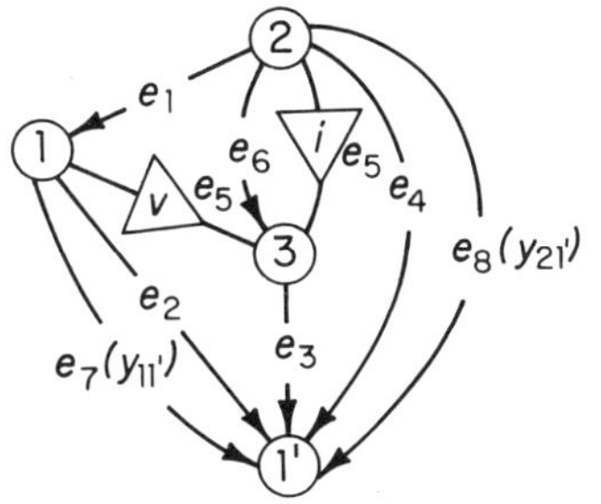

Figure 8-48 Graph $\underline{G}'$.

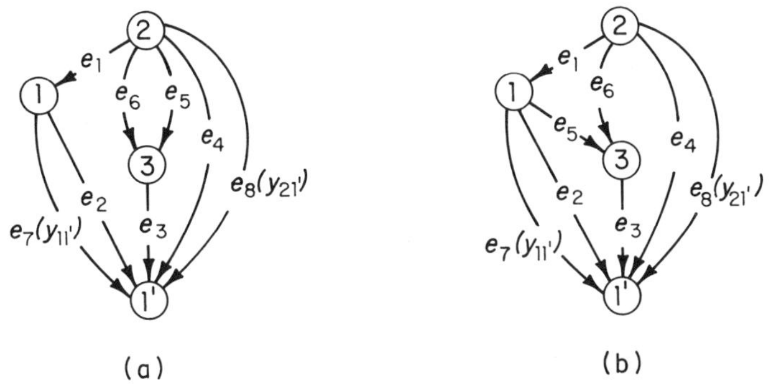

Figure 8-49 (a) Current graph G_i'. (b) Voltage graph G_v' of $\underline{G}$.

The corresponding current graph G_i' and voltage graph G_v' are shown in Figure 8-49. We next find all the common trees that contain both $y_{11'}$ and $y_{21'}$. They are shown in *i-v* pairs in Figure 8-50 with their corresponding admittance products. The signs for $y_3 y_{11'} y_{21'}$ and $g_p y_{11'} y_{21'}$ are both positive, since both common trees contain only passive elements. The sign of $g_m y_{11'} y_{21'}$ is determined by the value of the sign permutation as

$$\epsilon = \begin{matrix} \\ i \\ v \end{matrix} \begin{pmatrix} 1 & 2 & 3 \\ y_{11'} & y_{21'} & \bar{g}_m \\ y_{11'} & y_{21'} & \bar{g}_m \end{pmatrix} = (-1)^{0+2} = +1$$

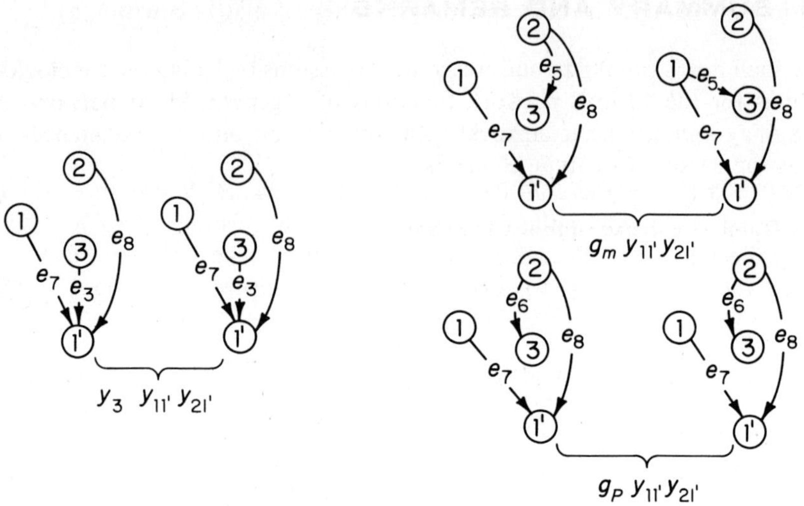

Figure 8-50 Common trees of $\underline{G}'$ that contain both $y_{11'}$ and $y_{22'}$.

Finally, by Equation (8-65), we get

$$U_{1,1';2,1'} = y_3 + g_m + g_p \tag{8-68}$$

(c) Other Network Functions

The voltage-ratio transfer function is defined as V_{mn}/V_{ij} between any two node-pair voltages V_{mn} and V_{ij}. In particular, we write

$$\frac{V_{22'}}{V_{11'}} \equiv \frac{V_o}{V_i} = \frac{V_o/I_1}{V_i/I_1}$$

where V_o/I_1 is the reciprocal of the open-circuit transfer admittance at node pair (1,1′), and V_i/I_1 is the reciprocal of the open-circuit driving-point admittance between node pairs (1,1′) and (2,2′). Thus, from Equation (8-59), we write immediately the following result:

$$\left.\frac{V_o}{V_i}\right|_{I_2=0} = \frac{W_{1,1';2,2'}/V}{W_{1,1'}/V} = \frac{W_{1,1';2,2'}}{W_{1,1'}} \tag{8-69}$$

Similarly, the current-ratio transfer function can be expressed in topological terms as

$$\left.\frac{I_2}{I_1}\right|_{V_1=0} = -\frac{W_{1,1';2,2'}}{W_{2,2'}} \tag{8-70}$$

The derivation of Equation (8-70) is straightforward and, hence is left as a problem.

8-8 SUMMARY AND REMARKS

We shall now summarize our preceding discussions by listing the topological formulas for the various network functions of a general linear network N, which may contain nonreciprocal elements and/or mutual inductances as well as the passive RLC-type elements.

(a) Open-circuit Admittance Functions

$$Y_{o_{i,j}} \triangleq -y_{ij} = \frac{V}{W_{ij}} \tag{8-49}$$

$$Y_{o_{ij,mn}} \triangleq -y_{ij,mn} = \frac{V}{W_{i,j;m,n}} \tag{8-54}$$

$$\left.\frac{I_1}{V_i}\right|_{I_2=0} \triangleq y_{11'} = \frac{\Delta}{\Delta_{11}} = \frac{V}{W_{1,1'}}$$

$$\left.\frac{I_2}{V_i}\right|_{I_1=0} \triangleq y_{22',11'} = \frac{\Delta}{\Delta_{21} - \Delta_{2'1}} = \frac{V}{W_{2,2';1,1'}}$$

(8-58) and

$$\left.\frac{I_1}{V_o}\right|_{I_2=0} \triangleq y_{11',22'} = \frac{\Delta}{\Delta_{12} - \Delta_{12'}} = \frac{V}{W_{1,1';2,2'}}$$

$$\left.\frac{I_2}{V_o}\right|_{I_1=0} \triangleq y_{22'} = \frac{\Delta}{\Delta_{22} + \Delta_{2'2'} - \Delta_{22'} - \Delta_{2'2}} = \frac{V}{W_{2,2'}}$$

(8-59)

(b) Short-circuit Admittance Functions

$$\begin{bmatrix} I_1 \\ I_2 \end{bmatrix} = \frac{1}{\Delta_{1122} + \Delta_{112'2'} - \Delta_{1122'} - \Delta_{112'2}} \times \begin{bmatrix} (\Delta_{22} + \Delta_{2'2'} - \Delta_{22'} - \Delta_{2'2}) & -(\Delta_{21} - \Delta_{2'1}) \\ -(\Delta_{12} - \Delta_{12'}) & (\Delta_{11}) \end{bmatrix} \begin{bmatrix} V_i \\ V_o \end{bmatrix} \tag{8-66}$$

$$= \frac{1}{U_{1,1';2,2'}} \begin{bmatrix} W_{2,2'} & -W_{2,2';1,1'} \\ -W_{1,1';2,2'} & W_{1,1'} \end{bmatrix} \begin{bmatrix} V_i \\ V_o \end{bmatrix} \tag{8-67}$$

(c) Other Network Functions

$$\text{Voltage ratio} \triangleq \left.\frac{V_o}{V_i}\right|_{I_2=0} = \frac{W_{1,1';2,2'}}{W_{1,1'}} \tag{8-69}$$

$$\text{Current ratio} \triangleq \left.\frac{I_2}{I_1}\right|_{V_1=0} = -\frac{W_{1,1';2,2'}}{W_{2,2'}} \tag{8-70}$$

In conclusion, we have presented two different topological methods of analysis of active networks — the digraph method of Chen and the two-graph technique of Mayeda. There are, in fact, four methods of developing topological formulas for general electrical networks — those of Mason [57], Coates [23], Mayeda [59], and Chen [20, 22]. Mason's method is conceptually quite different from the theory of topological analysis, which we have developed in Chapters 5 and 6 and, hence, is not included here. The methods of Coates and Mayeda are very similar in the computational techniques when applied to most practical networks. Both schemes are based on the two-graph concept. Although the theory of Coates is more general from a theoretical point of view, the method of Mayeda is considerably more convenient for practical computations and, hence selected over Coates' method for presentation. The digraph method of Chen is also included here because, as mentioned earlier, it is the latest development of the topological techniques for the analysis of active networks; and, more important, in that method only one graph instead of two is used to analyze a given active network. Furthermore, there is no sign problem, since the directed tree products will automatically give the correct signs for the terms in the formula. This gives the justification for our selection of the two methods presented here.

It is worth noting that the method of Mayeda is an extension of the conventional topological formulas derived from Maxwell's and Kirchhoff's rules in that the concepts of trees and k-trees are preserved; whereas, the theory of Chen is a mixture of the conventional k-tree approach and the flow-graph techniques. It is indeed our hope that the interested reader will recognize the strong points and drawbacks of each method presented (or even mentioned) in this chapter and understand the similarities or differences between them so that superior topological analysis techniques will be obtained as a result of this discussion.

Finally, as pointed out and further illustrated by the derivation of a new set of topological expressions in Chapter 6 (Section 6-8), one of the important features of the topological approach is the fact that the topological formulas not only preserve the topology of the network so as to give the reader a better insight into the system, but also are well suited for computer calculation. Thus, recognizing the fact that a wide adoption of topological techniques for solving practical problems in network analysis and design depends mainly upon the success in developing superior computer programs to be used with these topological expressions to save time and effort (and hence cut down the cost), one should, for example, work toward this goal by producing highly efficient algorithms for the listing of all trees and k-trees of a graph. In this way, effective computer programs can be generated from these algorithms for the analysis and design of general electrical networks.

PROBLEMS

8-1 For the equivalent network shown in Figure P8-1, find the digraph associated with it and illustrate one application of this digraph by means of an example.

8-2 Derive the digraph for a transistor. Show the transistor equivalent

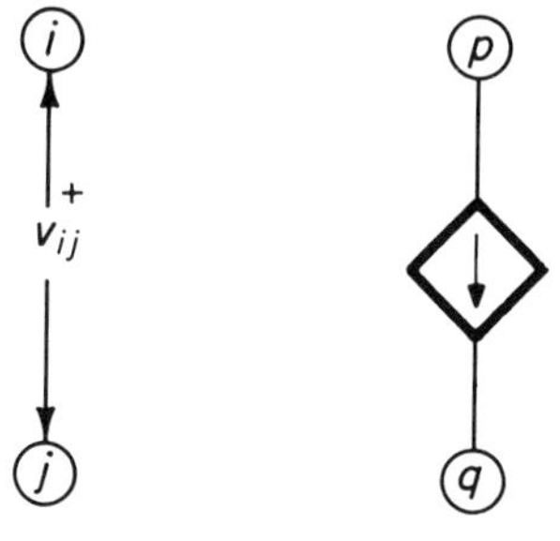

Figure P8-1

circuit used in the derivation and relate it to the digraph derived.

8-3 Refer to the digraph of Figure 8-21.
(a) Obtain the determinant $\Delta = \det Y_n$.
(b) Evaluate Δ_{1g}, Δ_{gk}, Δ_{gp}, and Δ_{pk} by the use of Equation (8-14).

8-4 Consider the potential feedback amplifier discussed in Example 8-7. The digraph of this amplifier is redrawn in Figure P8-4 with the nodes and edges relabelled for simplicity. The letters $a, b, \ldots, g$ may be considered as the weights of the edges. Determine the set of all directed trees of the graph taking as the reference node (a) node 1, (b) node 2, (c) node 3, and (d) node 4.

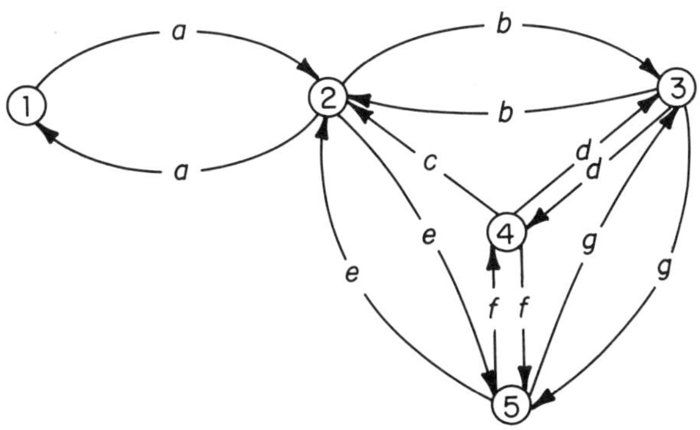

Figure P8-4

8-5 Refer to the digraph of Figure P8-4. Obtain the following sets of directed 2-trees: (a) $\{T_{2d}^{12}\}$. (b) $\{T_{2d}^{23}\}$. (c) $\{T_{2d}^{34}\}$. (d) $\{T_{2d}^{45}\}$.

8-6 The network diagram of a transistor feedback amplifier is shown in Figure P8-6. Using Chen's digraph analysis technique, find the current gain of the amplifier.

8-7 Derive a digraph representation of an ideal transformer. Show your work in detail.

8-8 Refer to the active network N with magnetic coupling shown in Figure

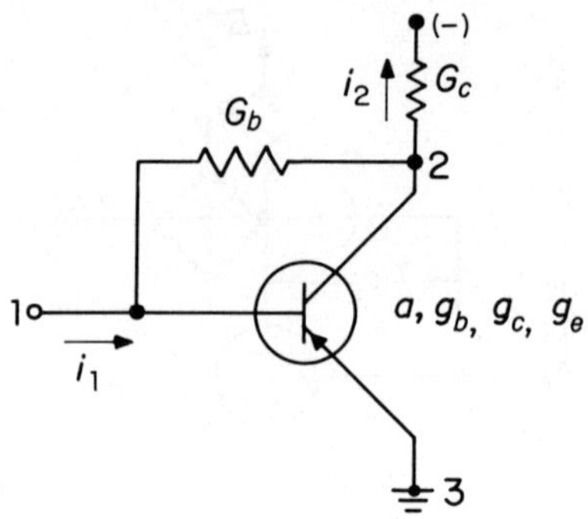

Figure P8-6

8-29. Using the result obtained in Problem 8-6, find the digraph G of N (Chen's method).

8-9 For the active network N of Figure 8-34(a),
(a) Obtain the digraph G of N.
(b) Evaluate the determinant Δ of the node admittance matrix Y_n of N using Chen's digraph technique.
(c) Compare the result obtained in (b) with that shown in Equation (8-32) and comment on the two methods used.

8-10 Find the voltage ratio e_2/e_1 of the triode feedback amplifier shown in Figure P8-10 using Chen's digraph analysis method.

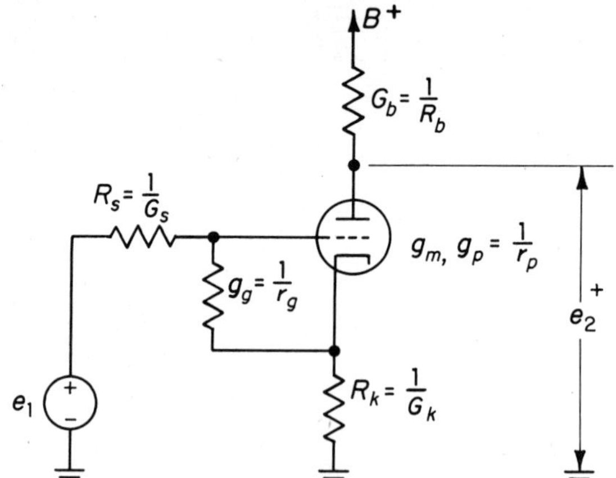

Figure P8-10

8-11 Determine the voltage ratio e_2/e_1 of the Miller integrator circuit shown in Figure P8-11 using Chen's digraph technique. Show your work in detail.

8-12 Repeat Problem 8-11 for the determination of e_3/e_1.

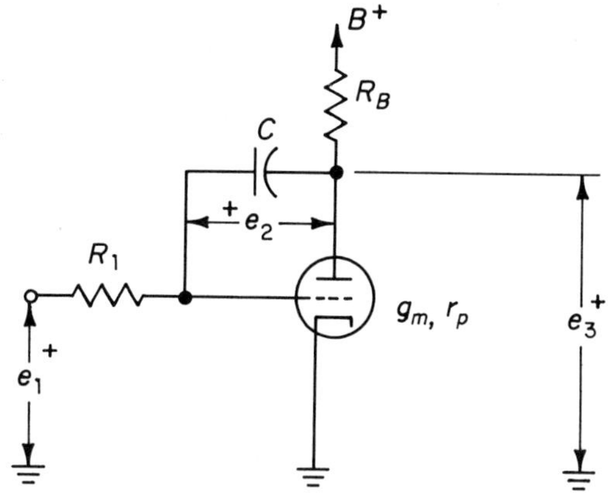

Figure P8-11

8-13 Using Chen's method, (a) find the driving-point impedance Z_{in} and (b) find the voltage gain E_2/E_1 of the general potential feedback amplifier shown in Figure P8-13.

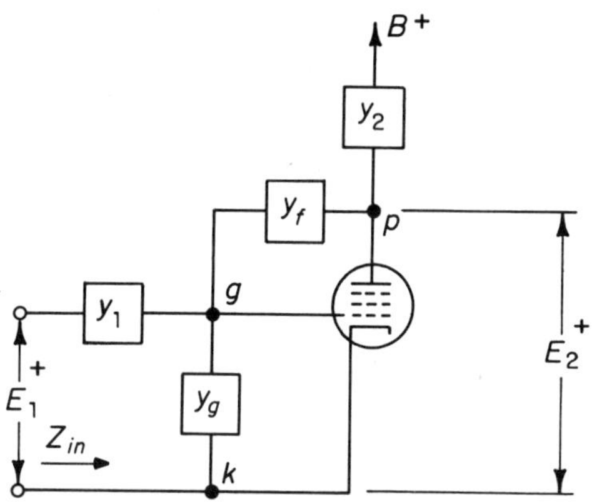

Figure P8-13

8-14 Refer to the potential amplifier of Figure 8-20(a).
(a) Obtain the Mayeda (resultant) graph $\underline{G}$ of the amplifier.
(b) Determine the set of all common trees of $\underline{G}$. Show all the common trees and identify the voltage graph G_v, the current graph G_i, and the edges and vertices in each common tree.

8-15 Find the voltage gain E_2/E_1 of the potential feedback amplifier of Figure 8-20(a) by Mayeda's two-graph analysis method making use of the results obtained in Problem 8-14.

8-16 Refer to Figure 8-20(a). When the admittance y_k is removed by short-circuiting node k to node r, the network becomes identical to that of

Figure P8-13. Modify your answer obtained in Problem 8-15 for the voltage gain E_2/E_1 and compare this modified expression with the answer obtained for part (b) of Problem 8-13.

8-17 Repeat Problem 8-6 using Mayeda's two-graph technique. Compare your answer with that obtained in Problem 8-6 and comment on the two methods used.

8-18 Repeat Problem 8-10 using Mayeda's two-graph technique. Compare your answer with that obtained in Problem 7-10 and comment on the two methods used.

8-19 Repeat Problem 8-11 using Mayeda's two-graph method. Compare your answer with that obtained in Problem 8-11 and comment on the two methods used.

8-20 Repeat Problem 8-12 using Mayeda's method. Compare the answers obtained by these two methods and comment on the methods.

8-21 Refer to the networks of Figure P7-13. Solve each network problem as stated in Problem 7-13 using (a) Chen's digraph method and (b) Mayeda's two-graph method. Compare the methods used for each network problem and list the advantages and disadvantages of each method over the other.

8-22 Repeat Problem 8-21 for the networks of Figure P7-15.

8-23 For the network of Figure P7-20, determine the voltage gain using (a) Chen's digraph method and (b) Mayeda's two-graph method. Compare your results with those obtained in Problem 7-20 and comment on the methods used.

8-24 Repeat Problem 8-23 for the network of Figure P7-22.

8-25 Determine Z_{in} of the network shown in Figure P7-24 using (a) Chen's digraph method and (b) Mayeda's two-graph method. Compare your answer with that obtained in Problem 7-24 and comment on the methods used.

8-26 Repeat Problem 7-25 using (a) Chen's digraph method and (b) Mayeda's two-graph method. Compare your answer with that obtained in Problem 7-25 and comment on the various methods used.

8-27 Repeat Problem 7-27 using (a) Chen's digraph method and (b) Mayeda's two-graph method. Compare your answers with the corresponding ones obtained in Problem 7-27 and comment on the methods used.

8-28 Comment on the flow-graph techniques as a whole versus the "k-tree" methods presented in this chapter. Make your comparison as complete as possible.

TOPOLOGICAL APPLICATIONS TO SWITCHING NETWORKS

CHAPTER 9

9-1 INTRODUCTION

As implied by the title, the main objective of this chapter is to discuss some of the properties of switching networks through the application of graph theory. In order to make our discussions self-contained, a brief examination of the mathematics of switching will be given first.[1] Then, we shall study the topological properties of switching networks and present the relationship between conventional electrical networks and switching networks based on the work of Mayeda[2] and Seshu.[3] Because of their importance with respect to minimality in realization in terms of number of contacts, single-contact networks will be studied at length in this chapter. The discussions of the properties and the synthesis of single-contact networks are based mainly on the work of Chan [14, 15].

9-2 BASIC DEFINITIONS

A switching network may be described as an electrical network that consists of two-state or bistate elements such as switches, contacts, relays, etc. Thus, we define, formally, the following terms:

[1]For a further study of this subject, see F. E. Hohn [43].
[2]See W. Mayeda [60], Chapter 4; [61].
[3]See S. Seshu [74].

DEFINITION 9-1. Contact. A *contact* is a two-terminal electric device that has two states — open state and closed state. In the open state, there exists no conductive path between the terminals; whereas, in the closed state, there exists between the terminals a path through which a current may flow in either direction.

We shall use "1" to represent a short circuit and "0" to represent an open circuit. Also, "+" stands for Boolean addition (thus $x_i + x_j$ signifies the union of x_i and x_j) and "·" stands for Boolean multiplication (thus $x_i \cdot x_j$, or simply $x_i x_j$) represents the intersection of the two terms. Also, with each variable x, we associate a variable x', which is the complement of x, so that x' is opened ($x' = 0$) when x is closed ($x = 1$), and x' is closed ($x' = 1$) when x is opened ($x = 0$).[4]

DEFINITION 9-2. Switching (Contact) Network. A *switching* (*contact*) *network* is a nonoriented network of switches (contacts) in which a binary variable (which has only two values — 0 and 1) is associated with each switch.

We shall consider the two terms "switching networks" and "contact networks" synonymous, and shall use them interchangeably throughout our discussions in this chapter.

EXAMPLE 9-1. A two-terminal switching network consisting of a single switch with a driving voltage (battery) E and a resistor R is shown in Figure 9-1. In Figure 9-2, we depict a two-terminal switching network with four variables — x_i and x_j and their complements x_i' and x_j'. The source and the load are not explicitly shown in the figures.

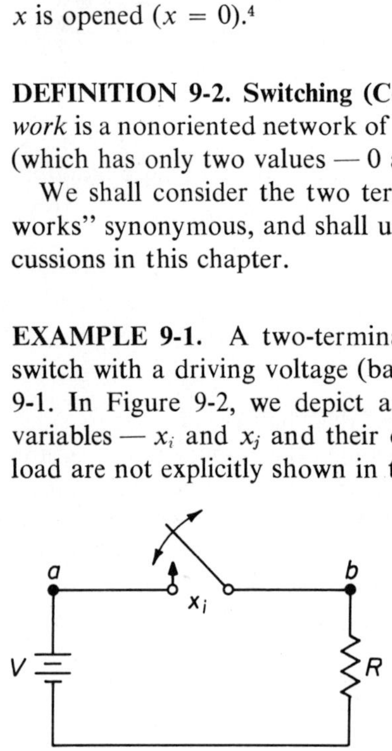

Figure 9-1 A two-terminal switching network with only one switch x_i.

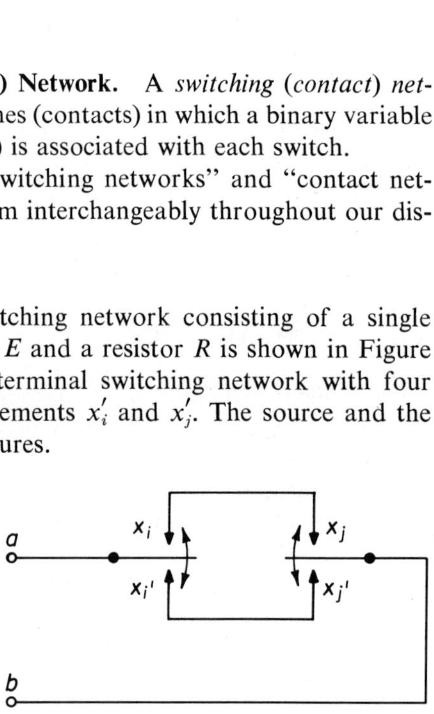

Figure 9-2 A two-terminal network with four variables: x_i, x_j, x_i, and x_j.

DEFINITION 9-3. Path Set and Path Product. A *path set* is a set of variables associated with the edges of a path between the terminals of the contact network. A *path product* is the product of the variables in a path set.

[4]A brief discussion of elementary switching algebra will be presented in Section 9-3.

DEFINITION 9-4. Switching Function. The *switching function* F_{ij} between vertices v_i and v_j of a contact network is the union of all path products between the two vertices of the network. Thus,

$$F_{ij} = \sum_{\substack{\text{all} \\ \text{paths}}} P_{ij}$$

DEFINITION 9-5. Natural Form. When a switching function is expressed as a sum of products of variables it is said to be in *natural form*. Similarly, any function, when expressed as a sum of products of variables, is said to be in natural form no matter whether it is a switching function or not.

DEFINITION 9-6. Minimality. A contact network is *minimal* if it realizes a given switching function using a minimum number of contacts.

EXAMPLE 9-2. Consider the switching network N of Figure 9-3(a). A simplified drawing of N is shown in Figure 9-3(b), and the switching function F_{ij} of N between terminals v_i and v_j is as follows:

$$F_{ij} = (x_1x_4 + x_2x_3)x_5 \tag{9-1}$$

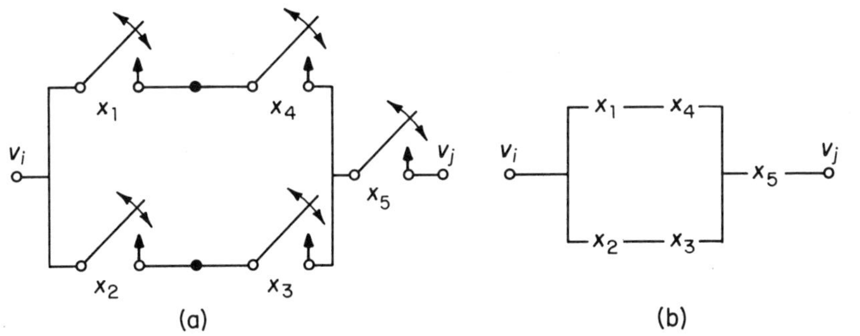

Figure 9-3 (a) A switching network N. (b) A simplified representation of N.

The natural form of Equation (9-1) is

$$F_{ij} = x_1x_4x_5 + x_2x_3x_5 \tag{9-2}$$

Since it needs at least n contacts to realize a given switching function of n variables, the switching network N of Figure 9-3 is a minimal realization of the switching function F_{ij} given by Equation (9-2).

DEFINITION 9-7. Single-contact (SC) Network and SC Function. A *single-contact network* (or an *SC network*, for brevity) is a switching network in which each binary variable x_i associated with contact i is distinct from the others. An *SC function* is the switching function (between two prescribed terminals) of an SC network.

As an example, the switching network N of Figure 9-3 is an SC network and, consequently, the switching function F_{ij} between terminals v_i and v_j of N as expressed in Equation (9-2) is an SC function. If, on the other hand, contact x_5 in N (Figure 9-3) is replaced by x_1 so that there are two contacts identified by the same variable x_1, the resultant network is a non-SC network.

DEFINITION 9-8. Nonseparable Two-terminal Graph. A two-terminal graph G is *nonseparable* if the resulting graph G_0 is nonseparable when an additional edge x_0 is connected between the two terminals of G.

Definition 9-8 is useful in studying the properties of switching networks. It should be noted, however, that the graph of a nonseparable two-terminal switching network may be separable, as illustrated by the following example.

EXAMPLE 9-3. The graph G of the switching network N of Figure 9-3 is shown in Figure 9-4(a). Obviously, G is a separable graph with v_a as a cut-vertex. But it is nonseparable in the "*two-terminal*" sense with respect to its terminals v_i and v_j, according to Definition 9-8, since the resulting graph G_0, from the addition of edge x_0 across terminals v_i and v_j of G, is nonseparable (in the "*general*" sense, as stated in Definition 1-9). Graph G_0 is shown in Figure 9-4(b).

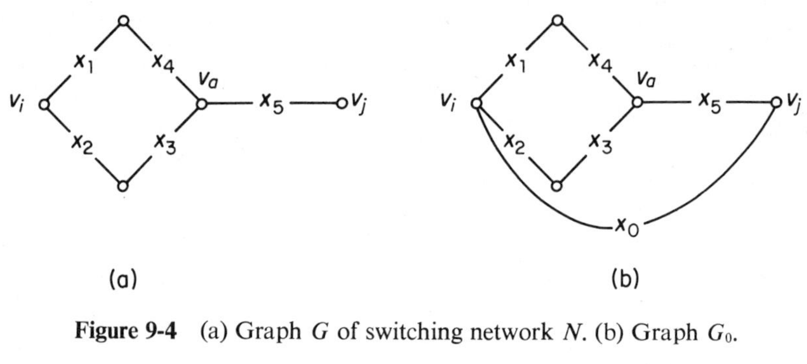

Figure 9-4 (a) Graph G of switching network N. (b) Graph G_0.

9-3 ELEMENTARY SWITCHING ALGEBRA

The so-called switching algebra we shall discuss in this section is known to the mathematician as *Boolean algebra*[5] or the *algebra of logic*. We prefer to call it *switching algebra* because we shall apply it to the theory of switching throughout the remainder of this chapter.

[5]It is called *Boolean algebra* because G. Boole developed the theory in about 1850. It is also referred to as the *algebra of sets* or the *algebra of classes*.

The variables denoted by a, b, c, x_i, x_j, etc., are the *network variables* that identify the contacts in the switching network. As mentioned in Section 9-2, each variable takes on the value 0 when the contact is open, and the value 1 when it is closed. Also, the symbol x_i' is used to denote the complement of a variable x_i so that $x_i' = 0$ when $x_i = 1$, and $x_i' = 1$ when $x_i = 0$. We define the addition and the multiplication between the two states, 0 and 1, as follows:

$$\begin{aligned} 0 + 0 &= 0 \\ 0 + 1 &= 1 \\ 1 + 0 &= 1 \\ 1 + 1 &= 1 \end{aligned} \tag{9-3}$$

$$\begin{aligned} 0 \cdot 0 &= 0 \\ 0 \cdot 1 &= 0 \\ 1 \cdot 0 &= 0 \\ 1 \cdot 1 &= 1 \end{aligned} \tag{9-4}$$

Physically, the addition of two variables indicates the parallel connection between the two variables, while the multiplication, on the other hand, represents the series connection between them. Thus, for any two variables a and b (and their complements a' and b'), we summarize Equations (9-3) and (9-4) in Tables 9-1 and 9-2, respectively.

TABLE 9-1

The Addition Table

Variables				*Addition Table*					
a	b	a'	b'	$a+b$	$a+a'$	$a+b'$	$b+a'$	$b+b'$	$a'+b'$
0	0	1	1	0	1	1	1	1	1
0	1	1	0	1	1	0	1	1	1
1	0	0	1	1	1	1	0	1	1
1	1	0	0	1	1	1	1	1	0

From Tables 9-1 and 9-2, we see that for any variable a and its complement a'

$$a + a' = 1 \tag{9-5}$$

and

$$a \cdot a' = 0 \tag{9-6}$$

This should be quite obvious from the physical point of view because, for example, if contacts a and a' are connected in parallel (as in the case of

TABLE 9-2

The Multiplication Table

Variables				*Multiplication Table*					
a	b	a'	b'	$a\cdot b$	$a\cdot a'$	$a\cdot b'$	$b\cdot a'$	$b\cdot b'$	$a\cdot b'$
0	0	1	1	0	0	0	0	0	1
0	1	1	0	0	0	0	1	0	0
1	0	0	1	0	0	1	0	0	0
1	1	0	0	1	0	0	0	0	0

addition), it is evident that a conducting path exists between the two terminals when either of the contacts is closed. Similarly, if the two contacts are connected in series and if either one of the contacts is open, there will be no conducting path between the terminals.

As indicated in Equations (9-3) and (9-4), the commutative law holds for addition and multiplication. That is, for any two variables a and b, we write

$$a + b = b + a \tag{9-7}$$

$$a\cdot b = b\cdot a \tag{9-8}$$

When there are three or more contacts involved in a switching network, more properties must be considered. For example, the associative law holds for both addition and multiplication for any three variables a, b, and c. Thus,

$$a + (b + c) = (a + b) + c \qquad \text{and} \qquad a\cdot(b\cdot c) = (a\cdot b)\cdot c$$

For the sake of compactness, all the postulates just discussed, as well as some others, are summarized in Table 9-3.

It is interesting to note that, except for the involution law (Postulate 9), all the postulates in Table 9-3 appear in dual pairs, and each postulate in a pair can be obtained by the replacement of each of the operations "+" and "·" by the other and each of the symbols "0" and "1" by the other wherever applicable. This property is known as the *principle of duality*.

It should also be noted that although the symbols a, b, and c are represented as three contacts in the network diagram in Table 9-3, they can very well be interpreted as three switching functions and the postulates will be equally valid.

Finally, we should point out that the importance of the duality principle lies in the fact that once a theorem is derived using Postulates 1 through 8, a dual theorem can be readily obtained by the interchange of the operations "+" and "·" and of the symbols "0" and "1" wherever they appear in the theorem.

TABLE 9-3

Postulates (Laws) of Switching Algebra

Postulate Number		*Name of Postulate*	*Algebraic Statement*	*Network Equivalents*
1	(a)	Commutative	$a + b = b + a$	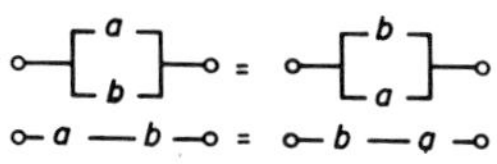
	(b)		$a \cdot b = b \cdot a$	
2	(a)	Associative	$a + (b + c) = (a + b) + c$	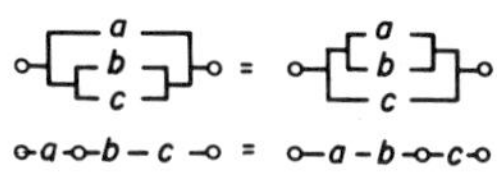
	(b)		$a \cdot (b \cdot c) = (a \cdot b) \cdot c$	
3	(a)	Distributive	$a \cdot (b + c) = ab + ac$	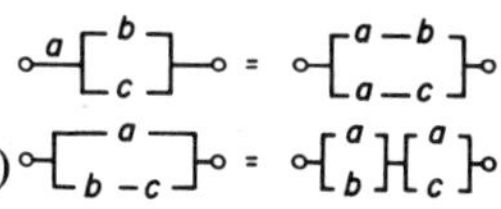
	(b)		$a + (b \cdot c) = (a + b) \cdot (a + c)$	
4	(a)	Idempotent	$a + a = a$	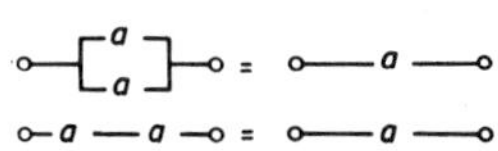
	(b)		$a \cdot a = a$	
5	(a)	Operation with 0 and 1	$a + 0 = a$	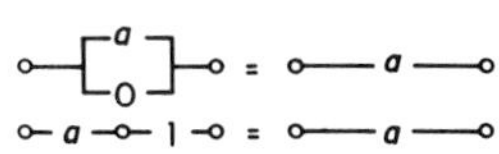
	(b)		$a \cdot 1 = a$	
6	(a)	Operation with 0 and 1	$a + 1 = 1$	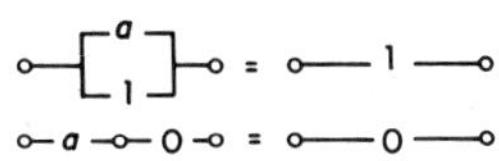
	(b)		$a \cdot 0 = 0$	
7	(a)	Complementarity	$a + a' = 1$	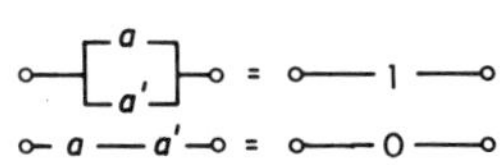
	(b)		$a \cdot a' = 0$	
8	(a)	Dualization	$(a + b)' = a'b'$	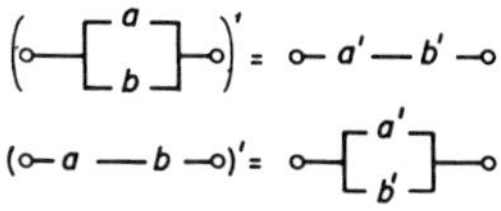
	(b)		$(a \cdot b)' = a' + b'$	
9		Involution	$(a')' = a$	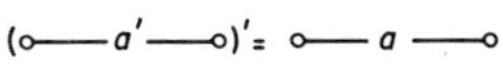

9-4 PATH MATRIX AND ITS PROPERTIES

The concept of the path matrix in the theory of linear graphs was first introduced by Wing and Kim [84]. The discussion given in this section is based on that paper.

DEFINITION 9-9. Path Matrix P. The *path matrix* P of an SC network of v vertices and e edges is the matrix $P = [p_{ij}]$ of a finite number of rows (corresponding to all the paths between vertices v_i and v_j) and e columns (corresponding to the e edges) where

$$p_{ij} = 1 \text{ if edge } x_j \text{ is in path } i$$

$$p_{ij} = 0 \text{ if edge } x_j \text{ is not in path } i$$

Since there are only $e - v + 2$ independent paths between the terminals of an SC network of v vertices and e edges, as pointed out by Mayeda,[6] the following theorem is self-evident.

THEOREM 9-1. The rank of the path matrix of an SC network of v vertices and e edges is equal to $e - v + 2$.

DEFINITION 9-10. Fundamental Path Matrix P_f. A *fundamental path matrix* is a matrix of $e - v + 2$ rows which correspond to a set of $e - v + 2$ independent paths between the two terminals of the SC network.

It is therefore obvious that P_f is a submatrix of the path matrix, and the rank of P_f is also $e - v + 2$. The other basic properties of P_f may be summarized as follows:

(a) The ring sum[6] of an even number of rows of P_f is either a circuit or an edge-disjoint union[7] of circuits.

(b) The ring sum of an odd number of rows of P_f is either a path or a redundant path (that is, a path and a circuit, or a path and an edge-disjoint union of circuits).

(c) Any row in the path matrix P is either a row of P_f or an odd ring sum (that is, the ring sum of an odd number of rows) of the rows of P_f.

These three properties have also been proved by Mayeda.[8]

DEFINITION 9-11. Basic and Nonbasic Cutset. A *basic cutset* of an SC network is a cutset such that the removal of its elements will separate the remainder of the network into two disjoint subgraphs, each of which contains one terminal of the network. All other cutsets which do not have this property are *nonbasic cutsets.*

Since each row of the path matrix P corresponds to a term of the SC function F, and to a path product of the SC network N that realizes P and F, it is obvious that an SC network can be described by either its SC function or its path matrix. Finally, each term in the SC function corresponds to either a row or the ring sum of a number of rows of a fundamental path matrix of the SC network.

[6]The ring-sum operation was defined in Chapter 1 (Definition 1-18).

[7]The union operation was defined in Definition 1-16. For the definition of disjoint subgraphs, see Definition 1-10. The term "edge-disjoint" was discussed in Example 1-2.

[8]See W. Mayeda [60], Chapter 4, for a proof of this statement. It is also listed as a problem in this chapter (Problem 9-5).

EXAMPLE 9-4. Consider the SC function F:

$$F = x_1x_2x_5 + x_1x_4x_8 + x_2x_3x_6 + x_3x_4x_7 + x_1x_2x_6x_7x_8 + x_1x_4x_5x_6x_7 + x_2x_3x_5x_7x_8 + x_3x_4x_5x_6x_8$$

which corresponds to the SC network N with terminals v_1 and v_2 as shown in Figure 9-5. The path matrix of F

$$P = \begin{array}{c} \\ 1 \\ 2 \\ 3 \\ 4 \\ 5 \\ 6 \\ 7 \\ 8 \end{array}\begin{array}{c} \begin{array}{cccccccc} x_1 & x_2 & x_3 & x_4 & x_5 & x_6 & x_7 & x_8 \end{array} \\ \begin{bmatrix} 1 & 1 & 0 & 0 & 1 & 0 & 0 & 0 \\ 1 & 0 & 0 & 1 & 0 & 0 & 0 & 1 \\ 0 & 1 & 1 & 0 & 0 & 1 & 0 & 0 \\ 0 & 0 & 1 & 1 & 0 & 0 & 1 & 0 \\ 1 & 1 & 0 & 0 & 0 & 1 & 1 & 1 \\ 1 & 0 & 0 & 1 & 1 & 1 & 1 & 0 \\ 0 & 1 & 1 & 0 & 1 & 0 & 1 & 1 \\ 0 & 0 & 1 & 1 & 1 & 1 & 0 & 1 \end{bmatrix} \end{array}$$

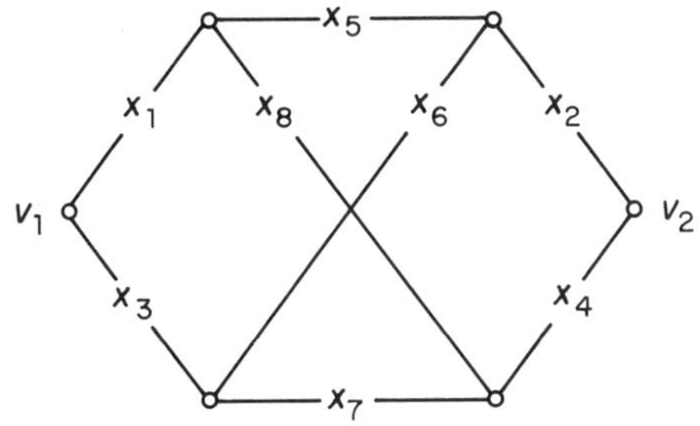

Figure 9-5 The SC network N.

has a rank $r = 4$, which can be readily verified. Hence F has four independent terms; but, by inspection of the SC network N of Figure 9-5, we see that $e = 8$ and $v = 6$, so that $e - v + 2 = 4$, which is indeed equal to the rank of P.

9-5 SINGLE-CONTACT NETWORKS AND SC FUNCTIONS

As defined in Section 9-4, an SC function is the switching function of an SC network. SC functions are also known as *network functions* or *graph functions* since they uniquely specify the corresponding switching network,[9] as will be seen later in this section.

[9]This means that for any given SC function there exists a unique network realization between two specified terminals. The realization is unique to within a 2-isomorphism. The interested reader is referred to a paper by R. L. Ashenhurst [2].

Because of its single-contact nature, an SC network is always minimal in the sense that it contains a minimum number of contacts necessary to realize the given switching function. For this reason, we devote this entire section to the study of the SC function and derive some of its properties.

Since in the theory of switching

$$A + AB = A$$

where A and B denote single (binary) variables or products of variables, we have the following definition:

DEFINITION 9-12. Redundancy. If A and AB are two terms of a switching function where A and B denote single variables or products of variables, then AB is a *redundant term* with respect to A in the sense that $A + B = 1$ (or 0) whenever $A = 1$ (or 0).

Obviously, an SC function contains no redundant terms due to its single-contact nature.

In an SC network N, if an edge x_i is opened, all the paths containing x_i between the terminals will be destroyed and, hence, the terms containing x_i in the SC function F will vanish. The new function F',[10] which is formed by eliminating all the terms containing x_i in F, is the SC function of the network with x_i removed (opened). It can be seen that the new function F' may be obtained from F simply by setting $x_i = 0$ in F; that is, $F' = F(x_i = 0)$.

Now we shall define the edge-opening transformation as follows:

DEFINITION 9-13. Edge-opening Transformation. Let $F(x_i = 0)$ be the function obtained from a function F by setting $x_i = 0$. Then $F(x_i = 0)$ is the *edge-opening transformation* of F with respect to x_i. Alternatively, we say that $F(x_i = 0)$ is obtained from F by "opening" x_i in F.

The following theorem is self-evident.

THEOREM 9-2. Any edge-opening transformation of an SC function is again an SC function.

Proof: Let the given function F be the SC function of some network N. Then, by Definition 9-13, the edge-opening transformation $F(x_i = 0)$ of F with respect to edge x_i is simply the SC function of the new network N' resulting from N by opening edge x_i. Hence the theorem.

Now, suppose edge x_i is shorted in N. Then the endpoints of x_i will coincide and all the paths between the terminals of N containing x_i will no longer contain x_i. Each of these paths will become either a path between the terminals of the resultant network $N(x_i = 1)$ or a "redundant path" in $N(x_i = 1)$ in

[10] Note that F' as used in this chapter *does not* represent the complement of F. It simply denotes a new function as defined in the context.

the sense that the edges contained in it form a path and a circuit in $N(x_i = 1)$. In the latter case, the term corresponding to the path becomes a redundant term in the new function $F(x_i = 1)$ that is realized by $N(x_i = 1)$. Therefore, the new function $F(x_i = 1)$, which is obtained by removing from F the variable x_i in each term containing it and by also removing the redundant terms, is the SC function of the resultant network $N(x_i = 1)$ where edge x_i has been short-circuited. Therefore, $F(x_i = 1)$ can be obtained by setting $x_i = 1$ and removing the redundant terms in F. The next definition and theorem are the consequences of this discussion.

DEFINITION 9-14. Edge-shorting Transformation. Let $F(x_i = 1)$ be the function which is obtained from a function F by first setting $x_i = 1$ and then removing the redundant terms, if any, from the function. Then, $F(x_i = 1)$ is the *edge-shorting transformation* of F with respect to x_i. Alternatively, we say that $F(x_i = 1)$ is obtained from F by "shorting" the variable x_i in F.

THEOREM 9-3. Any edge-shorting transformation of an SC function is again an SC function.

Proof: If the given function is the SC function of some SC network, then, from the discussion preceding and from Definition 9-12, the theorem follows immediately.

Finally, we shall give an example to illustrate the concepts introduced in this section.

EXAMPLE 9-5. Consider the SC function F:

$$F = x_1x_2x_5 + x_1x_4x_8 + x_2x_3x_6 + x_3x_4x_7 + x_1x_2x_6x_7x_8 + x_1x_4x_5x_6x_7 + x_2x_3x_5x_7x_8 + x_3x_4x_5x_6x_8$$

which corresponds to the SC network N as shown in Figure 9-6. Now, obtain the edge-opening transformation $F(x_3 = 0)$ of F:

$$F(x_3 = 0) = x_1x_2x_5 + x_1x_4x_8 + x_1x_2x_6x_7x_8 + x_1x_4x_5x_6x_7$$

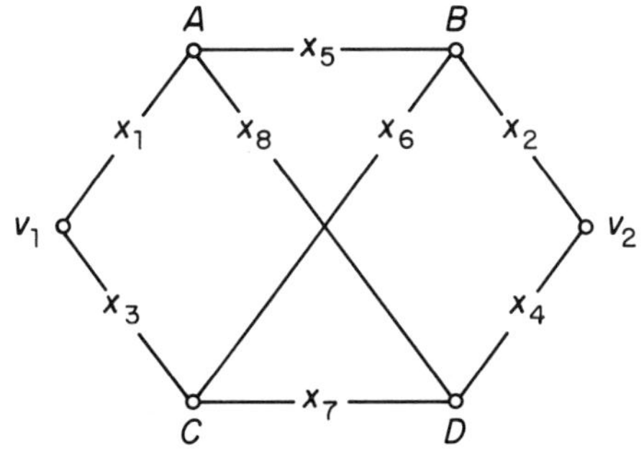

Figure 9-6 Network N realizing F between vertices v_1 and v_2.

$F(x_3 = 0)$ is realized by the network $N(x_3 = 0)$, which is a subnetwork of N, as shown in Figure 9-7. Next, obtain the edge-shorting transformation $F(x_7 = 1)$ of F:

$$F(x_7 = 1) = x_1x_2x_5 + x_1x_4x_8 + x_2x_3x_6 + x_3x_4 + x_1x_2x_6x_8 + x_1x_4x_5x_6 + x_2x_3x_5x_8 + \cancel{x_3x_4x_5x_6x_8}$$

(redundant)

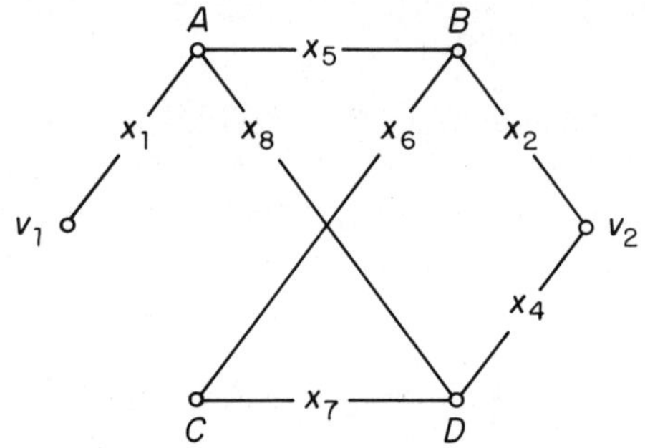

Figure 9-7 Network $N(x_3 = 0)$ realizing $F(x_3 = 0)$ between v_1 and v_2.

where the term $x_3x_4x_5x_6x_8$ is redundant due to the presence of the term x_3x_4 in F. The network $N(x_7 = 1)$ realizing $F(x_7 = 1)$ is shown in Figure 9-8.

Before introducing a new transformation, we first state and prove the following theorem:

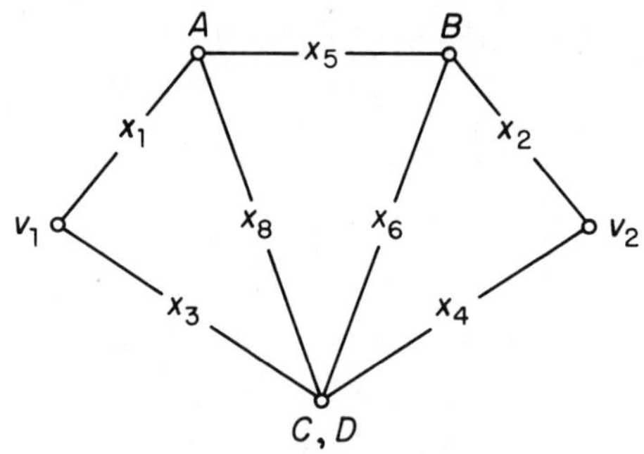

Figure 9-8 Network $N(x_7 = 1)$ realizing $F(x_7 = 1)$ between v_1 and v_2.

THEOREM 9-4. Let F be the SC function of an SC network N of v vertices and e edges between terminals v_i and v_j. Let F' be a sum (union) of n product terms, each of which is a path product between two vertices, v_i' and v_j', which are the endpoints of a given edge x_k of the e edges of N where n is the number of all such paths. If the number of variables contained in F' is denoted by e', then

(a) $e' \leqq e$.

(b) If $e' = e$, N is nonseparable in the general sense (in contrast to the nonseparability in the two-terminal sense (Definition 9-8), and F' is the SC function of N with respect to v_i' and v_j'.

(c) If $e' < e$, N is separable (in the general sense) and F' is the SC function of some subnetwork N' of N.

Proof:

(a) Since each term in F corresponds to a path between v_i' and v_j', and the variables contained in each term of F' are the edges contained in some path between v_i' and v_j', and since the total number of edges in N is e, the number of variables, e', of F' cannot be greater than e; that is, $e' \leqq e$.

(b) If $e' = e$, all the e edges (and hence all the v vertices) of N are contained in the paths between v_i' and v_j'. By definition, F' is the SC function of N with respect to v_i' and v_j'. Furthermore, since all the edges of N are included in F', every edge must be in some path between v_i' and v_j'. This implies that any two of the e edges are contained in some circuit, the reason being as follows: If x_a and x_b denote any two edges of N, then either x_a and x_b are in the same path between v_i' and v_j' or they are not. If they are in the same path, the path forms a circuit with x_k; if not, then the path containing x_a and that containing x_b form a circuit. Therefore, any two vertices of N are contained in some circuit and, by a theorem of Whitney,[11] N is nonseparable.

(c) If $e' < e$, some of the variables in F will not be contained in F'. This means that the edges corresponding to these variables cannot be placed in any path between v_i' and v_j'. Any edge x_0 of these edges and the edge x_k cannot be contained in any circuit of N; if they are contained in some circuit, say C, then C becomes a path between v_i' and v_j' containing x_0 when x_k is removed from it, which contradicts the fact that x_0 cannot be placed in any path between v_i' and v_j'. Thus, by a theorem of Ashenhurst,[12] N is a separable graph. The fact that F' is the SC function of a subnetwork N' of N is obvious since N' is obtained by opening from N the edges which are not contained in any of the n path products between v_i' and v_j', and the terms of F' correspond to all the path products between v_i' and v_j' in N. The theorem is now proved.

DEFINITION 9-15. Terminal Transformation (With Respect to Some Variable x_k). Let the function F of e variables be the SC function between vertices v_i and v_j of an SC network N of v vertices and e edges. Also, let $S_1 = \{f_{1i}/i = 1, 2, \cdots\}$ be the set of all terms in F containing the variable x_k; and let $S_2 = \{f_{2j}/j = 1, 2, \cdots\}$ be the set of all terms in F not containing x_k. Then the function $F_v(x_k)$, which is obtained step by step by the application of the

[11]This theorem states that a connected graph G is nonseparable if and only if every pair of vertices of G is contained in some circuit. See H. Whitney [81].

[12]This theorem states that if G is a nonseparable graph and if e_a and e_b are any two edges of G, then e_a and e_b are contained in some circuit of G. See R. L. Ashenhurst [2].

following operations, is called the *terminal transformation* of F with respect to x_k.[13]

(a) Obtain the function F', where

$$F' = x_k + \sum_{i,j} f_{1i} \oplus f_{2j} \oplus x_k \qquad \text{when } S_2 \neq \phi$$

$$F' = x_k \qquad \text{when } S_2 = \phi$$

(b) Remove all repeated redundant terms in F', and the resultant function is $F_v(x_k)$.

THEOREM 9-5. Any terminal transformation of an SC function is again an SC function.

Proof: We shall prove that (a) any term contained in $F_v(x_k)$ is a path between the x_k endpoints, v_i' and v_j', and that (b) every path between v_i' and v_j' in the SC network N that realizes the given SC function F between v_i and v_j is contained in $F_v(x_k)$.

(a) Consider any two paths between v_i and v_j, f_{1i} and f_{2j} (see Definition 9-15). Then, f_{1i} and f_{2j} either have some variables in common or none at all. If they do have common variables, $f_{1i} \oplus f_{2j}$ is either a circuit containing x_k [Figure 9-9(a)] or an edge-disjoint union of circuits, one of which contains x_k [Figure 9-9(b)]. If $f_{1i} \oplus f_{2j}$ is a circuit, then $(f_{1i} \oplus f_{2j}) \oplus x_k$ is a path between v_i' and v_j'. If $f_{1i} \oplus f_{2j}$ is an edge-disjoint union of circuits, let the one containing x_k be C. Then C is the union of two paths, say C_1 and C_2, where C_1 is a proper subpath of f_{1i} and contains x_k, and C_2 is a proper subpath of f_{2j} and does not contain x_k; that is, $C = C_1 + C_2$. But there exists a third path, f_j', between v_i and v_j such that f_j' does not contain x_k (and, hence, $f_j' \in S_2$) and that

$$f_j' = (f_{1i} \oplus C_1) + C_2$$

so that

$$\begin{aligned} f_j' \oplus f_{1i} &= f_{1i} \oplus f_j' = f_{1i} \oplus [(f_{1i} \oplus C_1) + C_2] \\ &= [(f_{1i} \oplus f_{1i}) \oplus C_1] + C_2 = C_1 + C_2 = C \end{aligned}$$

Hence, $(f_j' \oplus f_{1i}) \oplus x_k = (f_{1i} \oplus x_k) \oplus f_j'$ is a path between x_i' and x_j'. This shows that if $f_{1i} \oplus f_{2j}$ is an edge-disjoint union of circuits, $(f_{1i} \oplus f_{2j}) \oplus x_k = (f_{1i} \oplus x_k) \oplus f_{2j}$ will then be an edge-disjoint union of a path between v_i' and v_j' and a circuit (or circuits), and is a redundant term when compared to $(f_j' \oplus f_{1i}) \oplus x_k$, which is the path contained in $(f_{1i} \oplus f_{2j}) \oplus x_k$ (along with some circuits), and therefore will be eliminated in Step (b) of Definition 9-15. If f_{1i} and f_{2j} have no variables in common, $f_{1i} \oplus f_{2j}$ is either a circuit containing x_k and vertices v_i and v_j — in which case f_{1i} and f_{2j} are two disjoint paths between v_i and v_j as shown in Figure 9-9(c) — or an edge-disjoint union of

[13]Note that in (a) the addition symbol "+" denotes the Boolean addition defined in Section 9-3, and the symbol "$\oplus$" is the ring-sum operation defined in Definition 1-18.

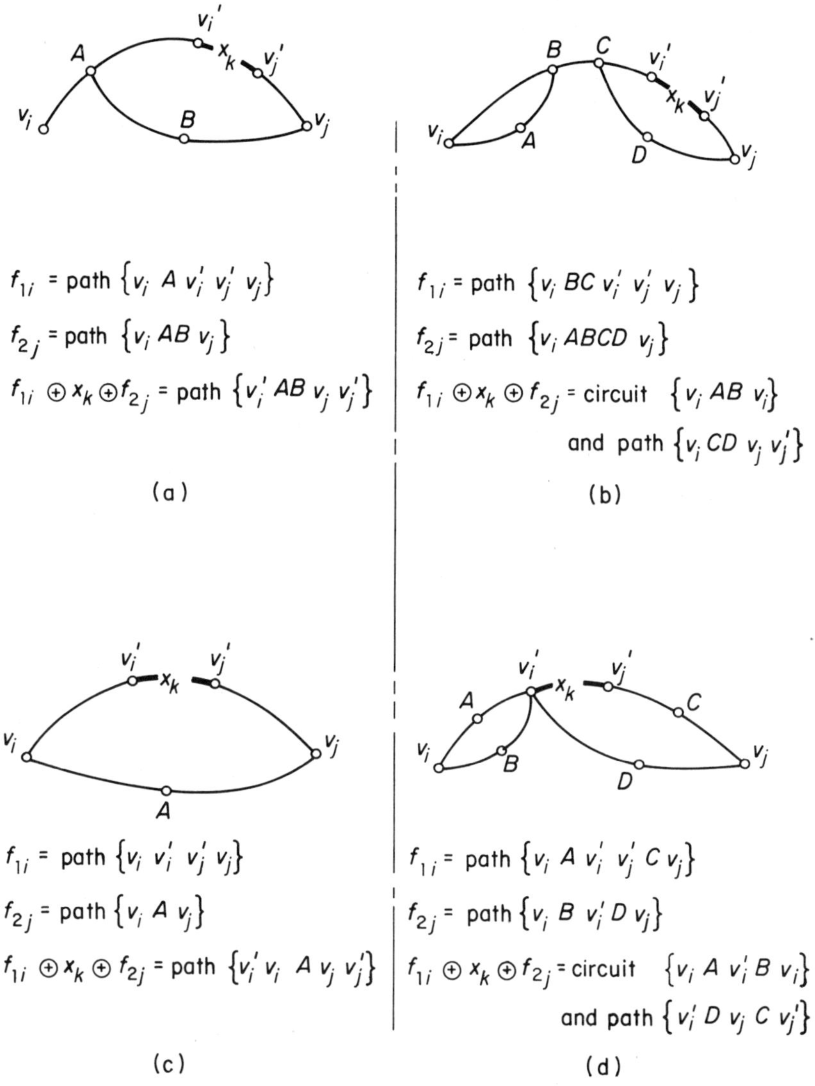

Figure 9-9 Various cases for the proof of part (a) of Theorem 9-5.

circuits connected in "series" [Figure 9-9(d)], one of which contains x_k. It is seen that the same arguments applied in the two previous cases may very well be applied here. Therefore, we conclude that every term contained in $F_v(x_k)$ is indeed a path between v_i' and v_j'.

(b) Let p' ($\neq x_k$) be any path between v_i' and v_j'. Then $p' \oplus x_k$ is a circuit. Since $S_1 \neq \phi$, there exists a path from v_i to one of the vertices in $p' \oplus x_k$, which contains no edges in $p' \oplus x_k$ (or otherwise v_i is a vertex in $p' \oplus x_k$). Let this path be denoted by p_{v_i} and let the vertex in $p' \oplus x_k$ at which this path is terminated be denoted by v_i''. Similarly, let p_{v_j} be the path from v_j to one of the vertices in $p' \oplus x_k$, and let v_j'' be the vertex in $p' \oplus x_k$ at which p_{v_j} is terminated. Then it is clear that there exists an f_{1i} in S_1 and an f_{2j} in S_2 such that $f_{1i} \oplus f_{2j} \oplus x_k = p'$. This completes the proof.

THEOREM 9-6. Let the function $F_v(x_k)$ of e' variables be the terminal transformation of a given SC function F of e variables of some SC network N. Then,

(a) $e' \leq e$.
(b) N is nonseparable if $e' = e$.
(c) N is separable if $e' < e$.

Proof: By Theorem 9-5, F_v is the SC function of some subnetwork N' of N. (Note that N is the subnetwork of N itself.) The rest of the proof is identical to that of Theorem 9-4.

A few examples seem to be appropriate at this point.

EXAMPLE 9-6. Consider the same SC function F of Example 9-5. To obtain the terminal-transformation $F_v(x_5)$ of F with respect to x_k, we let $x_k = x_5$. Then the terms in S_1 are those in F containing x_5, and the terms in S_2 are those in F not containing x_5. Thus,

$$S_1: \begin{cases} f_{11} = x_1x_2x_5 \\ f_{12} = x_1x_4x_5x_6x_7 \\ f_{13} = x_2x_3x_5x_7x_8 \\ f_{14} = x_3x_4x_5x_6x_8 \end{cases} \qquad S_2: \begin{cases} f_{21} = x_1x_4x_8 \\ f_{22} = x_2x_3x_6 \\ f_{23} = x_3x_4x_7 \\ f_{24} = x_1x_2x_6x_7x_8 \end{cases}$$

Hence,

$$\sum_{i,j} f_{1i} \oplus f_{2j} \oplus x_k$$

is the sum of the following terms:[14]

$$\begin{aligned} f_{11} \oplus f_{21} \oplus x_5 &= x_2x_4x_8 \\ f_{11} \oplus f_{22} \oplus x_5 &= x_1x_3x_6 \\ f_{11} \oplus f_{23} \oplus x_5 &= x_1x_2x_3x_4x_7 \\ f_{11} \oplus f_{24} \oplus x_5 &= x_6x_7x_8 \end{aligned} \qquad \begin{aligned} f_{13} \oplus f_{21} \oplus x_5 &= x_1x_2x_3x_4x_7 \\ f_{13} \oplus f_{22} \oplus x_5 &= x_6x_7x_8 \\ f_{13} \oplus f_{23} \oplus x_5 &= x_2x_4x_8 \\ f_{13} \oplus f_{24} \oplus x_5 &= x_1x_3x_6 \end{aligned}$$

$$\begin{aligned} f_{12} \oplus f_{21} \oplus x_5 &= x_6x_7x_8 \\ f_{12} \oplus f_{22} \oplus x_5 &= x_1x_2x_3x_4x_7 \\ f_{12} \oplus f_{23} \oplus x_5 &= x_1x_3x_6 \\ f_{12} \oplus f_{24} \oplus x_5 &= x_2x_4x_8 \end{aligned} \qquad \begin{aligned} f_{14} \oplus f_{21} \oplus x_5 &= x_1x_3x_6 \\ f_{14} \oplus f_{22} \oplus x_5 &= x_2x_4x_8 \\ f_{14} \oplus f_{23} \oplus x_5 &= x_6x_7x_8 \\ f_{14} \oplus f_{24} \oplus x_5 &= x_1x_2x_3x_4x_7 \end{aligned}$$

[14]Note that since the ring-sum operation is a binary operation, each of the following terms is obtained by taking the ring sum $f_{1i} \oplus f_{2j}$ of the first two functions f_{1i} and f_{2j} and then taking the ring sum $(f_{1i} \oplus f_{2j}) \oplus x_k$ of $(f_{1i} \oplus f_{2j})$ and x_k. Thus the first result is obtained in the following manner: $f_{11} \oplus f_{21} \oplus x_5 = [(x_1x_2x_5) \oplus (x_1x_4x_8)] \oplus x_5 = (x_2x_4x_5x_8) \oplus (x_5) = x_2x_4x_8$ for example.

and

$$F' = x_5 + \sum_{i,j} f_{1i} \oplus f_{2j} \oplus x_5$$

Finally, eliminating the repeated and redundant terms from F' [Step (b)], we find

$$F_v(x_5) = x_5 + x_1x_3x_6 + x_2x_4x_8 + x_6x_7x_8 + x_1x_2x_3x_4x_7$$

which is the SC function of the same network N as shown in Figure 9-6 between vertices A and B, which are the endpoints of x_5.

EXAMPLE 9-7. Consider the following SC function:

$$\begin{aligned} F = {} & x_1x_5x_6x_7 + x_1x_5x_6x_8 + x_2x_5x_7x_9 + x_2x_5x_8x_9 + x_3x_6x_7x_{10} \\ & + x_3x_6x_8x_{10} + x_4x_7x_9x_{10} + x_4x_8x_9x_{10} + x_1x_2x_3x_7x_9x_{10} \\ & + x_1x_2x_3x_8x_9x_{10} + x_1x_2x_4x_6x_7x_{10} + x_1x_2x_4x_6x_8x_{10} \\ & + x_1x_3x_4x_5x_7x_9 + x_1x_3x_4x_5x_8x_9 + x_2x_3x_4x_5x_6x_7 \\ & + x_2x_3x_4x_5x_6x_8 \end{aligned}$$

The network realizing F is shown in Figure 9-10. Now, the terminal transformation $F_v(x_1)$ of F with respect to x_1 is

$$F_v(x_1) = x_1 + x_2x_3x_4 + x_2x_6x_9 + x_3x_5x_{10} + x_4x_5x_6x_9x_{10}$$

from which we get $e' = 8 < e = 10$. Hence, by Theorem 9-6(c), N is separable, as seen in Figure 9-10.

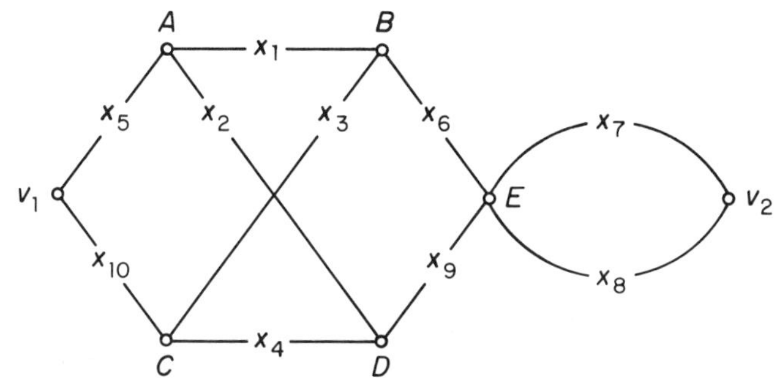

Figure 9-10 Network N realizing F between v_1 and v_2.

The network N' realizing $F_v(x_1)$ is shown in Figure 9-11. We see from Example 9-7 that the terminal transformation $F_v(x_k)$ is useful in that it can be applied to a given SC function F to detect whether F is separable or nonseparable. If it is separable, the function will be simplified as a result of such a transformation. This is summarized in the next theorem.

THEOREM 9-7. Let the function $F_v(x_k)$ of e' variables be the terminal transformation of a given SC function F of e variables of some SC network N. Then N is nonseparable if and only if $e' = e$.

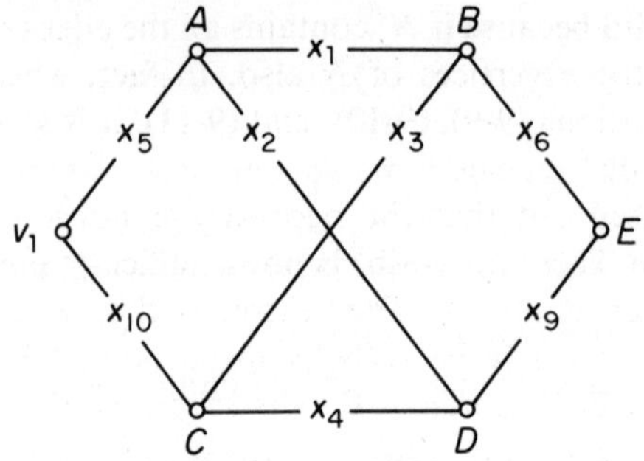

Figure 9-11 Network N' realizing $F_v(x_1)$ between vertices A and B.

Proof: The sufficiency part of the theorem has been established in Theorem 9-6. We shall now prove the necessity part. If N is nonseparable, then, by a theorem of Ashenhurst, any two edges of N may be placed in some circuit.[15] Consider two edges, x_m and x_n. Let x_m be x_k [the reference edge of $F_v(x_k)$], and let x_n be any of the $e - 1$ edges in N. Then, the fact that x_m and x_n may be placed in some circuit implies that x_n be contained in some path between v_i' and v_j', the endpoints of x_k. Since x_n is any of the $e - 1$ edges of N, all the $e - 1$ edges must be contained in the paths between v_i' and v_j' and the edge, x_k $(= x_m)$, is, by itself, a path between v_i' and v_j'. Therefore, all the e variables of F are contained in $F_v(x_k)$; that is, $e' = e$. This completes the proof.

THEOREM 9-8. Let the function $F_v(x_k)$ of e' variables be the terminal transformation of a given SC function F of e variables of some SC network N. Then, (a) $r' \leq r$ always, and (b) $r' = r$ if N is nonseparable, where r and r' are the ranks of the path matrices corresponding to F and $F_v(x_i)$, respectively.

Proof:

(a) By Theorem 9-6(a), $e' \leq e$, which implies that $v' = v$, or $v - v' = e - e'$. (Why?) But, $e' - v' + 2 = r'$ and $e - v + 2 = r$. It follows that $r' \leq r$.

(b) By Theorem 9-7, $e' = e$ if N is nonseparable. But,

$$r' = e' - v' + 2 \tag{9-9}$$

and

$$r = e = v + 2 \tag{9-10}$$

Furthermore,

$$v' = v \qquad \text{if } e' = e \tag{9-11}$$

[15] See R. L. Ashenhurst [2] for the proof of this statement.

Equation (9-11) is valid because if N' contains all the edges of N, it is necessary that N' contain all the v vertices of N also. In fact, when $e' = e$, $N' = N$. Therefore, from Equations (9-9), (9-10), and (9-11), it is seen that $r' = r$. Thus the theorem is proved.

It should be pointed out that the necessary condition for N to be nonseparable as stated in Theorem 9-8(b) is not a sufficient condition, for N may still be separable even if $r' = r$. The reason is that when $e - e' = v - v'$, then $r' = r$, which can easily be seen by inspection of Equations (9-9) and (9-10). Note also that by Theorem 9-6, N is separable if $e' \neq e$.

We shall now summarize the results of this section, which shows the effects of the edge transformations and terminal transformations of an SC function on the original SC network.

Let N be any given SC network whose function between vertices v_i and v_j is denoted by F, which is of rank r. Also, let F' stand for any of the functional transformations, and let N' be the SC network realizing F'. Finally, let N have v vertices and e edges, and let N' have v' vertices and e' edges. Then, we have the following statements:

(a) If F' is an edge-opening transformation of F, then N' is a proper subnetwork of N. Also, $v' \leq v$, $e' < e$, and $r' < r$.

(b) If F' is an edge-shorting transformation of F, then N' may or may not be (but generally is not) a subnetwork of N.

(c) Moreover, if F' is a terminal transformation with respect to x_i (v_i', v_j'), and

(i) if $e' \leq e$, then N' is a subnetwork of N, $v' \leq v$, and $r' \leq r$;

(ii) if $e' < e$, then N is separable, N' is a proper subnetwork of N containing x_i, $v' < v$, and $r' \leq r$;

(iii) if $e' = e$, then $N' = N$; that is, N' and N are identical.

Before stating an important realizability condition of an SC function, we define the following.

DEFINITION 9-16. Odd Ring Sum and Even Ring Sum. Let $S = \{S_1, S_2, \ldots, S_n\}$ be a collection of n sets of elements, each set of which may or may not have elements in common (that is, $S_1, S_2, \ldots, S_n$ may or may not be disjoint sets). Then the ring sum of an odd number p of sets of S is called an *odd ring sum* of the n sets and is denoted by

$$\text{odd} \oplus \{S_{i1}, S_{i2}, \ldots, S_{ip}\} \equiv S_{i1} \oplus S_{i2} \oplus \cdots \oplus S_{ip}$$

where $S_{i1}, S_{i2}, \ldots, S_{ip}$ are the p sets of S involved in the odd ring sum. Similarly, the ring sum of an even number q of sets of the n sets is called an *even ring sum* of the n sets and is denoted by

$$\text{even} \oplus \{S_{j1}, S_{j2}, \ldots, S_{jq}\} \equiv S_{j1} \oplus S_{j2} \oplus \cdots \oplus S_{jq}$$

where $S_{j1}, S_{j2}, \ldots, S_{jq}$ are the q sets of S involved in the even ring sum.

DEFINITION 9-17. The Odd-ring-sum Condition. Let r be the rank of a matrix M of n rows. If all the irredundant odd ring sums of any r independent rows of M ($r \leq n$), together with the r rows, form the n rows of M, then M is said to satisfy the *odd-ring-sum condition.*

One of the properties of a path matrix P is that any odd ring sum of the rows of P is either a row of P or a row that contains all the nonzero elements of a row of P. In the latter case, the row is called a redundant row of P since it corresponds to a path plus either a circuit or an edge-disjoint union of circuits, as pointed out in Section 9-4. It is therefore obvious that a set of all the r (independent) rows of any fundamental path matrix P_f and all the non-redundant odd ring sums of the r rows of P_f, where r is the rank of P_f, together form a path matrix.

It is known that an odd ring sum of the paths between the terminals v_i and v_j of an SC network is either a path between v_i and v_j or an edge-disjoint union of a path between v_i and v_j and circuits. This gives the following theorem:

THEOREM 9-9. A necessary condition for a matrix M to be a path matrix is that M satisfy the odd-ring-sum condition.

DEFINITION 9-18. Correspondence Between a Function and a Matrix. Let F be a function of n variables expressed as a sum of m product terms. Let M be the matrix of order $m \times n$ whose m rows correspond to the m terms of F and whose n columns correspond to the n variables of F. Then M is said to be the *corresponding matrix* of F; likewise, F is said to be the *corresponding function* of M.

COROLLARY 9-1. A necessary condition for F to be an SC function is that the corresponding matrix M of F satisfy the odd-ring-sum condition.

Corollary 9-1 follows directly from Theorem 9-9 and Definition 9-18.

EXAMPLE 9-8. Consider the same SC function as in Example 9-5:

$$\begin{aligned} F = {} & x_1x_2x_5 + x_1x_4x_8 + x_2x_3x_6 + x_3x_4x_7 + x_1x_2x_6x_7x_8 \\ & + x_1x_4x_5x_6x_7 + x_2x_3x_5x_7x_8 + x_3x_4x_5x_6x_8 \end{aligned}$$

which has four independent terms ($r = 4 = e - v + 2$). Choose the four independent terms from F as follows:

$$f_1 = x_1x_2x_5 \qquad f_2 = x_1x_4x_8 \qquad f_3 = x_2x_3x_6 \qquad f_4 = x_3x_4x_7$$

Now, obtain all the possible odd ring sums:

$$\begin{aligned} f_1 \oplus f_2 \oplus f_3 &= x_3x_4x_5x_6x_8 & \qquad f_1 \oplus f_2 \oplus f_4 &= x_2x_3x_5x_7x_8 \\ f_1 \oplus f_3 \oplus f_4 &= x_1x_4x_5x_6x_7 & \qquad f_2 \oplus f_3 \oplus f_4 &= x_1x_2x_6x_7x_8 \end{aligned}$$

We see that all the odd ring sums are the dependent terms contained in F. Therefore, F satisfies the odd-ring-sum condition.

It should be pointed out that the odd-ring-sum condition stated in Corollary 9-1 is just one of the (many) necessary realizability conditions of SC networks. Some other realizability conditions may be derived from the properties of SC networks. Such conditions may be used to formulate a testing procedure for the determination of the realizability (or nonrealizability) of a given function as an SC function.[16]

9-6 TOPOLOGICAL SYNTHESIS OF SC NETWORKS

The main problem in the synthesis of SC networks is the development of a method of constructing an SC network from the prescribed function. In this section, it is assumed that the given function is an SC function; that is, we assume that the function is realizable as an SC network.

There exist several synthesis procedures, each of which gives an algorithm leading to the realization of the same SC network (within 2-isomorphism, of course). Mayeda's method [60], for example, is a systematic procedure for constructing the SC network from the given SC function by first obtaining the cutset matrix and then the incidence matrix. The method of Gould [36] is one that approaches the network realization by a step-by-step procedure for first realizing a tree of the graph from which the network is finally obtained. Löfgren's procedure [52] is a part-by-part method in which the fundamental circuits are constructed from the prescribed function. The method devised by Seshu [72] is the formation of the circuit matrix from the given SC function by adding a "driver element" between the input vertices.

Of these four methods, Seshu's procedure is believed to be the most straightforward in theory and the easiest to apply. Based on Seshu's method and with the help of Mayeda's algorithm for the construction of the incidence matrix from the cutset matrix, Chan develops a synthesis procedure[17] that allows one to realize the SC network from the SC function very easily because all the steps in the procedure not only are clearly defined, but are very simple in nature. We shall first give a brief discussion of the theory behind the method; then the entire procedure will be presented with all the steps listed in proper sequence. Finally, an example will be given to illustrate the application of this synthesis algorithm.

(a) A General Discussion

Let the function F_{AB} be the SC function of a certain SC network N with

[16]For further study, refer to S. P. Chan [15].

[17]For a more detailed discussion on the topological analysis and synthesis of SC networks, the interested reader is referred to S. P. Chan [14].

vertices A and B as the terminal vertices of N. Let F_{AB} be expressed in its natural form — that is, a sum of products of the variables

$$F_{AB} = p_1 + p_2 + \cdots + p_k = \sum_{j=1}^{k} p_j$$

Let an edge, called the "driver element" and denoted by x_0, be added across vertices A and B, the terminals of N; and let the resultant network be denoted by N^*. Now, if F_{AB} is multiplied by x_0,

$$x_0F_{AB} = x_0p_1 + x_0p_2 + \cdots + x_0p_k + \sum_{j=1}^{k} x_0p_j$$

each term x_0p_j in the product x_0F_{AB} will represent a circuit in N^*. Since p_j is a path between A and B, and the driver element x_0 is connected between these two vertices, p_j and x_0 together must form a circuit in the new network N^*. Therefore, if P_a is the path matrix of the original SC network N, the matrix B_p, which is obtained by adding a column of ones to P_a, will then be a circuit matrix of the new resultant network N^*. Using a theorem of Ashenhurst [2], which states that, in any nonseparable network, any circuit that does not contain a given edge can be expressed as the ring sum of some two paths between the endpoints of that element, one can easily see that every circuit in N^* is the ring sum of some two paths (two rows) of P_a of N. It follows directly that the rank of B_p is equal to $e - v + 2$, the nullity of N^*, where e and v are the number of edges and vertices, respectively, of N. This guarantees that a fundamental circuit matrix B_f of N^* can be obtained from B_p. Knowing B_f, we can then obtain corresponding cutset matrix Q_f by employing the familiar relationship between B_f and Q_f (see Theorem 2-15):

$$Q_f = [B_{f12}^T \quad U]$$

where

$$B_f = [U \quad B_{f_{12}}]$$

and

$$Q_f = [Q_{f_{11}} \quad U]$$

with the understanding that B_f and Q_f have the same column sequence.

Next, an incidence matrix A can be found from Q_f by performing the ring-sum operations on the rows of Q_f until each column of the resultant matrix has no more than two 1's, this being the necessary and sufficient condition of the incidence matrix. Usually, such a transformation from Q_f to A can be made quite easily and sometimes even by inspection if the order of Q_f is sufficiently small. However, if this is not the case, Mayeda's method [61], which gives a systematic procedure to transform Q_f into A, can be used here to determine A from Q_f.

(b) The Synthesis Procedure

Let F_{AB} be a given SC function with variables $x_1, x_2, \ldots, x_n$.

Step 1: Form the path matrix P_a with $x_1, x_2, \ldots, x_n$ as its columns and $p_1, p_2, \ldots, p_k$ as its rows, where p_j is the jth term of F_{AB}.

Step 2: Add a column of ones to P_a to form the circuit matrix B_p and identify the added column by x_0.

Step 3: Find the rank of B_p by reducing it to the normal form using row operations with possibly rearrangement of columns. This will produce a fundamental circuit matrix B_f:

$$B_f = \begin{array}{c} \\ [\end{array}\begin{array}{cccccccc} x_1' & x_2' & \ldots & x_r' & x_{r+1}' & \ldots & x_n' & x_{n+1}' \\ & & U_R & \vdots & & B_{f_{12}} & & \end{array}\begin{array}{c} \\ ] \end{array}$$

where the order $x_1', x_2', \ldots, x_{n+1}'$ denotes the new sequence of the columns after B_f is obtained. (Note that the set $\{x_1', x_2', \ldots, x_r'\}$ constitutes a chord set, and the set $\{x_{r+1}', x_{r+2}', \ldots, x_n', x_{n+1}'\}$ constitutes a tree-branch set.)

Step 4: Form the fundamental cutset matrix Q_f:

$$Q_f = [B_{f12}^T \quad U]$$

where B_{f12}^T denotes the transpose of $B_{f_{12}}$.

Step 5: Form an incidence submatrix A:

$$A = D \cdot Q_f$$

where D is nonsingular.[18] This corresponds to the application of ring-sum operations on the rows of Q_f until there are no more than two 1's in each column of the transformed matrix A. If the order of Q_f is large, apply Mayeda's method to transform Q_f into A.

Step 6: Form the incidence matrix A_a by adding the last row to A so that each column of A_a has exactly two 1's.

Step 7: Determine the network N^* from A_a and then remove the edge identified by x_0. This gives the SC network N that realizes F_{AB} as its switching function, and the two terminals A and B of N are the endpoints of the removed edge x_0.

(c) An Example and Remarks

EXAMPLE 9-9. Let F be the SC function of a certain SC network N, which is to be determined.

$$\begin{aligned} F_{AB} &= x_1x_2x_3x_5x_7x_9 + x_1x_2x_5x_7x_8 + x_1x_4x_5 + x_1x_5x_6 + x_2x_3x_4x_8x_{10} \\ &\quad + x_2x_3x_6x_8x_{10} + x_2x_4x_9x_{10} + x_2x_6x_9x_{10} + x_3x_7x_{10} + x_7x_8x_9x_{10} \\ &= \sum_{j=1}^{10} p_j \end{aligned}$$

[18]The proof of this statement was listed as a problem at the end of Chapter 2 (Problem 2-32).

Step 1: The path matrix P_a is obtained as shown:

$$
P_a = \begin{array}{c} p_1 \\ p_2 \\ \\ \cdot \\ \\ \cdot \\ \\ \cdot \\ \\ p_{10} \end{array}
\overset{\begin{array}{cccccccccc} x_1 & x_2 & x_3 & x_4 & x_5 & x_6 & x_7 & x_8 & x_9 & x_{10} \end{array}}{\begin{bmatrix}
1 & 1 & 1 & 0 & 1 & 0 & 1 & 0 & 1 & 0 \\
1 & 1 & 0 & 0 & 1 & 0 & 1 & 1 & 0 & 0 \\
1 & 0 & 0 & 1 & 1 & 0 & 0 & 0 & 0 & 0 \\
1 & 0 & 0 & 0 & 1 & 1 & 0 & 0 & 0 & 0 \\
0 & 1 & 1 & 1 & 0 & 0 & 0 & 1 & 0 & 1 \\
0 & 1 & 1 & 0 & 0 & 1 & 0 & 1 & 0 & 1 \\
0 & 1 & 0 & 1 & 0 & 0 & 0 & 0 & 1 & 1 \\
0 & 1 & 0 & 0 & 0 & 1 & 0 & 0 & 1 & 1 \\
0 & 0 & 1 & 0 & 0 & 0 & 1 & 0 & 0 & 1 \\
0 & 0 & 0 & 0 & 0 & 0 & 1 & 1 & 1 & 1
\end{bmatrix}}
$$

Step 2: A column of ones is added to P_a to form B_p. This added column is identified by x_0.

$$
B_p = \begin{array}{c} p_1 \\ p_2 \\ \cdot \\ \cdot \\ \cdot \\ p_{10} \end{array}
\overset{\begin{array}{ccccc} x_1 & x_2 & \cdots & x_{10} & x_0 \end{array}}{\left[\begin{array}{cc} & 1 \\ & 1 \\ & \cdot \\ P_a & \cdot \\ & \cdot \\ & 1 \end{array}\right]}
$$

Step 3: B is found to be of rank 5 and is reduced to the normal form by the usual ring-sum operations on the rows. The reduced matrix denoted by B_N is the following:

$$
B_p \sim \begin{array}{c} p_1 \\ p_2 \\ p_3 \\ p_4 \\ p_5 \\ p_6 \\ \cdot \\ \cdot \\ \cdot \\ p_{10} \end{array}
\overset{\begin{array}{ccccccccccc} x_1 & x_2 & x_3 & x_4 & x_5 & x_6 & x_7 & x_8 & x_9 & x_{10} & x_0 \end{array}}{\left[\begin{array}{ccccccccccc}
1 & 1 & 1 & 0 & 1 & 0 & 1 & 0 & 1 & 0 & 1 \\
0 & 1 & 1 & 1 & 0 & 0 & 1 & 0 & 1 & 0 & 0 \\
0 & 0 & 1 & 0 & 0 & 0 & 0 & 1 & 1 & 0 & 0 \\
0 & 0 & 0 & 1 & 0 & 1 & 0 & 0 & 0 & 0 & 0 \\
0 & 0 & 0 & 0 & 0 & 0 & 1 & 1 & 1 & 1 & 1 \\
\hdashline
\multicolumn{11}{c}{\text{every row from } p_6 \text{ to } p_{10} \text{ is the same}} \\
\multicolumn{11}{c}{\text{as row } p_5} \\
 \\
 \\
 \\
\end{array}\right]} = B_N
$$

Next, B is obtained by first taking the first five rows of B_N to form a new matrix of order 5×10 (and of rank 5, of course). Then, rearrange the column sequence as $x_1, x_2, x_3, x_4, x_7, x_5, x_6, x_8, x_9, x_{10}, x_0$ so that a triangular matrix of order 5 appears in the (1,1) position. Finally, ring-sum operations are performed on the rows so that the leading triangular matrix is transformed into a unit matrix. This process also transforms B_N into B_f as follows:

$$
B_f = \begin{array}{c} \\ p_1 \\ p_2 \\ p_3 \\ p_4 \\ p_5 \end{array}
\begin{array}{c}
\begin{array}{ccccccccccc} x_1 & x_2 & x_3 & x_4 & x_7 & \cdot\ x_5 & x_6 & x_8 & x_9 & x_{10} & x_0 \end{array} \\
\left[\begin{array}{ccccc|cccccc}
1 & 0 & 0 & 0 & 0 & 1 & 1 & 0 & 0 & 0 & 1 \\
0 & 1 & 0 & 0 & 0 & 0 & 1 & 0 & 1 & 1 & 1 \\
0 & 0 & 1 & 0 & 0 & 0 & 0 & 1 & 1 & 0 & 0 \\
0 & 0 & 0 & 1 & 0 & 0 & 1 & 0 & 0 & 0 & 0 \\
0 & 0 & 0 & 0 & 1 & 0 & 0 & 1 & 1 & 1 & 1
\end{array}\right]
\end{array} = [U_5 \cdot B_{f12}]
$$

Step 4: The fundamental cutset matrix is formed by

$$Q_f = [B^T_{f\,12} \cdot U_6]$$

as follows:

$$
Q_f = \begin{array}{c}
\begin{array}{ccccccccccc} x_1 & x_2 & x_3 & x_4 & x_7 & \cdot\ x_5 & x_6 & x_8 & x_9 & x_{10} & x_0 \end{array} \\
\left[\begin{array}{ccccc|cccccc}
1 & 0 & 0 & 0 & 0 & 1 & 0 & 0 & 0 & 0 & 0 \\
1 & 1 & 0 & 1 & 0 & 0 & 1 & 0 & 0 & 0 & 0 \\
0 & 0 & 1 & 0 & 1 & 0 & 0 & 1 & 0 & 0 & 0 \\
0 & 1 & 1 & 0 & 1 & 0 & 0 & 0 & 1 & 0 & 0 \\
0 & 1 & 0 & 0 & 1 & 0 & 0 & 0 & 0 & 1 & 0 \\
1 & 1 & 0 & 0 & 1 & 0 & 0 & 0 & 0 & 0 & 1
\end{array}\right]
\end{array} = [B^T_{f\,12} \cdot U_6]
$$

Step 5: Form A from Q_f by performing the following ring-sum operations on the rows of Q_f:

(a) Add (mod 2) row 5 to row 6.
(b) Add row 1 to row 6.
(c) Add row 4 to row 5.
(d) Add row 3 to row 4.

The resultant matrix is the incidence submatrix A:

$$
A = \begin{array}{c} \\ a \\ b \\ c \\ d \\ e \\ f \end{array}
\begin{array}{c}
\begin{array}{ccccccccccc} x_1 & x_2 & x_3 & x_4 & x_7 & x_5 & x_6 & x_8 & x_9 & x_{10} & x_0 \end{array} \\
\left[\begin{array}{ccccccccccc}
1 & 0 & 0 & 0 & 0 & 1 & 0 & 0 & 0 & 0 & 0 \\
1 & 1 & 0 & 1 & 0 & 0 & 1 & 0 & 0 & 0 & 0 \\
0 & 0 & 1 & 0 & 1 & 0 & 0 & 1 & 0 & 0 & 0 \\
0 & 1 & 0 & 0 & 0 & 0 & 0 & 1 & 1 & 0 & 0 \\
0 & 0 & 1 & 0 & 0 & 0 & 0 & 0 & 1 & 1 & 0 \\
0 & 0 & 0 & 0 & 0 & 1 & 0 & 0 & 0 & 1 & 1
\end{array}\right]
\end{array}
$$

Note that in each column of A there are no more than two 1's, this being the necessary and sufficient condition of an incidence matrix.

Step 6: The complete incidence matrix A_a is obtained by adding the last row, g (the reference node), to A of Step 5 so that each column of the resultant matrix A_a has exactly two 1's. Thus,

$$
A = \begin{array}{c} \\ a \\ b \\ c \\ d \\ e \\ f \\ g \end{array}
\begin{array}{c}
\begin{array}{ccccccccccc} x_1 & x_2 & x_3 & x_4 & x_7 & x_5 & x_6 & x_8 & x_9 & x_{10} & x_0 \end{array} \\
\left[\begin{array}{ccccccccccc}
1 & 0 & 0 & 0 & 0 & 1 & 0 & 0 & 0 & 0 & 0 \\
1 & 1 & 0 & 1 & 0 & 0 & 1 & 0 & 0 & 0 & 0 \\
0 & 0 & 1 & 0 & 1 & 0 & 0 & 1 & 0 & 0 & 0 \\
0 & 1 & 0 & 0 & 0 & 0 & 0 & 1 & 1 & 0 & 0 \\
0 & 0 & 1 & 0 & 0 & 0 & 0 & 0 & 1 & 1 & 0 \\
0 & 0 & 0 & 0 & 0 & 1 & 0 & 0 & 0 & 1 & 1 \\
0 & 0 & 0 & 1 & 1 & 0 & 1 & 0 & 0 & 0 & 1
\end{array}\right]
\end{array}
$$

Step 7: Network N^* is constructed from A_a and the network is shown in Figure 9-12. Finally, with the driver element x_0 removed from N^*, the desired SC network is obtained with its terminal vertices identified as f and g, which are the endpoints of x_0 in N^*. The final realization is shown in Figure 9-13.

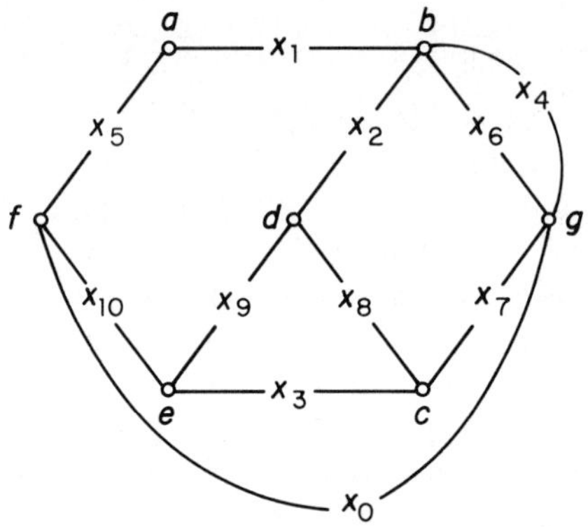

Figure 9-12 Network N^*.

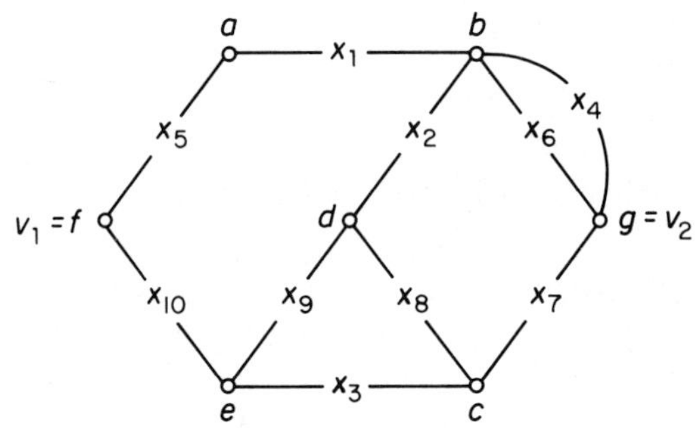

Figure 9-13 Network N, the desired SC network.

It should be noted that in Steps 3, 4, 5, and 6, the matrices B_f, Q_f, A, and A_a all have the same column sequence: x_1, x_2, x_3, x_4, x_7, x_5, x_6, x_8, x_9, x_{10}, x_0. This is very important because if this sequence is not kept during the entire procedure, the final network obtained will not yield the same SC function as prescribed, because the interrelationships between the matrices are derived on the assumption that the columns of these matrices are arranged in the same edge sequence.

Finally, we should point out that the synthesis procedure always leads to an SC network realization if the given function is an SC function. On the other hand, if the function is not an SC function, the procedure fails at Step 5; then Q_f cannot be transformed into A through row operations or Mayeda's method, thus indicating the nonrealizability of the function as an SC function.

9-7 REMARKS

The main objective of this chapter has been to introduce the topological applications to the analysis and synthesis of switching networks. In Section 9-5 we derived some of the functional transformations that can be used to analyze a given SC function. Among these, the terminal transformation has proved to be the most useful one since it can be applied to determine the separability of corresponding SC networks, as well as to obtain new SC functions of the same network with respect to a different pair of terminals (which are, of course, the endpoints of some specified edge). Other functional transformations can also be derived[19] from different properties of the SC networks.

We have also shown how realizability conditions may be obtained for SC networks. As mentioned earlier, a selected set of such conditions may be formulated to simplify the realization process if the function in question is realizable as an SC network — or, perhaps even more important, to reveal the nonrealizability of the function should it be non-SC. The usefulness of such a set of realizability conditions lies mainly in whether or not these conditions are simple and easy to apply so that considerable time and effort may be saved as a result of such applications to the realization process.[20]

Our emphasis has been placed on a special class of switching networks — namely, the SC networks — because if a switching function can be realized as an SC network, the realization is always minimal. Obviously, if the given function is non-SC, other synthesis procedures must be employed.[21] We have not attempted to study the non-SC networks; such discussions are beyond the scope of this text.

PROBLEMS

9-1 Evaluate the following expressions by switching algebra:
(a) $1\cdot\{[(0+1)\cdot 1+1]+1\cdot 0\}+(0+1)$
(b) $1\cdot 1\cdot\{1\cdot 0\cdot[(0+1)+(1+0)+1+(0\cdot 1)]+1\}$
(c) $\{[1\cdot 1+0\cdot 0\cdot(0\cdot 1)\cdot 1+(1+0)]\cdot(1\cdot 0)+(1+0)+(0+1)\cdot 1\}$
$+[(1+0)+(0+0)+(0+1)\cdot 1]\cdot 1$

9-2 Assume that the complements of variables a and b are a' and b', respectively. Evaluate the following expressions by switching algebra:
(a) $b\{(a'+a)a+[(a\cdot a+a'\cdot b')(a'+b')+b]a'\}$ for $a=0$, $b=1$
(b) $a'b+a\{(a'+b)a+b'[b'+a+a'b'(a+b)+b]\}$ for $a=1, b=0$

9-3 If the complements of the variables a, b, and c are a', b', and c', show that

[19]See S. P. Chan [14] for further study of functional transformations of SC networks.

[20]For further discussions, see S. P. Chan [14].

[21]The interested reader is referred to R. Gould [36], L. Löfgren [52], W. Mayeda [60], and S. Seshu [72].

(a) $(a + b + c)' = a' \cdot b' \cdot c'$
(b) $[(a + b) \cdot c]' = a' \cdot b' + c'$
(c) $[a + (b \cdot c)]' = a' + b' \cdot c'$
by its network equivalents.

9-4 Find the path matrix P_{12} between the vertices v_1 and v_2 of the graphs shown in Figure P9-4.

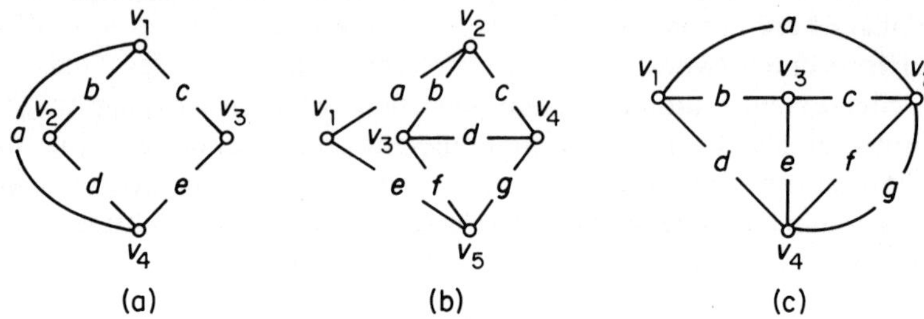

Figure P9-4

9-5 Prove that the number of independent paths between an SC network of v vertices and e edges is $e - v + 2$.

9-6 Prove that the ring sum of an even number of rows of a fundamental path matrix P_f is either a circuit or a disjoint union of circuits.

9-7 Prove that the ring sum of an odd number of rows of P_f is either a path or a redundant path (that is, a path and a circuit or a disjoint union of circuits).

9-8 Refer to the single-contact (SC) network N shown in Figure P9-8 with terminals (A,B).

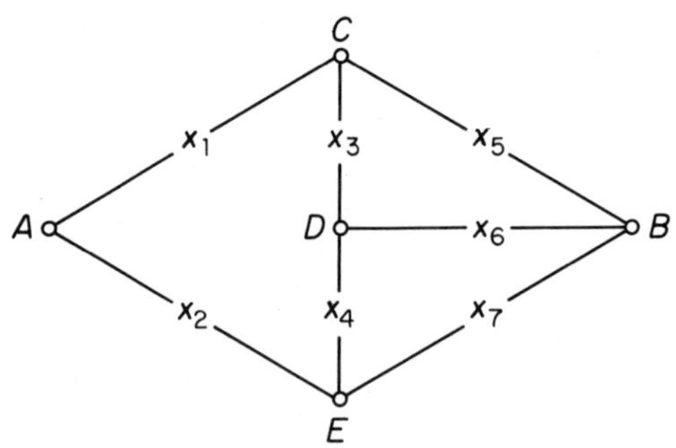

Figure P9-8

(a) Obtain the SC function F_{AB} of N.
(b) Write the path matrix P from F_{AB} obtained in part (a) and determine the rank of P. Does your result verify Theorem 9-1?

9-9 Repeat Problem 9-8 but with terminals (C,E) (so that F_{CE} is the function of interest).

9-10 Refer to Example 9-5. Obtain the edge-shorting transformations:
(a) $F(x_1 = 1)$ (c) $F(x_3 = 1)$
(b) $F(x_2 = 1)$ (d) $F(x_4 = 1)$

9-11 Refer to Example 9-5. Obtain the terminal transformations:
(a) $F_v(x_1)$ (c) $F_v(x_3)$
(b) $F_v(x_2)$ (d) $F_v(x_4)$

9-12 Refer to Example 9-7. Obtain the terminal transformations:
(a) $F_v(x_2)$ (c) $F_v(x_6)$
(b) $F_v(x_4)$ (d) $F_v(x_8)$

9-13 Develop at least two functional transformations (original ones) not discussed in this chapter. State clearly your definitions and discuss the properties of each of the new transformations. Also point out some possible applications of these transformations.

9-14 Given as SC function F_{AB}:

$$\begin{aligned} F_{AB} = {} & x_1x_2x_4x_5 + x_1x_2x_6x_7x_8 + x_1x_2x_6x_9 + x_1x_4x_6 + x_1x_5x_7x_8 \\ & + x_1x_5x_9 + x_2x_3x_4x_5x_7 + x_2x_3x_6x_7x_9 + x_2x_3x_6x_8 \\ & + x_3x_4x_6x_7 + x_3x_4x_6x_8x_9 + x_3x_5x_7x_9 + x_3x_5x_8 \end{aligned}$$

Perform the following functional transformations:
(a) $F(x_1 = 0)$, edge-opening of x_1.
(b) $F(x_2 = 1)$, edge-shorting of x_2.
(c) $F_v(x_3)$, terminal transformation with respect to x_3.

9-15 Check to see whether the following matrices satisfy the odd-ring-sum condition:

(a)
$$M_1 = \begin{bmatrix} 1 & 0 & 1 & 0 & 1 & 0 & 0 \\ 0 & 1 & 1 & 0 & 0 & 1 & 0 \\ 0 & 0 & 0 & 1 & 0 & 1 & 0 \\ 1 & 1 & 0 & 1 & 1 & 0 & 0 \\ 0 & 0 & 0 & 0 & 0 & 0 & 1 \end{bmatrix}$$

(b)
$$M_2 = \begin{bmatrix} 1 & 0 & 1 & 0 & 0 & 0 & 0 \\ 1 & 1 & 0 & 1 & 0 & 0 & 0 \\ 1 & 1 & 0 & 0 & 0 & 1 & 1 \\ 0 & 0 & 0 & 0 & 1 & 0 & 1 \\ 0 & 0 & 0 & 1 & 1 & 1 & 0 \\ 0 & 1 & 1 & 0 & 1 & 1 & 0 \end{bmatrix}$$

9-16 Obtain the SC networks from the following SC functions:
(a) $F_{AB} = x_1 + x_2x_3 + x_2x_4x_5 + x_2x_4x_6x_7$
(b) $F_{CD} = x_1x_3x_5 + x_2x_3x_6 + x_4x_6 + x_1x_2x_4x_5 + x_7$
(c) $F_{MN} = x_1x_3 + x_1x_2x_4 + x_1x_2x_6x_7 + x_5x_7 + x_4x_5x_6 + x_2x_3x_5x_6$

9-17 Let

$$\begin{aligned} F = {} & x_1x_2x_8 + x_1x_4x_5 + x_2x_6x_8 + x_4x_5x_6 + x_1x_2x_3x_4x_7 \\ & + x_1x_3x_5x_7x_8 + x_2x_3x_4x_6x_7 + x_3x_5x_6x_7x_8 \end{aligned}$$

be the SC function of an SC network N.
(a) Without realizing the network, determine whether or not it is separable. Show your work in detail.
(b) Find N using any systematic approach. Show your work in detail.

9-18 Construct a function F, expressed as a sum of products of a set of variables $x_1, x_2, \ldots, x_k$, such that the odd-ring-sum condition defined in Definition 9-17 is not satisfied by F. (This shows that F is not realizable as an SC function.)

9-19 Construct a function F, expressed as a sum of products of a set of variables $x_1, x_2, \ldots, x_k$ such that F satisfies the odd-ring-sum condition (Definition 9-17) but fails one of the other realizability conditions discussed in this chapter (so that F is not an SC function).

9-20 Refer to the SC network N of Figure P9-8. Add an edge, say x_0 across the terminals (A,B) of N and find the dual N' of N by the dot method discussed in Chapter 3 (Section 3-4). Identify the endpoints of edge x_0' as terminals (A',B') of N' where x_0' in N' is the edge corresponding to edge x_0 of N. Next, remove x_0 from N and x_0' from N' and then observe the relationship between the set of all edges contained in a path between terminals (A,B) of N and the set of corresponding edges in N'. Draw as many conclusions as you can from this observation and make your statements as general as possible.

9-21 Derive a set of (at least two) necessary realizability conditions for an SC function that have not been discussed in this chapter.

9-22 State any new ideas (theorems, properties, conjectures, and so on) about the analysis and synthesis of SC networks.

9-23 Obtain an SC network realization for the SC function given in Problem 9-14 using the synthesis procedure discussed in Section 9-6.

9-24 Repeat Problem 9-23 for the SC network given in Problem 9-17.

9-25 Determine the realizability of each of the following functions as an SC function. If it is realizable, find its SC network realization and show your work in detail. If it is not realizable, explain fully how you arrive at this conclusion.

(a) $F_1 = x_1x_2x_5 + x_1x_3 + x_2x_3x_4 + x_4x_5$

(b) $F_2 = x_1x_2x_4 + x_1x_3x_6 + x_2x_3x_5 + x_4x_5x_6$

(c) $F_3 = x_1x_2x_6 + x_1x_3x_8x_9x_{10} + x_1x_4x_6x_7x_8 + x_1x_5x_{10} + x_2x_3x_7 + x_2x_4x_6x_9x_{10} + x_2x_5x_7x_8x_9 + x_3x_4x_8 + x_3x_5x_6x_7x_{10} + x_4x_5x_9 + x_6x_7x_8x_9x_{10}$

(d) $F_4 = x_1x_3x_5x_6 + x_1x_3x_5x_7x_9 + x_1x_4x_6 + x_1x_4x_7x_9 + x_2x_3x_4x_6 + x_2x_3x_4x_7x_9 + x_2x_5x_6 + x_2x_5x_7x_9 + x_6x_7x_8 + x_8x_9$

(e) $F_5 = x_1x_5x_{10} + x_1x_5x_{12} + x_2x_3x_4x_6x_9 + x_2x_3x_{13} + x_2x_4x_7x_9x_{13} + x_2x_6x_7 + x_3x_4x_6x_8x_9 + x_3x_8x_{13} + x_4x_7x_8x_9x_{13} + x_5x_{11} + x_6x_7x_8 + x_{10}x_{12}$

9-26 After gaining some experience in determining the realizability of functions (such as those given in Problem 9-25) as SC functions, can you formulate a systematic testing procedure? Show your work in detail in deriving the procedure.

9-27 The method for obtaining a set of $e - v + 1$ independent circuits of a given network using loop analysis is straightforward. One simply picks a tree from the network and forms a fundamental circuit with every chord with possibly some tree branches so that $e - v + 1$ independent circuits are obtained. For a given SC network, can you suggest a simple algorithm for determining a set of $e - v + 2$ independent paths between the specified terminals? (See Theorem 9-1.)

9-28 Prove each of the three properties, (a), (b), and (c), of a fundamental path matrix P_f described in the text directly following Definition 9-2.

9-29 Refer to the synthesis procedure discussed in Section 9-6. Prove that, in Step 3, when B_p is transformed into B_f by performing the ring-sum operations on the rows of B_p, each resultant row of B_f corresponds to a circuit but *never* a disjoint union of circuits.

9-30 It has been pointed out in Section 9-6 that the synthesis procedure derived always leads to an SC network realization N if the given function F is realizable as such. But, if F is not an SC function, the non-realizability of F will not be revealed until Step 5 has been reached, where Q_f cannot be transformed into an incidence submatrix A. In such a case, a great deal of time and effort will be wasted by following this procedure. Can you suggest a better procedure which will enable one to determine easily (hopefully by inspection with simple criteria) the realizability of a given function F as an SC network? Show your work in detail in deriving the procedure.

INTRODUCTION TO TOPOLOGICAL SYNTHESIS AND OTHER APPLICATIONS

CHAPTER 10

10-1 INTRODUCTION

As mentioned in the Preface, this book is intended to be a self-contained text on topological analysis of electrical networks. In the preceding chapters we presented the basic theory of network topology and its application to the analysis of passive networks with no magnetic coupling. We then extended its application to the analysis of active networks, including such elements as vacuum tubes and transistors, as well as mutual inductances, and finally to the analysis and synthesis of switching networks, with the emphasis on single-contact networks. In this final chapter, we shall conclude our discussions by introducing the idea of topological synthesis of electrical networks as well as some other topological applications.

Since certain prerequisites are necessary before a rigorous theory can be presented in each area, and since this text is designed to be at an introductory level with the assumption of a minimum prerequisite, we shall be contented here with merely a concise introduction to the topological synthesis of passive networks without mutual inductances and a brief survey of some of the other topological applications as implied by the title of this chapter. Thus, most of the theorems included in the following sections will be stated without proof. However, references will be made in each theorem so that a proof can always be found, if desired. Our main objective is merely to acknowledge the existence of some of the other topological applications, and to describe a few unsolved problems in various fields so that the interested reader may do further research in any of these areas at a future date.

10-2 TOPOLOGICAL DETERMINATION OF THE ORDER OF COMPLEXITY

Let N be a passive network without mutual inductances. We define the following:

DEFINITION 10-1. Natural Frequency and Order of Complexity.[1] The *natural frequencies* of a network are the zeros of the network determinant (either node determinant Δ or loop determinant Δ') of N. The *order of complexity* of N is the number of finite nonvanishing zeros of the network determinant (node or loop) of N with each passive element (R, L, C) considered as a single element. (That is, if the number of R, L, and C is e, the order of the (square) element impedance matrix Z_e is e.)

One of the topological problems in network synthesis is to determine the order of complexity of N by inspection. It can be shown[2] that the determinants Δ and Δ' are related by

$$\Delta' = (\det Z_e)\,\Delta = Ks^k\Delta \tag{10-1}$$

where K is a positive constant and k is an integer (which may be positive, negative, or zero). Thus, both node and loop determinants have the same zeros if the *external* natural frequencies[3] ($s = 0$, $s = \infty$) are excluded. Consequently, it makes no difference whether Δ or Δ' is considered to determine the order of complexity of a given network. We next state the following theorem:

THEOREM 10-1. The order of complexity of a passive network without mutual inductances is given by

$$\alpha = R(G_{C_o}) + R(G_{L_o}) - R(G_{C_s}) - R(G_{L_s}) \tag{10-2}$$

where:

G_{C_o} is the graph obtained from the network by open-circuiting all noncapacitor elements,

G_{L_o} is the graph obtained by open-circuiting all noninductor elements,

G_{C_s} is the graph obtained by short-circuiting all noncapacitor elements,

G_{L_s} is the graph obtained by short-circuiting all noninductor elements, and

R stands for the rank of the graph.[4]

A proof of this theorem can be found in Seshu and Reed [76].

[1]For a detailed discussion on the order of complexity, see P. R. Bryant [8].

[2]It is listed as a problem in Chapter 6 (Problem 6-25).

[3]Natural frequencies at $s = 0$ and $s = \infty$ are referred to as external natural frequencies. All other natural frequencies are called *internal* natural frequencies.

[4]Refer to Definition 2-10 for the definition of the rank of a graph.

EXAMPLE 10-1. Consider the network shown in Figure 10-1, which is the same network used in Example 5-1. The node determinant Δ with node $1'$ as reference is, from Equation (5-9),

$$\Delta = \frac{24s^3 + 42s^2 + 18s + 1}{60s} \tag{10-3}$$

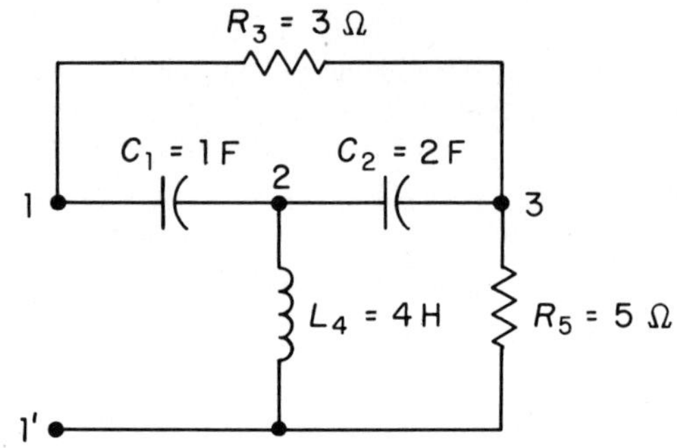

Figure 10-1 A passive network without mutual inductances.

which implies that the order of complexity of the given network is $\alpha = 3$. To verify this result topologically, we obtain the four graphs G_{C_o}, G_{L_o}, G_{C_s}, G_{L_s}, as illustrated in Figure 10-2. Thus, noting the fact that for any graph G with v vertices and p maximally connected subgraphs, $R(G) = v - p$, we have

$$R(G_{C_o}) = r - 2 = 2 \qquad R(G_{C_s}) = 1 - 1 = 0$$
$$R(G_{L_o}) = 4 - 3 = 1 \qquad R(G_{L_s}) = 1 - 1 = 0$$

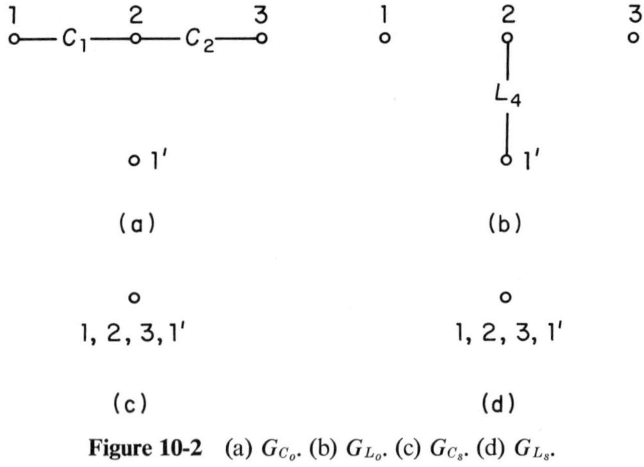

Figure 10-2 (a) G_{C_o}. (b) G_{L_o}. (c) G_{C_s}. (d) G_{L_s}.

Finally, by Equation (10-2),

$$\alpha = 2 + 1 - 0 - 0 = 3$$

which agrees with the previous result.

10-3 TOPOLOGICAL REALIZATION OF $Y_d(s)$

We have shown in Chapter 5 that the driving-point admittance $Y_d(s)$ of a passive one-port without mutual inductances may be expressed in topological terms, Equation (5-31), as

$$Y_d(s) = \frac{V}{W_{1,1'}} \tag{10-4}$$

where V and $W_{1,1'}$ are homogeneous polynomials of the element admittances $y_1, y_2, \ldots, y_e$, linear in any y_j, with the coefficient of each term equal to $+1$, and the degree of V is one more than that of $W_{1,1'}$. Also, we know from our discussions in Chapter 5 that each term in V is a tree admittance product and each term in $W_{1,1'}$ a 2-tree admittance product with node 1 in one component and the reference node 1′ in the other. Thus, each term in V is a combination of the e admittances $y_1, y_2, \ldots, y_e$ taken $v - 1$ at a time, and is in the form $K_p s^q$ where K_p is a positive constant determined by the $v - 1$ admittance values in the combination, and q is an integer, which can be positive, zero, or negative, depending on whether the number of capacitors is larger than, equal to, or less than the number of inductors, respectively, in the combination. This presents the following problem in topological synthesis.

How can each term, $K_p s^q$, in the homogeneous polynomial V (or $W_{1,1'}$) be decomposed into $v - 1$ of e admittances, $y_1, y_2, \ldots, y_e$, each of which represents the admittance of a capacitor (Cs), a resistor ($1/R$), or an inductor $[(1/L)s^{-1}]$ such that $Y_d(s) = V/W_{1,1'}$ can be realized? The following example serves to illustrate some of the points we have discussed.

EXAMPLE 10-2. Consider the same network N that was used in Example 5-1. The network was shown again in Figure 10-1. From Example 5-1, the driving-point admittance $Y_d(s)$ is given by Equation (5-12) as

$$Y_d(s) = \frac{\Delta}{\Delta_{11}} = \frac{24s^3 + 42s^2 + 18s + 1}{120s^3 + 96s^2 + 30s + 8} \tag{10-5}$$

which was obtained by direct expansion of the ratio Δ/Δ_{11}. But, from Example 5-8, the determinant Δ and its cofactor Δ_{11} are obtained topologically as

$$\begin{aligned}\Delta = V &= y_1y_2y_4 + y_1y_2y_5 + y_1y_3y_4 + y_1y_3y_5 + y_1y_4y_5 + y_2y_3y_4 + y_2y_3y_5 + y_2y_4y_5 \\ &= \frac{2}{5}s^2 + \frac{21}{30}s + \frac{18}{60} + \frac{1}{60s}\end{aligned} \tag{10-6}$$

and

$$\begin{aligned}\Delta_{11} = W_{1,1'} &= y_1y_2 + y_1y_3 + y_1y_5 + y_2y_3 + y_2y_4 + y_2y_5 + y_3y_4 + y_4y_5 \\ &= 2s^2 + \frac{24}{15}s + \frac{1}{2} + \frac{8}{60s}\end{aligned} \tag{10-7}$$

where

$$y_1 = s \qquad y_2 = 2s \qquad y_3 = \frac{1}{3} \qquad y_4 = \frac{1}{4s} \qquad y_5 = \frac{1}{5} \tag{10-8}$$

Thus, taking the ratio Δ/Δ_{11}, we get

$$Y_d(s) = \frac{\Delta}{\Delta_{11}} = \frac{\frac{2}{5}s^2 + \frac{21}{30}s + \frac{18}{60} + \frac{1}{60s}}{2s^2 + \frac{24}{15}s + \frac{1}{2} + \frac{8}{60s}} = \frac{24s^3 + 42s^2 + 18s + 1}{120s^3 + 96s^2 + 30s + 8} \tag{10-9}$$

which is identical to Equation (10-5).

From Equations (10-6) and (10-7), we realize that it is not easy to "decompose" the polynomial $V(W_{1,1'})$ into a sum of admittance products each of which will contain exactly three (two) of the five element admittances [as given by Equation (10-8)] such that $Y(s)$ as given by Equation (10-9) will be realized by the network N shown in Figure 5-3.

Before discussing the topological synthesis methods of $Y_d(s)$, we define the following:

DEFINITION 10-2. Positive Real (PR) Function. A function $F(s)$ of a complex variables, $s = \sigma + j\omega$, is called *positive real* (pr) if

(a) $F(s)$ is real when s is real.

(b) Re $F(s) \geq 0$ when Re $(s) \geq 0$.

The following theorem is well known in network synthesis:

THEOREM 10-2.[5] A rational function $F(s)$ is realizable as the immittance[6] of a passive one-port if and only if $F(s)$ is pr.

There are several conventional synthesis methods[5] for the realization of a given pr function. Our interest here is not in these conventional techniques but, rather, in the topological ones.

In order to develop a topological synthesis procedure for the realization of

$$Y_d(s) = \frac{V}{W_{1,1'}} = \frac{\Delta}{\Delta_{11}}$$

we shall first state some of the elementary properties of Δ/Δ_{11}.

THEOREM 10-3.[7] Let a pr rational function be represented by the admittance

$$\frac{P(s)}{Q(s)} = Y_d(s) = \frac{\Delta}{\Delta_{11}} = \frac{V(y_1, y_2, \ldots, y_e)}{W_{1,1'}(y_1, y_2, \ldots, y_e)} = \frac{V}{W_{1,1'}} \tag{10-10}$$

where $\Delta = \det(AY_eA^T)$; Δ_{11} = (1,1) cofactor of Δ; P and Q are polynomials in s; V and $W_{1,1'}$ are homogeneous polynomials in $y_1, y_2, \ldots, y_e$; and the symbol y_i stands both for the element y_i in the graph corresponding to the

[5] For a proof of Theorem 10-2 and the discussion of some of the synthesis techniques, refer to M. E. Van Valkenburg [79], for example.

[6] An immittance means that the function is either an impedance or an admittance. It is the combination of the two words IMpedance and adMITTANCE.

[7] Proofs of the four properties in Theorem 10-3 can be found in Seshu and Reed [76].

network and the admittance y_i of that element. Then some of the properties of V and $W_{1,1'}$ and, hence, of

$$\frac{V}{W_{1,1'}} = \frac{\Delta}{\Delta_{11}}$$

are as follows:

(a) If the elements $y_{j1}, y_{j2}, \ldots, y_{jk}$ from the set $\{y_1, y_2, \ldots, y_e\}$ $(k \leq e)$ are connected in parallel across nodes 1 and 1′, then,

$$\frac{V}{W_{1,1'}} = \left(\sum_{p=1}^{k} y_{jp}\right) + \frac{V_1}{W_{1,1'}} = y_0 + \frac{V_1}{W_{1,1'}} \tag{10-11}$$

where V_1 and $W_{1,1'}$ are not functions of $y_{j1}, y_{j2}, \ldots, y_{jk}$.

(b) Every element admittance y_i is contained in V.

(c) An element admittance y_i is contained in $W_{1,1'}$ if and only if y_i does not have nodes 1 and 1′ as its endpoints.

(d) Every product which appears in $W_{1,1'}$ also appears in V, multiplied by some y_k not in this product in $W_{1,1'}$.

We next introduce the following definitions:

DEFINITION 10-3. Reduced Admittance Function. If $\{y_{j1}, y_{j2}, \ldots, y_{jk}\}$ is a subset of the set $\{y_1, y_2, \cdots, y_e\}$ such that

$$\begin{aligned} V &= \left(\sum_{p \leq k} y_{jp}\right) V_1 + V_2 \\ W_{1,1'} &= \left(\sum_{p \leq k} y_{jp}\right) W_1 + W_2 \end{aligned} \tag{10-12}$$

where V_1, V_2, W_1, and W_2 are not functions of $y_{j1}, y_{j2}, \ldots, y_{jk}$, then

$$\sum_{p \leq k} y_{jp} = y_j$$

is called the *reduced admittance function* corresponding to $y_{j1}, y_{j2}, \ldots, y_{jk}$.

DEFINITION 10-4.[8] **Product Π_{kj}.** *Product Π_{kj}* is a product of an element y_j not contained in a specified (tree admittance) product $T^{(k)}(y)$ by some or all of the elements in $T^{(k)}(y)$ that do not appear as a factor of any product in V.

More properties of V are stated in the next theorem.

THEOREM 10-4. With the same notation as in Theorem 8-2, the following are some more of the properties of V:

(a) If V is realized as the node determinant of a network N and if e_r is the number of reduced element admittances in V, then

$$e_r \leq \frac{v(v-1)}{2}$$

[8]This definition as well as Theorems 10-4 through 10-7 can be found in Seshu and Reed [76], Chapter 8.

(b) For every chosen term $T^{(k)}(y)$ and every element admittance y_j not in $T^{(k)}(y)$, there exists at least one Π_{kj}.

(c) Every Π_{kj} corresponds to a subnetwork of N containing exactly one circuit — namely, a fundamental circuit with respect to tree $T^{(k)}$.

(d) The sum

$$\sum_i \Pi_{jk}^{(i)}$$

interpreted as a Boolean function,[9] can be reduced to a single product for every $T^{(k)}$ and every y_j not in $T^{(k)}$.

(e) With all the $e - v + 1$ fundamental circuits obtained and the fundamental circuit matrix $B_f = [U \quad B_{f_{12}}]$ formed with respect to tree $T^{(k)}$ by the procedure of (c), then {sets of elements corresponding to complements of nonvanishing majors of B_f} $\equiv$ {sets of elements corresponding to tree admittance products in V}.

The next theorem gives the realizability conditions for a matrix $B = [U \quad B_{12}]$ of order $(e - v + 1) \times e$ to be a fundamental circuit matrix of a graph.

THEOREM 10-5. The matrix $B = [U \quad B_{12}]$ of order $(e - v + 1) \times e$ with all entries equal to 0 or 1 is the fundamental circuit of a graph G if and only if there exists a matrix Q of order $(v - 1) \times e$ with all entries equal to 0 or 1 such that

(a) $R(Q) = v - 1$,

(b) Each of the columns of Q contains at least one, but no more than two, nonzero entries, and

(c) $B \cdot Q^T = 0$ (modulo 2).

Combining (d) and (e) of Theorem 10-4 and Theorem 10-5, we form the necessary and sufficient reason for $V(y_1, y_2, \ldots, y_e)$ to be realized as the node determinant of a graph, as stated in the following theorem.

THEOREM 10-6. A homogeneous polynomial $V(y_1, y_2, \ldots, y_e)$, linear in any y_j, with all coefficients equal to 1, is realizable as the node determinant of a graph if and only if

(a) Every sum

$$\sum_i \Pi_{jk}^{(i)}$$

interpreted as a Boolean function, can be reduced to a single product for every product $T^{(k)}$ in V and every y_j not in $T^{(k)}$,

(b) {Sets of elements corresponding to products in V} $\equiv$ {sets of elements corresponding to complements of nonvanishing majors of B (of maximum order, by definition)}, where B is derived from V by the procedure of (c) in Theorem 10-4, and

[9]Refer to Chapter 9 for the definition of a Boolean function and the Boolean operations.

(c) There exists a matrix Q of order $(v - 1) \times e$ and rank $v - 1$ with all entries equal to 0 or 1 such that each of its columns contains at least one but no more than two nonzero entries and $BQ^T = 0$.

Note that Equation (10-12) is obtained by multiplying Equation (10-11) by $W_{1,1'}$. This implies that if V is expressible in the form of Equation (10-12), there exist parallel elements across nodes 1 and 1′, and these elements, expressed as a sum of element admittances, are reduced to y_j. If there exists no such y_j in V, it is always possible to introduce a new element y_0 such that

$$Y_{d(y_0)} = \frac{V_0}{W_{1,1'}} = \frac{y_0 W_{1,1'} + V}{W_{1,1'}} \tag{10-13}$$

where $Y_{d(y_0)}$ is the driving-point admittance of the resultant network with y_0 connected across nodes 1 and 1′; V and $W_{1,1'}$ are defined as before [see Equation (10-12)]; and V_0 is the node determinant of the resultant network (with y_0 across 1 and 1′).

The next theorem gives the topological relationship between V and $W_{1,1'}$.

THEOREM 10-7. Let $V(y_0, y_1, \ldots, y_e)$ be the node determinant of a graph G; T be a tree of G containing y_0; and B_f be the fundamental circuit matrix with respect to T. Also, let G_1 be the graph obtained from G by short-circuiting the endpoints (1,1′) of y_0, where y_0 may be a single element or a reduced element. Then the subgraph T_1, obtained from T by deleting y_0, is a tree of G_1. Furthermore, the necessary and sufficient conditions for $W_{1,1'}(y_1, y_2, \ldots, y_e)$ to be the node determinant of G_1 are:

(a) B_{f_1}, obtained from $W_{1,1'}$ with respect to T_1, is identical to B_f with the column for y_0 deleted, and

(b) {Sets of elements corresponding to complements of nonvanishing majors of B_{f_1}} $\equiv$ {sets of elements corresponding to products in $W_{1,1'}$}.

We shall now outline the topological synthesis procedure for the realization of $V/W_{1,1'}$ in the following steps:

(1) Assume a set of element admittances $y_1, y_2, \ldots, y_e$, each of which is in the form $K_j s^k$ representing the admittance of an R, L, or C ($K_j > 0$ and $k = 0$ or ± 1).

(2) Express the given function $Y_d(s) = V/W_{1,1'}$ in terms of the y's in (1) such that the numerator V is expressible in the form

$$V = \left(\sum_{p \le k} y_{jp}\right) W_{1,1'} + V_1 = y_j W_{1,1'} + V_1$$

If no such y_j exists, then introduce y_0 and write

$$V_0 = y_0 W_{1,1'} + V \qquad \text{and} \qquad Y_d(y_0) = \frac{y_0 W_{1,1'} + V}{W_{1,1'}}$$

according to Equation (10-13).

(3) Choose a product T in V (or V_0) containing y_j (or y_0) and obtain the matrix $B_f = [U \quad B_{f_{12}}]$ by the procedure of (c) in Theorem 10-4.

(4) Form the matrix $Q = [B_{f12}^T \quad U]$.

(5) Obtain the matrix A from Q by performing the ring-sum operation on the rows of Q until each column of the resulting matrix A contains one or two nonzero entries. A is the incidence submatrix of the graph G to be realized.

(6) Identify G from A, and the endpoints of y_j identify the input nodes 1 and 1′. If y_0 has been introduced in step (2), then remove y_0 to obtain the desired graph and the endpoints of y_0 identify 1 and 1′.

EXAMPLE 10-3. Let us realize topologically the driving-point admittance function $Y(s)$ following the steps outlined in the procedure above:

$$Y_d(s) = \frac{2s^3 + 2s^2 + 3s + 1}{s^3 + s^2 + 2s + 1} \tag{10-14}$$

(1) Multiplying Equation (10-14) by s^2/s^2, we get

$$Y_d(s) = \frac{2s + 2 + \dfrac{3}{s} + \dfrac{1}{s^2}}{s + 1 + \dfrac{2}{s} + \dfrac{1}{s^2}} \tag{10-15}$$

and then assume the following element admittances:

$$y_1 = 1 \qquad y_2 = 1 \qquad y_3 = \frac{1}{s} \qquad y_4 = s \qquad y_5 = s \tag{10-16}$$

(2) Express $Y_d(s)$ in terms of the y's in Equation (10-16):

$$Y_d(s) = \frac{y_1y_2y_3 + y_1y_2y_4 + y_1y_3y_4 + y_1y_3y_5 + y_1y_4y_5 + y_2y_3y_5 + y_2y_4y_5 + y_3y_4y_5}{y_2y_3 + y_2y_4 + y_3y_4 + y_3y_5 + y_4y_5} \tag{10-17}$$

which can be expressed in the form

$$\begin{aligned} Y_d(s) &= \frac{V}{W_{1,1'}} = \frac{y_1(y_2y_3 + y_2y_4 + y_3y_4 + y_3y_5 + y_4y_5) + y_2y_3y_5 + y_2y_4y_5 + y_3y_4y_5}{y_2y_3 + y_2y_4 + y_3y_4 + y_3y_5 + y_4y_5} \\ &= \frac{y_1W_{1,1'} + V_1}{W_{1,1'}} \end{aligned} \tag{10-18}$$

where

$$W_{1,1'} = y_2y_3 + y_2y_4 + y_3y_4 + y_3y_5 + y_4y_5$$

and

$$V_1 = y_2y_3y_5 + y_2y_4y_5 + y_3y_4y_5$$

clearly are not functions of y_1. Thus, y_1 satisfies the condition for y_j as given by Equation (10-12).

(3) Choose the product $y_1y_2y_3$ in the numerator of Equation (10-18) and obtain the sums

$$\sum_i \Pi_{kj}^{(i)}$$

for the elements of V. They are given by

$$y_4 = y_4y_2y_3 + y_4y_1y_2y_3 = y_2y_3y_4 \text{ (Boolean)} \qquad (10\text{-}19)$$
$$y_5 = y_5y_1y_2 + y_5y_1y_2y_3 = y_1y_2y_5 \text{ (Boolean)}$$

Hence we get

$$B = \begin{bmatrix} \overset{y_1}{0} & \overset{y_2}{1} & \overset{y_3}{1} & \overset{y_4}{1} & \overset{y_5}{0} \\ 1 & 1 & 0 & 0 & 1 \end{bmatrix} \sim \left[\begin{array}{cc|ccc} \overset{y_4}{1} & \overset{y_5}{0} & \overset{y_1}{0} & \overset{y_2}{1} & \overset{y_3}{1} \\ 0 & 1 & 1 & 1 & 0 \end{array}\right] = [U_2 \quad B_{f_{12}}] = B_f \qquad (10\text{-}20)$$

where

$$B_{f_{12}} = \begin{bmatrix} 0 & 1 & 1 \\ 1 & 1 & 0 \end{bmatrix}$$

(4) Next we write

$$Q = [B_{f_{12}}^T \quad U_3] = \begin{bmatrix} \overset{y_4}{0} & \overset{y_5}{1} & \overset{y_1}{1} & \overset{y_2}{0} & \overset{y_3}{0} \\ 1 & 1 & 0 & 1 & 0 \\ 1 & 0 & 0 & 0 & 1 \end{bmatrix} \qquad (10\text{-}21)$$

(5) By inspection of Equation (10-21), we see that Q is already in the form of A since each of its columns contains one or two nonzero entries. Hence, we write

$$A = Q = \begin{matrix} V_1 \\ V_2 \\ V_3 \end{matrix} \begin{bmatrix} \overset{y_4}{0} & \overset{y_5}{1} & \overset{y_1}{1} & \overset{y_2}{0} & \overset{y_3}{0} \\ 1 & 1 & 0 & 1 & 0 \\ 1 & 0 & 0 & 0 & 1 \end{bmatrix} \qquad (10\text{-}22)$$

(6) The incidence matrix A_a can be readily obtained from Equation (10-22) by completing the row that corresponds to the reference vertex. This is done by adding the fourth row to A such that each column of the resultant matrix A_a contains exactly two 1's. Thus,

$$A_a = \begin{matrix} V_1 \\ V_2 \\ V_3 \\ V_4 \end{matrix} \begin{bmatrix} \overset{y_4}{0} & \overset{y_5}{1} & \overset{y_1}{1} & \overset{y_2}{0} & \overset{y_3}{0} \\ 1 & 1 & 0 & 1 & 0 \\ 1 & 0 & 0 & 0 & 1 \\ 0 & 0 & 1 & 1 & 1 \end{bmatrix}$$

Finally, the graph G and the network N corresponding to A_a are shown in Figure 10-3(a) and (b), respectively, with the endpoints of y_1 identifying the input nodes 1 and 1′.

EXAMPLE 10-4. Consider the following driving-point admittance:

$$Y_d(s) = \frac{2s^2 + 16s}{2s^2 + 9s + 8} \qquad (10\text{-}23)$$

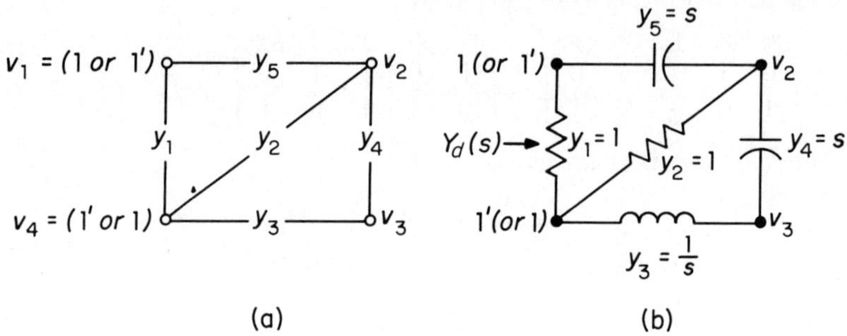

Figure 10-3 (a) Graph G. (b) Network N.

We shall synthesize $Y_d(s)$ in steps as in Example 10-3:

(1) Multiply $Y_d(s)$ by s/s:

$$Y_d(s) = \frac{2s + 16}{2s + 9 + \dfrac{8}{s}} \tag{10-24}$$

and assume the following element admittances:

$$y_1 = 1 \quad y_2 = 1 \quad y_3 = \frac{4}{s} \quad y_4 = 2 \tag{10-25}$$

(2) Next express $Y_d(s)$ in terms of the y's in Equation (10-25):

$$Y_d(s) = \frac{V}{W_{1,1'}} = \frac{(y_1y_2y_4) + (y_1y_3y_4 + y_2y_3y_4)}{(y_1y_4) + (y_1y_2 + y_3y_4) + (y_1y_3 + y_2y_3)} \tag{10-26}$$

We see that V in Equation (10-26) cannot be expressed in the form $V = y_jW_{1,1'} + V_1$. Thus we introduce y_0, so that

$$\begin{aligned} Y_{d(y_0)} &= \frac{V_0}{W_{1,1'}} = \frac{y_0W_{1,1'} + V}{W_{1,1'}} \\ &= \frac{y_0(y_1y_2 + y_1y_3 + y_1y_4 + y_2y_3 + y_3y_4) + (y_1y_2y_4 + y_1y_3y_4 + y_2y_3y_4)}{y_1y_2 + y_1y_3 + y_1y_4 + y_2y_3 + y_3y_4} \end{aligned} \tag{10-27}$$

(3) Choose the product $y_0y_1y_2$ in V_0 and obtain the matrix B_f by the procedure of (c) in Theorem 10-4 as follows:

The sums

$$\sum_i \Pi_{kj}^{(i)}$$

for the elements of V are:

$$\left.\begin{aligned} y_3 &= y_3y_1y_2 + y_3y_0y_1y_2 = y_1y_2y_3 \\ y_4 &= y_4y_0y_2 + y_4y_0y_1y_2 = y_0y_2y_4 \end{aligned}\right\}\text{(Boolean)}$$

Then we get

$$B = \begin{bmatrix} \overset{y_0}{0} & \overset{y_1}{1} & \overset{y_2}{1} & \overset{y_3}{1} & \overset{y_4}{0} \\ 1 & 0 & 1 & 0 & 1 \end{bmatrix} \sim \left[\begin{array}{cc|ccc} \overset{y_1}{1} & \overset{y_0}{0} & \overset{y_2}{1} & \overset{y_3}{1} & \overset{y_4}{0} \\ 0 & 1 & 1 & 0 & 1 \end{array}\right] = [U_2 \quad B_{f_{12}}] = B_f$$

(4) The matrix Q is given by

$$Q = \begin{bmatrix} \overset{y_1}{1} & \overset{y_0}{1} & \overset{y_2}{1} & \overset{y_3}{0} & \overset{y_4}{0} \\ 1 & 0 & 0 & 1 & 0 \\ 0 & 1 & 0 & 0 & 1 \end{bmatrix}$$

which is equal to A since it satisfies the (necessary and sufficient) condition that each column of A contains either one or two 1's.

(5) Next, A_a is obtained:

$$A_a = \begin{array}{c} V_1 \\ V_2 \\ V_3 \\ V_4 \end{array}\begin{bmatrix} \overset{y_1}{1} & \overset{y_0}{1} & \overset{y_2}{1} & \overset{y_3}{0} & \overset{y_4}{0} \\ 1 & 0 & 0 & 1 & 0 \\ 0 & 1 & 0 & 0 & 1 \\ 0 & 0 & 1 & 1 & 1 \end{bmatrix}$$

and the corresponding graph G_0 and the network N_0 are shown in Figure 10-4(a) and (b), respectively, with nodes 1 and 1′ identified by the endpoints of y_0. Obviously, the final realization is the network N derived from N_0 by removing the added element y_0 as illustrated in Figure 10-4(c).

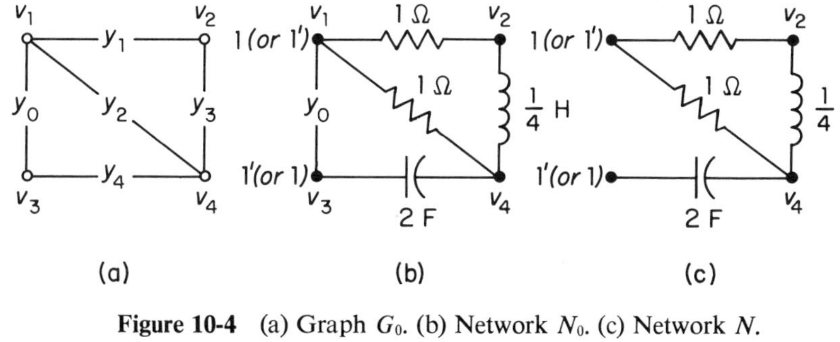

Figure 10-4 (a) Graph G_0. (b) Network N_0. (c) Network N.

With the main ideas in the theory and the steps in the synthesis procedure illustrated by Examples 10-3 and 10-4, a few comments seem appropriate at this point.

(1) A major difference between the topological synthesis procedure presented here and any conventional synthesis techniques, such as Brune's and the Bott-Duffin *RLC* one-port synthesis methods, to name just two, discussed in ordinary network synthesis texts,[10] is that the former is algebraic, whereas the latter are analytic.

[10]See, for example, M. E. Van Valkenburg [79]

(2) From the two examples, it should be evident that the topological method does not preassume any type of network structure; whereas, in any of the conventional techniques, a form of topology has been prescribed (for example, a ladder, as in the Cauer ladder development[10]).

(3) In the topological procedure presented above, once the B_f matrix is obtained, the realization is unique to within a 2-isomorphism.[11] In fact, the problem of deriving the graph from B_f has been discussed in Chapter 9 (compare Examples 10-3 and 10-4 with Example 9-9).

(4) The main drawback of this topological method is that an algorithm has not been formulated in Step (1) of the procedure to express $Y_d(s)$ as the ratio of V to $W_{1,1'}$. In other words, for a given V specification in $Y_d(s) = V/W_{1,1'}$, a network realization N can be readily determined; and, for a given $W_{1,1'}$, an N can also be found; but there exists no *general* method for determining an N that *simultaneously* realizes V and $W_{1,1'}$. This problem[12] must then be solved before the present topological method can be of effective and practical use.

(5) Another future problem naturally is to extend the present synthesis technique to the topological synthesis of transfer functions of a passive two-port and even to the synthesis of active networks.

10-4 THE PROBLEMS OF MINIMALITY

We start this section with the following definitions:

DEFINITION 10-5. Minimality in Reactive Elements. An *RLC* realization without mutual inductances of a positive real (pr) function

$$Z(s) = K\frac{s^n + a_{n-1}s^{n-1} + \cdots + a_1 s + a_0}{s^m + b_{m-1}s^{m-1} + \cdots + b_1 s + b_1} \tag{10-28}$$

with $K > 0$, not both a_0, $b_0 = 0$ and no common factor in numerator and denominator, is *minimal in reactive elements* if either or both of the following two conditions are satisfied:

$$n = n_L + n_C \qquad m = n_L + n_C \tag{10-29}$$

where n_L and n_C are the numbers of inductors and capacitors, respectively, in the realization.

DEFINITION 10-6. Minimum PR Function. A pr function is called *minimum* if

(a) It has no poles or zeros on the $j\omega$-axis, and

[11]Refer to Definition 1-24 for a formal definition of 2-isomorphism.

[12]It is listed as a research problem at the end of this chapter (Problem 10-24).

(b) It has a zero real part at some finite nonzero frequency $s = j\omega_i$ ($\omega_i \neq 0, \infty$) on the $j\omega$-axis.

The following theorem states the necessary and sufficient conditions for a passive network to be minimal in reactive elements.

THEOREM 10-8. A passive one-port N without mutual inductances that realizes the pr function $Z(s) = \Delta_{11}/\Delta$ of Equation (10-28) where Δ and Δ_{11} have no common factors, is minimal in reactive elements if and only if

(a) There are no loops consisting of capacitors alone or inductors alone in N.

(b) There are no cutsets consisting of capacitors alone or inductors alone in N other than the basic cutsets[13] with respect to terminals 1 and 1′.

(c) Because of (b), there are no cutsets with all capacitors or all inductors in N', where N' is derived from N by short-circuiting nodes 1 and 1′.

A formal proof of Theorem 10-8 is rather lengthy,[14] but, from Theorem 10-1, it should be evident that loops and cutsets consisting of capacitors alone or inductors alone must not be allowed either in N or in N', which is obtained from N by short-circuiting terminals 1 and 1′ at the input if N is minimal in reactive elements.

EXAMPLE 10-5. Consider the network N used in Example 5-1 and again in Example 10-2. This network N is repeated in Figure 10-5 for convenience. In this network,

$$Z_d(s) = \frac{1}{Y_d(s)} = \frac{\Delta_{11}}{\Delta} = \frac{120s^3 + 96s^2 + 30s + 8}{24s^3 + 42s^2 + 18s + 1}$$

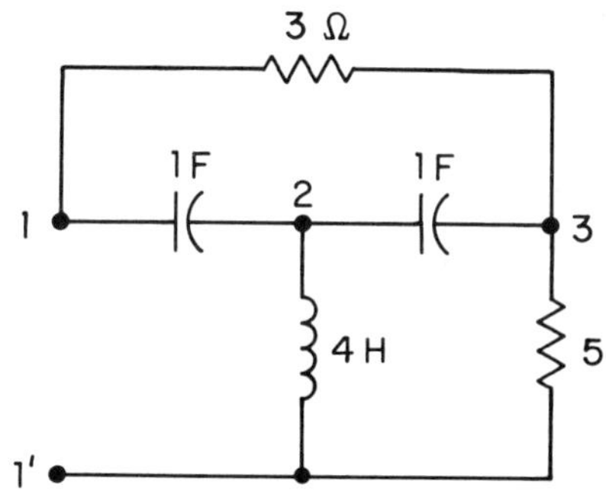

Figure 10-5 A passive one-port N.

which indicates that $n = m = 3$. However, by Definition 10-5 and by inspection of N in Figure 10-5, it is evident that N is minimal in reactive elements since

$$n_L = 1 \qquad \text{and} \qquad n_C = 2$$

[13]See Definition 9-11.

[14]See Seshu and Reed [76], Chapter 8, for a formal proof.

so that

$$n = m = n_L + n_C = 3$$

Also, we see that both conditions (a) and (b) stated in Theorem 10-8 are satisfied by N.

EXAMPLE 10-6. Consider the network N shown in Figure 10-6. The node determinant Δ of N is

$$\Delta = \begin{vmatrix} sC_1 + \dfrac{1}{sL_1} & -\left(sC_1 + \dfrac{1}{sL_1}\right) \\ -\left(sC_1 + \dfrac{1}{sL_1}\right) & s(C_1 + C_2) + \dfrac{L_1 + L_2}{sL_1L_2} \end{vmatrix}$$

$$= \begin{vmatrix} s + \dfrac{1}{s} & -\left(s + \dfrac{1}{s}\right) \\ -\left(s + \dfrac{1}{s}\right) & 2\left(s + \dfrac{1}{s}\right) \end{vmatrix}$$

$$= 2\left(s + \frac{1}{s}\right)^2 - \left(s + \frac{1}{s}\right)^2$$

$$= \left(s + \frac{1}{s}\right)^2$$

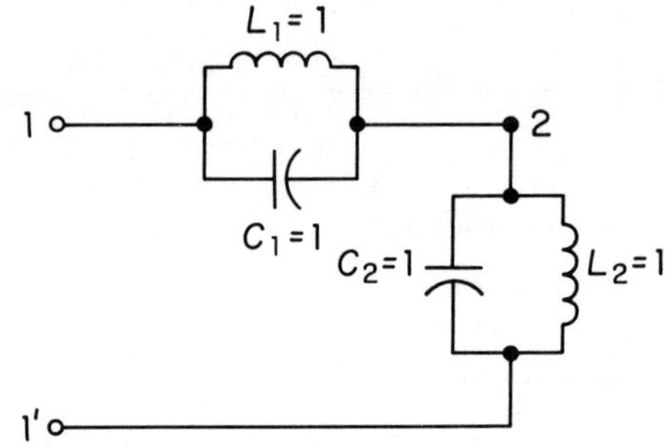

Figure 10-6 Network N.

Similarly, the cofactor $\Delta_{11} = 2(s + 1/s)$; thus

$$Z_d(s) = \frac{\Delta_{11}}{\Delta} = \frac{2\left(s + \dfrac{1}{s}\right)}{\left(s + \dfrac{1}{s}\right)^2} = \frac{2}{s + \dfrac{1}{s}} = \frac{2s}{s^2 + 1}$$

which shows that $n = 1$ and $m = 2$. But, from the network N, we see that

$$n_L = n_C = 2$$

so that

$$n_L + n_C = 4$$

Hence, $n \neq n_L + n_C$ and $m \neq n_L + n_C$, and, according to Definition 10-5, we conclude that N is not minimal in reactive elements.

We note that in N there exists a loop of all inductors (L_1 and L_2) and a loop of all capacitors (C_1 and C_2) when terminals 1 and 1′ are short-circuited. Thus, condition (c) in Theorem 10-8 is violated by N. The fact that N is not minimal in reactive elements is due to the cancellation of the common factor $(s + 1/s)$ of Δ and Δ_{11}. Such a network is known as a *degenerate* network in network theory.[15]

The next theorem states the effect on the realization of $Z(s)$ by resistors.

THEOREM 10-9. Let $Z(s)$ be a pr analytic function such that

$$0 \neq Z(0) \neq Z(\infty) \neq 0 \tag{10-30}$$

Then, any realization N of $Z(s)$ without mutual inductances contains at least two resistors. The proof of this theorem is quite evident since all inductors and capacitors in N become open- and short-circuited at $s = 0$ and $s = \infty$, respectively. Thus the theorem follows directly from Equation (10-30).

We now state some of the topological properties of a transformerless realization of a minimum pr function in the following theorems.[16]

THEOREM 10-10. Let N be a transformerless realization of a minimum pr function $Z(s)$. Then N contains between its terminals 1 and 1′

(a) No paths or basic cutsets of the one-element kind (all-R, all-L, or all-C),

(b) Paths and basic cutsets consisting of any of the two-element kind (RL, RC, or LC),

(c) No single-element path or basic cutset.

THEOREM 10-11. With the same notation as in Theorem 10-10,

(a) N cannot be a combination of a two-element parallel branch and another network. As a result,

(b) N must contain at least five elements with the additional restriction that only bridge networks are allowed on the five-element case.

THEOREM 10-12. Let $Z(s)$ be a *biquadratic*[17] minimum pr function and let N be a passive transformerless realization of $Z(s)$. Then,

(a) At least two resistors are contained in N if $Z(0) \neq Z(\infty)$,

[15]Refer to L. Weinberg [80], Section 1-9, and Seshu and Balabanian [75], Section 4-3, for example.

[16]Proofs of Theorems 10-10 through 10-12 can be found in Seshu and Reed [76], Chapter 8.

[17]A biquadratic function in s is a rational function in s of the form

$$\frac{a_2 s^2 + a_1 s + a_0}{b_2 s^2 + b_1 s + b_0}$$

where $a_2 \neq 0$ and $b_2 \neq 0$.

(b) At least three reactive elements are contained in N, and, consequently,
(c) The node determinant Δ, as well as its cofactor Δ_{11}, is of the form

$$\Delta \text{ (and } \Delta_{11} \text{ also)} = K\frac{s^k + a_{k-1}s^{k-1} + \cdots + a_1 s + a_0}{s^j}$$

where $K > 0$ and $k \geq 3$ (k being an integer, of course).

Proof: Part (a) of this theorem is obvious from the physical point of view since $Z(s)$ is resistive at $s = 0$ and $s = \infty$. The formal proof of this statement is left as an exercise.

Statement (b) can be proved as follows: From Theorem 10-10(c), there can be no single element between 1 and 1′. Hence, a path between 1 and 1′ must contain two or more elements. But, by Theoerm 10-10(b), there exists at least one L-C path between 1 and 1′ and, hence, N contains at least two reactive elements. If there were only two, they must be an L and a C connected in series between 1 and 1′. But, according to Theorem 10-10(b), there must also be an L-C basic cutset between 1 and 1′, also. Hence, at least three reactive elements are required in N.

The proof of (c) follows from (b) and Theorem 10-1 and is left as a problem.

The theorems presented here on minimality give us some insight, from the topological point of view, of the problem concerning the minimal realization of a minimum pr function $Z(s)$. For example, to obtain a minimal realization of a given minimum pr $Z(s)$, we begin with a five-element bridge, in light of Theorem 10-10(d). If none of the five-element bridge network realizes $Z(s)$, we then proceed to networks with six elements. As before, the series-parallel structure described in Theorem 10-11(b) will first be eliminated by virtue of the theorem. Finally, by exhausting all possibilities, it can be shown[18] that the two seven-element realizations of a biquadratic minimum function, illustrated in Figure 10-7 are minimal provided that the five-element bridges fail to realize the function.

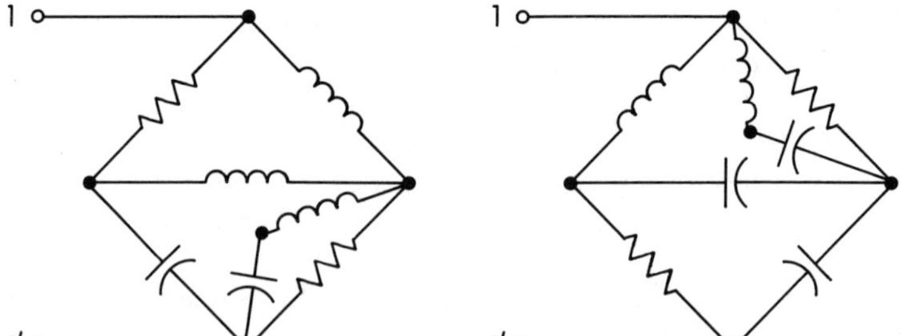

Figure 10-7 Seven-element minimal realizations of a biquadratic minimum function $Z(s)$.

[18]See S. Seshu [73] for further discussion of minimal realization of the biquadratic minimum function.

10-5 TOPOLOGICAL APPLICATIONS IN TWO-PORT SYNTHESIS

In this section, we shall discuss some of the topological applications in the synthesis of passive two-ports without mutual inductances. In particular, we shall show how some of the well-known theorems in two-port synthesis may be proved using topological means.

DEFINITION 10-7. Zero of Transmission (or Transmission Zero). A *zero of transmission* is a zero of a transfer function (of the form of output over input) of a given network. Thus, transmission zeros are zeros of z_{12}, y_{12}, and G_{12} (defined in Chapter 6). They are frequencies at which zero output occurs for finite input.[19]

DEFINITION 10-8. Minimum-phase and Nonminimum-phase Networks. A two-port network is called a *minimum-phase network* if it has no transmission zero in the open right half of the s-plane; otherwise, it is a *nonminimum-phase network*. Similarly, a transfer function without (open) right half s-plane zeros is referred to as a *minimum-phase function;* otherwise it is a *nonminimum-phase function.*

DEFINITION 10-9. Common-ground (Common-terminal) Network. A *common-ground* (*common-terminal*) *network* is a two-port with one ($1'$) of the two input terminals (1 and $1'$) common to one ($2'$) of the two output terminals (2 and $2'$) as shown in Figure 10-8. Common-terminal networks are also called *three-terminal* or *unbalanced networks.*

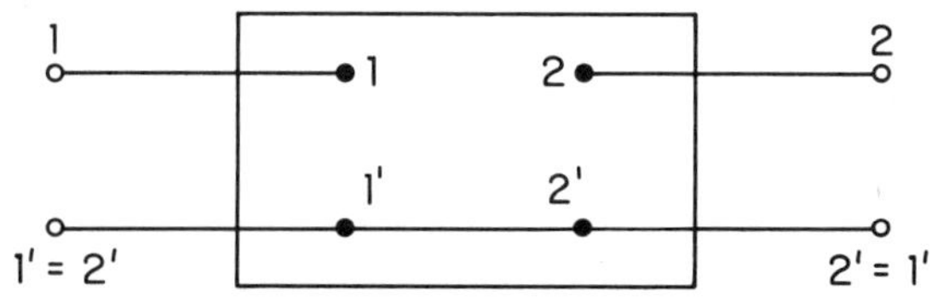

Figure 10-8 A common-ground network.

DEFINITION 10-10. Ladder Network. A *ladder network* is a passive series-parallel network without mutual coupling of the form shown in Figure 10-9 where each Z_i and Y_j represents a one-port (which may be a single element) and either, or both, of Z_1 and Z_{2n+1} may be absent. Thus, a ladder network is a special type of common-ground network. The following theorem deals with common-ground networks.

[19]Here, the words "output" and "input" refer to either voltage or current.

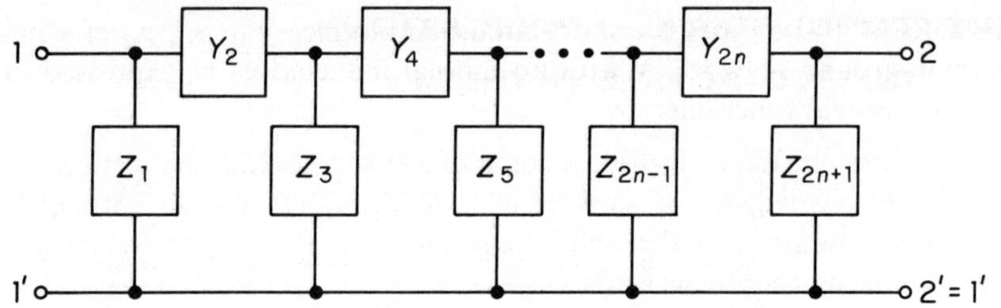

Figure 10-9 A ladder network.

THEOREM 10-13. Let N be a common-ground network without transformers. Then,

(a) The coefficients of the numerator and denominator polynomials of z_{12} of N are necessarily all real and non-negative if common factors in Δ and Δ_{12} (if any) are not canceled.

(b) None of the transmission zeros of N can be on the positive real axis.

(c) If N is a ladder network, all its transmission zeros are contained among the zeros of Z_i and Y_j ($i = 1, 2, \ldots, 2n + 1$; $j = 2, 4, \ldots, 2n$).

Proof: To prove property (a), we recall the topological formula for z_{12} derived in Chapter 6 [namely, Equation (6-126)] for a passive two-port with no mutual inductances:

$$z_{12} = \frac{W_{12,1'2'} - W_{12',1'2}}{V} \tag{10-31}$$

This expression becomes simply

$$z_{12} = \frac{W_{12,1'2'}}{V} \tag{10-32}$$

for a common-ground network, since terminals $1'$ and $2'$ are the same terminal and

$$W_{12',1'2} = 0$$

Furthermore, since N is a passive network consisting only of R, L, and C elements, each term in V and W represents a k-tree admittance product with a positive coefficient determined by the values of the passive elements contained in the product. Hence, property (a) is true.

Property (b) follows from the fact that a polynomial with positive coefficients cannot have any positive roots.

The proof of (c) is simple, but interesting, and hence is left to the reader as an exercise [Problem 10-28(b)].

The next theorem, well known in network synthesis, is attributable to Fialkow and Gerst.[20]

[20]This theorem is known in network synthesis as the Fialkow-Gerst coefficient condition; see, for example, M. E. Van Valkenburg [79] pp. 254–256.

THEOREM 10-14. Let the short-circuit admittances y_{11}, y_{12}, y_{22} of a passive common-ground network N with no mutual inductances be expressed in the form of rational functions:

$$-y_{12} = \frac{p_1(s)}{q(s)} = \frac{a_0 + a_1 s + a_2 s^2 + \cdots + a_{k-1}s^{k-1} + a_k s^k}{q(s)} \tag{10-33a}$$

$$y_{11} = \frac{p_2(s)}{q(s)} = \frac{b_0 + b_1 s + b_2 s^2 + \cdots + b_{m-1}s^{m-1} + b_m s^m}{q(s)} \tag{10-33b}$$

$$y_{22} = \frac{p_3(s)}{q(s)} = \frac{c_0 + c_1 s + c_2 s^2 + \cdots + c_{n-1}s^{n-1} + c_n s^n}{q(s)} \tag{10-33c}$$

Then,

$$a_i \geq 0 \tag{10-34a}$$

$$b_i \geq a_i \tag{10-34b}$$

$$c_i \geq a_i \tag{10-34c}$$

for $i = 0, 1, 2. \ldots$

Proof: By comparing Equations (10-33a), (10-33b), and (10-33c) with Equations (6-118), (6-117), and (6-119), respectively, we get immediately

$$p_1(s) = W_{12,1'2'} - W_{12',1'2} \tag{10-35a}$$

$$p_2(s) = W_{2,2'} \tag{10-35b}$$

$$p_3(s) = W_{1,1'} \tag{10-35c}$$

$$q(s) = \Sigma\, U \tag{10-35d}$$

Since, by hypothesis, N is a passive common-ground network without magnetic coupling, we have

$$W_{12',1'2} = 0$$

so that

$$p_1(s) = W_{12,1'2'} \tag{10-36}$$

which implies that

$$a_i \geq 0$$

as asserted in (a) of the theorem (and in Theorem 10-31). Next, from discussions in Chapter 6, it can easily be shown that

$$W_{1,1'} = W_{12,1'} + W_{1,1'2} \tag{10-37}$$

and

$$W_{2,2'} = W_{12,2'} + W_{2,12'} \tag{10-38}$$

But, since N is a common-ground network, Equations (10-37) and (10-38) reduce to

$$W_{1,1'} = W_{12,1'2'} + W_{1,1'22'} = p_3(s) \tag{10-39}$$

and

$$W_{2,2'} = W_{12,1'2'} + W_{2,11'2'} - p_2(s) \tag{10-40}$$

respectively. Now, comparing Equations (10-39) and (10-40) with (10-36), we see immediately that every term (2-tree admittance product) that appears in $p_1(s)$ also appears in $p_2(s)$ and $p_3(s)$. More precisely,

$$p_2(s) = p_1(s) + W_{2,11'2'} \tag{10-41}$$

and

$$p_3(s) = p_1(s) + W_{1,1'22'} \tag{10-42}$$

Since the coefficients in $W_{11'2',2}$ and $W_{1,1'22'}$ are necessarily non-negative, (b) and (c) of the theorem follow immediately, and the proof is now complete.

It is worth noting that, as implied in Equation (10-33), we have assumed in Theorem 10-32 that all the three functions have the same denominator $q(s)$. This is a very important point, for if this condition of common denominators is not satisfied by $-y_{12}$, y_{11}, and y_{22}, Equations (10-34b) and (10-34c) in the theorem will not be true, as the following example will illustrate.

EXAMPLE 10-7. Consider the passive transformerless common-ground network N shown in Figure 10-10. The short-circuit admittances of N are given by

$$-y_{12} = \frac{s}{s+1} \tag{10-43a}$$

$$y_{11} = \frac{2s+1}{s+1} \tag{10-43b}$$

$$y_{22} = 1 \tag{10-43c}$$

as can be easily verified.

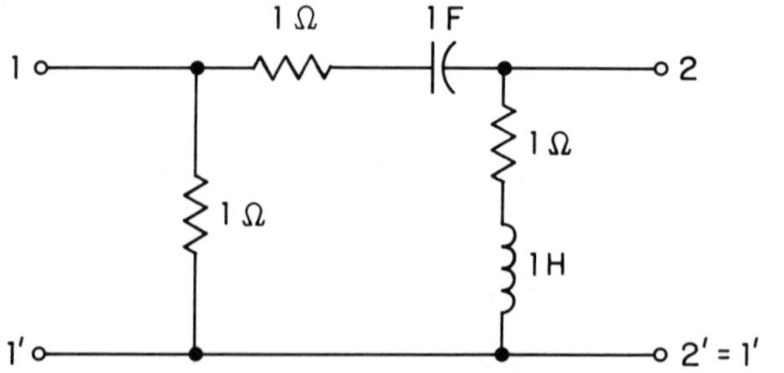

Figure 10-10 A passive transformerless common-ground network N.

It is seen that y_{22} in Equation (10-43b) does not have the same denominator as the other two admittances. It is also evident that Equation (10-33c) is not satisfied by y_{22} and $-y_{12}$. But if y_{22} is rewritten in the form

$$y_{22} = \frac{s+1}{s+1}$$

so that all the three admittances have the same denominator $q(s) = s + 1$, we then have the following values:

$$\begin{aligned} a_1 &= 1 & a_0 &= 0 \\ b_1 &= 2 & b_0 &= 1 \\ c_1 &= 1 & c_0 &= 1 \end{aligned}$$

so that all the three inequalities in Equation (10-34) are now satisfied, as asserted in Theorem 10-14.

We should perhaps note also that although the Fialkow-Gerst coefficient conditions [Equation (10-34)] are stated in terms of the short-circuit admittance functions, they are equally valid for the open-circuit impedance function as well. That is, if, in Equation (10-33), $-y_{12}$ is replaced by z_{12}, y_{11} by z_{11}, and y_{22} by z_{22}, inequalities (10-34a) through (10-34c) also hold, and the derivations for these inequalities can be obtained following a similar development used in the theorem.

EXAMPLE 10-8. Consider the same network N shown in Figure 10-10. It is easily verified that

$$z_{12} = \frac{s+1}{s^2+3s+1} \qquad z_{11} = \frac{s^2+2s+1}{s^2+3s+1} \qquad z_{22} = \frac{2s^2+3s+1}{s^2+3s+1}$$

Thus,

$$\begin{aligned} a_2 &= 0 & a_1 &= 1 & a_0 &= 1 \\ b_2 &= 1 & b_1 &= 2 & b_0 &= 1 \\ c_2 &= 2 & c_1 &= 3 & c_0 &= 1 \end{aligned}$$

so that all the three inequalities in Equation (10-34) hold for the three open-circuit impedances of N.

Finally, we should point out that the main purpose of this present section has been to illustrate how *some* of the theorems in network synthesis may be proved from the topological point of view. Needless to say, it has not been our intent to exhaust such examples of proving theorems in synthesis by topological methods. Our primary concern here has been to introduce to the reader the idea of topological synthesis and some of the topological applications in this regard. It is interesting to note that although the theory of network topology can be dated back as far as the nineteenth century,[21] the application

[21]Refer to Section 1-1.

of topology to network synthesis is a relatively recent development, and has thus far proved to be very challenging. It will undoubtedly play an increasingly significant role in the theory and application of network synthesis in the near future.

10-6 TOPOLOGICAL APPLICATIONS IN COMMUNICATION THEORY

One of the applications of topology to communication theory is the concept of a communication net, which may be defined as follows.

DEFINITION 10-11. Communication Net and Associated Terms. A *communication net* is a graph representing a communication system, with its edges signifying the communication channels and its vertices representing the stations of the system. The weights of the edges are called *channel capacities* since they specify the limited capacity of the corresponding channels that are capable of transferring information within the limit in either direction between stations. Between any pair of vertices, there exists a finite capacity of transferring information, called a *terminal capacity*, through one or more paths between the two vertices.

One of the problems in the theory of communication nets is the so-called maximum-flow problem, which is concerned with the determination of the maximum rate of flow of information between a pair of stations of a communication net. This problem has been attacked by a number of investigators.[22] Elias, Feinstein, and Shannon [29] proved that the maximum rate of flow between a pair of vertices (stations) in the net is equal to the minimum value of all basic cutsets with respect to the two vertices, where the *value* of a basic cutset is defined as the sum of the channel capacities of the edges of that cutset. Ford and Fulkerson [31, 32] independently solved the same problem with a different method of approach. A natural extension of the maximum flow problem between a pair of vertices is the generalization of this problem to simultaneous flows between k pairs of vertices through a given communication net. To this end, Hakimi [38] established the necessary and sufficient conditions for a communication net to realize a specified set of constant simultaneous flows. Tang [78] then extended Hakimi's problem by considering communication nets with time-varying branch capacities.

Another topological application in communication theory is the idea of a probabilistic net which we shall define next.

DEFINITION 10-12. Probabilistic Net. *A probabilistic net* is a communication net in which each edge is assigned a probability that the specified channel

[22]See references [4], [25], [29], [31], and [32].

exists and can perform its purpose (transfer information from one station to another, for example) successfully. The probability assigned to an edge (that is, the weight of that edge) is called the *reliability* of that edge.

As in the case of communication nets discussed here, a problem of probabilistic nets is the analysis of realization of such nets with a set of specification requirements. Problems concerning the pairwise reliabilities between any two stations have been discussed in several papers.[23] Wing and Demetriou [83] discussed the analysis of probabilistic nets in a recent paper. Most recently, Fu [34] made a generalization by introducing the idea of the overall reliability of the simultaneous communication among all (or part of) the stations of a given probabilistic net, where the *overall reliability* is defined as the probability that simultaneous communication can be secured among all the stations of the net. In his paper, Fu presented several algorithms for the analysis of a given probabilistic net, but little work has been done in the area of synthesis of such nets.

10-7 DIFFERENTIATION OF COMPOSITE FUNCTIONS

In this section, we shall show how topology may be used as an aid to obtain correct expressions for the differentiation of composite functions. The idea is introduced by means of a few examples.

EXAMPLE 10-9. Let us assume the following system of equations:

$$\begin{aligned} W &= f(u,v) \\ u &= g(x,y) \\ v &= h(x) \end{aligned} \tag{10-44}$$

so that the function W may be expressed as a function of x and y; that is,

$$W = F(x,y) \tag{10-45}$$

Now let us represent the system of Equation (10-44) by a directed graph such that each vertex corresponds to a variable in the system and an edge exists between two variables i and j, with orientation directed toward i from j if i is a function of j. Thus the graph corresponding to Equation (10-44) is the one shown in Figure 10-11. From differential calculus, we know that the differentiations of W with respect to x and y are given by

$$\frac{\partial W}{\partial x} = \frac{\partial W}{\partial u}\frac{\partial u}{\partial x} + \frac{\partial W}{\partial v}\frac{\partial v}{\partial x} \tag{10-46}$$

$$\frac{\partial W}{\partial y} = \frac{\partial W}{\partial u}\frac{\partial u}{\partial y} + \frac{\partial W}{\partial v}\frac{\partial v}{\partial y} \tag{10-47}$$

[23]See references [33], [35], [82], and [83].

By inspection of the graph in Figure 10-11 and Equations (10-46) and (10-47), we observe the obvious relationships between the graph and the equations. For example, in Equation (10-46) the product terms on the right side correspond to the paths from x to W, and the partial differentiations in each product term correspond to the edges in the path corresponding to that term.

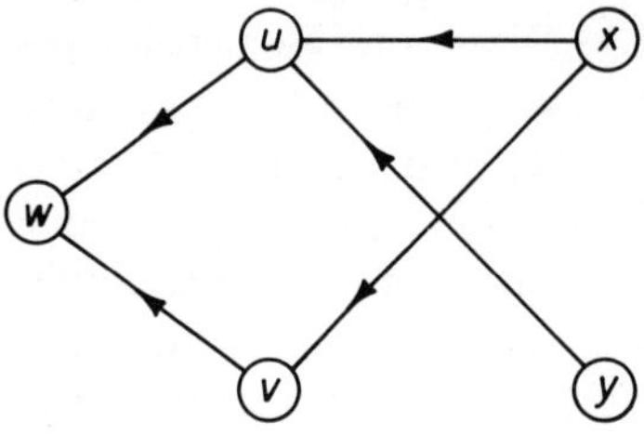

Figure 10-11 The graph corresponding to Equation (10-44).

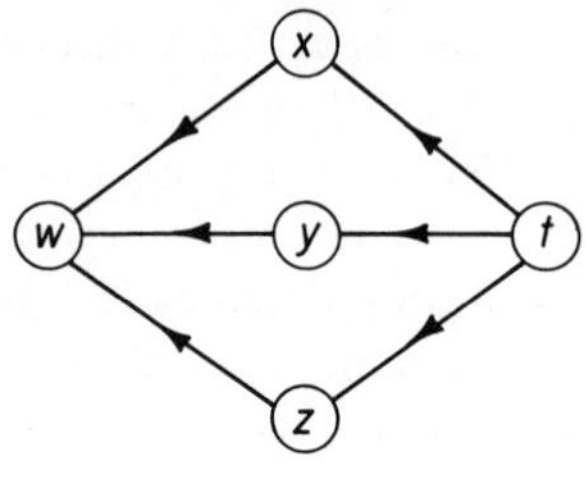

Figure 10-12 The graph corresponding to Equations (10-48).

EXAMPLE 10-10. Consider the system of equations

$$\begin{aligned} W &= f(x,y,z) \qquad & y &= h(t) \\ x &= g(t) & z &= k(t) \end{aligned} \tag{10-48}$$

From Equation (10-48) we see that W can be written as a function of t alone:

$$W = F(t) \tag{10-49}$$

The graph corresponding to Equation (10-48) is depicted in Figure 10-12. The formula for the *total* derivative of W with respect to t is obtained directly from the graph and is given by

$$\frac{dW}{dt} = \frac{\partial W}{\partial x}\frac{\partial x}{\partial t} + \frac{\partial W}{\partial y}\frac{\partial y}{\partial t} + \frac{\partial W}{\partial z}\frac{\partial z}{\partial t} \tag{10-50}$$

EXAMPLE 10-11. Let us consider a more complicated system of equations

$$W = f(x,y,u,v) \qquad u = g(x,v,t) \qquad v = h(x,y,t) \tag{10-51}$$

Thus, W may be expressed as a function of x, y, and t:

$$W = F(x,y,t) \tag{10-52}$$

The graph corresponding to Equation (10-51) is illustrated in Figure 10-13. From this graph we write

$$\frac{\partial W}{\partial x} = \frac{\partial W}{\partial x} + \frac{\partial W}{\partial u}\frac{\partial u}{\partial x} + \frac{\partial W}{\partial v}\frac{\partial v}{\partial x} + \frac{\partial W}{\partial u}\frac{\partial u}{\partial v}\frac{\partial v}{\partial x} \tag{10-53a}$$

$$\frac{\partial W}{\partial y} = \frac{\partial W}{\partial y} + \frac{\partial W}{\partial v}\frac{\partial v}{\partial y} + \frac{\partial W}{\partial u}\frac{\partial u}{\partial v}\frac{\partial v}{\partial y} \tag{10-53b}$$

$$\frac{\partial W}{\partial t} = \frac{\partial W}{\partial u}\frac{\partial u}{\partial t} + \frac{\partial W}{\partial v}\frac{\partial v}{\partial t} + \frac{\partial W}{\partial u}\frac{\partial u}{\partial v}\frac{\partial v}{\partial t} \tag{10-53c}$$

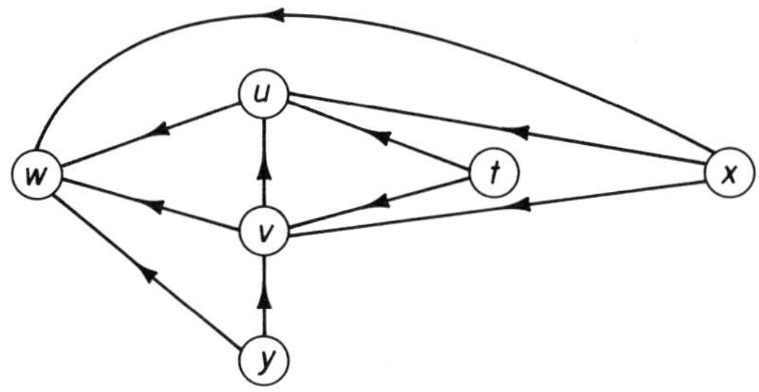

Figure 10-13 The graph corresponding to Equation (10-50).

In Equation (10-53a), the term $\partial W/\partial x$ on the left-hand side signifies the partial derivative of W as a function of the independent variables x, y, and t; that is, it is the partial derivative of W with respect to x with the other independent variables held constant. On the other hand, the same term, $\partial W/\partial x$, on the right-hand side of Equation (10-53a) refers to the partial derivative of W as a function of the independent variables x, u, and v. Similar remarks may be made about $\partial W/\partial y$ in Equation (10-53b).

10-8 BINARY RELATIONS

Our main objective in this section is to present a brief discussion of some topological applications in the calculus of binary relations. We shall first give a very brief review of the idea of relations and then illustrate how the idea may be treated topologically.

Let x and y be two objects, which may be either mathematical objects (such as numbers or theorems) or nonmathematical objects (such as people or animals). Then x and y may be "related" to each other in various ways, as the following examples will illustrate.

EXAMPLE 10-12. If x and y are two integers, then some of the mathematical relations are

$$x = y \tag{10-54a}$$

$$x > y \tag{10-54b}$$

$$x < y \tag{10-54c}$$

In Equation (10-54a) the symbol denoting "is equal to" specifies the relation between x and y; in (10-54b) the relation between the two integers is "is greater than"; and similarly, in (10-54c) the relation is "is smaller than."

EXAMPLE 10-13. Let x and y stand for two people. Then, some of the nonmathematical relations may be stated as

x is the father of y (10-55a)

x is the brother of y (10-55b)

x is older than y (10-55c)

where the relations in each of the statements in Equation (10-55) is underlined.

From the two examples, some of the properties are evident. In Equation (10-54a), we see that the positions of x and y in the mathematical statement may be interchanged and the resulting statement remains valid. Thus we write

$$y = x \tag{10-56}$$

However, we observe also that the positions of x and y cannot be interchanged in Equation (10-54b) to produce a valid statement; that is, if "$x > y$" is valid, it is obvious that "$y > x$" cannot be valid. In other words, the two statements, $x > y$ and $y > x$, cannot be satisfied simultaneously. The property that the positions of the two objects may be interchanged with respect to the same relationship to produce another valid statement is known as the *symmetric* property of that relation. We say that the relation is *a symmetric relation.* Other laws that classify relations, in addition to congruence ($\equiv$), and equality ($=$) are stated in the following definition:

DEFINITION 10-13. Reflexive, Symmetric, and Transitive Laws; Equivalence Relations. Let x, y, and z be three elements in a set S, and let R be a specified relation between any two of them. Let the symbol xRy be used to relate x to y by R. Then the following laws are defined as follows:

(a) *Reflexive law:* xRx for all x in S.

(b) *Symmetric law:* xRy implies ($\rightarrow$) yRx for all x, y in S.

(c) *Transitive law:* (xRy and yRz) $\rightarrow$ (xRz) for all x, y, and z in S.

Relations which are simultaneously reflexive, symmetric, and transitive are called *equivalence relations.*

EXAMPLE 10-14. Using the new terms just defined, it is evident that the relation "is equal to" is a reflexive relation since, for any integer x, we have always $x = x$. It is also symmetric because, for any two integers x and y, it is always true that

$$(x = y) \rightarrow (y = x)$$

This relation is also transitive since, for any three integers x, y, and z, we have

$$(x = y \text{ and } y = z) \rightarrow (x = z)$$

Similarly, it is easily verified that the relations "is greater than" or "is less then" are transitive but are neither reflexive nor symmetric. In like manner, the reader can verify that the relation "is the father of" does not satisfy any one of the three laws defined in Definition 10-13; the relation "is the brother of" satisfies both the symmetric and transitive law but does not satisfy the reflexive law; and the relation "is older than" satisfies the transitive law but does not satisfy the other two.

With the concept of binary relations briefly introduced, we are now ready to discuss some simple topological applications in this connection. First, we shall introduce the following definitions:

DEFINITION 10-14. Relation Graph G_R. Let S be a set of n objects $S = \{1, 2, \ldots, i, \ldots, j, \ldots, n\}$ with the binary relations $\{R_k\}$ defined on them. Then the *relation graph* G_R of S with respect to these relations is a weighted directed graph in which a vertex marked i corresponds to an object i in S and there exists an edge with a weight R_k from vertex i to vertex j in G_R if iR_kj is true.

DEFINITION 10-15. Relative Product. Let R_1 and R_2 be two relations for the set S. Then R_1R_2 is the *relative product* such that iR_1R_2k is true if iR_1j and jR_2k are true for some i, j, and k in S.

Definitions 10-14 and 10-15, as well as some topological applications in the calculus of binary relations, are illustrated by means of the following examples:

EXAMPLE 10-15. Let $S_1 = \{x,y,u,v\}$ be a set of four people, x, y, u, and v, and let the following relations be specified:

$$\begin{array}{l} x \text{ is the father of } y \\ y \text{ is the wife of } v \\ u \text{ is the teacher of } x \\ v \text{ is the employer of } u \\ u \text{ is younger than } y \end{array} \tag{10-57}$$

Suppose we want to describe the relations of (a) x to v, (b) v to y, and (c) y to y, First, Equation (10-57) is rewritten in the form

$$xR_1y \tag{10-58a}$$

$$yR_2v \tag{10-58b}$$

$$uR_3x \tag{10-58c}$$

$$vR_4u \tag{10-58d}$$

$$uR_5y \tag{10-58e}$$

where

$$R_1 = \text{“is the father of”} \tag{10-59a}$$

$$R_2 = \text{“is the wife of”} \tag{10-59b}$$

$$R_3 = \text{“is the teacher of”} \tag{10-59c}$$

$$R_4 = \text{“is the employer of”} \tag{10-59d}$$

$$R_5 = \text{“is younger than”} \tag{10-59e}$$

The graph G_R is obtained from Equation (10-58) as shown in Figure 10-14, with the weights and orientations of the edges properly assigned.

From G_R, we immediately obtain the following:

$$xR_1R_2v \tag{10-60}$$

$$vR_4R_5y \tag{10-61}$$

$$vR_4R_3R_1y \tag{10-62}$$

$$yR_2R_4R_5y \tag{10-63}$$

Finally, substituting Equation (10-59) into (10-60), (10-61), (10-62), and (10-63) we find the following relations:

(a) x is the father of the wife of v;
(b$_1$) v is the employer of u who is younger than y; also,
(b$_2$) v is the employer of the teacher of the father of y; and
(c) y is the wife of the employer of u who is younger than y herself.

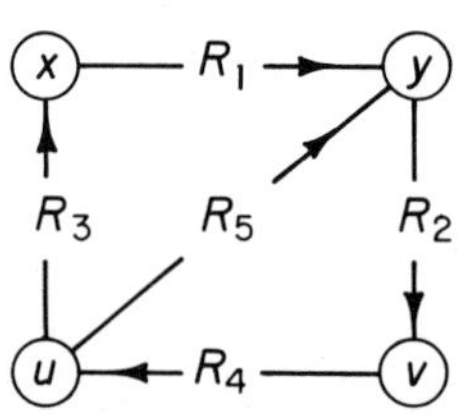

Figure 10-14 The relation graph G_R corresponding to Equation (10-58).

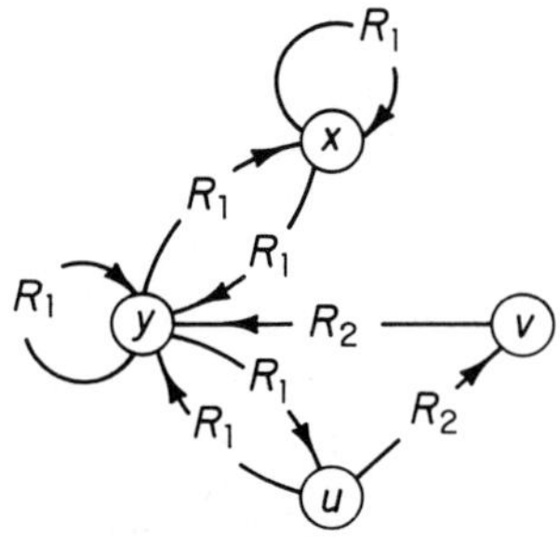

Figure 10-15 The relation graph G_R for Example 10-16.

EXAMPLE 10-16. Given the following relations among four people, x, y, u, and v:

$$xR_1y \qquad yR_1u \qquad uR_2v \qquad v_2R_2y$$

where R_1 = "is as old as" and R_2 = "is the brother of." Then we get the relation graph G_R as depicted in Figure 10-15. Note that since R_1 satisfies all the three laws defined in Definition 10-13, it is necessary to introduce the two self-loops at vertices x and y and the two pairs of edges between vertices x and y and vertices y and u as illustrated in Figure 10-15. From G_R, we obtain the following relations of x to the other three people (as well as to himself):

x is as old as himself (10-64a)

x is as old as y (10-64b)

x is as old as u (10-64c)

x is as old as the brother of v (10-64d)

x is as old as the brother of y (10-64e)

Finally, we should point out that Equation (10-64c) is obtained by the following steps in light of the transitive law:

$$(xR_1R_1u) \rightarrow (xR_1u).$$

Similarly, for Equation (10-64e) we write

$$(xR_1R_1R_2R_2y) \rightarrow (xR_1R_2y)$$

10-9 OTHER TOPOLOGICAL APPLICATIONS

In this final section, we shall mention only very briefly some of the topological applications which have not been discussed in the previous sections. The topological applications to (a) the theory of games, (b) the problem of the shortest route, (c) transport networks, and (d) the Euler problem, have been discussed by Berge [3]. Marshall [54] has presented topological applications in other areas such as (e) operational research, (f) social science, (g) psychology, (h) biology, and (i) physical science. The interested reader is also referred to Seshu and Reed [75], Kim and Chien [45], Harary, Norman, and Cartwright [42], Ore [66], and Busacker and Saaty [10] for still other possible applications of topology.

Finally, we should emphasize the fact that our brief discussion of other applications presented here is by no means exhaustive. It is meant as no more than a brief survey of some of the topological applications in other fields. The main purpose of such a partial list given here is to acquaint the reader with the existence of such applications.

PROBLEMS

10-1 Determine the order of complexity of each of the networks shown in Figure P10-1.

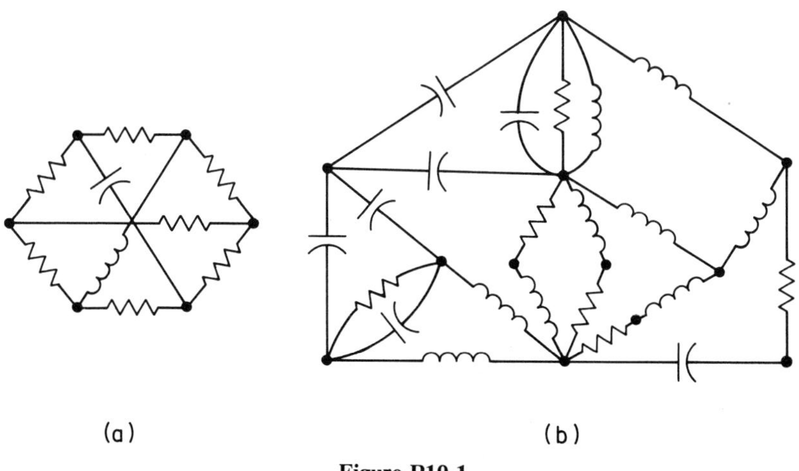

Figure P10-1

10-2 For each of the networks shown in Figure P10-1, determine the number of finite nonzero natural frequencies (not necessarily distinct).

10-3 Repeat Problems 10-1 and 10-2 for the network of Figure P10-3.

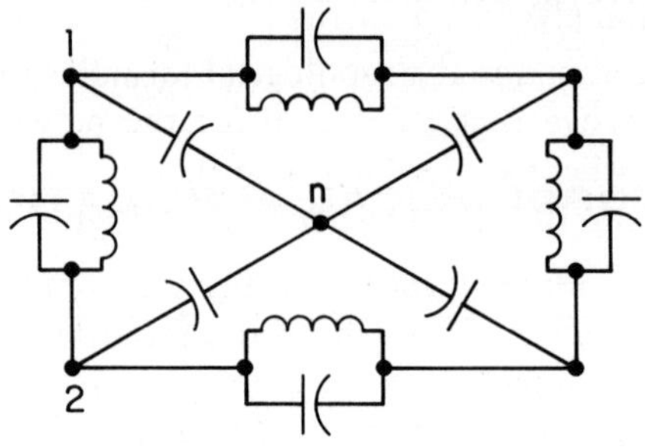

Figure P10-3

10-4 Using as few elements as possible, construct an LC one-port whose driving-point impedance is potentially equal to the impedance measured across nodes 1 and n of the network of Figure P10-3.

10-5 Using as few elements as possible, construct an LC one-port whose driving-point impedance is potentially equal to the impedance measured across nodes 1 and 2 of the network of Figure P10-3.

10-6 If every inductance is one henry and every capacitance one farad, find the natural frequencies of the network shown in Figure P10-3. (*Hint:* Exploit the symmetry of the network.)

10-7 Show that the determinant Δ of the node-admittance matrix Y_n of an RLC network N is an even (or odd) polynomial if the resistors in N form a separable subnetwork of N.
(*Hint:* Make use of Theorem 5-2.)

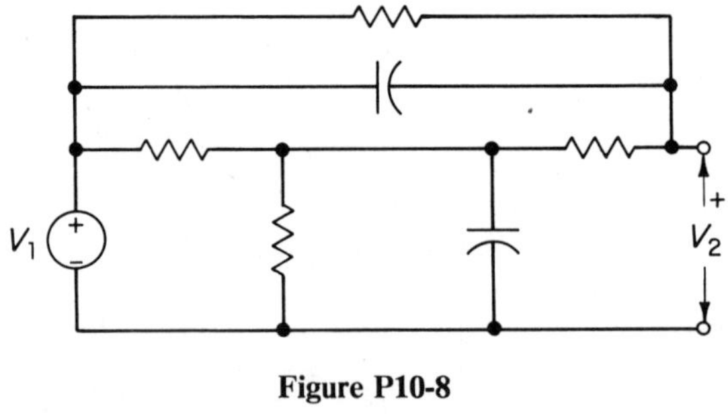

Figure P10-8

10-8 Given the following theorem (H. Poincaré):
A polynomial $p(s)$ of degree n ($n > 0$) with non-negative coefficients ($a \geq 0, j = 1, 2, \cdots, n$)

$$p(s) = a_n s^n + a_{n-1} s^{n-1} + \cdots + a_1 s + a_0$$

has no zeros in the sector $|\arg s| < \pi/n$, and

$$p(s) = a_n s^n + a_0$$

if $p(s)$ has a zero s_k such that $|\arg s_k| = \pi/n$.

Using this theorem and the topological formulas for transfer functions (Chapter 6), prove that $V_2/V_1 \neq 0$ for the network of Figure P10-8 for $|\theta| \leq 90°$, where $s = \rho\epsilon^{j\theta}$ and $\rho \geq 0$.

10-9 Repeat Problem 10-8 for the network of Figure P10-9 for $|\theta| \leq 45°$.

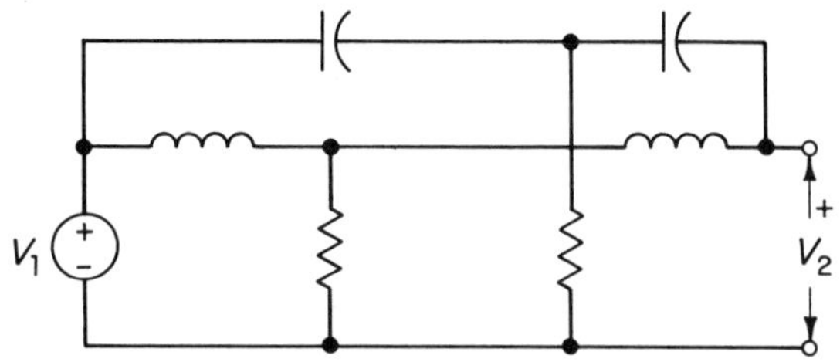

Figure P10-9

10-10 Repeat Problem 10-8 for the network shown in Figure P10-10 for $|\theta| \leq 60°$.

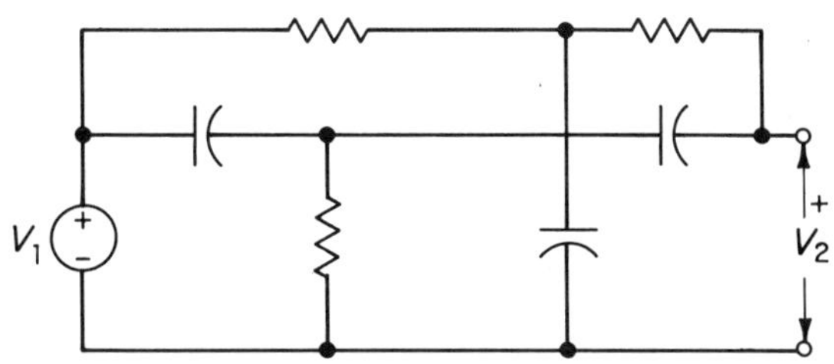

Figure P10-10

10-11 Repeat Problem 10-8 for the network of Figure P10-11 for $\theta = 0$.

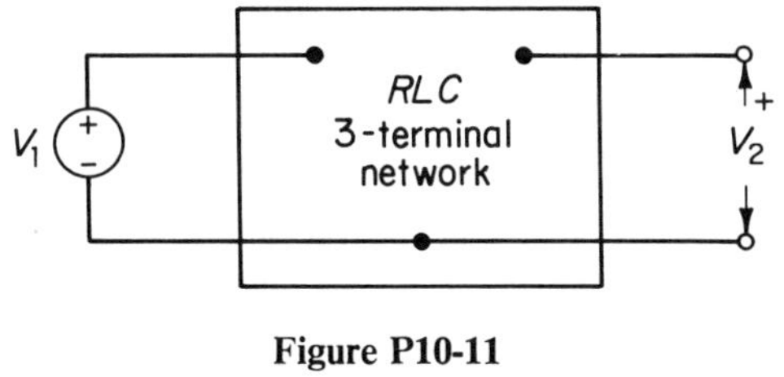

Figure P10-11

10-12 Prove that for an *RC* 3-terminal network (Figure P10-12), the polynomial $W_{12,0}(s)$ has no missing powers.

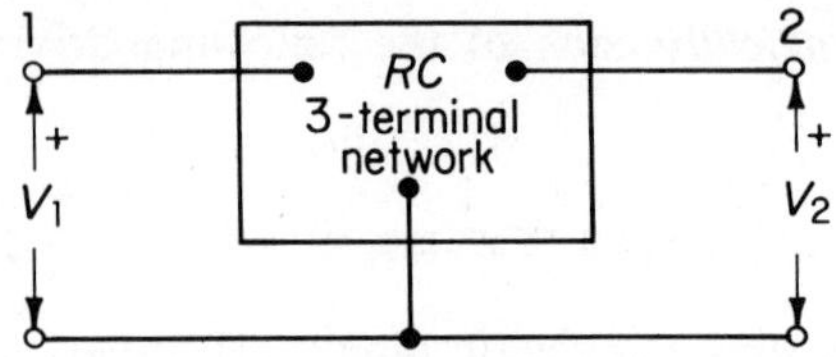

Figure P10-12

10-13 Using the statement of Problem 10-12, determine the minimum number of resistors and capacitors required to realize the following voltage ratio transfer function as an RC 3-terminal network (Figure P10-12):

$$\frac{V_2}{V_1} = \frac{4s^4 + 2}{s(s + 1)(s + 2)(s + 3)(s + 4)}$$

10-14 There are a number of different RC grounded two-ports with a minimum number of resistors and capacitors having the following $W_{12,0}$ form:

$$W_{12,0} = a_3s^3 + a_2s^2 + a_1s + a_0$$

where $a_i \neq 0$, $i = 0, 1, 2$, and 3. Find as many as you can and explain why each of the networks found satisfies the required $W_{12,0}$ form.

10-15 Determine *topologically* the number of poles and zeros of the driving-point impedance $Z_d(s)$ of the network shown in Figure P10-15. Show your work in detail.

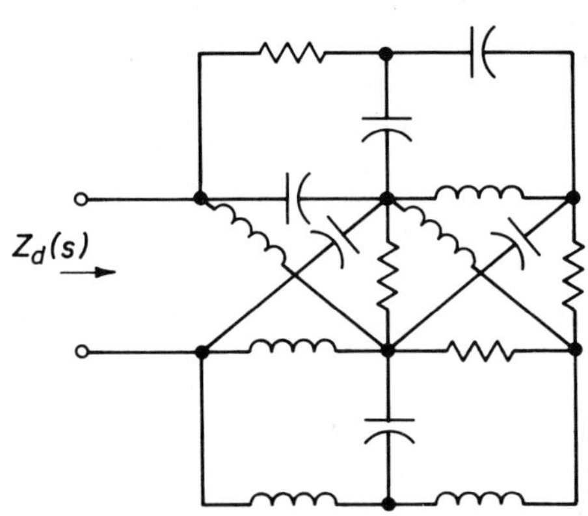

Figure P10-15

10-16 Given a set of 2-trees $\{T_{2i,j}^{(k)}\}$. How can a graph G be determined such that G realizes each $T_{2i,j}^{(k)}$ in the set as a 2-tree? What restrictions does the set $\{T_{2i,j}^{(i)}\}$ impose upon G? Give a detailed discussion on each of such restrictions.

10-17 Realize *topologically* each of the following driving-point impedance functions:

$$\text{(a) } Z_a = \frac{2s^2 + 2s + 1}{s^2 + s + 1} \qquad \text{(b) } Z_b = \frac{s^3 + 3s^2 + 2s + 1}{2s^2 + s + 1}$$

Indicate every step involved in your work. Also, identify the terminals in each network obtained.

10-18 Determine whether or not each of the networks shown in Figure P10-18 is minimal in reactive elements assuming a value of 1 (with a correct unit) for each element in the network.

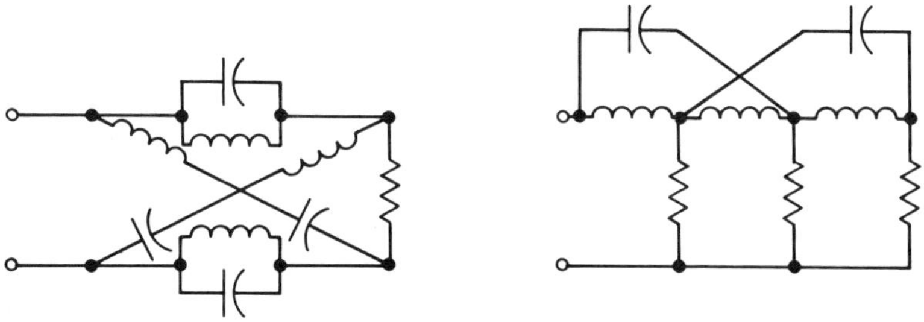

(a) (b)

Figure P10-18

10-19 Repeat Problem 10-17 for each of the networks shown in Figure P10-19, assuming unit values for all elements.

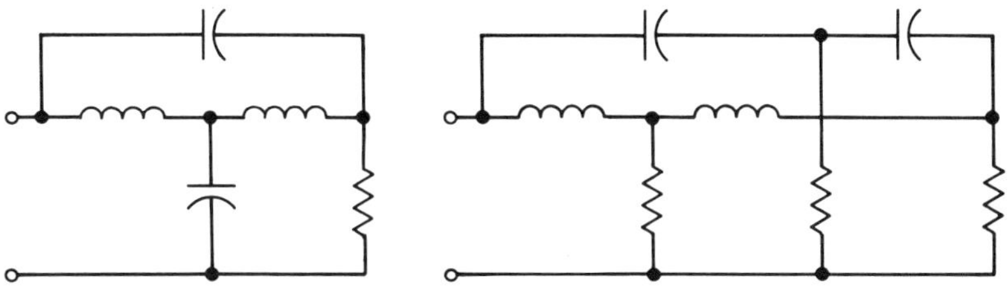

(a) (b)

Figure P10-19

10-20 With the aid of topological formulas for network functions, construct an RC two-port such that $E_2/E_1 = 0$ at $s = 1$. Make the network as simple as you can.

10-21 Derive a set of necessary and sufficient conditions for the realizability of a given matrix as (a) a circuit matrix; (b) a cutset matrix of the graph corresponding to some network. (Also refer to Problem 2-33.)

10-22 Derive a set of necessary and sufficient realizability conditions for a given matrix as (a) the short-circuit admittance matrix Y_{sc} and (b) the open-circuit impedance matrix Z_{oc} of a certain two-port, where

$$Y_{sc} = \begin{bmatrix} y_{11} & y_{12} \\ y_{21} & y_{22} \end{bmatrix} \quad \text{and} \quad Z_{oc} = \begin{bmatrix} z_{11} & z_{12} \\ z_{21} & z_{22} \end{bmatrix}$$

10-23 Extend the results obtained in Problem 10-22 by repeating the problem for a general n-port.

10-24 Let F_1 be a rational function in S that represents the sum of all tree admittance products (of some passive one-port) and let F_2 be another function in S representing the sum of all 2-tree admittance products (of some passive one-port). Formulate a topological method for finding a passive one-port N that simultaneously realizes F_1 as V and F_2 as $W_{1,1'}$ of $Z_d(s)$, where

$$Z_d(s) = \frac{W_{1,1'}}{V}$$

is the driving-point impedance of N. Assume that N contains no mutual inductances. State all your assumptions.

10-25 Extend the topological method obtained in Problem 10-24 to the synthesis of a passive two-port with no magnetic coupling. State your assumptions.

10-26 Repeat Problem 10-25 for an active two-port. State your assumptions.

10-27 Repeat Problem 10-25 for an n-port. State your assumptions.

10-28 (a) Prove statement (a) of Theorem 10-12 using a topological approach. (b) Prove statement (c) of Theorem 10-13.

10-29 Write the system of differential equations for the partial derivatives of W with respect to t and z from the corresponding graph shown in Figure P10-29.

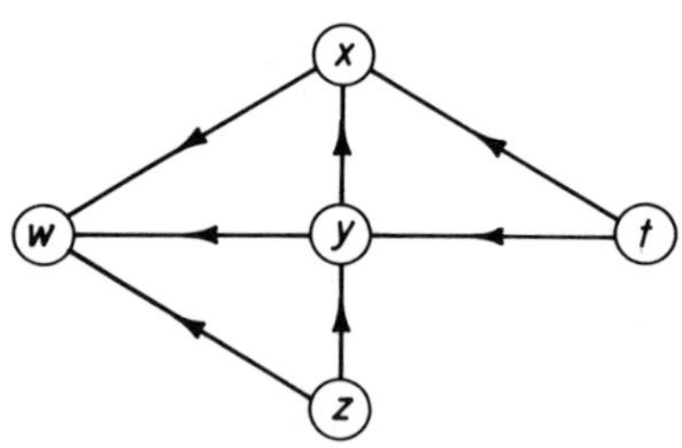

Figure P10-29

10-30 Consider the following system of equations:

$$\begin{aligned} W &= f(x,y,t,u,v,w) \\ x &= x(y,u,w) \\ y &= y(t,u,v,w) \\ u &= u(v) \end{aligned}$$

Following the procedure used in Example 10-11,
(a) Obtain the graph G corresponding to the given system.
(b) Write the partial differential equations from G obtained in part (a).

10-31 Compare the graph of Figure 10-12 with the SC network N of Figure P10-31 where nodes w and t are the terminals of N.
(a) List the similarities and differences between the two types of graphs.
(b) State the relationship between the properties of the two types of graphs.

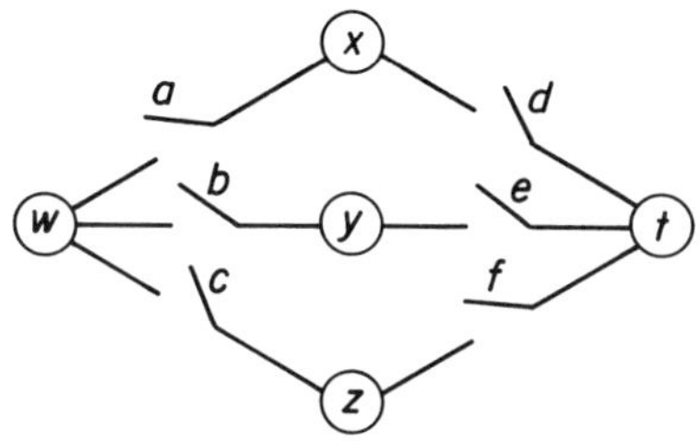

Figure P10-31

10-32 Let a, b, c, d, and e be five persons. The relationships between them are expressed by the relation graph shown in Figure P10-32, where

$$R_1 = \text{“the son of”}$$
$$R_2 = \text{“the father of”}$$
$$R_3 = \text{“the wife of”}$$
$$R_4 = \text{“the brother of”}$$

State the relation, in simplest terms, between (a) b and c; (b) b and e; (c) c and d.

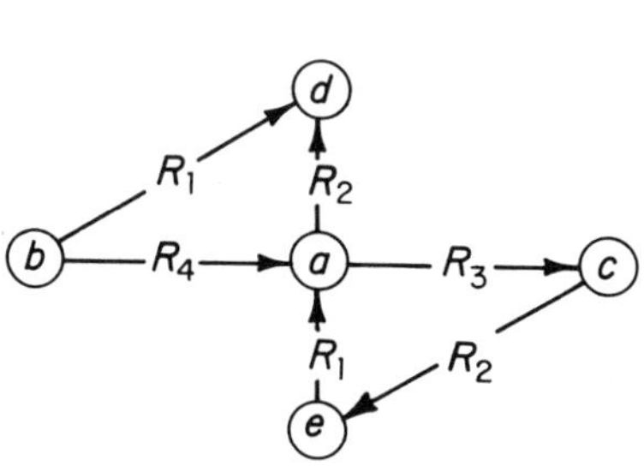

Figure P10-32

10-33 In Figure P10-33, each node represents an island and each line represents a bridge joining two islands. Determine whether it is possible to traverse each bridge precisely once (not necessarily coming back to the starting point).

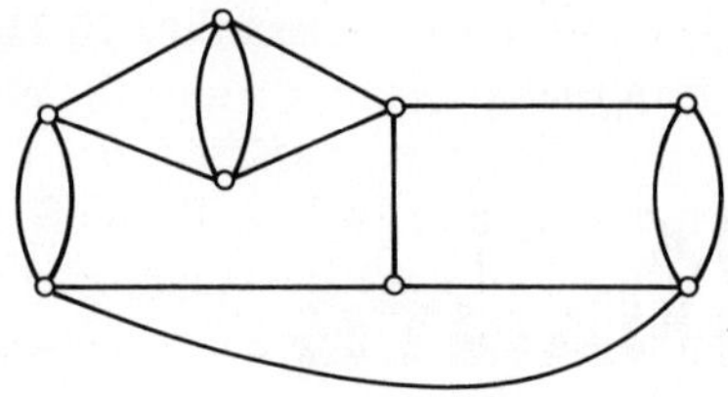

Figure P10-33

10-34 In Figure P10-34, each node represents a town and each line represents a road connecting two towns. Determine whether it is possible to visit each town exactly once *and* come back to the starting point. (This is the famous *Hamilton circuit problem.*)

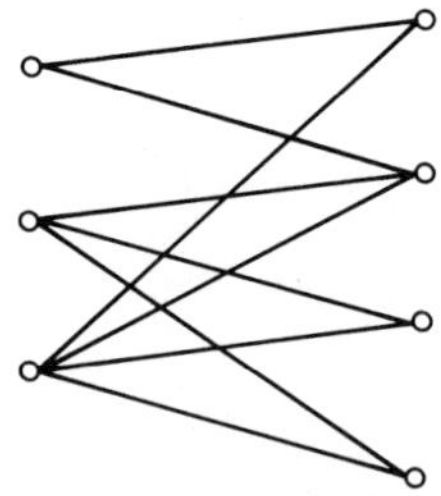

Figure P10-34

10-35 In Figure P10-35, vertices v_1 and v_2 represent the two banks of a river, vertices v_3 and v_4 denote two islands in the river, and each of the seven edges stands for a bridge. Determine whether it is possible to cross all the seven bridges, passing over each one only once. This is a classic problem in graph theory, known as the *Konigsberg bridge problem.*

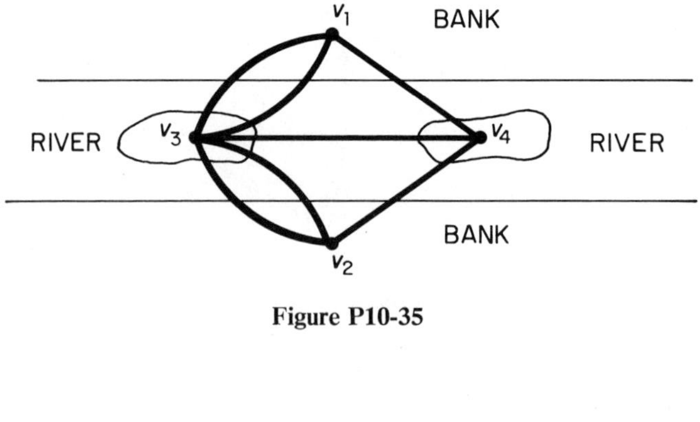

Figure P10-35

APPENDIX 1
DETERMINANTS

A1-1 INTRODUCTION

Determinants have their origin in connection with the problem of solving systems of simultaneous linear equations. The concept of a determinant is already familiar to the student from ordinary algebra where, in solving a system of n linear algebraic equations in n unknowns, it is quite convenient to express the solutions in compact form as ratios of determinants. Although the theory of determinants is widely applicable for the theoretical development as well as various numerical computational methods in different branches of engineering and physical sciences, our main concern in this chapter is to present only a brief discussion of certain properties of determinants that are useful to us in our later development of the theory of the main subject.

A1-2 DETERMINANTS OF ORDER 2

Consider the following system of two equations in two unknowns:

$$a_{11}x_1 + a_{12}x_2 = b_1 \quad \text{(A1-1)}$$

$$a_{21}x_1 + a_{22}x_2 = b_2 \quad \text{(A1-2)}$$

where a_{11}, a_{12}, a_{21}, and a_{22} are the coefficients; b_1 and b_2 are the constants; and x_1 and x_2 are the unknowns of the system. The a's and b's are assumed to be known constants here. The double-subscript notation is used for the coefficients to specify the position of each coefficient.

Multiplying Equation (A1-1) by a_{22} and (A1-2) by $-a_{12}$ and adding, we get

$$(a_{11}a_{22} - a_{21}a_{12})x_1 = a_{22}b_1 - a_{12}b_2$$

Thus,

$$x_1 = \frac{a_{22}b_1 - a_{12}b_2}{a_{11}a_{22} - a_{21}a_{12}} \tag{A1-3}$$

provided that

$$a_{11}a_{22} - a_{21}a_{12} \neq 0 \tag{A1-4}$$

Similarly, multiplying Equation (A1-1) by $-a_{21}$ and (A1-2) by a_{11} and adding, we get

$$(a_{11}a_{22} - a_{21}a_{12})x_2 = a_{11}b_2 - a_{21}b_1$$

Hence,

$$x_2 = \frac{a_{11}b_2 - a_{21}b_1}{a_{11}a_{22} - a_{21}a_{12}} \tag{A1-5}$$

with the same assumption that $a_{11}a_{22} - a_{21}a_{12} \neq 0$ as stated in Equation (A1-4).

Next we introduce the symbol

$$\det A = \begin{vmatrix} a_{11} & a_{12} \\ a_{21} & a_{22} \end{vmatrix}$$

which we shall call a determinant of order 2 having four elements, a_{11}, a_{12}, a_{21}, and a_{22}, arranged in two horizontal and two vertical lines. Furthermore, we define the expression $a_{11}a_{22} - a_{21}a_{22}$ to be the expanded value of det A. Then Equations (A1-3) and (A1-4) may be written in compact form as

$$x_1 = \frac{\begin{vmatrix} b_1 & a_{12} \\ b_2 & a_{22} \end{vmatrix}}{\begin{vmatrix} a_{11} & a_{12} \\ a_{21} & a_{22} \end{vmatrix}} \tag{A1-6}$$

and

$$x_2 = \frac{\begin{vmatrix} a_{11} & b_1 \\ a_{21} & b_2 \end{vmatrix}}{\begin{vmatrix} a_{11} & a_{12} \\ a_{21} & a_{22} \end{vmatrix}} \tag{A1-7}$$

provided that

$$\begin{vmatrix} a_{11} & a_{12} \\ a_{21} & a_{22} \end{vmatrix} \neq 0 \tag{A1-8}$$

It should be noted that the determinant in Equation (A1-8) is called the system determinant, or the determinant of the system. Its value is obtained by taking the difference between the two "cross products" $a_{11}a_{22}$ and $a_{12}a_{21}$ in that order. In general, the expanded value of a determinant of order 2 is obtained by adding the product of the diagonal elements in the upper left and lower right corners to the *negative* of the product of the diagonal elements in the upper right and low left corners.

With the above introduction, we now formally define a second-order determinant and its expanded value.

DEFINITION A1-1. Second-order Determinant and Its Value. A *second-order determinant* of the form

$$\det A = \begin{vmatrix} a_{11} & a_{12} \\ a_{21} & a_{22} \end{vmatrix} = a_{11}a_{22} - a_{12}a_{21}$$

is defined as a square array of four *elements* a_{11}, a_{12}, a_{21}, a_{22}, arranged in two rows (horizontal lines) and two columns (vertical lines) with an *expanded value* (or simply value) obtained by taking the sum of the product of the diagonal elements in the upper left and lower right corners and the *negative* of the product of the diagonal elements in the upper right and lower left corners.

The diagonal from upper left to lower right is usually referred to as the *major* (*or main*) *diagonal*, whereas the one from upper right to lower left is called the minor diagonal of the determinant. Using these terms, the value of a second-order determinant may be defined as the sum of the product of the major diagonal elements and the negative of the product of the minor diagonal elements.

EXAMPLE A1-1. Consider the system

$$\begin{aligned} 2x_1 + \ x_2 &= -1 \\ -x_1 + 3x_2 &= 2 \end{aligned}$$

By Equations (A1-6) and (A1-7), we get

$$x_1 = \frac{\begin{vmatrix} -1 & 1 \\ 2 & 3 \end{vmatrix}}{\begin{vmatrix} 2 & 1 \\ -1 & 3 \end{vmatrix}} \qquad x_2 = \frac{\begin{vmatrix} 2 & -1 \\ -1 & 2 \end{vmatrix}}{\begin{vmatrix} 2 & 1 \\ -1 & 3 \end{vmatrix}}$$

since

$$\begin{vmatrix} 2 & 1 \\ -1 & 3 \end{vmatrix} \neq 0$$

Now, evaluating the four determinants, we get

$$x_1 = \frac{(-1)(3) - (1)(2)}{(2)(3)-(1)(-1)} = \frac{-3 - 2}{6 + 1} = \frac{-5}{7}$$

$$x_2 = \frac{(2)(2) - (-1)(-1)}{(2)(3) - (1)(-1)} = \frac{4 - 1}{6 + 1} = \frac{3}{7}$$

which are the solutions of the system. Note that in the system determinant

$$\begin{vmatrix} 2 & 1 \\ -1 & 3 \end{vmatrix}$$

2 and 3 are the major diagonal elements, and 1 and -1 are the minor diagonal elements. The value of the determinant is 7.

We shall define and discuss a determinant of order 3 (or a third-order determinant) in the next section.

A1-3 DETERMINANTS OF ORDER 3

If the solution of a linear system of two simultaneous equations in two unknowns involves determinants of order 2, systems of three equations in three unknowns naturally suggest the introduction of third-order determinants. We shall define a third-order determinant as follows.

DEFINITION A1-2. Determinant of Order 3. A *third-order determinant*, is a

$$\det A = \begin{vmatrix} a_{11} & a_{12} & a_{13} \\ a_{21} & a_{22} & a_{23} \\ a_{31} & a_{32} & a_{33} \end{vmatrix} \qquad \text{(A1-9)}$$

square array containing nine elements arranged in three rows and three columns flanked by two vertical lines as shown in Equation (A1-9).

DEFINITION A1-3. Minor. The second-order determinant M_{ij}, obtained by deleting the ith row and the jth column from det A, is called the *minor* of the element a_{ij} of det A, where $i = 1, 2, 3$; $j = 1, 2, 3$.

DEFINITION A1-4. Cofactor. The *cofactor* C_{ij} of the element a_{ij} of det A is defined as

$$C_{ij} = (-1)^{i+j} M_{ij} \qquad \text{(A1-10)}$$

Now we are ready to define the value of det A.

DEFINITION A1-5. The Value of Det A. The *value* of det A is defined as

$$\det A = a_{i1}C_{i1} + a_{i2}C_{i2} + a_{i3}C_{i3} \qquad i = 1, 2, 3 \qquad \text{(A1-11)}$$

or

$$\det A = a_{1j}C_{1j} + a_{2j}C_{2j} + a_{3j}C_{3j} \qquad j = 1, 2, 3 \qquad \text{(A1-12)}$$

EXAMPLE A1-2. Let us expand the third-order determinant

$$\det A = \begin{vmatrix} 1 & 1 & 2 \\ -1 & 3 & 1 \\ 3 & -1 & -3 \end{vmatrix}$$

Here,

$$
\begin{array}{lll}
a_{11} = 1 & a_{12} = 1 & a_{13} = 2 \\
a_{21} = -1 & a_{22} = 3 & a_{23} = 1 \\
a_{31} = 3 & a_{32} = -1 & a_{33} = -3
\end{array}
$$

By Equation (A1-11), for $i = 2$, we get

$$\det A = a_{21}C_{21} + a_{22}C_{22} + a_{23}C_{23} \tag{A1-13}$$

But, by Equation (A1-10), we get

$$C_{21} = (-1)^{2+1}M_{21} = (-1)^3\begin{vmatrix} 1 & 2 \\ -1 & -3 \end{vmatrix} = -\begin{vmatrix} 1 & 2 \\ -1 & -3 \end{vmatrix}$$

$$= -(1)(-3) - (2)(-1) = -(-3 + 2) = 1$$

Similarly,

$$C_{22} = (-1)^{2+2}M_{22} = (-1)^4\begin{vmatrix} 1 & 2 \\ 3 & -3 \end{vmatrix} = (-3 - 6) = -9$$

$$C_{23} = (-1)^{2+3}M_{23} = (-1)^5\begin{vmatrix} 1 & 1 \\ 3 & -1 \end{vmatrix} = -(-1 - 3) = 4$$

Substituting these values into Equation (A1-11) we get

$$\det A = (-1)(1) + (3)(-9) + (1)(4)$$
$$= -1 - 27 + 4$$

or

$$\det A = -24$$

We should note that by evaluating the determinant using Equations (A1-11) and (A1-12) with different i and j values one will always arrive at the same result each time.

We recall, in the second-order case, the value of the determinant is obtained by adding the product of the major diagonal elements to the negative of the product of the minor diagonal elements:

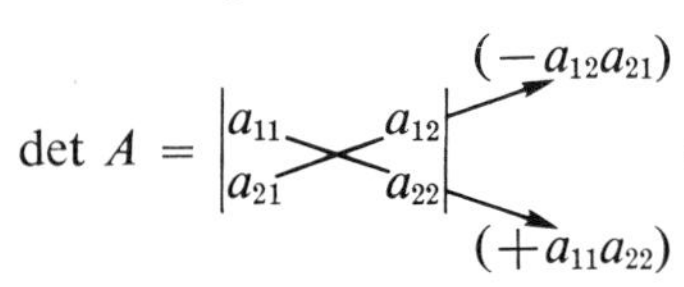

$$= a_{11}a_{22} - a_{12}a_{21}$$

Similarly, the third-order determinant may be evaluated with a slight modification:

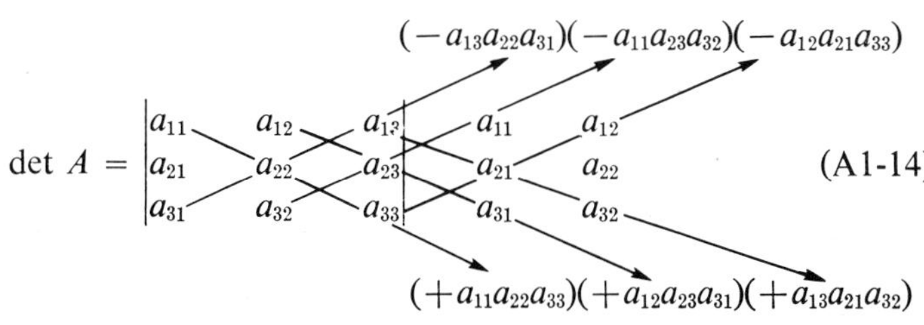

(A1-14)

Thus,

$$\det A = a_{11}a_{22}a_{33} + a_{12}a_{23}a_{31} + a_{13}a_{21}a_{32} - a_{13}a_{22}a_{31} - a_{11}a_{23}a_{32} - a_{12}a_{21}a_{33}$$

However, we should emphasize that this scheme applies only to the evaluation of a determinant of *order 3 or less*. It *will not* give the correct value to a determinant of higher order!

EXAMPLE A1-3. Let us evaluate the same determinant as given in Example A1-2 using this "cross product" expansion. We have

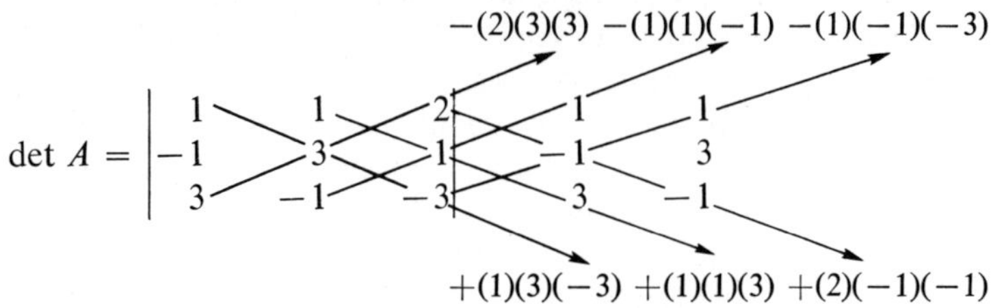

$$\begin{aligned} &= (1)(3)(-3) + (1)(1)(3) + (2)(-1)(-1) \\ &\quad -(2)(3)(3) - (1)(1)(-1) - (1)(-1)(-3) \\ &= -9 + 3 + 2 - 18 + 1 - 3 \end{aligned}$$

or

$$\det A = -24$$

which agrees with the value obtained by the application of the more general definition in Example A1-2.

A1-4 DETERMINANTS OF ORDER *n*

DEFINITION A1-6. Determinant of Order *n*. A *determinant of order n*

$$\det A = \begin{vmatrix} a_{11} & a_{12} & \cdots & a_{1n} \\ a_{21} & a_{22} & \cdots & a_{2n} \\ \cdot & & & \\ \cdot & & & \\ \cdot & & & \\ a_{n1} & a_{n2} & \cdots & a_{nn} \end{vmatrix} \tag{A1-15}$$

is a square array of $n \times n$ elements arranged in n rows and n columns as shown in Equation (A1-15). The value of the determinant is given by

$$\det A = a_{k1}C_{k1} + a_{k2}C_{k2} + \cdots + a_{kn}C_{kn} = \sum_{i=1}^{n} a_{ki}C_{ki}$$

$$k \text{ being a positive integer} \leq n \tag{A1-16}$$

or

$$\det A = a_{1i}C_{1i} + a_{2i}C_{2i} + \cdots + a_{ni}C_{ni} = \sum_{j=1}^{n} a_{ji}C_{ji}$$

$$i \text{ being a positive integer} \leq n \tag{A1-17}$$

where a_{ki} (k,i are any two positive integers $\leq n$) is the element of det A_n located at the intersection of the kth and the ith column, and C_{ki} is the cofactor of a_{ki}.

A1-5 PROPERTIES OF DETERMINANTS

The properties discussed here are a selected set of properties which are used frequently in this text. They are stated here without proof. Examples, however, will be given to illustrate the ideas involved.

Property 1: If the determinant contains a row (or a column) of zeros, the value of the determinant is zero.

Property 2: If, in a determinant, two rows (or two columns) are interchanged, the value of the resulting determinant is equal to the negative of the value of the original determinant.

Property 3: If any two rows (or two columns) are proportional, the value of the determinant is zero.

Property 4: If each element in a row (or a column) is multiplied by the same constant k, the value of the resulting determinant is k times the value of the original one.

Property 5: If each element in a row (or a column) is multiplied by the same constant k and then added to the corresponding element of some other row (or column), the value of the resultant determinant is the same as that of the original one.

Property 6: If the 1st, 2d, . . ., nth rows of a determinant are taken to be the 1st, 2d, . . ., nth columns, respectively, of a new determinant, the values of the two determinants are equal.

EXAMPLE A1-4. To demonstrate Property 1, let us consider the third-order determinant:

$$\det A = \begin{vmatrix} 1 & -3 & 1 \\ 0 & 0 & 0 \\ -1 & 2 & -2 \end{vmatrix}$$

Expanding it by the cross-product scheme, we get

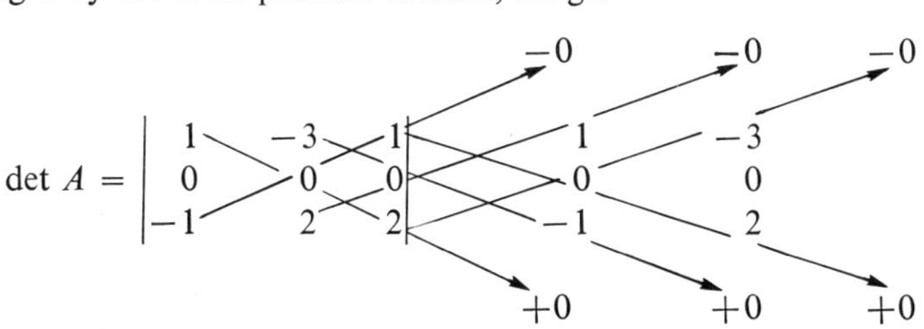

which is seen to be zero since each term in the expansion contains a zero in the product.

EXAMPLE A1-5. Let us take the same determinant as that given in Example A1-2:

$$\det A = \begin{vmatrix} 1 & 1 & 2 \\ -1 & 3 & 1 \\ 3 & -1 & -3 \end{vmatrix}$$

the value of which is -24. Now, form a new determinant, det B, by interchanging the first and third column:

$$\det B = \begin{vmatrix} 2 & 1 & 1 \\ 1 & 3 & -1 \\ -3 & -1 & 3 \end{vmatrix}$$

Evaluating it using Equation (A1-16) and taking $k = 1$,

$$\begin{aligned} \det B &= a_{11}C_{11} + a_{12}C_{12} + a_{13}C_{13} \\ &= (2)(-1)^{1+1}\begin{vmatrix} 3 & -1 \\ -1 & 3 \end{vmatrix} + (1)(-1)^{1+2}\begin{vmatrix} 1 & -1 \\ -3 & 3 \end{vmatrix} + (1)(-1)^{1+3}\begin{vmatrix} 1 & 3 \\ -3 & -1 \end{vmatrix} \\ &= (2)(1)(9 - 1) + (1)(-1)(3 - 3) + (1)(1)(-1 + 9) \\ &= 16 - 0 + 8 \end{aligned}$$

or

$$\det B = 24$$

which is the negative of det A, as predicted by Property 2.

EXAMPLE A1-6. Consider

$$\det A = \begin{vmatrix} 1 & -1 & 2 \\ -1 & 3 & 1 \\ -2 & 6 & 2 \end{vmatrix}$$

Expanding it by the cross-product method, we get

$$\begin{aligned} \det A &= (1)(3)(2) + (-1)(1)(-2) + (2)(-1)(6) \\ &\quad -(-2)(3)(2) - (6)(1)(1) - (2)(-1)(-1) \\ &= 6 + 2 - 12 + 12 - 6 - 2 \end{aligned}$$

or

$$\det A = 0.$$

Note that the corresponding elements of the second and third row of det A are proportional. Thus, by Property 3, the value of the determinant is expected to be zero.

EXAMPLE A1-7. To illustrate Property 4, we shall use the determinant A of Example A1-2 again and multiply the second row by 2, yielding

$$\det B = \begin{vmatrix} 1 & 1 & 2 \\ -2 & 6 & 2 \\ 3 & -1 & -3 \end{vmatrix}$$

Next, evaluating it by the cross-product method,

$$\det B = -18 + 6 + 4 - 36 + 2 - 6$$

or

$$\det B = -48$$

which is indeed twice the value of det A, as expected.

EXAMPLE A1-8. Once again, consider the determinant of Example A1-2:

$$\det A = \begin{vmatrix} 1 & 1 & 2 \\ -1 & 3 & 1 \\ 3 & -1 & -3 \end{vmatrix}$$

Multiplying the first row by -3 and then adding it to the third row, we get

$$\det B = \begin{vmatrix} 1 & 1 & 2 \\ -1 & 3 & 1 \\ 0 & -4 & -9 \end{vmatrix}$$

Expanding det B by the cross-product method,

$$\det B = -27 + 0 + 8 - 0 + 4 - 9$$

or,

$$\det B = -24$$

This demonstrates Property 5.

EXAMPLE A1-9. Finally, let us obtain a new determinant, det C, by taking the first, second, and third rows of det A of Example A1-8 as the first, second, and third columns of det C. Thus,

$$\det C = \begin{vmatrix} 1 & -1 & 3 \\ 1 & 3 & -1 \\ 2 & 1 & -3 \end{vmatrix}$$

Expanding it, we get

$$\det C = -9 + 2 + 3 - 18 + 1 - 3$$

or

$$\det C = -24,$$

which is the same as det A, as stated in Property 6.

This concludes our discussions on determinants.

APPENDIX A2 MATRICES

A2-1 INTRODUCTION

Both matrices and determinants are closely related in connection with the solving of simultaneous linear equations. Although they are quite different in many ways, there exist sufficient similarities between them to cause confusion for those who encounter the two topics for the first time. For this reason, we shall be careful in pointing out the differences between determinants and matrices in the following discussions.

A2-2 BASIC CONCEPTS, EXAMPLES OF MATRICES, AND DEFINITIONS

Consider a linear system of m simultaneous equations in n unknowns:

$$\begin{aligned}
a_{11}x_1 + a_{12}x_2 + \cdots + a_{1n}x_n &= b_1 \\
a_{21}x_1 + a_{22}x_2 + \cdots + a_{2n}x_n &= b_2 \\
&\vdots \\
a_{m1}x_1 + a_{m2}x_2 + \cdots + a_{mn}x_n &= b_m
\end{aligned}$$

where $a_{11}, a_{12}, \ldots, a_{mn}$ are the coefficients associated with the unknown variables $x_1, x_2, \ldots, x_n$ and $b_1, b_2, \ldots, b_m$ are the known constants. To be more specific,

a_{ij} is the jth coefficient which is associated with variable x_j in the ith equation of the system, and b_i is the known constant which appears on the right-hand side of that equation. If the coefficients are arranged in the same way as they occur in the system, there results a rectangular array of the form

$$\begin{bmatrix} a_{11} & a_{12} & \cdots & a_{1n} \\ a_{21} & a_{22} & \cdots & a_{2n} \\ \cdot & & & \\ \cdot & & & \\ \cdot & & & \\ a_{m1} & a_{m2} & \cdots & a_{mn} \end{bmatrix}$$

which is known as the *coefficient matrix* of the system.

Other examples of matrices are the so-called unknown column matrix and the constant column matrix,

$$\begin{bmatrix} x_1 \\ x_2 \\ \cdot \\ \cdot \\ \cdot \\ x_n \end{bmatrix} \quad \text{and} \quad \begin{bmatrix} b_1 \\ b_2 \\ \cdot \\ \cdot \\ \cdot \\ b_m \end{bmatrix}$$

respectively. Obviously the unknown column matrix is an array of all the n unknowns of the system arranged in a column in exactly the same order as they appear in the left-hand member of each equation of the system. Similarly, the constant column matrix is an array of all the m constants arranged in a column in the same sequence as they appear on the right-hand sides of the equations of the given system. If the constant column matrix is "merged" with the coefficient matrix, we obtain a new matrix of the form

$$\begin{bmatrix} a_{11} & a_{12} & \cdots & a_{1n} & b_1 \\ a_{21} & a_{22} & \cdots & a_{2n} & b_2 \\ \cdot & & & & \\ \cdot & & & & \\ \cdot & & & & \\ a_{m1} & a_{m2} & \cdots & a_{mn} & b_m \end{bmatrix}$$

which is called the *augmented matrix* of the system.

All these matrices will be discussed again in more detail when the topics of linear dependence and linear equations are discussed in the next chapter. We shall now define a matrix formally.

DEFINITION A2-1. Matrix of Order $m \times n$. A *matrix*, $A_{m\times n}$, *of order* $m \times n$ (read "m by n") is a rectangular array of $m \times n$ numbers (or functions) called *elements* arranged in m rows and n columns subject to a set of rules of operation (to be defined) and is denoted by the symbol

$$A_{m\times n} = \begin{bmatrix} a_{11} & a_{12} & \cdots & a_{1n} \\ a_{21} & a_{22} & \cdots & a_{2n} \\ \cdot & & & \\ \cdot & & & \\ \cdot & & & \\ a_{m1} & a_{m2} & \cdots & a_{mn} \end{bmatrix} \tag{A2-1}$$

where a_{ij} is the element of $A_{m\times n}$ located in the ith row and the jth column.

We shall use upper-case letters to represent matrices. When the order is immaterial in the discussion, a single letter A, without subscripts, will be used to denote a matrix. We should point out also that when the order is the only information required, expression (A2-1) may then be abbreviated as

$$A = [a_{ij}]_{m\times n} \tag{A2-2}$$

which gives essentially the same information as that in Equation (A2-1). The element a_{ij} is called the typical element or the (i,j) element. As in the case of a determinant, the first subscript i of a_{ij} indicates its row position, and the second subscript, j, shows its column position.

The reader should note that a determinant is always a square array; whereas, a matrix, say $A_{m\times n}$, is, in general, rectangular in shape unlesss $m = n$; that is, unless the number of rows equal the number of columns. In this case, $A_{m\times n}$ becomes a square matrix and a single-subscripted letter A_n may be used to indicate such information.

A2-3 EQUALITY, ADDITION, AND SUBTRACTION OF MATRICES

Some of the rules of operation for matrices are quite similar to those for the algebra of scalars. However, since matrices are of elements and not single elements, it should not be surprising if some of the rules are rather unfamiliar to the reader and are even completely different from the corresponding ones in ordinary algebra. After all, we are not dealing with the same kind of animals!

DEFINITION A2-2. Equality of Matrices. Two matrices, $A = [a_{ij}]_{m\times n}$ and $B = [b_{ij}]_{p\times q}$, are said to be *equal* if and only if

(a) A and B are of the same order: $m = p$, $n = q$; and

(b) $a_{ij} = b_{ij}$ for every pair of i and j.

In short, A and B are equal if and only if they are of the same order and their corresponding elements are equal one by one.

Before defining addition, we shall first define conformability of two matrices for addition.

DEFINITION A2-3. Conformability for Addition. Two matrices A and B are said to be *conformable* for addition if they both have the same order.

DEFINITION A2-4. Addition of Matrices. If two matrices A and B are conformable for addition and if $A = [a_{ij}]_{m \times n}$ and $B = [b_{ij}]_{m \times n}$, then the *sum*, $A + B$, is defined as $A + B = [(a_{ij} + b_{ij})]_{m \times n}$.

EXAMPLE A2-1. Consider the two matrices

$$A = \begin{bmatrix} 2 & -1 & 3 \\ -1 & 0 & 1 \end{bmatrix} \quad \text{and} \quad B = \begin{bmatrix} 1 & 3 & 0 \\ 0 & -1 & -2 \end{bmatrix}$$

The sum $A + B$ is

$$A + B = \begin{bmatrix} 2+1 & -1+3 & 3+0 \\ -1+0 & 0+(-1) & 1+(-2) \end{bmatrix}$$

or

$$A + B = \begin{bmatrix} 3 & 2 & 3 \\ -1 & -1 & -1 \end{bmatrix}$$

We shall next state the following definitions before defining subtraction of matrices.

DEFINITION A2-5. Negative of a Matrix. If $A = [a_{ij}]_{m \times n}$, then the *negative* of A, denoted by $-A$, is defined as $-A = [-a_{ij}]_{m \times n}$; that is, the negative of a matrix is obtained by taking the negative of every element of the matrix.

DEFINITION A2-6. Zero (Null) Matrix. A matrix of order $m \times n$ is called a *zero matrix* (or *null matrix*) if all its elements are equal to zero.

A zero matrix is denoted by $0_{m \times n}$; or 0_n, if it is square; or simply 0 if the order of the matrix is immaterial in the discussion.

Now we shall define subtraction of matrices.

DEFINITION A2-7. Subtraction of Matrices. If $A = [a_{ij}]_{m \times n}$ and $B = [b_{ij}]_{m \times n}$, then the *difference*, $A - B$, is defined as $A = B = A + (-B) = [(a_{ij} - b_{ij})]_{m \times n}$.

Since subtraction of matrices is defined in terms of addition, the conformability for subtraction will be defined exactly as that for addition; that is, two matrices are said to be conformable for subtraction if they are of the same order.

EXAMPLE A2-2. Let us consider once again the same two matrices, A and B, of Example A2-1. We have

$$A - B = A + (-B) = \begin{bmatrix} 2-1 & -1-3 & 3-0 \\ -1-0 & 0-(-1) & 1-(-2) \end{bmatrix}$$

or

$$A - B = \begin{bmatrix} 1 & -4 & 3 \\ -1 & 1 & 3 \end{bmatrix}$$

Similarly,

$$B - A = \begin{bmatrix} 1-2 & 3-(-1) & 0-3 \\ 0-(-1) & -1-0 & -2-1 \end{bmatrix}$$

or

$$B - A = \begin{bmatrix} -1 & 4 & -3 \\ 1 & -1 & -3 \end{bmatrix}$$

From this example, we see that

$$A - B = -(B - A) \tag{A2-3}$$

which is indeed a true statement for any two matrices A and B, provided that they are of the same order.

THEOREM A2-1. If A, B, and C are three matrices of the same order, then both the commutative and associative laws for addition hold for the three matrices; that is,

$$A + B = B + A \quad \text{(commutative law)} \tag{A2-4}$$

$$A + (B + C) = (A + B) + C \quad \text{(associative law)} \tag{A2-5}$$

A2-4 MULTIPLICATION OF MATRICES

DEFINITION A2-8. Scalar Multiple of a Matrix. The *product of a matrix* $A = [a_{ij}]_{m\times n}$ *and a scalar* k is the matrix $kA = [ka_{ij}]_{m\times n}$; that is, when a matrix A is multiplied by a scalar k, every element of A is multiplied by k.

DEFINITION A2-9. Conformability for Multiplication. Two matrices, $A = [a_{ij}]_{m\times n}$ and $B = [b_{ij}]_{p\times q}$, are said to be *conformable for multiplication* in the order AB if and only if $n = p$. In other words, A and B are conformable for multiplication in the order AB if and only if the number of columns in A is equal to the number of rows in B.

DEFINITION A2-10. Product of Two Matrices. *The product C of two matrices, $A = [a_{ij}]_{m \times n}$ and $B = [b_{ij}]_{n \times q}$,* which are conformable for multiplication in the order AB is defined by

$$A \times B = C = [c_{ij}]_{m \times q}$$

where

$$c_{ij} = a_{i1}b_{1j} + a_{i2}b_{2j} + \cdots + a_{in}b_{nj} = \sum_{k=1}^{n} a_{ik}b_{kj}$$

In other words, we say that the (i,j) element c_{ij} of the product matrix $C = A \times B$ is the sum of the products of the elements in the ith row of A and the corresponding elements in the jth column of B.

Note that the order $m \times q$ of the product C of the two matrices, A and B, in the order AB is defined by the number of rows of A and the number of columns of B.

It is worth noting that when we talk about the multiplication of two matrices, we first should examine them to see if they are conformable in the order specified. If this requirement is not satisfied, the product will not exist. We should also note that even if the two matrices are conformable in one order, they may not be conformable in the reversed order. This leads to the following theorem.

THEOREM A2-2. The commutative law for multiplication of two matrices generally does not hold; that is, in general,

$$A \times B \neq B \times A \tag{A2-6}$$

It can be shown that the commutative law for multiplication holds for the two matrices, $A = [a_{ij}]_{m \times n}$ and $B = [b_{ij}]_{p \times q}$, if and only if

(a) $m = q$ and $n = p$, and

(b) $\sum_{k=1}^{n} a_{ik}b_{kj} = \sum_{k=1}^{n} a_{kj}b_{ik}$ for every pair of i and j.

Since the commutative law, in general, does not hold for multiplication of matrices, the statement "Matrix A is multiplied by matrix B" will be very confusing because one does not know whether it means the product AB or the product BA. Thus, the following definition will be necessary.

DEFINITION A2-11. Premultiplication and Postmultiplication. In the product AB, matrix B is said to be *premultiplied* by matrix A; or A is *postmultiplied* by B.

Finally, we shall list the following properties of matrices with respect to the operation of multiplication including the result of our preceding discussion on the commutative law for multiplication. Let A, B, C be three matrices.

Property 1: The commutative law, in general, does not hold for multiplication of matrices; that is,

$$AB \neq BA$$

Property 2: The associative law holds for multiplication of matrices:

$$(AB)C = A(BC)$$

Property 3: The distributive law holds for multiplication of matrices with respect to addition:

$$A(B + C) = AB + AC$$

Property 4: The cancellation law, in general, does not hold for multiplication of matrices; that is, from

$$AB = AC \qquad \text{or} \qquad BA = CA$$

we *may not* conclude that $B = C$ even if $A \neq 0$.

Property 5: The equation, $A \cdot B = 0$, does not imply $A = 0$ or $B = 0$ (or both $A = 0$, $B = 0$) in contrast with that in ordinary algebra of scalars.

Needless to say, we have assumed that all the products and sums involved in these properties exist.

EXAMPLE A2-3. Let A and B be the following matrices:

$$A = [a_{ij}]_{2\times 3} = \begin{bmatrix} 1 & -2 & 0 \\ -1 & 3 & 2 \end{bmatrix} \qquad B = [b_{ij}]_{3\times 2} = \begin{bmatrix} 1 & -2 \\ -1 & 0 \\ 2 & 3 \end{bmatrix}$$

First, we multiply matrix A by a constant, say $k = -2$, to illustrate Definition A2-8, the scalar multiple of a matrix. From this definition, $kA = [ka_{ij}]$, we have

$$kA = (-2)\begin{bmatrix} 1 & -2 & 0 \\ -1 & 3 & 2 \end{bmatrix} = \begin{bmatrix} (-2)(1) & (-2)(-2) & (-2)(0) \\ (-2)(-1) & (-2)(3) & (-2)(2) \end{bmatrix}$$

or

$$kA = \begin{bmatrix} -2 & 4 & 0 \\ 2 & -6 & -4 \end{bmatrix}$$

We shall next demonstrate Property 1. Let $P = A \times B$ and $Q = B \times A$. Then,

$$P = [p_{ij}]_{2\times 2} = \begin{bmatrix} 1 & -2 & 0 \\ -1 & 3 & 2 \end{bmatrix} \begin{bmatrix} 1 & -2 \\ -1 & 0 \\ 2 & 3 \end{bmatrix}$$

$$= \begin{bmatrix} (1)(1) + (-2)(-1) + (0)(2) & (1)(-2) + (-2)(0) + (0)(3) \\ (-1)(1) + (3)(-1) + (2)(2) & (-1)(-2) + (3)(0) + (2)(3) \end{bmatrix}$$

or

$$P = \begin{bmatrix} 3 & -2 \\ 0 & 8 \end{bmatrix}$$

Next,

$$Q = [q_{ij}]_{3\times 3} = \begin{bmatrix} 1 & -2 \\ -1 & 0 \\ 2 & 3 \end{bmatrix}\begin{bmatrix} 1 & -2 & 0 \\ -1 & 3 & 2 \end{bmatrix}$$

$$= \begin{bmatrix} (1)(1) + (-2)(-1) & (1)(-2) + (-2)(3) & (1)(0) + (-2)(2) \\ (-1)(1) + (0)(-1) & (-1)(-2) + (0)(3) & (-1)(0) + (0)(2) \\ (2)(1) + (3)(-1) & (2)(-2) + (3)(3) & (2)(0) + (3)(2) \end{bmatrix}$$

or

$$Q = \begin{bmatrix} 3 & -8 & -4 \\ -1 & 2 & 0 \\ -1 & 5 & 6 \end{bmatrix}$$

We see that the products P and Q do not even have the same order. This clearly demonstrates that the commutative law generally does not apply for multiplication of matrices.

EXAMPLE A2-4. To illustrate the associative law (Property 2), let A and B be the same two matrices as given in Example A2-3, and let

$$C = [c_{ij}]_{2\times 3} = \begin{bmatrix} -2 & 1 & -1 \\ 1 & 3 & 0 \end{bmatrix}$$

From Example A2-3, we have

$$A \times B = P = \begin{bmatrix} 3 & -2 \\ 0 & 8 \end{bmatrix}$$

so that

$$(AB)C = P \times C = \begin{bmatrix} 3 & -2 \\ 0 & 8 \end{bmatrix}\begin{bmatrix} -2 & 1 & -1 \\ 1 & 3 & 0 \end{bmatrix}$$

$$= \begin{bmatrix} (-6 - 2) & (3 - 6) & (-3 + 0) \\ (0 + 8) & (0 + 24) & (0 + 0) \end{bmatrix}$$

or

$$(AB)C = \begin{bmatrix} -8 & -3 & -3 \\ 8 & 24 & 0 \end{bmatrix}$$

Next, we form $B \times C$:

$$B \times C = \begin{bmatrix} 1 & -2 \\ -1 & 0 \\ 2 & 3 \end{bmatrix}\begin{bmatrix} -2 & 1 & -1 \\ 1 & 3 & 0 \end{bmatrix} = \begin{bmatrix} (-2 - 2) & (1 - 6) & (-1 + 0) \\ (2 + 0) & (-1 + 0) & (1 + 0) \\ (-4 + 3) & (2 + 9) & (-2 + 0) \end{bmatrix}$$

or

$$B \times C = \begin{bmatrix} -4 & -5 & -1 \\ 2 & -1 & 1 \\ -1 & 11 & -2 \end{bmatrix}$$

Then

$$A(BC) = \begin{bmatrix} 1 & -2 & 0 \\ -1 & 3 & 2 \end{bmatrix} \begin{bmatrix} -4 & -5 & -1 \\ 2 & -1 & 1 \\ -1 & 11 & -2 \end{bmatrix}$$
$$= \begin{bmatrix} (-4 - 4 + 0) & (-5 + 2 + 0) & (-1 - 2 + 0) \\ (4 + 6 - 2) & (5 - 3 + 22) & (1 + 3 - 4) \end{bmatrix}$$

or

$$A(BC) = \begin{bmatrix} -8 & -3 & -3 \\ 8 & 24 & 0 \end{bmatrix}$$

Thus we see clearly from this example that

$$(AB)C = A(BC)$$

EXAMPLE A2-5. Let A, B, and C be the following matrices of order 2:

$$A = \begin{bmatrix} 1 & -1 \\ -2 & 1 \end{bmatrix} \quad B = \begin{bmatrix} -1 & 3 \\ 0 & 2 \end{bmatrix} \quad C = \begin{bmatrix} 2 & 0 \\ 3 & -1 \end{bmatrix}$$

Then,

$$B + C = \begin{bmatrix} (-1 + 2) & (3 + 0) \\ (0 + 3) & (2 - 1) \end{bmatrix} = \begin{bmatrix} 1 & 3 \\ 3 & 1 \end{bmatrix}$$

and

$$A(B + C) = \begin{bmatrix} 1 & -1 \\ -2 & 1 \end{bmatrix} \begin{bmatrix} 1 & 3 \\ 3 & 1 \end{bmatrix} = \begin{bmatrix} (1 - 3) & (3 - 1) \\ (-2 + 3) & (-6 + 1) \end{bmatrix}$$

or

$$A(B + C) = \begin{bmatrix} -2 & 2 \\ 1 & -5 \end{bmatrix}$$

Next we form the products AB and AC:

$$AB = \begin{bmatrix} 1 & -1 \\ -2 & 1 \end{bmatrix} \begin{bmatrix} -1 & 3 \\ 0 & 2 \end{bmatrix} = \begin{bmatrix} (-1 + 0) & (3 - 2) \\ (2 + 0) & (-6 + 2) \end{bmatrix} = \begin{bmatrix} -1 & 1 \\ 2 & -4 \end{bmatrix}$$

$$AC = \begin{bmatrix} 1 & -1 \\ -2 & 1 \end{bmatrix} \begin{bmatrix} 2 & 0 \\ 3 & -1 \end{bmatrix} = \begin{bmatrix} (2 - 3) & (0 + 1) \\ (-4 + 3) & (0 - 1) \end{bmatrix} = \begin{bmatrix} -1 & 1 \\ -1 & -1 \end{bmatrix}$$

Finally, we have

$$AB + AC = \begin{bmatrix} -1 & 1 \\ 2 & -4 \end{bmatrix} + \begin{bmatrix} -1 & 1 \\ -1 & -1 \end{bmatrix}$$

$$= \begin{bmatrix} (-1 - 1) & (1 + 1) \\ (2 - 1) & (-4 - 1) \end{bmatrix} = \begin{bmatrix} -2 & 2 \\ 1 & -5 \end{bmatrix}$$

which indeed is equal to $A(B + C)$, shown previously.

EXAMPLE A2-6. To show that the cancellation law generally does not hold for the multiplication of matrices (Property 4), let A, B, C be the following matrices:

$$A = \begin{bmatrix} 1 & 1 \\ 1 & 1 \end{bmatrix} \qquad B = \begin{bmatrix} 1 & -1 \\ -1 & 1 \end{bmatrix} \qquad C = \begin{bmatrix} -1 & 1 \\ 1 & -1 \end{bmatrix}$$

so that

$$AB = \begin{bmatrix} 1 & 1 \\ 1 & 1 \end{bmatrix}\begin{bmatrix} 1 & -1 \\ -1 & 1 \end{bmatrix} = \begin{bmatrix} 0 & 0 \\ 0 & 0 \end{bmatrix}$$

and

$$AC = \begin{bmatrix} 1 & 1 \\ 1 & 1 \end{bmatrix}\begin{bmatrix} -1 & 1 \\ 1 & -1 \end{bmatrix} = \begin{bmatrix} 0 & 0 \\ 0 & 0 \end{bmatrix}$$

We see clearly that $B \neq C$ even though the products AB and AC are equal.

The fact that $AB = 0$ does not imply $A = 0$ or $B = 0$ — or both $A = 0$ and $B = 0$ — (Property 5) has already been illustrated by observing the two products AB and AC in this example. Note that none of the three matrices is a zero matrix, and yet a zero product can result by multiplying two nonzero matrices. This property is one of the examples that reveal the differences between the matrix algebra and the ordinary algebra of scalars. We may mention that there are special matrices for which the commutative law for multiplication holds.

EXAMPLE A2-7. Let

$$A = \begin{bmatrix} 1 & 2 \\ -1 & 1 \end{bmatrix} \qquad B = \begin{bmatrix} 1 & 0 \\ 0 & 1 \end{bmatrix}$$

so that

$$AB = \begin{bmatrix} 1 & 2 \\ -1 & 1 \end{bmatrix}\begin{bmatrix} 1 & 0 \\ 0 & 1 \end{bmatrix} = \begin{bmatrix} 1 & 2 \\ -1 & 1 \end{bmatrix}$$

and

$$BA = \begin{bmatrix} 1 & 0 \\ 0 & 1 \end{bmatrix}\begin{bmatrix} 1 & 2 \\ -1 & 1 \end{bmatrix} = \begin{bmatrix} 1 & 2 \\ -1 & 1 \end{bmatrix}$$

Obviously, $AB = BA$ and the commutative law indeed holds for these two matrices! In fact, matrix B is a very special matrix called the *unit* (or *identity*) *matrix*.

A2-5 TRANSPOSE AND DETERMINANT OF A MATRIX

DEFINITION A2-12. Transpose of a Matrix. The matrix A^T of order $n \times m$, obtained by interchanging the rows and columns of an $m \times n$ matrix A, is defined as the *transpose* of A. In symbolic terms, if $A = [a_{ij}]_{m\times n}$, then $A^T = [a_{ji}]_{n\times m}$.

EXAMPLE A2-8. Let A, B, and C be the following matrices:

$$A = \begin{bmatrix} 3 & 1 & -2 \\ 0 & 2 & 4 \end{bmatrix} \quad B = \begin{bmatrix} -1 & 0 & 1 \\ 2 & -1 & 0 \end{bmatrix} \quad C = \begin{bmatrix} 1 & -1 \\ 0 & 1 \\ 2 & 3 \end{bmatrix}$$

The transposes of the three matrices are

$$A^T = \begin{bmatrix} 3 & 0 \\ 1 & 2 \\ -2 & 4 \end{bmatrix} \quad B^T = \begin{bmatrix} -1 & 2 \\ 0 & -1 \\ 1 & 0 \end{bmatrix} \quad C^T = \begin{bmatrix} 1 & 0 & 2 \\ -1 & 1 & 3 \end{bmatrix}$$

Now, taking the transpose of A^T, we get

$$(A^T)^T = \begin{bmatrix} 3 & 1 & -2 \\ 0 & 2 & 4 \end{bmatrix}$$

which becomes the original matrix A.

Let us multiply A and A^T by a constant, say -2:

$$-2A = \begin{bmatrix} -6 & -2 & 4 \\ 0 & -4 & -8 \end{bmatrix} \quad \text{and} \quad -2A^T = \begin{bmatrix} -6 & 0 \\ -2 & -4 \\ 4 & -8 \end{bmatrix}$$

Taking the transpose of $-2A$ we have

$$(-2A)^T = \begin{bmatrix} -6 & 0 \\ -2 & -4 \\ 4 & -8 \end{bmatrix}$$

which is seen to be equal to $-2A^T$.

Next we form the sum of A and B:

$$A + B = \begin{bmatrix} 2 & 1 & -1 \\ 2 & 1 & 4 \end{bmatrix}$$

the transpose of which is

$$(A + B)^T = \begin{bmatrix} 2 & 2 \\ 1 & 1 \\ -1 & 4 \end{bmatrix}$$

But, if we take the sum of A^T and B^T we get

$$A^T + B^T = \begin{bmatrix} 3 & 0 \\ 1 & 2 \\ -2 & 4 \end{bmatrix} + \begin{bmatrix} -1 & 2 \\ 0 & -1 \\ 1 & 0 \end{bmatrix} = \begin{bmatrix} 2 & 2 \\ 1 & 1 \\ -1 & 4 \end{bmatrix}$$

which is the same as $(A + B)^T$. Finally, premultiplying C by A, we have

$$AC = \begin{bmatrix} 3 & 1 & -2 \\ 0 & 2 & 4 \end{bmatrix} \begin{bmatrix} 1 & -1 \\ 0 & 1 \\ 2 & 3 \end{bmatrix} = \begin{bmatrix} -1 & -8 \\ 8 & 14 \end{bmatrix}$$

the transpose of which is

$$(AC)^T = \begin{bmatrix} -1 & 8 \\ -8 & 14 \end{bmatrix}$$

Now, premultiplying C^T by A^T we get

$$A^T C^T = \begin{bmatrix} 3 & 0 \\ 1 & 2 \\ -2 & 4 \end{bmatrix} \begin{bmatrix} 1 & 0 & 2 \\ -1 & 1 & 3 \end{bmatrix} = \begin{bmatrix} 3 & 0 & 6 \\ -1 & 2 & 8 \\ -6 & 4 & 8 \end{bmatrix}$$

which obviously is *not* equal to the matrix $(AC)^T$ for the two matrices are not even of the same order. In fact, the product, $A^T C^T$, will not even exist if the number of rows of A is not equal to the number of columns of C. But, if C^T is postmultiplied by A^T, we have

$$C^T A^T = \begin{bmatrix} 1 & 0 & 2 \\ -1 & 1 & 3 \end{bmatrix} \begin{bmatrix} 3 & 0 \\ 1 & 2 \\ -2 & 4 \end{bmatrix} = \begin{bmatrix} -1 & 8 \\ -8 & 14 \end{bmatrix}$$

which is indeed equal to $(AC)^T$.

Example A2-8 suggests some of the properties of the transpose of matrices which are listed as follows:

Property 1: The transpose of a matrix always exists.

Property 2: The orders of a matrix and its transpose are the same if and only if the matrix is a square matrix.

Property 3: The transpose of the transpose of a matrix is the matrix itself; that is, $(A^T)^T = A$.

Property 4: The transpose of the scalar multiple of a matrix is equal to the scalar multiple of its transpose; that is, $(kA)^T = kA^T$.

Property 5: The transpose of the sum of two matrices is equal to the sum of their transposes. Thus, if A and B are two matrices of the same order, we have $(A + B)^T = A^T + B^T$.

Property 6: The transpose of the product of two matrices is equal to the product of the two transposes *in the reverse order*. That is, if A and B are conformable for multiplication, we write $(AB)^T = B^T A^T$.

Most of the proofs of these properties are simple and straightforward. For example, to prove Property 5, let $A = [a_{ij}]_{m\times n}$ and $B = [b_{ij}]_{m\times n}$. Then, by Definition A2-12, the transposes are $A^T = [a_{ji}]_{n\times m}$, $B^T = [b_{ji}]_{n\times m}$.

The sum $C = A + B$ is given by

$$C = A + B = [a_{ij}]_{m\times n} + [b_{ij}]_{m\times n} = [(a_{ij} + b_{ij})]_{m\times n} = [c_{ij}]_{m\times n}$$

where $c_{ij} = a_{ij} + b_{ij}$ for all i and j.

The transpose of C is

$$C^T = [c_{ij}]^T_{m\times n} = [c_{ji}]_{n\times m}$$

where $c_{ji} = a_{ji} + b_{ji}$ for all j and i.

Now, the sum of the transposes is

$$A^T + B^T = [a_{ji}]_{n\times m} + [b_{ji}]_{n\times m} = [(a_{ji} + b_{ji})]_{n\times m}$$

which is seen to be equal to C^T. Therefore,

$$C^T = (A + B)^T = A^T + B^T$$

DEFINITION A2-13. Determinant of a Square Matrix. Let A be a square matrix of order n. Then the *determinant of A*, denoted by det A, is the determinant of n^2 elements arranged in n rows and n columns in exactly the same way as the elements appear in A; that is, if

$$\mathrm{A} = [a_{ij}]_{n\times n} = \begin{bmatrix} a_{11} & a_{12} & \cdots & a_{1n} \\ a_{21} & a_{22} & \cdots & a_{2n} \\ \cdot & & & \\ \cdot & & & \\ \cdot & & & \\ a_{n1} & a_{n2} & \cdots & a_{nn} \end{bmatrix}$$

then,

$$\det A = \det\, [a_{ij}]_{n\times n} = \begin{bmatrix} a_{11} & a_{12} & \cdots & a_{1n} \\ a_{21} & a_{22} & \cdots & a_{2n} \\ \cdot & & & \\ \cdot & & & \\ \cdot & & & \\ a_{n1} & a_{n2} & \cdots & a_{nn} \end{bmatrix}$$

It should be pointed out that only square matrices have corresponding determinants since all determinants are, by definition, square arrays. More important, a determinant of numbers, after being evaluated, can be reduced to a numerical value, whereas a matrix never has a value as such. This is a very important point, which can be rather confusing to those who encounter matrix algebra and the theory of determinants for the first time.

A2-6 SPECIAL TYPES OF MATRICES

DEFINITION A2-14. Symmetric Matrix. A *square* matrix $S_n = [s_{ij}]_n$ of order n is called a *symmetric* matrix if and only if $s_{ij} = s_{ji}$ for every pair of i and j.

From this definition it is seen that a matrix A is symmetric if and only if $A = A^T$.

DEFINITION A2-15. Diagonal Matrix. A *square* matrix $D_n = [d_{ij}]_n$ of order n is called a *diagonal* matrix if and only if $d_{ij} = 0$ for $i \neq j$.

Note that in this definition there are no restrictions on d_{ii} and, hence, the elements on the major diagonal can take in any value, including zero. However, the elements which are not on the major diagonal must *all* be zero in a diagonal matrix.

DEFINITION A2-16. Unit (or Identity) Matrix. A *square* matrix $U_n = [u_{ij}]_n$ of order n is defined as a *unit matrix* if and only if $u_{ii} = 1$ for all i, and $u_{ij} = 0$ for all i and j where $i \neq j$. (Note that I_n is used to denote an identity matrix. Thus $I_n \equiv U_n$, and the two symbols represent the same matrix.)

Thus, a unit matrix is a special type of diagonal matrix in which the major diagonal elements are all equal to 1.

The identity matrix has the similar property as the identity element in ordinary algebra since, for any matrix A, it is true that $AI = IA$ provided, of course, that the product AI and IA exist. Because of the possible difference in their orders, two identity matrices, I_m and I_n, may not be equal unless $m = n$. The following example will illustrate the points discussed in this section thus far.

EXAMPLE A2-9. Consider the following matrices:

$$A = \begin{bmatrix} 1 & -1 & 0 \\ -1 & 2 & 3 \\ 0 & 3 & -1 \end{bmatrix} \quad B = \begin{bmatrix} 1 & 0 & 0 \\ 0 & 0 & 0 \\ 0 & 0 & -2 \end{bmatrix} \quad I = \begin{bmatrix} 1 & 0 & 0 \\ 0 & 1 & 0 \\ 0 & 0 & 1 \end{bmatrix}$$

A is a symmetric matrix, B is a diagonal matrix, and I is a unit matrix — all of order 3. Note that in B, $b_{11} = 1$, $b_{22} = 0$, $b_{33} = -2$ and all the other elements are equal to zero, illustrating the fact that the major diagonal elements may be of any value in a diagonal matrix.

Next, we form the products AI and IA:

$$AI = \begin{bmatrix} 1 & -1 & 0 \\ -1 & 2 & 3 \\ 0 & 3 & -1 \end{bmatrix} \begin{bmatrix} 1 & 0 & 0 \\ 0 & 1 & 0 \\ 0 & 0 & 1 \end{bmatrix} = \begin{bmatrix} 1 & -1 & 0 \\ -1 & 2 & 3 \\ 0 & 3 & -1 \end{bmatrix} = A$$

$$IA = \begin{bmatrix} 1 & 0 & 0 \\ 0 & 1 & 0 \\ 0 & 0 & 1 \end{bmatrix} \begin{bmatrix} 1 & -1 & 0 \\ -1 & 2 & 3 \\ 0 & 3 & -1 \end{bmatrix} = \begin{bmatrix} 1 & -1 & 0 \\ -1 & 2 & 3 \\ 0 & 3 & -1 \end{bmatrix} = A$$

We see that $AI = IA = A$, resembling the property $a \cdot 1 = 1 \cdot a = a$ for any number a in ordinary algebra. In general, equation $AI = IA = A$ is valid for any square matrix A of an arbitrary order n as long as the identity matrix is of the same order so that the products AI and IA exist. If A is not a square matrix and if the order of A is mn, we can still write

$$I_m A = A I_n = A$$

for the obvious reason.

DEFINITION A2-17. Real Matrix. A matrix $R = [r_{ij}]_{m \times n}$ is called a *real matrix* if and only if r_{ij} is real for every pair of i and j; that is, R is a real matrix if and only if *all* its elements are real.

In general, the elements of a matrix may be real or complex (or even functions). If one of its elements is complex, the matrix is a "complex" matrix.

DEFINITION A2-18. Conjugate of a Matrix. Let $A = [a_{ij}]_{m \times n}$ be a matrix of complex elements of order mn. If every element $a_{ij} = \alpha_{ij} + \beta_{ij}\sqrt{-1}$ (α, β real) of A is replaced by its complex conjugate $\bar{a}_{ij} = \alpha_{ij} - \beta_{ij}\sqrt{-1}$, the resulting matrix, denoted by $\bar{A}$, is defined as the *conjugate of A*.

With the help of Definition A2-18, a real matrix may be redefined as a matrix R such that $R = \bar{R}$.

DEFINITION A2-19. Hermitian Matrix. A *square* matrix $H = [h_{ij}]$ is called a *Hermitian matrix* if and only if $h_{ij} = \bar{h}_{ji}$ for every pair of i and j. Stated in a different way, a Hermitian matrix H is one such that $H = (\bar{H})^T$.

It is interesting to note that when all the elements are real, a Hermitian matrix is reduced to a real symmetric matrix, which can easily be verified. Real symmetric matrices will be encountered quite frequently in later developments.

DEFINITION A2-20. Scalar Matrix. A *scalar matrix* is defined as a diagonal matrix such that all the elements in the major diagonal are equal to a scalar. In other words, a square matrix $K = [k_{ij}]_n$ of order n is called a diagonal matrix if and only if $k_{ii} = \alpha$, a scalar for all i; and $k_{ij} = 0$ for all $i \neq j$.

DEFINITION A2-21. Row and Column Matrices. A *row matrix* of m elements is a matrix of order $1 \times m$; and a *column matrix* of n elements is a matrix of order $n \times 1$ where m and n may be any positive integer. A column matrix of n elements is also called an *n-vector*.

EXAMPLE A2-10. Let A be the following matrix:

$$A = \begin{bmatrix} a + jb & c + jd \\ g + jh & p + jq \end{bmatrix}$$

where a, b, c, d, g, h, p, q are real numbers and $j = \sqrt{-1}$.
The conjugate of A is the following matrix:

$$\bar{A} = \begin{bmatrix} a - jb & c - jd \\ g - jh & p - jq \end{bmatrix}$$

Case 1: Let $b = d = h = q = 0$ so that

$$A = \bar{A} = \begin{bmatrix} a & c \\ g & p \end{bmatrix}$$

which is a real matrix.

Case 2: Let $b = q = 0$, $c = g$ and $d = -h$ so that a_{11} and a_{22} are real and a_{12} and a_{21} are complex conjugates. Under these conditions, $A = (\bar{A}^T)$ and thus becomes a Hermitian matrix. To verify this, let $a = 1$, $p = 2$, $b = q = 0$, $c = g = -1$ and $d = -h = 3$, so that

$$A = \begin{bmatrix} 1 & -1 + j3 \\ -1 - j3 & 2 \end{bmatrix} \qquad A^T = \begin{bmatrix} 1 & -1 - j3 \\ -1 + j3 & 2 \end{bmatrix}$$

$$(\bar{A})^T = \begin{bmatrix} 1 & -1 + j3 \\ -1 - j3 & 2 \end{bmatrix}$$

Hence, $A = (\bar{A})^T$ and, by definition, it is a Hermitian matrix.

EXAMPLE A2-11. Let A, B, and C be the following matrices:

$$A = [-1 \quad 2 \quad 0 \quad 1], \qquad B = \begin{bmatrix} 2 \\ 3 \\ 1 \\ -1 \end{bmatrix} \qquad C = \begin{bmatrix} 0 \\ 1 \\ 2 \\ -1 \end{bmatrix}$$

By Definition A2-21, A is a row matrix of four elements and B and C are column matrices of four elements, or simply 4-vectors. Now let us obtain the products AB and B^TC:

$$AB = [-1 \quad 2 \quad 0 \quad 1]\begin{bmatrix} 2 \\ 3 \\ 1 \\ -1 \end{bmatrix} = [3]$$

$$B^TC = [2 \quad 3 \quad 1 \quad -1]\begin{bmatrix} 0 \\ 1 \\ 2 \\ -1 \end{bmatrix} = [6]$$

Both products are 1×1 matrices or 1-vectors. In the case of $B^T C$, we say that 6 is the *scalar product* of the two 4-vectors B and C.

DEFINITION A2-22. Singular and Nonsingular Matrices. A *singular matrix* is a square matrix whose determinant is equal to zero, whereas a *nonsingular matrix* is a square matrix whose determinant is different from zero.

The most obvious examples of nonsingular and singular matrices are the *unit* matrices and the *square* zero matrices, respectively.

We shall devote the next section to the discussion of the inverse of a matrix.

A2-7 INVERSE OF A MATRIX

In many ways matrix algebra and the algebra of scalar are quite similar, as witnessed in the previous discussions. Once again, we shall define a new matrix, called the *inverse matrix*, which will take the part of the reciprocal, or inverse, of a given matrix in matrix algebra. Thus we give the following definition:

DEFINITION A2-23. Inverse. The *inverse* of a given matrix A is defined as a matrix A^{-1} such that the products of the two matrices in both orders yield the identity matrix I; that is,

$$A \cdot A^{-1} = A^{-1} \cdot A = I \tag{A2-7}$$

From this defining equation, (A2-7), we see that a matrix must be square in order to have an inverse. (Why?) Also, it is not difficult to establish that

$$A^{-1} = \left(\frac{1}{\det A}\right) \cdot [A_{ij}]^T = \left[\frac{A_{ij}}{\det A}\right]^T = \begin{bmatrix} \dfrac{A_{11}}{\det A} & \dfrac{A_{21}}{\det A} & \cdots & \dfrac{A_{n1}}{\det A} \\ \dfrac{A_{12}}{\det A} & \dfrac{A_{22}}{\det A} & \cdots & \dfrac{A_{n2}}{\det A} \\ \cdot & & & \\ \cdot & & & \\ \cdot & & & \\ \dfrac{A_{1n}}{\det A} & \dfrac{A_{2n}}{\det A} & \cdots & \dfrac{A_{nn}}{\det A} \end{bmatrix} \tag{A2-8}$$

where A_{ij} is the cofactor of a_{ij}, so that

$$[A_{ij}]^T = \begin{bmatrix} A_{11} & A_{12} & \cdots & A_{1n} \\ A_{21} & A_{22} & \cdots & A_{2n} \\ \cdot & & & \\ \cdot & & & \\ \cdot & & & \\ A_{n1} & A_{n2} & \cdots & A_{nn} \end{bmatrix}^T = \begin{bmatrix} A_{11} & A_{21} & \cdots & A_{n1} \\ A_{12} & A_{22} & \cdots & A_{n2} \\ \cdot & & & \\ \cdot & & & \\ \cdot & & & \\ A_{1n} & A_{2n} & \cdots & A_{nn} \end{bmatrix} \tag{A2-9}$$

The matrix $[A_{ij}]^T$ defined in Equation (A2-9) is known as the *adjoint* of the given matrix A, which we shall define next.

DEFINITION A2-24. Adjoint. The *adjoint* of a given matrix A is the matrix $\mathcal{A}$, which is the transpose of a matrix obtained by replacing every element a_{ij} of A by its cofactor A_{ij}; that is,

$$\mathcal{A} = [A_{ij}]^T = \begin{bmatrix} A_{11} & A_{21} & \cdots & A_{n1} \\ A_{12} & A_{22} & \cdots & A_{n2} \\ \cdot & & & \\ \cdot & & & \\ \cdot & & & \\ A_{1n} & A_{2n} & \cdots & A_{nn} \end{bmatrix} \tag{A2-10}$$

From Equations (A2-8) and (A2-10) it is evident that

$$A^{-1} = \frac{1}{\det A} \cdot \mathcal{A} \tag{A2-11}$$

which gives the relationship between the inverse and the adjoint of a given matrix.

EXAMPLE A2-12. Consider the following matrix:

$$A = \begin{bmatrix} 1 & 0 & -1 \\ -1 & 1 & 0 \\ 0 & 1 & 2 \end{bmatrix}$$

We shall find the inverse of A using Equations (A2-10) and (A2-11):

$$\det A = \begin{vmatrix} 1 & 0 & -1 \\ -1 & 1 & 0 \\ 0 & 1 & 2 \end{vmatrix} = 2 + 1 + 0 - 0 - 0 - 0 = 3$$

$$A_{11} = \begin{vmatrix} 1 & 0 \\ 1 & 2 \end{vmatrix} = 2 - 0 = 2 \qquad A_{21} = -\begin{vmatrix} 0 & -1 \\ 1 & 2 \end{vmatrix} = -(0 + 1) = -1$$

$$A_{31} = \begin{vmatrix} 0 & -1 \\ 1 & 0 \end{vmatrix} = 0 + 1 = 1$$

$$A_{12} = -\begin{vmatrix} -1 & 0 \\ 0 & 2 \end{vmatrix} = -(-2 - 0) = 2 \qquad A_{22} = \begin{vmatrix} 1 & -1 \\ 0 & 2 \end{vmatrix} = 2 - 0 = 2$$

$$A_{32} = -\begin{vmatrix} 1 & -1 \\ -1 & 0 \end{vmatrix} = -(0 - 1) = 1$$

$$A_{13} = \begin{vmatrix} -1 & 1 \\ 0 & 1 \end{vmatrix} = -1 - 0 = -1 \qquad A_{23} = -\begin{vmatrix} 1 & 0 \\ 0 & 1 \end{vmatrix} = -(1 - 0) = -1$$

$$A_{33} = \begin{vmatrix} 1 & 0 \\ -1 & 1 \end{vmatrix} = 1 - 0 = 1$$

The adjoint of A is, by Equation (A2-10),

$$\mathcal{A} = \begin{bmatrix} A_{11} & A_{21} & A_{31} \\ A_{12} & A_{22} & A_{32} \\ A_{13} & A_{23} & A_{33} \end{bmatrix} = \begin{bmatrix} 2 & -1 & 1 \\ 2 & 2 & 1 \\ -1 & -1 & 1 \end{bmatrix}$$

We shall list some of the properties of matrices concerning the inverse matrices just discussed.

Property 1: The inverse of a matrix, if it exists, is unique.

Property 2: The inverse of a matrix exists if and only if (a) the matrix is square and (b) the determinant of the matrix is different from zero; that is, the matrix is nonsingular.

Of course, the second condition implies the first since, by definition, both singular and nonsingular matrices must be square matrices.[1]

Property 3: If A is a nonsingular matrix (so that A^{-1} exists), then

$$(A^T)^{-1} = (A^{-1})^T \tag{A2-12}$$

Property 4: If A and B are two nonsingular matrices of the same order, then

$$(A \cdot B)^{-1} = B^{-1} \cdot A^{-1} \tag{A2-13}$$

Some of these above properties are quite evident. For example, to show that Property 1 is valid, let us assume that both matrices, B and C, are the inverses of a given matrix A. Then, by Definition A2-23,

$$AB = BA = I$$
$$AC = CA = I$$

so that we may conclude that $AB = AC$ since both products are equal to the same matrix I. Now, premultiplying this last equation by B, we get

$$B(AB) = B(AC)$$

or

$$(BA)B = (BA)C \qquad \text{(due to associative law)}$$

or

$$IB = IC \qquad \text{(due to definition of an inverse)}$$

or

$$B = C \qquad \text{(due to the property of the identity matrix)}$$

which shows the uniqueness of the inverse. Property 2 follows directly from Equation (A2-8). Obviously, if the given matrix A is not a square matrix, det

A will not exist. Furthermore, even if A is square, but if det $A = 0$, then the inverse, A^{-1} in Equation (A2-8), will clearly be undefined, because the adjoint will be divided by 0, which is not allowed.

A2-8 RANKS, ELEMENTARY TRANSFORMATIONS, AND EQUIVALENCE

DEFINITION A2-25. Submatrix of a Matrix. A *submatrix* of a given matrix is the resulting matrix that is obtained by deleting certain rows and/or columns of the given matrix.

DEFINITION A2-26. Rank of a Matrix. The *rank* of a matrix A of order $m \times n$ is defined to be the order of the determinant of a highest-order nonsingular submatrix A_r of A. We denote the rank of A by $R(A) = r$ and say that matrix A is of rank r.

We should remark that there may be *more than one* nonsingular submatrix of A that are of the same maximum order r. But, by this definition, there should be no nonsingular submatrix of A which is of order higher than r if A is of rank r.

EXAMPLE A2-13. Consider the following matrix:

$$A = \begin{bmatrix} 1 & -3 & -8 & -2 \\ 2 & 5 & 6 & -4 \\ 3 & 1 & -4 & -6 \end{bmatrix}$$

The four submatrices of order 3 of A are

$$\begin{bmatrix} 1 & -3 & -8 \\ 2 & 5 & 6 \\ 3 & 1 & -4 \end{bmatrix} \qquad \begin{bmatrix} 1 & -3 & -2 \\ 2 & 5 & -4 \\ 3 & 1 & -6 \end{bmatrix}$$

$$\begin{bmatrix} 1 & -8 & -2 \\ 2 & 6 & -4 \\ 3 & -4 & -6 \end{bmatrix} \qquad \begin{bmatrix} -3 & -8 & -2 \\ 5 & 6 & -4 \\ 1 & -4 & -6 \end{bmatrix}$$

It can be easily verified that the determinant of each of the four matrices above is zero. Hence, $R(A) \leq 2$. But the determinant of the submatrix, obtained by deleting row 3 and columns 3 and 4 of A, is

$$\begin{vmatrix} 1 & -3 \\ 2 & 5 \end{vmatrix} = 5 + 6 = 11$$

Since 2 is the order of the determinant of a highest-order, nonsingular submatrix of A, namely,

$$\begin{bmatrix} 1 & -3 \\ 2 & 5 \end{bmatrix}$$

the rank of A is 2, or $R(A) = 2$.

EXAMPLE A2-14. Let B be the following matrix:

$$B = \begin{bmatrix} 1 & 0 & 0 \\ 0 & 2 & -1 \\ -1 & 0 & 1 \end{bmatrix}$$

The determinant of B is

$$\det B = \begin{vmatrix} 1 & 0 & 0 \\ 0 & 2 & -1 \\ -1 & 0 & 1 \end{vmatrix} = 2$$

Since $B \neq 0$, we have, by Definition A2-26, $R(B) = 3$.

From the two examples, we see that if A is a matrix of order $m \times n$, the rank of A is, at most, equal to the smaller of the two numbers, m and n. Also, another point should be obvious to the reader from the foregoing examples: If m and n are large numbers — that is, if there are quite a few rows and columns in A — it could be extremely laborious to find out the rank of A by definition because it might involve the evaluation of a large number of determinants, possibly of different orders. For this reason, we need to seek a better way of determining the rank of a given matrix in the sense that it will be more systematic and less time-consuming. That is exactly what we are going to do; but, first, let us examine the following definition:

DEFINITION A2-27. Elementary Transformations of a Matrix. The following three operations are called the *elementary transformations* of a matrix:

(1) Interchange of any two rows (or columns).

(2) Multiplication of all the elements of any row (or column) by a nonzero constant.

(3) Addition of the elements of a row (or column) multiplied by a nonzero constant to the corresponding elements of another row (or column).

THEOREM A2-3. The rank of a matrix will remain unaltered under (repeated applications of) any or all of the three operations defined in Definition A2-27.

DEFINITION A2-28. Equivalence of Matrices. Two matrices, A and B, are said to be *equivalent* if and only if they are of the same order and same rank. We write $A \sim B$ if they are equivalent.

It should be noted that the statement, $A \sim B$, is quite different from the statement $A = B$ because, for A and B to be equivalent, it requires only that they be of the same order and rank. On the other hand, for A and B to be equal, each pair of elements, a_{ij} and b_{ij}, in the corresponding position between A and B must be equal; namely, $a_{ij} = b_{ij}$ for all i and j.

Next we shall define the normal form of a matrix.

DEFINITION A2-29. Normal Form of a Matrix. A matrix A of order $m \times n$ and of rank r is said to be in its *normal form* if under (repeated operations of) the elementary transformations it is reduced to one and only one of the following four forms:

$$\begin{bmatrix} u_r & 0 \\ 0 & 0 \end{bmatrix}_{m\times n} \qquad [u_r \quad 0]_{m\times n} \qquad \begin{bmatrix} u_r \\ 0 \end{bmatrix}_{m\times n} \qquad [u_r]_{m\times n}$$

where u_r is a unit matrix of order r and the zeros in the matrices denote zero submatrices (possibly of different orders).

THEOREM A2-4. The rank of a given matrix is equal to the order of the unit submatrix in its normal form.

Theorem A2-4 implies that in order to find the rank of a matrix, we shall first reduce it to its normal form through the applications of the elementary transformations of Definition A2-27. Then we find $R(A) = R(U_r) = r$.

EXAMPLE A2-15. Let us consider the matrix A of Example A2-13:

$$A = \begin{bmatrix} 1 & -3 & -8 & -2 \\ 2 & 5 & 6 & -4 \\ 3 & 1 & -4 & -6 \end{bmatrix}$$

To reduce A to its normal form A_N, we shall multiply row 1 by -2 and add it to row 2 (operation 3) and multiply row 1 by -3 and add it to row 3 (operation 3) to obtain A_1 as the first step. Thus,

$$A_1 = \begin{bmatrix} 1 & -3 & -8 & -2 \\ 0 & 11 & 22 & 0 \\ 0 & 10 & 20 & 0 \end{bmatrix}$$

Next, multiplying row 2 of A_1 by $1/11$ (operation 2) and multiplying row 3 by $1/10$ (operation 2), we get A_2:

$$A_2 = \begin{bmatrix} 1 & -3 & -8 & -2 \\ 0 & 1 & 2 & 0 \\ 0 & 1 & 2 & 0 \end{bmatrix}$$

Now, multiplying row 2 of A_2 by -1 and adding it to row 3 (operation 3), we get A_3:

$$A_3 = \begin{bmatrix} 1 & -3 & -8 & -2 \\ 0 & 1 & 2 & 0 \\ 0 & 0 & 0 & 0 \end{bmatrix}$$

Then, multiplying column 1 by 3 and adding it to column 2 (operation 3); multiplying column 1 by 8 and adding it to column 3 (operation 3); and multiplying column 1 by 2 and adding it to column 4 (operation 3), we obtain A_4:

$$A_4 = \begin{bmatrix} 1 & 0 & 0 & 0 \\ 0 & 1 & 2 & 0 \\ 0 & 0 & 0 & 0 \end{bmatrix}$$

Finally, multiplying column 2 of A by -2 and adding it to column 3 (operation 3), we get A_N:

$$A_N = \left[\begin{array}{cc|cc} 1 & 0 & 0 & 0 \\ 0 & 1 & 0 & 0 \\ \hline 0 & 0 & 0 & 0 \end{array}\right] = \begin{bmatrix} U_2 & 0 \\ 0 & 0 \end{bmatrix}$$

which, of course, is identified by the first form in Definition A2-29. Thus, $R(A) = 2$, as expected.

DEFINITION A2-30. Elementary Matrix. An *elementary matrix* is a square matrix of order r that is obtained from an identity matrix of the same order to perform elementary transformations by matrix multiplication.

EXAMPLE A2-16. Consider the following matrices:

$$A = \begin{bmatrix} 1 & 0 & -2 \\ -1 & 2 & 1 \\ 3 & 1 & -1 \end{bmatrix} \qquad B = \begin{bmatrix} 1 & 0 & -2 \\ -1 & 2 & 1 \\ 1 & 5 & 1 \end{bmatrix} \qquad C = \begin{bmatrix} 5 & 0 & -2 \\ -3 & 2 & 1 \\ 5 & 1 & -1 \end{bmatrix}$$

Note that B is obtained from A by multiplying the second row of A by 2 and then adding it to the third row. Similarly, C is formed from A by multiplying the third column of A by -2 and then adding it to the first column.

Next, let us perform the very same elementary transformations on a third-order identity matrix as we just did on A in obtaining B and C. We have the elementary matrices E_1 and E_2, respectively:

$$E_1 = \begin{bmatrix} 1 & 0 & 0 \\ 0 & 1 & 0 \\ 0 & 2 & 1 \end{bmatrix} \qquad E_2 = \begin{bmatrix} 1 & 0 & 0 \\ 0 & 1 & 0 \\ -2 & 0 & 1 \end{bmatrix}$$

Note carefully that E_1 is obtained from the third-order identity matrix by multiplying its second row by 2 and adding it to the third row. E_2 is obtained by multiplying the third column of the identity matrix by -2 and adding it to the first column. Now, observe that if A is *premultiplied* by E_1, we have

$$E_1A = \begin{bmatrix} 1 & 0 & 0 \\ 0 & 1 & 0 \\ 0 & 2 & 1 \end{bmatrix} \begin{bmatrix} 1 & 0 & -2 \\ -1 & 2 & 1 \\ 3 & 1 & -1 \end{bmatrix} = \begin{bmatrix} 1 & 0 & -2 \\ -1 & 2 & 1 \\ 1 & 5 & 1 \end{bmatrix}$$

which is equal to B.

Finally, if we *postmultiply* A by E_2, we find

$$AE_2 = \begin{bmatrix} 1 & 0 & -2 \\ -1 & 2 & 1 \\ 3 & 1 & -1 \end{bmatrix} \begin{bmatrix} 1 & 0 & 0 \\ 0 & 1 & 0 \\ -2 & 0 & 1 \end{bmatrix} = \begin{bmatrix} 5 & 0 & -2 \\ -3 & 2 & 1 \\ 5 & 1 & -1 \end{bmatrix}$$

which is equal to C.

Example A2-16 suggests the following interesting properties of elementary matrices as summarized in Theorem A2-5.

THEOREM A2-5. A specified set of elementary transformations on the rows (columns) of a given matrix can be accomplished by premultiplying (postmultiplying) the matrix by an elementary matrix that is obtained from an identity matrix by performing the same set of elementary transformations on the identity matrix.

Since performing elementary transformations to a given matrix will not alter the rank of the matrix, Theorems A2-6 and A2-7 follow immediately.

THEOREM A2-6. All elementary matrices are nonsingular.

THEOREM A2-7. Multiplication of a given matrix by any number of elementary matrices will result in an equivalent matrix. In other words, multiplication by an elementary matrix will not alter the rank of a matrix.

A2-9 PARTITIONING OF MATRICES

When we defined a normal form of a matrix in Definition A2-29, we introduced the four possible forms as

$$\left[\begin{array}{c|c} U_r & 0 \\ \hline 0 & 0 \end{array}\right] \qquad [U_r \mid 0] \qquad \left[\begin{array}{c} U_r \\ \hline 0 \end{array}\right] \qquad [U_r] \qquad \text{(A2-14)}$$

The first three matrices, in fact, have been "partitioned" into submatrices. Thus we define matrix partitioning as follows.

DEFINITION A2-31. Matrix Partitioning. When a given matrix is subdivided into submatrices, the matrix is said to be *partitioned* into submatrices.

Matrix partitioning finds its usefulness in the multiplication of matrices and in working with a system of equations when the system is subdivided into subsets of equations in order to improve effectiveness in handling the equations.

EXAMPLE A2-17. Consider the following two matrices:

$$A = \begin{bmatrix} 1 & 0 & -1 & -2 \\ 0 & 1 & 0 & 0 \\ -1 & 0 & 1 & 0 \\ 3 & 0 & 0 & 1 \end{bmatrix} \qquad B = \begin{bmatrix} -1 & 1 \\ 0 & 2 \\ 1 & 0 \\ 0 & 1 \end{bmatrix}$$

Next, let us partition the two matrices as follows:

$$A = \left[\begin{array}{c|ccc} 1 & 0 & -1 & -2 \\ \hline 0 & 1 & 0 & 0 \\ -1 & 0 & 1 & 0 \\ 3 & 0 & 0 & 1 \end{array}\right] = \begin{bmatrix} A_{11} & A_{12} \\ A_{21} & A_{22} \end{bmatrix} \qquad B = \left[\begin{array}{cc} -1 & 1 \\ \hline 0 & 2 \\ 1 & 0 \\ 0 & 1 \end{array}\right] = \begin{bmatrix} B_{11} \\ B_{21} \end{bmatrix}$$

where

$$A_{11} = [1] \qquad A_{12} = [0 \quad -1 \quad -2]$$

$$A_{21} = \begin{bmatrix} 0 \\ -1 \\ 3 \end{bmatrix} \qquad A_{22} = \begin{bmatrix} 1 & 0 & 0 \\ 0 & 1 & 0 \\ 0 & 0 & 1 \end{bmatrix}$$

$$B_{11} = [-1 \quad 1] \qquad B_{21} = \begin{bmatrix} 0 & 2 \\ 1 & 0 \\ 0 & 1 \end{bmatrix}$$

Now form the product AB by partitioning:

$$A \cdot B = \begin{bmatrix} A_{11} & A_{12} \\ A_{21} & A_{22} \end{bmatrix} \begin{bmatrix} B_{11} \\ B_{21} \end{bmatrix} = \begin{bmatrix} A_{11}B_{11} + A_{12}B_{21} \\ A_{21}B_{11} + A_{22}B_{21} \end{bmatrix} \tag{A2-15}$$

But

$$A_{11}B_{11} = [1][-1 \quad 1] = [-1 \quad 1] \tag{A2-16}$$

$$A_{12}B_{21} = [0 \quad -1 \quad -2] \begin{bmatrix} 0 & 2 \\ 1 & 0 \\ 0 & 1 \end{bmatrix} = [-1 \quad -2] \tag{A2-17}$$

$$A_{21}B_{11} = \begin{bmatrix} 0 \\ -1 \\ 3 \end{bmatrix} [-1 \quad 1] = \begin{bmatrix} 0 & 0 \\ 1 & -1 \\ -3 & 3 \end{bmatrix} \tag{A2-18}$$

$$A_{22}B_{21} = \begin{bmatrix} 1 & 0 & 0 \\ 0 & 1 & 0 \\ 0 & 0 & 1 \end{bmatrix} \begin{bmatrix} 0 & 2 \\ 1 & 0 \\ 0 & 1 \end{bmatrix} = \begin{bmatrix} 0 & 2 \\ 1 & 0 \\ 0 & 1 \end{bmatrix} \tag{A2-19}$$

Substituting Equations (A2-16), (A2-17), (A2-18), and (A2-19) into Equation (A2-15), we obtain

$$AB = \begin{bmatrix} [-1 \quad 1] + [-1 \quad -2] \\ \begin{bmatrix} 0 & 0 \\ 1 & -1 \\ -3 & 3 \end{bmatrix} + \begin{bmatrix} 0 & 2 \\ 1 & 0 \\ 0 & 1 \end{bmatrix} \end{bmatrix} = \begin{bmatrix} -2 & -1 \\ 0 & 2 \\ 2 & -1 \\ -3 & 4 \end{bmatrix} \tag{A2-20}$$

Next we partition A and B differently as follows:

$$A = \left[\begin{array}{cc|cc} 1 & 0 & -1 & -2 \\ 0 & 1 & 0 & 0 \\ \hline -1 & 0 & 1 & 0 \\ 3 & 0 & 0 & 1 \end{array}\right] = \begin{bmatrix} A_{11} & A_{12} \\ A_{21} & A_{22} \end{bmatrix} \qquad B = \left[\begin{array}{cc} -1 & 1 \\ 0 & 2 \\ \hline 1 & 0 \\ 0 & 1 \end{array}\right] = \begin{bmatrix} B_{11} \\ B_{21} \end{bmatrix}$$

Multiplying these two partitioned matrices in the order AB, we still get the same equation as (A2-15) this time, however,

$$A_{11} = \begin{bmatrix} 1 & 0 \\ 0 & 1 \end{bmatrix} \qquad A_{12} = \begin{bmatrix} -1 & -2 \\ 0 & 0 \end{bmatrix}$$

$$A_{21} = \begin{bmatrix} -1 & 0 \\ 3 & 0 \end{bmatrix} \qquad A_{22} = \begin{bmatrix} 1 & 0 \\ 0 & 1 \end{bmatrix}$$

$$B_{11} = \begin{bmatrix} -1 & 1 \\ 0 & 2 \end{bmatrix} \qquad B_{21} = \begin{bmatrix} 1 & 0 \\ 0 & 1 \end{bmatrix}$$

so that

$$A_{11}B_{11} = \begin{bmatrix} 1 & 0 \\ 0 & 1 \end{bmatrix}\begin{bmatrix} -1 & 1 \\ 0 & 2 \end{bmatrix} = \begin{bmatrix} -1 & 1 \\ 0 & 2 \end{bmatrix} \tag{A2-21}$$

$$A_{12}B_{21} = \begin{bmatrix} -1 & -2 \\ 0 & 0 \end{bmatrix}\begin{bmatrix} 1 & 0 \\ 0 & 1 \end{bmatrix} = \begin{bmatrix} -1 & -2 \\ 0 & 0 \end{bmatrix} \tag{A2-22}$$

$$A_{21}B_{11} = \begin{bmatrix} -1 & 0 \\ 3 & 0 \end{bmatrix}\begin{bmatrix} -1 & 1 \\ 0 & 2 \end{bmatrix} = \begin{bmatrix} 1 & -1 \\ -3 & 3 \end{bmatrix} \tag{A2-23}$$

$$A_{22}B_{21} = \begin{bmatrix} 1 & 0 \\ 0 & 1 \end{bmatrix}\begin{bmatrix} 1 & 0 \\ 0 & 1 \end{bmatrix} = \begin{bmatrix} 1 & 0 \\ 0 & 1 \end{bmatrix} \tag{A2-24}$$

Substituting Equations (A2-21), (A2-22), (A2-23), and (A2-24) into Equation (A2-15), we find that

$$AB = \begin{bmatrix} \begin{bmatrix} -1 & 1 \\ 0 & 2 \end{bmatrix} + \begin{bmatrix} -1 & -2 \\ 0 & 0 \end{bmatrix} \\ \begin{bmatrix} 1 & -1 \\ -3 & 3 \end{bmatrix} + \begin{bmatrix} 1 & 0 \\ 0 & 1 \end{bmatrix} \end{bmatrix} = \begin{bmatrix} -2 & -1 \\ 0 & 2 \\ 2 & -1 \\ -3 & 4 \end{bmatrix} \tag{A2-25}$$

which is equal to the same product AB given in Equation (A2-20).

A2-10 REMARKS

We learn two important facts about matrix partitioning from this example:

(a) In partitioning two matrices for multiplication, it is important that the second matrix in the multiplication order be partitioned in such a way that all the products of the submatrices exist. In other words, the submatrices of the two given matrices must be partitioned correctly in their orders so that they be conformable in each product of submatrices as indicated in Equation (A2-15).

(b) Even though the product of two matrices is always unique, no matter if it is obtained by direct multiplication or by multiplication after partitioning (or no matter how the two matrices are properly partitioned before multiplication is carried out), the labor and time consumed may differ widely from one case to another. For example, in the first partitioning in Example A2-17, submatrices A_{11} and A_{22} are identity matrices of order 1 and 3, respectively, and hence the operations in Equations (A2-16) and (A2-19) can be performed by inspection. However, by the second partitioning, there are three identity submatrices (A_{11}, A_{22}, and B_{21}) and, in addition, all the submatrices in this partitioning are of order 2; therefore all the operations involved are simplified. Finally, if the product AB is to be carried out by direct multiplication, we will not be able to take advantage of the simplicity of an identity matrix in matrix multiplication at all!

APPENDIX A3 SYSTEMS OF LINEAR EQUATIONS

A3-1 INTRODUCTION

The systems of equations we will be dealing with here are only the systems of m linear algebraic equations in n unknowns of the general form

$$\begin{aligned} a_{11}x_1 + a_{12}x_2 + \cdots + a_{1n}x_n &= b_1 \\ a_{21}x_1 + a_{22}x_2 + \cdots + a_{2n}x_n &= b_2 \\ &\vdots \\ a_{m1}x_1 + a_{m2}x_2 + \cdots + a_{mn}x_n &= b_m \end{aligned} \tag{A3-1}$$

where the a_{ij} are constant coefficients and the x_j are the variables and the b_i are known constants. When the system of Equation (A3-1) is written in matrix form, it becomes

$$\begin{bmatrix} a_{11} & a_{12} & \cdots & a_{1n} \\ a_{21} & a_{22} & \cdots & a_{2n} \\ \cdot & & & \\ \cdot & & & \\ \cdot & & & \\ a_{m1} & a_{m2} & \cdots & a_{mn} \end{bmatrix} \begin{bmatrix} x_1 \\ x_2 \\ \cdot \\ \cdot \\ \cdot \\ x_n \end{bmatrix} = \begin{bmatrix} b_1 \\ b_2 \\ \cdot \\ \cdot \\ \cdot \\ b_m \end{bmatrix} \tag{A3-2}$$

or, in the abbreviated matrix notation, we write

$$AX = B \tag{A3-3}$$

where

$$A = [a_{ij}]_{m\times n} = \begin{bmatrix} a_{11} & a_{12} & \cdots & a_{1n} \\ a_{21} & a_{22} & \cdots & a_{2n} \\ \cdot & & & \\ \cdot & & & \\ \cdot & & & \\ a_{m1} & a_{m2} & \cdots & a_{mn} \end{bmatrix}$$

$$X = [x_j]_{n\times 1} = \begin{bmatrix} x_1 \\ x_2 \\ \cdot \\ \cdot \\ \cdot \\ x_n \end{bmatrix} \qquad B = [b_i]_{m\times 1} = \begin{bmatrix} b_1 \\ b_2 \\ \cdot \\ \cdot \\ \cdot \\ b_m \end{bmatrix}$$

As mentioned in Section A2-2, A is known as the coefficient matrix, X as the unknown vector, and B as the constant vector of the system of Equation (A3-1). When such a system of linear equations is given (where m may or may not be equal to n), the first problem to be studied is the conditions for solvability of the system. Such conditions may be established quite easily after a brief discussion on linear dependence and independence of equations.

A3-2 LINEAR DEPENDENCE AND INDEPENDENCE

The concept of linear dependence and independence of quantities, such as matrices, vectors, equations and others, plays a very important role in the study of linear systems of equations. In this section, we shall define linear dependence and independence in the most general sense and then apply the idea to solve various systems of linear equations with the help of the theory of determinants and matrices.

DEFINITION A3-1. Linear Dependence and Independence. A set of n quantities, $Q_1, Q_2, \ldots, Q_n$, are said to be *linearly dependent* if and only if there exists a set of n constants $k_1, k_2, \ldots, k_n$ (not all zero) such that the equation

$$k_1Q_1 + k_2Q_2 + \cdots + k_nQ_n = 0 \tag{A3-4}$$

holds identically. If no such set of constants k_j ($j = 1, 2, \ldots, n$) can be found, that is, if Equation (A3-4) can be satisfied only by the set $k_1 = k_2 = \cdots = k_n = 0$, the quantities Q_j ($j = 1, 2, \ldots, n$) are said to be *linearly independent*.

The fact that Equation (A3-4) is an identity should be emphasized because it implies that when the equation holds for a set of k_j's, it will remain valid for *any* set of values of the Q_j's whether they are constants or functions of some variables. This means, for example, that if the Q_j's are functions of a variable x, then Equation (A3-4) will hold independently of the value of x appearing in the n functions. The following examples should make the points clear.

EXAMPLE A3-1. Let us consider the following 3-vectors:

$$V_1 = \begin{bmatrix} 1 \\ 2 \\ -1 \end{bmatrix} \qquad V_2 = \begin{bmatrix} -1 \\ 0 \\ 2 \end{bmatrix} \qquad V_3 = \begin{bmatrix} 0 \\ 2 \\ 1 \end{bmatrix}$$

To determine whether the three 3-vectors are linearly dependent, we simply apply Definition A3-1 and see whether a set of k's can be found to satisfy the equation

$$k_1V_1 + k_2V_2 + k_3V_3 \equiv 0$$

Thus, we write

$$k_1\begin{bmatrix} 1 \\ 2 \\ -1 \end{bmatrix} + k_2\begin{bmatrix} -1 \\ 0 \\ 2 \end{bmatrix} + k_3\begin{bmatrix} 0 \\ 2 \\ 1 \end{bmatrix} = 0$$

or

$$\begin{bmatrix} k_1 & -k_2 & +0 \\ 2k_1 & +\ 0 & +2k_3 \\ -k_1 & +2k_2 & +k_3 \end{bmatrix} = 0$$

or

$$\begin{bmatrix} k_1 - \ k_2 \\ 2k_1 + 2k_3 \\ -k_1 + 2k_2 + k_3 \end{bmatrix} = 0 \tag{A3-5}$$

Since the left-hand member of Equation (A3-5) is a column matrix of three elements, or a 3-vector, the zero on the right side of this equation must also be a 3-vector with zero elements due to the definition of equality of two matrices. Hence, Equation (A3-5) implies that

$$\begin{aligned} k_1 - \ k_2 \qquad &= 0 \\ 2k_1 \qquad + 2k_3 &= 0 \\ -\ k_1 + 2k_2 + \ k_3 &= 0 \end{aligned} \tag{A3-6}$$

or, written in matrix form

$$\begin{bmatrix} 1 & -1 & 0 \\ 2 & 0 & 2 \\ -1 & 2 & 1 \end{bmatrix} \begin{bmatrix} k_1 \\ k_2 \\ k_3 \end{bmatrix} = \begin{bmatrix} 0 \\ 0 \\ 0 \end{bmatrix} \tag{A3-7}$$

This is a linear system of three equations in three unknowns. Such a system may be solved in many ways. For the time being, we shall not attempt to solve the systems, but, rather, give the following result as one of the (infinitely many) solutions to the system:

$$\begin{aligned} k_1 &= 1 \\ k_2 &= 1 \\ k_3 &= -1 \end{aligned} \tag{A3-8}$$

In fact, the reader will see by inspection that the sum $V_1 + V_2$ is indeed equal to V_3. Since a set of k's, not all zero, does exist, we conclude that the three vectors V_1, V_2, and V_3 are linearly dependent.

EXAMPLE A3-2. Let Q_1, Q_2, Q_3 be the following linear forms:

$$\begin{aligned} Q_1 &= x_1 + 2x_2 - 3x_3 \\ Q_2 &= x_1 + x_2 - x_3 \\ Q_3 &= x_2 - 2x_3 \end{aligned} \tag{A3-9}$$

Again, applying Definition A3-1, we get

$$k_1Q_1 + k_2Q_2 + k_3Q_3 = 0$$

or

$$k_1(x_1 + 2x_2 - 3x_3) + k_2(x_1 + x_2 - x_3) + k_3(x_2 - 2x_3) = 0$$

or

$$(k_1 + k_2)x_1 + (2k_1 + k_2 + k_3)x_2 + (-3k_1 - k_2 - 2k_3)x_3 = 0. \tag{A3-10}$$

Since this equation is an identity, it must hold for any set of values of x_1, x_2, x_3. This implies that all the three "coefficients" of the three variables in Equation (A3-10) must be zero. Thus,

$$\begin{aligned} k_1 + k_2 &= 0 \\ 2k_1 + k_2 + k_3 &= 0 \\ -3k_1 - k_2 - 2k_3 &= 0 \end{aligned} \tag{A3-11}$$

Our problem is now reduced to the solution of the system of three equations in three unknowns. One of the solutions of Equation (A3-11) is seen to be

$$\begin{aligned} k_1 &= 1 \\ k_2 &= -1 \\ k_3 &= -1 \end{aligned}$$

Hence, the three linear forms are linearly dependent.

EXAMPLE A3-3. Let $Q_1 = \sin x$ and let $Q_2 = \cos x$. To see whether Q_1 and Q_2 are linearly dependent, we write, by Equation (A3-4),

$$k_1Q_1 + k_2Q_2 = k_1 \sin x + k_2 \cos x = 0 \tag{A3-12}$$

Since Equation (A3-12) is an identity, it should hold for any values of x. So, if we let $x = 0$, this equation becomes

$$k_1 \cdot 0 + k_2(1) = 0$$

which implies that

$$k_2 = 0 \tag{A3-13}$$

Next, setting $x = \pi/2$ in Equation (A3-12), we get

$$k_1 \cdot (1) + k_2(0) = 0$$

or

$$k_1 = 0 \tag{A3-14}$$

Equations (A3-13) and (A3-14) together imply that (A3-12) can only be satisfied by $k_1 = k_2 = 0$ for all values of x. Hence, Q_1 and Q_2 are linearly independent, as expected from our knowledge of trigonometry.

Let us consider Equation (A3-7). This matrix equation, which represents a system of three simultaneous linear equations in three unknowns, may be rewritten in a more compact form:

$$A \cdot K = 0 \tag{A3-15}$$

where A is the coefficient matrix, K is the unknown column matrix, and zero is the constant column matrix. Note that the constant column matrix is a zero matrix for this system. Such a system is a special type of system, which has certain interesting properties.

DEFINITION A3-2. Homogeneous and Nonhomogeneous Systems. A system $AX = B$ of m linear equations in n unknowns is called a *homogeneous system* if the constant column matrix B is a null (zero) matrix. Otherwise, the system is a *nonhomogeneous system.*

Since matrix A in Equation (A3-15) is a square matrix, we see that if A^{-1} exists, that is, if det $A \neq 0$, we can premultiply (A3-15) by A^{-1} to get

$$A^{-1} \cdot A \cdot K = A^{-1} \cdot 0 = 0$$

or

$$I \cdot K = 0$$

or

$$K = 0$$

This suggests an important property of the homogeneous systems, which is stated in the following theorem:

THEOREM A3-1. In a homogeneous system of n equations in n unknowns, $AX = 0$ (where A is square) if the coefficient matrix A is nonsingular; the only solution to the system is the null matrix $X = 0$ where all the variables are equal to zero. (Such a solution is called the *trivial solution*, which we will discuss in more detail in the next section.)

THEOREM A3-2. A set of n n-vectors is a linearly independent set if and only if the (square) matrix, which is formed by taking all the n vectors as columns, is nonsingular.

Theorem A3-2 may be restated as follows:

THEOREM A3-3. A square matrix A is nonsingular (so that A^{-1} exists) if and only if its columns (or rows) are linearly independent.

THEOREM A3-4. A matrix A of order $m \times n$ is of rank r if and only if there exists (at least) one set of r linearly independent columns (rows).

EXAMPLE A3-4. Consider the following 3×4 matrix:

$$A = \begin{bmatrix} 1 & -1 & 0 & 2 \\ 2 & 0 & 2 & 0 \\ -1 & 2 & 1 & -4 \end{bmatrix} \tag{A3-16}$$

Let us reduce A to its normal form:

$$A \sim \begin{bmatrix} 1 & -1 & 0 & 2 \\ 0 & 2 & 2 & -4 \\ 0 & 1 & 1 & -2 \end{bmatrix} \sim \begin{bmatrix} 1 & -1 & 0 & 2 \\ 0 & 1 & 1 & -2 \\ 0 & 2 & 2 & -4 \end{bmatrix} \sim \begin{bmatrix} 1 & -1 & 0 & 2 \\ 0 & 1 & 1 & -2 \\ 0 & 0 & 0 & 0 \end{bmatrix}$$
$$\sim \begin{bmatrix} 1 & 0 & 0 & 0 \\ 0 & 1 & 0 & 0 \\ 0 & 0 & 0 & 0 \end{bmatrix} \tag{A3-17}$$

All the four 3×4 matrices are equivalent matrices of A. The last one, of course, is the normal form A_N:

$$A_N = \begin{bmatrix} 1 & 0 & 0 & 0 \\ 0 & 1 & 0 & 0 \\ 0 & 0 & 0 & 0 \end{bmatrix} = \begin{bmatrix} U_2 & 0 \\ 0 & 0 \end{bmatrix} \tag{A3-18}$$

indicating that $R(A) = 2$.

The first equivalent matrix of A in Equation (A3-17) is obtained by multiplying row 1 of A by -2 and adding it to row 2, and then adding row 1 to row 3. The second equivalent matrix is obtained from the first by interchanging the second and third rows. The third is obtained from the second by multiplying row 2 of the second equivalent matrix by -2 and adding it to row 3. Finally, the last one, which is the normal form A_N of A, is obtained by the obvious column operations.

Actually, once the original matrix is reduced to the form

$$\begin{bmatrix} T_r & \vdots & P \\ \cdots & & \cdots \\ 0_i & \vdots & 0_j \end{bmatrix}_{m \times n} \tag{A3-19}$$

where T_r is a square matrix having all elements below its major diagonal equal to zero; 0_i and 0_j are the zero matrices of order $(m - r) \times r$ and $(m - r) \times (n - r)$, respectively; and P is an $r \times (n - r)$ matrix, it can be reduced to normal form by replacing T_r by a unit matrix U_r of order r and replacing P by a zero matrix 0_k of order $r \times (n - r)$ since these replacements of T_r and P by U_r and 0_k, respectively, can always be accomplished by a number of column operations without affecting any other submatrices in Equation (A3-19).

To illustrate this point, we write the following matrices from the third matrix in Equation (A3-17):

$$T_r = \begin{bmatrix} 1 & -1 \\ 0 & 1 \end{bmatrix} \quad P = \begin{bmatrix} 0 & 2 \\ 1 & -2 \end{bmatrix} \quad 0_i = [0 \quad 0] \quad 0_j = [0 \quad 0].$$

Now, replacing T_r by U_r and P by 0_k, we get the same normal form A_N as shown in Equation (A3-18).

A3-3 SYSTEMS OF SIMULTANEOUS LINEAR EQUATIONS

In the preceding section, we defined the homogeneous and nonhomogeneous systems of linear equations. We shall begin our discussions on the solution of both types of systems of linear equations by stating the following definitions:

DEFINITION A3-3. Consistency and Inconsistency. A system of linear equations,

$$AX = B \tag{A3-20}$$

is *consistent* if there exists at least one solution to the system; otherwise, the system is said to be *inconsistent* (implying that the system has no solution at all).

It is interesting to note that in the homogeneous case where B is a zero column matrix in Equation (A3-20), the system is always consistent; for, if $B = 0$, Equation (A3-20) becomes

$$AX = 0 \tag{A3-21}$$

so that $X = 0$ is always *a solution* to the system.

DEFINITION A3-4. Trivial and Nontrivial Solutions. In a homogeneous system, $AX = 0$, the solution $X = 0$ is called the *trivial solution.* All other solutions, if any, are referred to as the *nontrivial solutions* to the system.

THEOREM A3-5. A homogeneous system of linear equations is always consistent.

Although the trivial solution $X = 0$ is always a solution to a homogeneous system $AX = 0$, it obviously will not be a solution to a nonhomogeneous system $AX = b$; for if $X = 0$, the product AX must be a zero matrix. But in the nonhomogeneous case, $B \neq 0$; hence, $X = 0$ will not be a solution to the matrix equation $AX = B$.

THEOREM A3-6. A system $AX = B$ of m linear equations in n unknowns is consistent if and only if the coefficient matrix A and the augmented matrix[1] $[A,B]$ of the system are of the same rank, namely,

$$R(A) = R([A,B]) \tag{A3-22}$$

Theorem A3-6 is known as the *Fundamental Theorem of Linear Equations.*

EXAMPLE A3-5. Consider the following system of three linear equations in three unknowns:

$$x_1 + 2x_2 - x_3 = 1 \tag{A3-23a}$$

$$-2x_1 + x_2 + x_3 = 0 \tag{A3-23b}$$

$$3x_1 + x_2 - 2x_3 = -1 \tag{A3-23c}$$

This system, when written in matrix form $AX = B$, becomes

$$\begin{bmatrix} 1 & 2 & -1 \\ -2 & 1 & 1 \\ 3 & 1 & -2 \end{bmatrix} \begin{bmatrix} x_1 \\ x_2 \\ x_3 \end{bmatrix} = \begin{bmatrix} 1 \\ 0 \\ -1 \end{bmatrix}$$

The augmented matrix $[A,B]$ of this system is given by

$$[A,B] = \begin{array}{c} \begin{matrix} x_1 & x_2 & x_3 & b \end{matrix} \\ \begin{bmatrix} 1 & 2 & -1 & 1 \\ -2 & 1 & 1 & 0 \\ 3 & 1 & -2 & -1 \end{bmatrix} \end{array} \tag{A3-24}$$

To determine the rank of $[A,B]$, we reduce it to normal form in the following steps:

$$[A,B] \sim \begin{array}{c} \begin{matrix} x_1 & x_2 & x_3 & b \end{matrix} \\ \begin{bmatrix} 1 & 2 & -1 & 1 \\ 0 & 5 & -1 & 2 \\ 0 & -5 & 1 & -4 \end{bmatrix} \end{array} \sim \begin{array}{c} \begin{matrix} x_1 & x_2 & x_3 & b \end{matrix} \\ \begin{bmatrix} 1 & 2 & -1 & 1 \\ 0 & 5 & -1 & 2 \\ 0 & 0 & 0 & -2 \end{bmatrix} \end{array}$$

$$\sim \begin{array}{c} \begin{matrix} x_1 & x_2 & b & x_3 \end{matrix} \\ \begin{bmatrix} 1 & 2 & 1 & -1 \\ 0 & 5 & 2 & -1 \\ 0 & 0 & -2 & 0 \end{bmatrix} \end{array} \sim \begin{bmatrix} 1 & 0 & 0 & 0 \\ 0 & 1 & 0 & 0 \\ 0 & 0 & 1 & 0 \end{bmatrix} \tag{A3-25}$$

[1]The definition of the augmented matrix $[A,B]$ of a system $AX = B$ of linear equations was given in Section A2-2.

The first two equivalent matrices of $[A,B]$ in Equation (A3-25) are obtained from $[A,B]$ by elementary row operations. The third equivalent matrix is obtained from the second by interchanging the third and the fourth columns of the second matrix as indicated by the letters x_1, x_2, x_3, and b, where x_1, x_2, and x_3 are the variables associated with the coefficients in the first, second, and third column, respectively, and b identifies the constant column (the B matrix in the matrix equation $AX = B$). Finally, the fourth equivalent matrix, the normal form of $[A,B]$, is obtained from the third by a series of elementary column operations. We see that $R([A,B]) = 3$; but from Equation (A3-24) and from the first and second matrices in (A3-25), we write by inspection that

$$A = \begin{bmatrix} 1 & 2 & -1 \\ -2 & 1 & 1 \\ 3 & 1 & -2 \end{bmatrix} \sim \begin{bmatrix} 1 & 2 & -1 \\ 0 & 5 & -1 \\ 0 & -5 & 1 \end{bmatrix} \sim \begin{bmatrix} 1 & 2 & -1 \\ 0 & 5 & -1 \\ 0 & 0 & 0 \end{bmatrix}$$

$$\sim \begin{bmatrix} 1 & 0 & 0 \\ 0 & 1 & 0 \\ 0 & 0 & 0 \end{bmatrix} \qquad \text{(A3-26)}$$

which is seen to be of rank 2. Hence,

$$R(A) \neq R([A,B])$$

and, by virtue of Theorem A3-6, the system is inconsistent.

We shall now verify it by the familiar method of elimination. Multiplying Equation (A3-23a) by 2 and adding it to (A3-23b) we get

$$5x_2 - x_3 = 2 \qquad \text{(A3-27)}$$

Next, multiplying Equation (A3-23a) by -3 and adding it to (A3-23c) we obtain

$$-5x_2 + x_3 = -4. \qquad \text{(A3-28)}$$

Finally, adding Equation (A3-27) to (A3-28), we have

$$0 = -2 \qquad \text{(A3-29)}$$

which shows the inconsistency of the system.

We might mention that this systematic procedure of elimination is known as the *Gauss reduction*, named after the famous mathematician Karl Friedrich Gauss (1777–1855).

Actually, in reducing $[A,B]$ to its normal form, we did exactly the same thing as we did by the Gauss reduction. If we observe the third row of the second equivalent matrix in Equation (A3-25), we see that this row represents the equation

$$0 \cdot x_1 + 0 \cdot x_2 + 0 \cdot x_3 = -2$$

or, simply,

$$0 = -2$$

which is identical to expression (A3-29), revealing the inconsistency of the system.

EXAMPLE A3-6. Let us consider the following system:

$$x_1 \qquad - x_3 = 2 \tag{A3-30a}$$

$$-x_1 + x_2 \qquad = 0 \tag{A3-30b}$$

$$2x_1 + x_2 - x_3 = 1 \tag{A3-30c}$$

which may be written in the matrix form:

$$\begin{bmatrix} 1 & 0 & -1 \\ -1 & 1 & 0 \\ 2 & 1 & -1 \end{bmatrix} \begin{bmatrix} x_1 \\ x_2 \\ x_3 \end{bmatrix} = \begin{bmatrix} 2 \\ 0 \\ 1 \end{bmatrix}$$

Next, we shall find the rank of the augmented matrix by reducing it to normal form:

$$[A,B] = \begin{array}{c} \begin{matrix} x_1 & x_2 & x_3 & b \end{matrix} \\ \begin{bmatrix} 1 & 0 & -1 & 2 \\ -1 & 1 & 0 & 0 \\ 2 & 1 & -1 & 1 \end{bmatrix} \end{array} \sim \begin{array}{c} \begin{matrix} x_1 & x_2 & x_3 & b \end{matrix} \\ \begin{bmatrix} 1 & 0 & -1 & 2 \\ 0 & 1 & -1 & 2 \\ 0 & 1 & 1 & -3 \end{bmatrix} \end{array}$$

$$\sim \begin{array}{c} \begin{matrix} x_1 & x_2 & x_3 & b \end{matrix} \\ \begin{bmatrix} 1 & 0 & -1 & 2 \\ 0 & 1 & -1 & 2 \\ 0 & 0 & 2 & -5 \end{bmatrix} \end{array} \sim \begin{bmatrix} 1 & 0 & 0 & 0 \\ 0 & 1 & 0 & 0 \\ 0 & 0 & 1 & 0 \end{bmatrix} \tag{A3-31}$$

Hence we have $R([A,B]) = 3$.

From the third equivalent matrix in Equation (A3-31), we also see that $R(A) = 3$. Therefore, $R(A) = R([A,B])$ and, according to Theorem A3-6, the system is consistent. In fact, from the third row of the second equivalent matrix in Equation (A3-31), we have

$$2x_3 = -5$$

or

$$x_3 = -\frac{5}{2} \tag{A3-32}$$

and from the second row of the same matrix we get

$$x_2 - x_3 = 2$$

or

$$x_2 = 2 + x_3 \tag{A3-33}$$

Substituting Equation (A3-32) into (A3-33), we obtain

$$x_2 = 2 - \frac{5}{2} = -\frac{1}{2} \tag{A3-34}$$

From the first row of the same matrix we get

$$x_1 - x_3 = 2$$

or

$$x_1 = 2 + x_3 \tag{A3-35}$$

Substituting Equation (A3-32) into (A3-35) we obtain

$$x_1 = 2 - \frac{5}{2} = -\frac{1}{2} \tag{A3-36}$$

Equations (A3-32), (A3-34), and (A3-36) together give the complete solution to the system:

$$\begin{aligned} x_1 &= -\frac{1}{2} \\ x_2 &= -\frac{1}{2} \\ x_3 &= -\frac{5}{2} \end{aligned} \tag{A3-37}$$

To check our answers, we can simply substitute the values of x_1, x_2, and x_3 of Equation (A3-37) into the original system [Equation (A3-30)] and see whether these values satisfy the given three equations, simultaneously. We shall leave this simple task to the reader as an exercise.

In a system $AX = B$ of n linear equations in n unknowns, if the coefficient matrix A is nonsingular such that its inverse A^{-1} exists, the system may be solved by premultiplying the matrix equation of the system by A^{-1}:

$$A^{-1} \cdot A \cdot X = A^{-1}B$$

or

$$I \cdot X = A^{-1}B$$

or

$$X = A^{-1}B \tag{A3-38}$$

Equation (A3-38) means that the solution vector X is equal to the product of the inverse of the coefficient matrix and the constant vector, in that order. This leads to the next well-known theorem.

THEOREM A3-7. In a system $AX = B$ of n linear equations in n unknowns, if the coefficient matrix A is nonsingular, the system has a unique solution given by

$$x_j = \frac{\det A_j}{\det A} \qquad j = 1, 2, \ldots, n \tag{A3-39}$$

where A is the matrix obtained from A by replacing its jth column by the constant column of the column matrix B.

Theorem A3-7 is called *Cramer's rule.*

EXAMPLE A3-7. Consider once again the system of Example A3-6. The determinant of A is

$$\det A = \begin{vmatrix} 1 & 0 & -1 \\ -1 & 1 & 0 \\ 2 & 1 & -1 \end{vmatrix} = -1 + 0 + 1 + 2 + 0 + 0 = 2$$

The three determinants, $\det A_1$, $\det A_2$, and $\det A_3$, are:

$$\det A_1 = \begin{vmatrix} 2 & 0 & -1 \\ 0 & 1 & 0 \\ 1 & 1 & -1 \end{vmatrix} = -2 + 0 + 0 + 1 + 0 + 0 = -1$$

$$\det A_2 = \begin{vmatrix} 1 & 2 & -1 \\ -1 & 0 & 0 \\ 2 & 1 & -1 \end{vmatrix} = 0 + 0 + 1 + 0 + 0 - 2 = -1$$

$$\det A_3 = \begin{vmatrix} 1 & 0 & 2 \\ -1 & 1 & 0 \\ 2 & 1 & 1 \end{vmatrix} = 1 + 0 - 2 - 4 + 0 + 0 = -5$$

Hence, by Cramer's rule, we get

$$x_1 = \frac{\det A_1}{\det A} = \frac{-1}{2} \qquad x_2 = \frac{\det A_2}{\det A} = \frac{-1}{2} \qquad x_3 = \frac{\det A_3}{\det A} = \frac{-5}{2}$$

These results agree with the values given by Equation (A3-37).

It should be noted that Theorem A3-7 is valid for the system $AX = B$ only when (a) $m = n$, the number of equations is equal to the number of unknowns and (b) the coefficient matrix A is nonsingular. If condition (a) is not satisfied, $\det A$ will not exist and the theorem will not hold. If condition (b) is not satisfied, the theorem again will fail because, when $\det A = 0$, the general expression (A3-39) for the unknowns x_j will not be defined.

We should also point out that when a given system $AX = B$ is known to be consistent, that is, when $R(A) = R([A,B])$, all we know about the system from the fundamental theorem is that there exists *at least* one solution to the system; but the solutions may or may not be unique. In other words, there may be more than one solution to the system. We shall demonstrate this point by means of the next example.

EXAMPLE A3-8. Consider the following homogeneous system $AX = 0$ of three equations in three unknowns:

$$\begin{bmatrix} 1 & 1 & 0 \\ 2 & 1 & 1 \\ -3 & -1 & -2 \end{bmatrix} \begin{bmatrix} x_1 \\ x_2 \\ x_3 \end{bmatrix} = \begin{bmatrix} 0 \\ 0 \\ 0 \end{bmatrix} \qquad \text{(A3-40)}$$

The system is consistent since it is a homogeneous system (Theorem A3-5). To find the nontrivial solutions, if any, let us reduce the augmented matrix to normal form:

$$[A,B] = \begin{array}{c} \begin{array}{cccc} x_1 & x_2 & x_3 & b \end{array} \\ \begin{bmatrix} 1 & 1 & 0 & 0 \\ 2 & 1 & 1 & 0 \\ -3 & -1 & -2 & 0 \end{bmatrix} \end{array} \sim \begin{array}{c} \begin{array}{cccc} x_1 & x_2 & x_3 & b \end{array} \\ \begin{bmatrix} 1 & 1 & 0 & 0 \\ 0 & -1 & 1 & 0 \\ 0 & 2 & -2 & 0 \end{bmatrix} \end{array} \sim \begin{array}{c} \begin{array}{cccc} x_1 & x_2 & x_3 & b \end{array} \\ \begin{bmatrix} 1 & 1 & 0 & 0 \\ 0 & 1 & -1 & 0 \\ 0 & 2 & -2 & 0 \end{bmatrix} \end{array}$$

$$\sim \begin{array}{c} \begin{array}{cccc} x_1 & x_2 & x_3 & b \end{array} \\ \begin{bmatrix} 1 & 1 & 0 & 0 \\ 0 & 1 & -1 & 0 \\ 0 & 0 & 0 & 0 \end{bmatrix} \end{array} \sim \left[\begin{array}{cc|cc} 1 & 0 & 0 & 0 \\ 0 & 1 & 0 & 0 \\ \hline 0 & 0 & 0 & 0 \end{array}\right] \qquad \text{(A3-41)}$$

The second matrix in Equation (A3-41) is obtained from $[A,B]$ by multiplying row 1 of $[A,B]$ by -2 and adding it to row 2, and multiplying row 1 by 3 and adding it to row 3. The third matrix is obtained by multiplying row 2 of the second matrix by -1. The fourth matrix is formed by multiplying row 2 of the third matrix by -2 and adding it to row 3; and the last matrix, which is the normal form of $[A,B]$ is obtained by the obvious elementary transformations on the columns of the fourth matrix.

From the second row of the fourth matrix in Equation (A3-41), we get

$$x_2 - x_3 = 0$$

or

$$x_2 = x_3 \qquad \text{(A3-42)}$$

and, from the first row of the same matrix, we have

$$x_1 + x_2 = 0$$

or

$$x_1 = -x_2 \qquad \text{(A3-43)}$$

Therefore, if we let $x_3 = t$, where t is an arbitrary constant, we get from Equations (A3-42) and (A3-43) the general solution to the system with all the three variables expressed in terms of t:

$$\begin{aligned} x_1 &= -t \\ x_2 &= t \\ x_3 &= t \end{aligned} \qquad \text{(A3-44)}$$

To check our results, let $t = 1$. Equation (A3-44) then becomes

$$\begin{aligned} x_1 &= -1 \\ x_2 &= 1 \\ x_3 &= 1 \end{aligned} \qquad \text{(A3-45)}$$

Substituting Equation (A3-45) back into the original system Equation (A3-40), we have

$$\begin{bmatrix} 1 & 1 & 0 \\ 2 & 1 & 1 \\ -3 & -1 & -2 \end{bmatrix} \begin{bmatrix} -1 \\ 1 \\ 1 \end{bmatrix} = \begin{bmatrix} 0 \\ 0 \\ 0 \end{bmatrix}$$

which is obviously a correct statement.

We might remark that the solution given by Equation (A3-44) is called a *complete solution;* whereas a solution such as (A3-45), which is obtained from (A3-44) by letting $t = 1$ (a particular value of t), is known as a *particular solution.*

From Example A3-8, we see that in this case

$$R(A) = R([A,B]) = 2$$

so that the necessary and sufficient condition stated in Theorem A3-6 is satisfied (as expected, since a homogeneous system is always consistent); and that the rank of the coefficient matrix is less than the number of unknowns, $r < n$ (or $2 < 3$, for the given system). Also, we see that $\det A = 0$; that is, A is singular. Furthermore, we observe from Equation (A3-44) that there are infinitely many nontrivial solutions to the given homogeneous system. And, finally, we note that if the Cramer's rule (Theorem A3-7) were used to find the solutions, we would have obtained

$$x_j = \frac{\det A_j}{\det A} = \frac{0}{0} \tag{A3-46}$$

since $\det A_j = b_1A_{1j} + b_2A_{2j} + \cdots + b_nA_{nj}$ and all the b's are equal to zero in case of a homogeneous system.

Expression (A3-46) is an indeterminate form indicating that $\det A = 0$ is the only possibility for nontrivial solutions to exist in a homogeneous system of n equations in n unknowns.

THEOREM A3-8. A homogeneous system $AX = 0$ of m linear equations in n unknowns

$$\sum_{j=1}^{n} a_{ij}x_j = 0 \qquad (i = 1, 2, \cdots, m) \tag{A3-47}$$

has nontrivial solutions if and only if the rank r of the coefficient matrix A is less than the number n of unknowns; that is, $r < n$.

THEOREM A3-9. The nontrivial solutions of a homogeneous system of linear equations are always infinitely many when they exist.

THEOREM A3-10. A homogeneous system $AX = 0$ of n linear equations in n unknowns has nontrivial solutions if and only if the coefficient matrix A is singular. (See Theorem A3-1, also.)

THEOREM A3-11. In a homogeneous system of m linear equations in n unknowns, if $m<n$, the system *always* has nontrivial solutions.

In the case of nonhomogeneous systems, it is also possible that a given system has infinitely many solutions as the following example will illustrate.

EXAMPLE A3-9. Consider the following nonhomogeneous system $AX = B$ of three equations in three unknowns:

$$\begin{aligned} x_1 + 2x_2 - 2x_3 &= 2 \\ 2x_1 + x_2 - 4x_3 &= 4 \\ -x_1 + 3x_2 + 2x_3 &= -2 \end{aligned} \tag{A3-48}$$

which may be written in matrix form:

$$\begin{bmatrix} 1 & 2 & -2 \\ 2 & 1 & -4 \\ -1 & 3 & 2 \end{bmatrix} \begin{bmatrix} x_1 \\ x_2 \\ x_3 \end{bmatrix} = \begin{bmatrix} 2 \\ 4 \\ -2 \end{bmatrix} \tag{A3-49}$$

The augmented matrix $[A,B]$ is reduced to normal form as follows:

$$[A,B] = \begin{array}{c} \begin{matrix} x_1 & x_2 & x_3 & b \end{matrix} \\ \begin{bmatrix} 1 & 2 & -2 & 2 \\ 2 & 1 & -4 & 4 \\ -1 & 3 & 2 & -2 \end{bmatrix} \end{array} \sim \begin{array}{c} \begin{matrix} x_1 & x_2 & x_3 & b \end{matrix} \\ \begin{bmatrix} 1 & 2 & -2 & 2 \\ 0 & -3 & 0 & 0 \\ 0 & 5 & 0 & 0 \end{bmatrix} \end{array}$$

$$\sim \begin{bmatrix} 1 & 0 & 0 & 0 \\ 0 & 1 & 0 & 0 \\ 0 & 0 & 0 & 0 \end{bmatrix} \tag{A3-50}$$

The first equivalent of $[A,B]$ in Equation (A3-50) is obtained from $[A,B]$ by the familiar row elementary transformations, whereas the normal form is obtained from the first equivalent by the column elementary transformations.

From the second row (or the third row) of the first equivalent, we see that

$$x_2 = 0 \tag{A3-51}$$

and, from the first row of the same matrix with $x_2 = 0$, we have

$$x_1 - 2x_3 = 2$$

or

$$x_1 = 2 + 2x_3 \tag{A3-52}$$

Now, if we let $x_3 = t$, an arbitrary constant, we have a complete solution in terms of t:

$$\begin{aligned} x_1 &= 2 + 2t \\ x_2 &= 0 \\ x_3 &= t \end{aligned} \tag{A3-53}$$

We shall obtain a particular solution from these three equations by letting $t = 1$. Thus,

$$\begin{aligned} x_1 &= 4 \\ x_2 &= 0 \\ x_3 &= 1 \end{aligned} \tag{A3-54}$$

Substituting these particular values into Equations (A3-48), we see that they indeed satisfy the original system.

The fact that a nonhomogeneous system may have more than one solution is not limited to the case in which the number of equations is equal to the number of unknowns, as Example A3-9 may have suggested. In fact, if the terms involving x_2 are eliminated from the system [Equation (A3-48)], we have a nonhomogeneous system of three equations in two unknowns:

$$\begin{aligned} x_1 - 2x_3 &= 2 \\ 2x_1 - 4x_3 &= 4 \\ -x_1 + 2x_3 &= -2 \end{aligned}$$

A complete solution to this system is evidently the following:

$$\begin{aligned} x_1 &= 2 + 2t \\ x_3 &= t \end{aligned}$$

which is the same as that in Equation (A3-53), except that the equation $x_2 = 0$ has been removed for the obvious reason.

Finally, we shall conclude this appendix by making a few remarks about linear dependence and independence as well as consistency and inconsistency of linear equations.

A3-4 REMARKS

In the previous sections we discussed linear dependence and independence among the linear equations within a given system. We also discussed consistency and inconsistency of homogeneous and nonhomogeneous systems of linear equations. From these discussions, we noted that the number m of equations, the number n of unknowns, the rank $R(A)$ of the coefficient matrix A and the rank $R([A,B])$ of the augmented matrix all play important roles in determining whether a given system is consistent or inconsistent, or whether the equations in the system are linearly dependent or independent. In this section, we shall point out the interrelationships between these characteristics. For convenience, let us define the following terms:

DEFINITION A3-5. Dependent and Independent Systems. When the equations of a given linear system are linearly dependent, we say the system is a *dependent system.* Similarly, if all the equations of the system are linearly independent, it is said to be an *independent system.*

We shall show that a system of linear equations can be simultaneously
(a) dependent and consistent,
(b) dependent and inconsistent,
(c) independent and consistent, or
(d) independent and inconsistent

by citing one example for each case. But, before constructing the examples, it may be wise to make clear the following point: What do we mean by saying that the equations in a system are linearly dependent (or independent)? To answer this question, let us recall our general definition for linear dependence and independence (Definition A3-1) which states that the n quantities Q_1, $Q_2, \ldots, Q_n$ are linearly dependent if and only if there exists a set of constants $k_1, k_2, \cdots, k_n$, not all zero, such that the equation

$$k_1Q_1 + k_2Q_2 + \cdots + k_nQ_n = 0 \tag{A3-55}$$

is satisfied identically. When the n quantities represent such things as vectors, linear forms, or others, the definition should be quite clear to the reader. But when the Q's represent n equations, it might require a careful interpretation, since there are two sides to an equation. This dilemma is easily solved if we agree to write every equation in a given system in the form $f_i = 0$ so that when we use Equation (3-55) to determine the dependence of the m equations, we simply write

$$k_1f_1 + k_2f_2 + \cdots + k_nf_n = 0 \tag{A3-56}$$

and see if a set of k_j's, not all zero, exists to satisfy this equation identically.

EXAMPLE A3-10. To show that a system can be dependent and consistent, let us consider the following system:

$$\begin{aligned} x_1 - 2x_2 &= 3 \\ -2x_1 + 4x_2 &= -6 \end{aligned} \tag{A3-57}$$

Here, $R(A) = R([A,B]) = 2$, which can be easily verified. Putting both equations in the form $f_i = 0$, we get

$$\begin{aligned} f_1 &= \quad x_1 - 2x_2 - 3 = 0 \\ f_2 &= -2x_1 + 4x_2 + 6 = 0 \end{aligned}$$

Substituting f_1 and f_2 into Equation (A3-56), we obtain

$$k_1(x_1 - 2x_2 - 3) + k_2(-2x_1 + 4x_2 + 6) = 0$$

or

$$(k_1 - 2k_2)x_1 + (-2k_1 + 4k_2)x_2 + (-3k_1 + 6k_2) = 0 \tag{A3-58}$$

For Equation (A3-58) to be satisfied identically, it is required that all the coefficients be zero; that is,

$$\begin{aligned} k_1 - 2k_2 &= 0 \\ -2k_1 + 4k_2 &= 0 \\ -3k_1 + 6k_2 &= 0 \end{aligned}$$

which, when written in matrix form, becomes

$$\begin{bmatrix} 1 & -2 \\ -2 & 4 \\ -3 & 6 \end{bmatrix} \begin{bmatrix} x_1 \\ x_2 \end{bmatrix} = \begin{bmatrix} 0 \\ 0 \\ 0 \end{bmatrix} \tag{A3-59}$$

This is a homogeneous system of three equations in two unknowns. The rank of the coefficient matrix is equal to 1 since

$$A = \begin{bmatrix} 1 & -2 \\ -2 & 4 \\ -3 & 6 \end{bmatrix} \sim \begin{bmatrix} 1 & -2 \\ 0 & 0 \\ 0 & 0 \end{bmatrix} \sim \begin{bmatrix} 1 & 0 \\ 0 & 0 \\ 0 & 0 \end{bmatrix}$$

Thus, by Theorem A3-8, the system has nontrivial solutions. This means that there exists a set of k_1 and k_2 not both zero such that Equation (A3-58) is satisfied identically. Hence, the system given by Equation (A3-57) is a dependent system.

Actually, we can tell by inspection that the system is dependent, since the second equation can be obtained from the first by multiplying the first equation by -2. We went through all this trouble for two important reasons: First, the work presented in this example clearly shows that even though one can always follow the conventional methods to solve a problem, it might not be the best approach to that particular problem, and sometimes it pays to think wisely and examine other possibilities instead of proceeding dogmatically to the end. Second, a beginner may gain some insight into the ideas presented by going through the conventional procedure at least once and thereby obtain a better understanding of the principles involved.

EXAMPLE A3-11. Consider the following system:

$$\begin{aligned} x_1 - 2x_2 &= 3 \\ 3x_1 - 6x_2 &= 4 \\ -2x_1 + 4x_2 &= -6 \end{aligned} \tag{A3-60}$$

Let us find the ranks of $[A,B]$ and A by reducing $[A,B]$ to normal form:

$$[A,B] = \begin{array}{c} \begin{matrix} x_1 & x_2 & b \end{matrix} \\ \begin{bmatrix} 1 & -2 & 3 \\ 3 & -6 & 4 \\ -2 & 4 & -6 \end{bmatrix} \end{array} \sim \begin{array}{c} \begin{matrix} x_1 & x_2 & b \end{matrix} \\ \begin{bmatrix} 1 & -2 & 3 \\ 0 & 0 & -5 \\ 0 & 0 & 0 \end{bmatrix} \end{array} \sim \begin{array}{c} \begin{matrix} x_1 & b & x_2 \end{matrix} \\ \begin{bmatrix} 1 & 3 & -2 \\ 0 & -5 & 0 \\ 0 & 0 & 0 \end{bmatrix} \end{array}$$

$$\sim \begin{bmatrix} 1 & 0 & 0 \\ 0 & 1 & 0 \\ 0 & 0 & 0 \end{bmatrix} \tag{A3-61}$$

From Equation (A3-61) we see that $R([A,B]) = 2$ and $R(A) = 1$. Hence, the system of Equation (A3-60) is inconsistent. The fact that the given system is dependent can be seen by noting that the third equation is dependent upon the first, since $f_3 = -2f_1$; thus we may let $k_1 = 2$, $k_2 = 0$, and $k_3 = 1$, and find that the equation

$$k_1 f_1 + k_2 f_2 + k_3 f_3 = 0$$

is satisfied identically.

EXAMPLE A3-12. For an independent and consistent system, consider the following:

$$\begin{aligned} x_1 - 2x_2 &= 3 \\ -x_1 + x_2 &= 1 \end{aligned} \tag{A3-62}$$

Since $\det A = -1 \neq 0$, it is immediately evident that

$$R([A,B]) = R(A) = 2$$

and hence the system is consistent.

To determine whether the system is dependent, we write

$$k_1 f_1 + k_2 f_2 = k_1(x_1 - 2x_2 - 3) + k_2(-x_1 + x_2 - 1) = 0$$

or

$$(k_1 - k_2)x_1 + (-2k_1 + k_2)x_2 + (-3k_1 - k_2) = 0$$

which implies that

$$\begin{aligned} k_1 - k_2 &= 0 \\ -2k_1 + k_2 &= 0 \\ -3k_1 - k_2 &= 0 \end{aligned} \tag{A3-63}$$

This is a homogeneous system of three equations in two unknowns. When written in matrix form, Equation (A3-63) becomes

$$\begin{bmatrix} 1 & -1 \\ -2 & 1 \\ -3 & -1 \end{bmatrix} \begin{bmatrix} k_1 \\ k_2 \end{bmatrix} = \begin{bmatrix} 0 \\ 0 \\ 0 \end{bmatrix} \tag{A3-64}$$

The coefficient matrix of Equation (A3-64) is of rank 2, which is equal to (and hence not less than) the number of unknowns. Then, according to Theorem A3-8, the homogeneous system [Equation (A3-63)] has no nontrivial solution and, consequently, the original system [Equation (A3-62)] is an independent system.

EXAMPLE A3-13. To form an independent and inconsistent system, let us take the first two equations from the system of Example A3-11:

$$\begin{aligned} x_1 - 2x_2 &= 3 \\ 3x_1 - 6x_2 &= 4 \end{aligned} \tag{A3-65}$$

The fact that Equation (A3-65) is inconsistent has already been revealed in

Example A3-13. Now we shall show that it is an independent system. Again we write

$$k_1f_1 + k_2f_2 = k_1(x_1 - 2x_2 - 3) + k_2(3x_1 - 6x_2 - 4) = 0$$

or

$$(k_1 + 3k_2)x_1 + (-2k_1 - 6k_2)x_2 + (-3k_1 - 4k_2) = 0$$

which implies that

$$\begin{aligned} k_1 + 3k_2 &= 0 \\ -2k_1 - 6k_2 &= 0 \\ -3k_1 - 4k_2 &= 0 \end{aligned}$$

or, in matrix form,

$$\begin{bmatrix} 1 & 3 \\ -2 & -6 \\ -3 & -4 \end{bmatrix} \begin{bmatrix} k_1 \\ k_2 \end{bmatrix} = \begin{bmatrix} 0 \\ 0 \\ 0 \end{bmatrix} \tag{A3-66}$$

The coefficient matrix of Equation (A3-66) is of rank 2 since the 2×2 submatrix, formed by taking row 1 and row 3, is nonsingular; and, by Theorem A3-8, the system has no nontrivial solutions. Hence, the given system is an independent system.

We have seen that Examples A3-10, A3-11, A3-12, and A3-13 are examples of systems of linear equations that are dependent and consistent, dependent and inconsistent, independent and consistent, and independent and inconsistent, respectively. Now we shall state the necessary and sufficient conditions for each of the four cases; but, first, let us introduce the following theorem:

THEOREM A3-12. A system $AX = B$ of m linear equations in n unknowns is a dependent system if and only if the rank of the augmented matrix is less than the number of equations; that is,

$$R([A,B]) < m$$

As a consequence of Theorem A3-12 and Theorem A3-13 (the fundamental theorem), we establish the following theorem.

THEOREM A3-13. For a system $AX = B$ of m linear equations in n unknowns, we have the following conditions:

(a) For consistency: $R(A) = R([A,B])$

(b) For inconsistency: $R(A) < R([A,B])$

(c) For dependence: $R([A,B]) < m$

(d) For independence: $R([A,B]) = m$

Now, if we re-examine Examples A3-10, A3-11, A3-12, and A3-13, we see why the systems are as they are with respect to consistency and dependence in light of Theorem A3-13.

APPENDIX B1 THE NECESSARY AND SUFFICIENT CONDITIONS

B1-1 THE "IF-THEN" STATEMENT

Let A and B represent two propositions and consider the statement:

$$\textit{If } A \text{ is true } \textit{then } B \text{ is true} \tag{B1-1}$$

This is called an "if-then" statement and is usually abbreviated as "If A then B." It has the following equivalent forms:

$$A \text{ implies } B \text{ (or simply } A \Rightarrow B) \tag{B1-2}$$

$$A \text{ is a } \textit{sufficient condition} \text{ for } B \tag{B1-3}$$

$$A \text{ is true } \textit{only if } B \text{ is true} \tag{B1-4}$$

$$B \text{ is true } \textit{if } A \text{ is true} \tag{B1-5}$$

$$B \text{ is a } \textit{necessary condition} \text{ for } A \tag{B1-6}$$

In statement (B1-1), proposition A is referred to as the *hypothesis* and proposition B is the *conclusion*.

A statement is said to be *valid* if the conclusion is a logical consequence of the hypothesis; otherwise, *invalid*. In other words, if the statement is valid, then the truth of the hypothesis *always guarantees* the truth of the conclusion. It should be noted, however, that the falsity of the hypothesis in the valid statement *does not guarantee* the falsity of the conclusion; that is, in a valid statement, if the hypothesis is false, the conclusion *may* still be true. But the falsity of the conclusion *always guarantees* the falsity of the hypothesis. In fact, the following statement is an equivalent to any of the statements (B1-1) through (B1-6):

$$\text{If } B \text{ is false then } A \text{ must be false} \tag{B1-7}$$

B1-2 THE CONVERSE OF A STATEMENT

The converse of a statement is formed by interchanging the hypothesis and conclusion. Thus, the converse of (B1-1) is

$$\text{If } B \text{ is true then } A \text{ is true} \tag{B1-8}$$

or, equivalently,

$$A \text{ is true if } B \text{ is true} \tag{B1-9}$$

which is the result of interchanging the hypothesis and conclusion in (B1-5). Obviously, interchanging the hypothesis and conclusion in any equivalent of (B1-1) constitutes an equivalent of (B1-8) which is the converse of (B1-1). Note that the converse of a valid statement may not necessarily be valid.

B1-3 THE "IF-AND-ONLY-IF" STATEMENT

If the converse of a statement is also valid, then we may combine the statement and its converse to form an *if-and-only-if statement* as follows: Take statement (B1-1) as an example. Consider its equivalent, (B1-4), and its converse, (B1-9). They read

$$A \text{ is true } \textit{only if } B \text{ is true} \tag{B1-10}$$

and

$$A \text{ is true } \textit{if } B \text{ is true} \tag{B1-11}$$

Combining (B1-10) and (B1-11) in the order of (B1-11) and (B1-12), we have

$$A \text{ is true } \textit{if and only if } B \text{ is true} \tag{B1-12}$$

Note that in (B1-10) B is a necessary condition for A so that the term "only if" *implies necessity;* and that in (B1-11) B is a sufficient condition for A [which is seen by obtaining the converse of (B1-3)], so that the phrase "if" *implies sufficiency*. Consequently, the term "if and only if" constitutes a *necessary and sufficient condition.*

Therefore, if a statement and its converse are both valid, then one serves as a necessary and sufficient condition of the other; and, using statement (B1-1) as an example, we represent this by writing

$$A \Leftrightarrow B \tag{B1-13}$$

meaning that (B1-13) can be read from left to right or from right to left. It is a "two-way" traffic!

B1-4 EXAMPLES

We shall now give a few examples to demonstrate the meaning of the terms introduced in the above discussions.

EXAMPLE B1-1. Let

(a) A be the hypothesis: $x = 4$, and $y = 6$.

(b) B be the conclusion: $x + y = 10$.

With the hypothesis and conclusion specified by (a) and (b), statement (B1-1) becomes

$$\text{If } x = 4 \text{ and } y = 6, \text{ then } x + y = 10 \tag{B1-14}$$

This is obviously a valid statement. The equivalent forms of (B1-14) are:

$$x = 4 \text{ and } y = 6 \text{ imply } x + y = 10 \tag{B1-15}$$

$$x = 4 \text{ and } y = 6 \text{ are a sufficient condition for } x + y = 10 \tag{B1-16}$$

$$x = 4 \text{ and } y = 6 \text{ only if } x + y = 10 \tag{B1-17}$$

$$x + y = 10 \text{ if } x = 4 \text{ and } y = 6 \tag{B1-18}$$

$$x + y = 10 \text{ is a necessary condition for } x = 4 \text{ and } y = 6 \tag{B1-19}$$

$$\text{If } x + y \neq 10, \text{ then the equations } x = 4 \text{ and } y = 6 \text{ cannot be satisfied simultaneously} \tag{B1-20}$$

Statements (B1-14) through (B1-20) are obtained by substituting (a) and (b) into statements (B1-1) through (B1-7), respectively.

All the statements (B1-14) through (B1-20) are valid because if $x = 4$ and $y = 6$, then the sum $x + y$ must be 10; that is, the conclusion, $x + y = 10$, follows inescapably from the hypothesis, $x = 4$, $y = 6$.

To demonstrate that the falsity of the hypothesis does not guarantee the falsity of the conclusion, we need only to observe the fact that even if $x \neq 4$ and $y \neq 6$, the conclusion, $x + y = 10$, can still be true. For example, let $x = 3$, $y = 7$.

However, if $x + y \neq 10$, then the equations $x = 4$ and $y = 6$ may not hold simultaneously. This shows that the falsity of the conclusion *does* indeed *guarantee* the falsity of the hypothesis.

Next we form the converse of (B1-14), which is

$$\text{If } x + y = 10, \text{ then } x = 4 \text{ and } y = 6 \text{ must hold simultaneously} \tag{B1-21}$$

Evidently the conclusion, $x = 4$ and $y = 6$, does not follow inescapably from the hypothesis, $x + y = 10$, because we may very well have $x = 2$ and $y = 8$ as another possibility for which the equation $x + y = 10$ holds.

Since the converse of (B1-14) is not valid, an if-and-only-if statement cannot be formed. Hence we shall need the next example to illustrate the points discussed in Section B1-3.

EXAMPLE B1-2. Let A be the statement: Jones is an atheist; and let B be the statement: Jones does not believe in God. Now, (B1-1) becomes

If Jones is an atheist *then* he does not believe in God (B1-22)

or, equivalently,

Jones is an atheist *only if* he does not believe in God (B1-23)

Statement (B1-22) is valid with A as its hypothesis and B as its conclusion.

The converse of (B1-22) is

If Jones does not believe in God, *then* he is an atheist (B1-24)

An equivalent of (B1-24) is given by

Jones is an atheist *if* he does not believe in God (B1-25)

Statement (B1-24), or (B1-25), is valid because the conclusion that Jones is an atheist follows irrevocably from the hypothesis that he does not believe in God.

Combining Equations (B1-23) and (B1-25), the if-and-only-if statement is formed:

Jones is an atheist *if* and *only if* he does not believe in God (B1-26)

or, equivalently,

Jones does not believe in God *if* and *only if* he is an atheist (B1-27)

Statements (B1-26) and (B1-27) may be abbreviated as

$$(A) \Leftrightarrow (B) \qquad \text{(B1-28)}$$

and

$$(B) \Leftrightarrow (A) \qquad \text{(B1-29)}$$

respectively. Obviously, (B1-28) and (B1-29) are equivalent statements.

B1-5 PROOF OF A THEOREM OF THE "IF-AND-ONLY-IF" CATEGORY

When a theorem is of an if-and-only-if type and stated in the form

$$A \Leftrightarrow B \qquad \text{(B1-30)}$$

the proof of the theorem will contain two parts: (a) *necessity* and (b) *sufficiency.*

(a) For the necessity, identified by "only if," or simply by ($\Rightarrow$), the statement to be proved is

$$A \text{ is true only if } B \text{ is true}$$

or, as an equivalent,

$$A \text{ implies } B \ (A \Rightarrow B) \tag{B1-31}$$

Therefore, to prove it, one begins with the hypothesis that A is true and shows that B must follow from A.

(b) For the sufficiency, denoted by "if" or ($\Leftarrow$), the statement to be proved is the converse of (B1-31); namely,

$$B \text{ implies } A \qquad (A \Leftarrow B) \tag{B1-32}$$

To prove (B1-32), one proceeds by assuming that B is true and arriving at A as a conclusion.

The methods of proof are discussed in Appendix B2.

APPENDIX B2
METHODS OF PROOF

B2-1 PRELIMINARY DISCUSSIONS

In a logical development, or any abstract mathematical science, a definition, when applied in mathematical sense, is simply an agreement to use an expression (such as a symbol or a phrase) as a substitute for another expression which usually is rather lengthy or complicated in its meaning. Thus the definition is conveniently used in the context as an equivalent to the more complicated expression which it defines. For example, in a trigonometry text, an "acute angle" is defined as "an angle whose measure is greater than 0°, but less than 90°," and the two phrases convey the same meaning in all the discussions of the text. Since a definition is merely a substitution, there obviously must be certain terms that are undefined to begin with in the development; that is, there must be some *undefined terms* as our starting point. With these undefined terms, we define new ones. The most familiar examples of undefined terms, perhaps, are the two terms "point" and "line" in geometry. Using these two undefined terms, we can define, for example, the term "two parallel lines" which means "two lines (in a plane) that have no points in common." Similarly, a statement that is accepted without proof is called an *axiom* or a *postulate*. A *theorem*, on the other hand, is a formula, proposition, or statement deduced from other propositions, postulates, definitions, and/or undefined terms.

In the proof of a theorem, one must establish that the conclusion be a logical consequence of the hypothesis (or hypotheses). For example, in a given theorem, if A is the hypothesis and B is the conclusion, one must show in the proof that if A is true, then B must be true as an inescapable consequence. Alternatively, one can prove the same theorem by showing that if B is false, A must also be false as a logical deduction. Three methods of proof will be discussed in the next two sections. They are the *direct proof* (or deductive proof), the *indirect proof* (or proof by contradiction), and the *mathematical induction.*

B2-2 DIRECT PROOF AND INDIRECT PROOF

Let A and B be two propositions and let us consider the following statement:

$$\text{If } A \text{ is true then } B \text{ is true} \tag{B2-1}$$

Here, A is the assumption or hypothesis, and B is the conclusion. In a direct proof, one starts by assuming the truth of A and proceeds to establish the truth of B as a logical consequence. In an indirect proof, one shows that the negation of the *conclusion* of the statement leads logically to a result which contradicts the hypothesis of the original statement. An illustration of the direct proof is given in Example B1-1 of Section B1-4. To compare the structure of these two methods, we shall first repeat the direct proof and then prove the same statement by the method of indirect proof.

$$\text{Let } A \text{ be the hypothesis: } x = 4 \text{ and } y = 6 \tag{B2-2}$$

$$\text{Let } B \text{ be the conclusion: } x + y = 10 \tag{B2-3}$$

Then (B2-1) becomes

$$\text{If } x = 4 \text{ and } y = 6, \text{ then } x + y = 10 \tag{B2-4}$$

To constitute a deductive proof, one simply substitutes the values of x and y in (B2-2) into (B2-3), yielding

$$x + y = 4 + 6$$

or

$$x + y = 10$$

which agrees with the conclusion in (B2-4).

In the indirect proof, one assumes that

$$x + y \neq 10 \tag{B2-5}$$

and shows that if the assumptions

$$x = 4 \qquad y = 6$$

are to be true simultaneously, a contradiction

$$10 \neq 10$$

will result by substituting (B2-2) into (B2-5). One concludes, therefore, that if $x = 4$ and $y = 6$, it must follow that $x + y = 10$; that is, (B2-4) is valid.

B2-3 MATHEMATICAL INDUCTION

The powerful method of proof, known as *mathematical induction*, is based on an axiom which may be stated as follows:

If a proposition P_n involving the positive integer n can be shown to have the following two properties:

Property A: P_n is true for $n = 1$

Property B: If k is any (positive integral) value of n for which P_n is true, then P_n is also true for the next larger (integral) value of n: $n = k + 1$

then P_n is true for all (positive integral) values of n.

EXAMPLE B2-1. Let P_n be the following formula:

$$1 + 2 + 3 + \cdots + (n - 1) + n = \frac{n(n + 1)}{2} \tag{B2-6}$$

We shall prove that Equation (B2-6) is correct for all positive integers n by two different methods.

Method 1: Let the left-hand member of Equation (B2-6) be represented by the symbol S:

$$S = 1 + 2 + 3 + \cdots + (n - 1) + n \tag{B2-7}$$

which can be rewritten in the descending order of the integers as follows:

$$S = n + (n - 1) + (n - 2) + \cdots + 2 + 1 \tag{B2-8}$$

Now, adding Equations (B2-7) and (B2-8) term by term, we get

$$S + S = \{1 + n\} + \{2 + (n - 1)\} + \{3 + (n - 2)\} + \cdots + \{(n - 1) + 2\} + \{n + 1\}$$

which, after simplification, becomes

$$2S = \underbrace{(n + 1) + (n + 1) + (n + 1) + \cdots + (n + 1) + (n + 1)}_{n \text{ terms}}$$

or

$$2S = n(n + 1) \tag{B2-9}$$

Dividing Equation (B2-9) by 2, we get

$$S = \frac{n(n + 1)}{2}$$

which, together with Equation (B2-7) yields (B2-6).

The second method of proof is the application of the axiom of mathematical induction.

Method 2: We shall show that P_n does have the two properties (A and B) stated in the axiom.

(A) To establish Property A, we note that the left-hand member of Equation (B2-6)

$$1 + 2 + \cdots + (n - 1) + n \tag{B2-10}$$

represents the sum of the first n positive integers. Hence, if $n = 1$, (B2-10) is simply equal to 1 because it is the "sum" of the first positive integer. Next, observe that the right-hand member of Equation (B2-6) is, for $n = 1$,

$$\frac{n(n + 1)}{2} = \frac{1(1 + 1)}{2} = 1$$

Therefore, for $n = 1$, (B2-6) becomes

$$1 = 1$$

which is, of course, a true statement. Hence, Property A is established.

(B) The establishment of Property B starts by assuming that P_n is true for $n = k$:

$$1 + 2 + 3 + \cdots + (k - 1) + k = \frac{k(k + 1)}{2} \tag{B2-11}$$

and we shall show that P_n must also be true for $n = k + 1$ whenever Equation (B2-11) holds. That is, we would like to show that the following equation is true:

$$1 + 2 + 3 + \cdots + k + (k + 1) = \frac{(k + 1)(k + 2)}{2} \tag{B2-12}$$

Equation (B2-12) is obtained by replacing k by $k + 1$ in (B2-11). To show that (B2-12) is true whenever (B2-11) is, let us add the term $(k + 1)$ to both sides of (B2-11). [Note that $(k + 1)$ would be the next term in the summation of the left-hand member of Equation (B2-11) if it were allowed to continue passing the value $n = k$.] Equation (B2-11) then becomes

$$1 + 2 + 3 + \cdots + (k - 1) + k + (k + 1) = \frac{k(k + 1)}{2} + (k + 1) \tag{B2-13}$$

The right-hand member of Equation (B2-13) may be simplified as follows:

$$\frac{k(k + 1)}{2} + (k + 1) = \frac{k^2 + k + 2k + 2}{2} = \frac{k^2 + 3k + 2}{2} = \frac{(k + 1)(k + 2)}{2}$$

Thus, Equation (B2-13) is seen to be identical to (B2-12) and Property B is established. Therefore, by virtue of mathematical induction, Equation (B2-6) is true for *all positive integers n*.

REFERENCES

[1] Aitken, A. C., *Determinants and Matrices*, 9th ed. New York. Interscience Publishers, 1956.

[2] Ashenhurst, R. L., "A Uniqueness Theorem for Abstract Two-terminal Switching Networks," Report No. BL-10, Harvard Computation Lab., pp. VI: 1-8, Mar. 1955.

[3] Berge, Claude, *The Theory of Graphs and Its Applications.* New York: John Wiley & Sons, Inc., 1962.

[4] Boldyreff, A. W., "Determination of Maximal Steady State Flow of Traffic through a Railroad Network," *Journal of the Operations Research Society of America*, pp. 443–465, Nov. 1955.

[5] Branin, F. H., "The Inverse of the Incidence Matrix of a Tree and the Formulation of the Algebraic-First-Order Differential Equations of an RLC Network," *IEEE Transactions on Circuit Theory*, vol. CT-10, pp. 543–544, Dec. 1963.

[6] Branin, F. H., "A New Method for Steady State A.C. Analysis of RLC Networks," 1966 *IEEE International Convention Record*, pt. 7, pp. 218–223.

[7] Brown, D. P., "New Topological Formulas for Linear Networks," *IEEE Transactions on Circuit Theory*, vol. CT-12, no. 3, pp. 358–365, Sept. 1965.

[8] Bryant, P. R., "Order of Complexity of Electrical Networks," *Proceedings of the IEE* (*London*), vol. 106C, pp. 174–188, 1959.

[9] Bryant, P. R., "The Explicit Form of Bashkow's *A* Matrix," *IRE Transactions on Circuit Theory*, vol. CT-9, no. 3, pp. 303–306, Sept. 1962.

[10] Busacker, Robert G., and Thomas L. Saaty, *Finite Graphs and Networks.* New York: McGraw-Hill Book Company, Inc., 1965.

[11] Calahan, D. A., "Computer Generation of Equivalent Networks," *1964 IEEE International Convention Record*, pt. 1, pp. 330–337.

[12] Calahan, D. A., "Computer Design of Linear Frequency Selective Networks," *Proceedings of the IEEE*, vol. 53, no. 11, pp. 1701–1706, Nov. 1965.

[13] Carpenter. R. M., and W. W. Happ, "Symbolic Approach to Computer-Aided Circuit Design," *Electronics*, vol. 39, no. 25, pp. 92–98, Dec. 1966.

[14] Chan, S. P., "Topological Analysis and Synthesis of Single-Contact Networks," Final Report, NSF Grant GP-2760, Electrical Engineering Department, University of Santa Clara, Santa Clara, Calif., July 1965.

[15] Chan, S. P., "Application of Graph Theory to the Synthesis of Single-Contact Networks," *Journal of the Franklin Institute*, vol. 280, no. 5, pp. 425–442 Nov. 1965.

[16] Chan S. P., and B. H. Bapna, "A Modification of the Coates Gain Formula for the Analysis of Linear Systems," *Inst. J. Control*, vol. 5, no. 5, pp. 483–495 1967.

[17] Chan, S. P., and S. G. Chan, "Modifications of Topological Formulas," *IEEE Transactions on Circuit Theory*, vol. CT-15, No. 1, pp. 84–86, Mar. 1968.

[18] Chan, S. P., and S. Y. Chan, "On Flow-Graph Techniques for the Analysis of Linear Networks and Systems," *Convention Record Tenth Midwest Symposium on Circuit Theory*, pp. XI-1-1 through XI-1-11, May 1967.

[19] Chan, S. P. and H. N. Mai, "A Flow-Graph Method for the Analysis of Linear Systems," *IEEE Transactions on Circuit Theory*, vol. CT-14, no. 3, pp. 350–354, Sept. 1967.

[20] Chen, W. K., "Topological Analysis for Active Networks," *IEEE Transactions on Circuit Theory*, vol. CT-12, no. 1, pp. 85–91, Mar. 1965.

[21] Chen, W. K., "Generation of Trees and *k*-Trees," *Proceedings of the Third Allerton Conference on Circuit and System Theory*, pp. 906–912, Oct. 1965.

[22] Chen, W. K., and R. C. Quisenberry, "On Network Topology," *Research Report EED 10-5*, Research Institute, Ohio University, Jan. 1966.

[23] Coates, C. L., "General Topological Formulas for Linear Network Functions," *IRE Transactions on Circuit Theory*, vol. CT-5, pp. 30–42, Mar. 1958.

[24] Coates, C. L., "Flow-Graph Solutions of Linear Algebraic Equations," *IRE Transactions on Circuit Theory*, vol. CT-6, pp. 170–187, June 1959.

[25] Dantzig, G. B., and D. R. Fulkerson, "On the Max-flow, Min-cut Theorem of Networks in Linear Inequalities," *Annals of Mathematical Studies*, no. 38, pp. 215–221, 1956.

[26] Desoer, C. A, "The Optimum Formula for the Gain of a Flow Graph or a Simple Derivation of Coates' Formula," *Proceedings of the IRE*, vol. 48, pp. 883–889, May 1960.

[27] Dunn, W. R., Jr., and S. P. Chan, "An Algorithm for Testing the Planarity of a Graph," *IEEE Transactions on Circuit Theory*, vol. CT-15, no. 2, pp. 166–168, June, 1968.

[28] Dunn, W. R. Jr., and S. P. Chan, "An Algorithm for Calculating the Gain of a Signal Flowgraph," *Proceedings of the Hawaii International Conference on System Sciences*, Jan. 1968.

[29] Elias, P., A. Feinstein, and C. E. Shannon, "A Note on the Maximal Flow through a Network," *IRE Transaction on Information Theory*, vol. IT-2, pp. 117–119, Dec. 1956.

[30] Euler, L., "Solutio Problematis and Geometriam Situs Pertinantis," *Academinae Petropolitanae*, 8, pp. 128–140, 1736.

[31] Ford, L. R., Jr., and D. R. Fulkerson, "Maximal Flow through a Network," *Canadian Journal of Mathematics*, vol. 8, pp. 399–404, 1956.

[32] Ford, L. R., Jr., and D. R. Fulkerson, "A Simple Algorithm for Finding Maximal Network Flows and an Application to the Hitchcock Problem," *Canadian Journal of Mathematics*, vol. 9, pp. 210–218, 1957.

[33] Fu, Y., "On Probabilistic Networks," Technical Report, Systems Laboratory Northwestern University, Evanston, Ill., 1963.

[34] Fu, Y., "Applications of Topological Methods to Probabilistic Communication Networks," *IEEE Transactions on Communication Technology*, vol. 13, no. 3, pp. 301–307, Sept. 1965.

[35] Fu, Y., and S. S. Yau, "A Note on the Reliability of Communication Networks," *Journal of the Society of Industrial and Applied Mathematics*, vol. 10, pp. 469–474, Sept. 1962.

[36] Gould, R., "Theory of Switching," Progress Report No. BL-18, Computation Laboratory, Harvard University, May 1957.

[37] Grigsby, L. L. and W. A. Blackwell, "Design of a Linear Multiport by State-space Techniques," *The Matrix and Tensor Quarterly*, vol. 16, no. 4, pp. 122–134, June 1966.

[38] Hakimi, S. L., "Simultaneous Flows through a Communication Network," *IRE Transactions on Circuit Theory*, vol. CT-9, no. 2, pp. 169–175, June 1962.

[39] Hakimi S. L. and N. Deo, "A Topological Approach to the Analysis of Linear Circuits," *Proceedings of the 4th Allerton Conference on Circuit and Systems*, pp. 197–206, Oct. 1966.

[40] Happ, W. W., "Flowgraph Techniques for Closed Systems," *IEEE Transactions on Aerospace and Electronic Systems*, vol. AES-2, no. 3, pp. 252–264, May 1966.

[41] Happ, W. W., "NASAP: Present Capabilities of a Maintained Program," Report ERC/CQ 67-631, National Aeronautics and Space Administration, 1967.

[42] Harary, Frank, Robert Z. Norman, and Dorwin Cartwright, *Structural Models: An Introduction to the Theory of Directed Graphs.* New York: John Wiley & Sons, Inc., 1965.

[43] Hohn, F. E., *Elementary Matrix Algebra*, 2d ed. New York: The Macmillan Company, 1964.

[44] Hohn, F. E., *Applied Boolean Algebra*, 2d ed. New York: The Macmillan Company, 1966.

[45] Kim, W. H., and R. T. Chien, *Topological Analysis and Synthesis of Communication Networks.* New York: Columbia University Press, 1962.

[46] Kirchhoff, G., "Über die Auflösung der Gleichungen, auf welche man bei der Untersuchung der linearen Vertheilung Galvanischer Ströme geführt wird," *Ann alen der Physik und Chemie*, vol. 72, pp. 497–508, 1847; English translation, *IRE Transactions on Circuit Theory*, vol. CT-5, no. 1, pp. 4–7, Mar. 1958.

[47] Koenig, H. E., and Y. Tokad, "State Models of Systems of Multiterminal Linear Components," *1964 IEEE International Convention Record*, pt. 1, pp. 318–329.

[48] Kuh, E. S., and Rohrer, R. A., "The State-variable Approach to Network Analysis," *Proceedings of the IEEE*, vol. 53, no. 7, pp. 672–686, July 1965.

[49] Kuo, F. F., "Network Analysis by Digital Computer," *Proceedings of the IEEE*, vol. 54, no. 6, pp. 820–829, June 1966.

[50] Kuratowski, C., "Sur le Probleme des Courbes Gauches en Topologie, *Fundamentae Mathematicae*, vol. 15, pp. 271–283, 1930.

[51] LePage, W. R., "State Variable Approach to a Second-Order System," *IEEE Transactions on Education*, vol. E-9, no. 2, pp. 80–90, June 1966.

[52] Löfgren, L., "Irredundant and Redundant Boolean Branch-Networks," *Transactions of the 1959 International Symposium on Circuit and Information Theory*, pp. 158–175 (issued as a special supplement to *IRE Transactions on Circuit Theory*, vol. CT-6, May 1959).

[53] Magnuson, W. G., Jr., "Computer-aided Design: Part 8, Picking Transient Analysis Programs," *Electronics*, pp. 84–87, Apr. 17, 1967.

[54] Marshall, C. W., "Applied Graph Theory," a publication of the Discrete System Concepts Project funded by the National Science Foundation under Grant GE-2546, Apr. 30, 1965.

[55] Mason, S. J., "Feedback Theory — Some Properties of Signal Flow Graphs," *Proceedings of the IRE*, vol. 41, pp. 1144–1156, Sept. 1953.

[56] Mason, S. J., "Feedback Theory — Further Properties of Signal Flow Graphs," *Proceedings of the IRE*, vol. 44, no. 7, pp. 920–926, July 1956.

[57] Mason, S. J., "Topological Analysis of Linear Nonreciprocal Networks," *Proceedings of the IRE*, vol. 45, pp. 829–838, June 1957.

[58] Maxwell, J. C., *Electricity and Magnetism*. Oxford: Clarendon Press, 1892.

[59] Meyeda, W., "Topological Formulas for Active Networks," *Interim Technical Report* No. 8, U.S. Army Contract DA-11-022-ORD-1938, University of Illinois, Urbana, Ill., Jan. 1958.

[60] Mayeda, W., "The Application of Mathematical Logic to Network Theory," *Interim Technical Report* No. 9, U.S. Army Contract DA-11-022-ORD-1983, University of Illinois, Urbana, Ill., Apr. 1958.

[61] Mayeda, W., "Synthesis of Switching Functions by Linear Graph Theory," *IBM Journal of Research and Development*, vol. 4, pp. 321–328 no. 3, July 1960.

[62] Mayeda, W., and S. Seshu, "Topological Formulas for Network Functions," Bulletin 446, University of Illinois Engineering Experiment Station, 1957.

[63] Mayeda, W., and S. Seshu, "Generation of Trees Without Duplications," *IEEE Transactions on Circuit Theory*, vol. CT-12, no. 2, pp. 181–185, June 1965.

[64] Minty, G. J., "A Simple Algorithm for Listing All the Trees of a Graph," *IEEE Transactions on Circuit Theory*, vol. CT-12, no. 1, p. 120, Mar. 1965.

[65] Nathan, A., "Topological Rules for Linear Networks," *IEEE Transactions on Circuit Theory*, vol. CT-12, no. 3, pp. 344–358, Sept. 1965.

[66] Ore, Oystein, *Graphs and Their Uses*. New York: Random House, Inc., 1963.

[67] Percival, W. S., "Solution of Passive Electrical Networks by Means of Mathematical Trees," *Proceedings of the IEE (London)*, vol. 100, pt. III, pp. 143–150, May 1953.

[68] Percival, W. S., "The Graphs of Active Networks," *Proceedings of the IEE (London)*, vol. 102, pt. C, pp. 270–278, Apr. 1955.

[69] Pottle, C., "Comprehensive Active Network Analysis by Digital Computer — a State-space Approach," *Proceedings of the 3rd Allerton Conference on Circuit and System Theory*, pp. 659–668, Oct. 1965.

[70] Pottle, C., "State-space Techniques for General Active Network Analysis," in *System Analysis by Digital Computer*. New York: John Wiley & Sons, Inc., 1966, pp. 59–98.

[71] Rider, P. R., *College Algebra*. New York: The Macmillan Company, 1947.

[72] Seshu, S., "On Electric Circuits and Switching Circuits," *IRE Transactions on Circuit Theory*, vol. CT-3, pp. 172–178, Sept. 1956.

[73] Seshu, S., "Minimal Realizations of the Biquadratic Minimum Function," *IRE Transactions on Circuit Theory*, vol. CT-6, no. 4, pp. 345–350, 1959.

[74] Seshu, S., "Network Applications of Graph Theory — A Survey," a paper presented at the Fifth Midwest Symposium on Circuit Theory, 1961.

[75] Seshu, S., and N. Balabanian, *Linear Network Analysis*. New York: John Wiley & Sons, Inc., 1959.

[76] Seshu, S., and M. B. Reed, *Linear Graphs and Electrical Networks*, Reading, Mass.: Addison-Wesley Publishing Company, Inc., 1961.

[77] Talbot, A., "Topological Analysis of General Linear Networks," *IEEE Transaction on Circuit Theory*, vol. CT-12, no. 2, pp. 170–180, June 1965.

[78] Tang, D. T., "Communication Networks with Simultaneous Flow Requirements," *IRE Transactions on Circuit Theory*, vol. CT-9, no. 2, pp. 176–182, June 1962.

[79] Van Valkenburg, M. E., *Introduction to Modern Network Synthesis*. New York: John Wiley & Sons, Inc., 1960.

[80] Weinberg, L., *Network Analysis and Synthesis*. New York: McGraw-Hill Book Company, Inc., 1962.

[81] Whitney, H., "Nonseparable and Planar Graphs," *Transactions of the American Mathematical Society*, vol. 3t, pp. 339–362, 1932.

[82] Williams, T. G. "The Design of Survivable Communications Networks," *IEEE Transactions on Communication Systems*, vol. CS-11, pp. 230–247, June 1963.

[83] Wing, O., and P. Demetriou, "Analysis of Probabilistic Networks," *IEEE Transactions on Communication Technology*, vol. COM-12, no. 3, pp. 38–40, Sept. 1964.

[84] Wing, O. and W. H. Kim, "The Path Matrix and Switching Functions," *Journal of the Franklin Institute*, vol. 268, no. 4, pp. 251–269, Oct. 1959.

[85] Zadeh, L. A., and C. A. Desoer, *Linear System Theory*. New York: McGraw-Hill Book Company, Inc., 1963.

INDEX